A New PROSPERITY

Building a Sustainable Energy Future

THE SERI SOLAR/CONSERVATION STUDY

Brick House Publishing
Andover, Massachusetts
1981

Library of Congress Cataloging in Publication Data

Main entry under title:

A New prosperity, building a sustainable energy
future.

Includes bibliographical references.
1. Energy conservation—United States.
2. Renewable energy sources—United States.
3. Energy policy—United States. I. Solar
Energy Research Institute.
TJ163.4.U6N49   333.79'0973   81-6089
ISBN 0-931790-27-1   AACR2
(ISBN 0-931790-53-0 pbk.)

*Edited by Nancy Irwin and
Designed by Mike Fender,
Cambridge, Massachusetts*

RICHARD L. OTTINGER
24TH DISTRICT, NEW YORK
———
COMMITTEES:
ENERGY AND COMMERCE
SCIENCE AND TECHNOLOGY

Congress of the United States
House of Representatives
Washington, D.C. 20515

REPLY, IF ANY TO:

☐ 2241 RAYBURN HOUSE OFFICE BUILDING
WASHINGTON, D.C. 20515
(202) 225-6506

☐ 100 STEVENS AVENUE, SUITE 203
MOUNT VERNON, NEW YORK 10550
(914) 699-2866

☐ 77 QUAKER RIDGE ROAD
NEW ROCHELLE, NEW YORK 10804
(914) 235-5600 OR 428-3040

March 30, 1981

During the four years of the Carter administration, I devoted considerable effort attempting to convince its officials that the United States could do more -- much more -- to increase the efficiency with which we use energy. Indeed, I pointed to a series of responsible studies showing that the United State could achieve independence from imported oil by the year 2000 by promoting energy efficiencies and renewables alone, and do so far quicker, cheaper and cleaner than any other means of oil displacement.

In July 1979, I asked John Sawhill, then Deputy Secretary of Energy, to address the problem of our inefficient use of energy, and to undertake an analysis of our potential for energy savings through conservation. The deputy secretary agreed and arranged to have the work done by the Solar Energy Research Institute. He asked SERI to document realistic energy savings through promoting efficiency in our buildings, transportation systems and industry, as well as the contribution achievable from renewable energy sources.

The study is now complete, after 18 months of hard work by some of the best minds in the country: coordinated by Henry Kelly and Karl Gawell of SERI, the report includes contributions by Art Rosenfeld of the Lawrence Berkeley Laboratory, Charles Gray of the Environmental Protection Agency, Marc Ross of the Energy Productivity Center at the Carnegie Mellon Institute, and Bob Williams and others at Princeton University.

The study tells us that the United States can enjoy our historic high economic growth rate while reducing our energy consumption by as much as 25 percent. It also tells us that the path we would follow to reach this goal is our cheapest energy future available.

These findings are startling, to say the least. They offer great promise for both our economic and energy futures. But I should note that the study does not pretend to be a prediction of our nations energy picture in the year 2000, but only what it can be if we take appropriate steps.

I should also note that the study does not presume that a net reduction in energy consumption will be cheap; the study estimates that investments over the next 20 years might reach $800 billion. While this sounds staggering, it must be weighed against a nuclear and coal alternative requiring investment estimated by the Department of Energy at over a trillion dollars, and the nuclear-electric segment only impacts the nine percent of our oil used to generate electricity.

Despite its cost, the program outlined by the report is by far our cheapest energy future. On the average, the recommended programs reduce energy consumption at a cost "per barrel saved" lower than today's world price for crude oil. An attempt to "produce our way to greatness" requires greater and greater energy expenditures, all at the cost of sacrificing investment in industrial modernization.

The findings of the report take on great significance when we consider our greatest economic conflict: our economy is hemorrhaging as we devote a larger and larger percentage of our Gross National Product to the purchase of imported oil, competing for scarce capital needed to rebuild our industrial and economic base.

Consider what is promised by the path outlined in the report:

- We can eliminate oil imports by the turn of the century, ending the greatest single threat to both our economy and national security.
- We can reach our goal of energy independence through efficiency at a cost far below that required to produce an equivalent amount of energy. Those savings will allow us to devote an increased percentage of our GNP to industrial capital investment, our greatest economic need.

Because of the huge economic benefits this report promises, I believe that our national security and economic interests demand that it be available for the public debate over the direction of federal fiscal policy. The earlier this report is published, the faster we can consider its invaluable information.

Sincerely,

Richard L. Ottinger
Chairman
House Subcommittee on Energy Conservation and Power

Contents

Acknowledgements

THE PUBLISHERS would like to thank all of the people at the Solar Energy Research Institute (SERI) in Golden, Colorado, who advised and guided us through the arduous task of editing this report, providing us with the final facts and figures necessary to get the report into the form you see here.

However, none of this would have been possible without the help and devotion of four people who worked tirelessly for two and a half weeks in Boulder, Colorado, designing, editing and typesetting the book for publication. They are Mike Fender and Nancy Irwin of Cambridge, Massachusetts; and Marylee Phillips of Thornton, Colo., and Carmen Echelmeier in Boulder. Ms. Irwin, who did all of the editing, and Mrs. Phillips, who managed the complete word-processing job, deserve special thanks. Each of them took time away from their regular jobs and their families to work night and day until the project was finished. Without their extremely competent and professional help, this report would never have been available to the general public.

Mike Fender would also like to thank Deborah Heard, of Horizon Graphics, in Boulder, for her speedy and accurate job in setting all the display type for the book and cover in one day. Finally, Mr. Fender would like to thank the management and staff of the Hotel Boulderado for their cheerful and uncomplaining support throughout the time he spent there in putting the final material together for the printers. The staff made endless trips to the fourth floor with all of the coffee and food necessary to fuel the editorial and production furnace, and operated their 1918 switchboard with great elan during the constant calls around the country that were necessary in arranging the production of the book.

The authors of the SERI Report would now like to take some space to acknowledge the contributions of some of the many people who devoted more than eighteen months of considerable effort in providing them with all that was necessary to produce this vital document.

PRIMARY CONTRIBUTORS

THIS REPORT was prepared with the assistance and cooperation of a large number of groups and individuals. The overall project was directed by Henry Kelly, Associate Director at SERI, and Karl Gawell, the project task leader. Robert Williams of the Princeton Center for Energy and Environmental Studies, made valuable contributions to many parts of the report--particularly in the analysis of industrial cogeneration, utilities, and industrial policy.

The work on buildings was conducted through a cooperative effort involving SERI, the Energy Efficient Buildings Program at the Lawrence Berkeley Laboratory, and the firm Gilford Derringer & Co. David Claridge and collaborators at SERI are responsible for the work on renewable energy in buildings. Arthur Rosenfeld, director of the Energy Efficient Buildings Program, and Jeff Harris, also of LBL, were responsible for developing the material on the potential for increasing the efficiency of residential and commercial buildings and policies affecting these areas. Jan Wright and Peter Cleary, of LBL, and David Goldstein, formerly of LBL and now with the Environmental Defense Fund (EDF), all made major contributions to the analysis of energy use in residential buildings. Joseph Derringer provided much of the material dealing with energy use in commercial buildings. Useful ideas and information was obtained from participants attending a summer study held in Santa Cruz, California, in August 1980, which reviewed an early draft of this study.

The work on industry was conducted under the direction of Marc Ross, of the University of Michigan, and the Mellon Institute's Energy Productivity Center, working with Kenneth Brown, formerly with SERI. Mark Gibson and Ruxton Villet, of SERI, supplied the bulk of the material dealing with biomass; Larry Flowers and Rainer Kern, of SERI, supplied the material on direct solar energy systems used in industry.

The material dealing with transportation is based on a task force report written under the direction of Charles Gray, Director of the Emission Control Technology Division of the U.S. Environmental Protection Agency, and J. Allson, also of EPA. Valuable contributions were also received from Frank Von Hippel, of Princeton, and Mark Gibson and Daniel Lashoff, of SERI.

The material dealing with the use of renewable resources in electric utilities was developed by Theresa Flaim and Roger Taylor, of SERI. The policy analysis was developed with the assistance of Michael Maguire.

Numerous other individuals and organizations contributed to the project or served as reviewers.

Note: The designer would like to thank Melissa Swanke and Kelley Hersey in Boston, Massachusetts for repairing and finishing up a word processing job that was complex, difficult and exhausting for everyone.

A New
PROSPERITY
INTRODUCTION

THE PAST HALF CENTURY has been a period of unprecedented economic growth for the United States. Much of this growth was fueled with cheap energy--primarily oil and gas, much of it imported. Events of the past few years, however, have called the stability of this economic foundation into doubt.

The SERI Study has redefined a stable foundation for growth in the American economy. The pillars of this new prosperity are more efficient energy consumption and economic use of renewable energy resources.

Specifically, SERI's findings show that through efficiency, the U.S. can achieve a full-employment economy and increase worker productivity, while reducing national energy consumption by nearly 25 percent. Table 1 breaks down U.S. energy demand under such a scenario.

Some 20 to 30 percent of this reduced demand could be supplied by renewable resources. Table 2 shows the potential contribution of each type of renewable resource. A strategy built around energy efficiency and the widespread use of renewable resources could result in the virtual elimination of oil imports. It must be emphasized that the numbers in Tables 1 and 2 are goals, not forecasts. But the benefits to the nation of attaining these goals are enormous. These figures must be given the serious examination they deserve.

SERI has not assumed that the U.S. must endure freezing homes, stalled traffic, or massive unemployment. Rather, the proposed energy strategy is based on highly optimistic assumptions about both social well-being and economic growth. Per capita income, for example, is assumed to increase by 45 percent over the next 20 years. Unemployment would fall from a current 7.4 percent to 4.0 percent by 1985 and would remain at that level through 2000. Labor productivity would grow more than 2 percent per year, as opposed to the 1.25 percent per year it has averaged since 1968. Table 3 presents a sampling of these growth rates.

Far from being incompatible with vigorous economic growth, the strategy proposed here may actually be essential for such growth. New supplies of oil, gas, and coal will almost certainly be more expensive than the energy that now fuels the American economy. Increased use of these sources would inevitably raise prices, accelerate inflation, and constrain growth. With an emphasis on efficiency, however, the cost of supplying basic energy services can increase much more slowly. For example, efficient industrial processes will allow growth to be relatively independent of unpredictable increases in energy prices. Similarly, development of energy-efficient buildings, both residential and commercial, will mean that energy prices need not be a barrier to the growth of construction.

The nation will spend more than 7 trillion dollars on energy during the next 20 years. Clearly, it is critical that the market be able to compare fairly investments in energy supplies with investments in efficiency and select the most productive use of available capital. The nation cannot afford federal programs that artificially prejudice this choice, encouraging inefficent investment.

The role of the federal government in achieving the proposed goals need not be costly or oppressive. Its role can be largely limited to the following. (1) It can ensure that investors have the information they need for decisions about energy investments. (2) It can revise existing programs that act to discourage capital investments in efficiency and subsidize or protect investments in energy supply. (3) It can ensure that investors have access to the capital they need for improvements that reduce energy demand. (4) It can maintain a vigorous research and development program in energy and related fields. (5) It can ensure that national energy investments are consistent with national interests in environment, equity, and security. The government cannot escape responsibility for a national energy policy. Its influence on energy

1

investments will be enormous, whether by design or by accident.

The strategy outlined in this report is rooted in the conservative sentiment that investments should be allowed to flow in the direction that produces the greater rate of return--rather than being channeled in less efficient directions through government action. This strategy therefore proposes to remedy many market distortions caused by public policy. It proposes that federal energy spending be both reduced and redirected.

The energy strategy proposed here has the enormous advantage that it does not rely on any unforeseen technological development or any draconian interference with the market process. It would provide protection immediately against the uncertainty that dominates the nation's near-term energy future. Yet, because the strategy proceeds in steps, it can also accommodate research triumphs as they occur. At a minimum, the proposed strategy is needed to supplement energy programs that hinge on possible--but unpredictable--technological breakthroughs.

SETTING GOALS AND GETTING THERE

SERI's analysis indicates that by the year 2000 national energy consumption could decrease in all sectors except industry, which shows a slight increase. The consumption of fuels can be reduced to a point where oil imports are eliminated by the end of the century. The consumption of electricity can be reduced to a point where, on a national basis, demands through the end of the century can be met with generating equipment now operating or in advanced stages of construction.

Renewable resources can play a major role. Hydroelectric facilities and energy derived from wood and other plant material, which contribute approximately 6 percent of the nation's energy today, are expected to continue to supply the bulk of the nation's solar energy by the end of the century. The potential contribution of direct solar thermal systems attached to buildings and industry is considerable, but the total contribution of such systems will be limited: increases in efficiency will reduce the total national demand for thermal energy.

Buildings

One would not commonly think of an investment in building insulation as an alternative to an investment in an oil well, but the two kinds of investments can have precisely the same outcome. For example, the equivalent of about 8.1 million barrels of oil per day (MBD) can be "produced" from existing and new residential and commercial buildings at an average cost that is about half of the cost of providing electricity,

oil, and/or gas to these buildings from new conventional sources.

New residential and commercial buildings can be built to use about a quarter of the energy required for heating and cooling required by the typical unit built in the U.S. today. This efficiency can be achieved with better insulation, tight construction, storm windows, daylighting, and efficient furnaces and air-conditioning equipment. Achieving these savings in standard practice, however, will require well-designed programs of applied research in new buildings and an effective technique for communicating the results of these programs to the nation's building industry and building owners. A total of 2 MBD could be saved in this way.

Given an aggressive retrofit program, it should be possible to reduce the demand for energy used by existing buildings, still standing in the year 2000, by 4.2 MBD. Programs to ensure that this potential is captured must be centered on the objective of creating profitable businesses, capable of delivering technically sound building retrofits. Aiding these emerging businesses will require a national program for applied research in building retrofits. Such a program would examine solar and energy-efficiency techniques and provide information and training materials.

The energy used in appliances to provide hot water, lighting, refrigeration, and other amenities for all residential buildings standing in the year 2000 could be over 1.5 MBD lower than anticipated in conventional forecasts, given an effective program of research and labeling for appliances. (This does not count the savings of 1.5 MBD resulting from efficiency improvements in heating and cooling equipment.) The cost would once again be much less than the cost of providing the additional energy from new conventional sources.

In addition to significantly reduced demand in buildings, renewable technologies could provide them with the equivalent of 2 to 2.5 MBD. Wood stoves, small wind generators, active and passive space heating, solar hot water, daylighting, and photovoltaics could provide up to 30 percent of the energy requirements of all buildings by the year 2000.

Industry

Industry is already moving rapidly to increase efficiency, but there is much room for improvement due to the relative inattention given to fuel efficiency before 1973. For example, the performance of industrial mechanical drives can be increased by 23 percent, the efficiency of delivering heat to industrial processes can be increased by 35 percent by using improved boilers and controls, and the efficiency of aluminum electrolysis can be improved by 33 percent.

These estimates are all relatively conservative because they are based solely on technologies already available. It is likely that much greater savings will be achieved by developing completely new industrial processes, such as the recently introduced float-glass process, which makes more efficient use of capital, labor, and materials, as well as energy. It is difficult to forecast such innovations, but they have always emerged and will certainly continue to occur now that more of the national engineering genius is being directed to problems of efficiency.

Programs designed to improve industrial efficiency should concentrate on ensuring that industries have adequate access to the capital needed to improve efficiency and that the marketplace receives the right signals to ensure that industries pay the full cost of the energy they use. This will require an adjustment in the federal tax code to encourage industrial investment. It is particularly important that capital be found to replace the obsolescent plants that now use a large fraction of the nation's industrial energy; a special "scrap and build" program to accelerate the rebuilding of American industrial plants could be funded from any windfall profits receipts available from energy decontrols or other sources. In addition, special attention also must be given to encouraging industrial research and developing new approaches to efficiency. Research is often the first victim of financial hard times, but industrial research is a vital national resource. Tax incentives and other programs are needed to promote industrial investments in new research and subsequent ventures based on such research.

Improved efficiency could to result in saving the equivalent of 5 MBD in industry by the year 2000. About half of these savings are already expected to occur; the other half would result from the new programs suggested. It would also be possible to provide the equivalent of 5 MBD from national biomass resources (although only about three-fourths of these resources are likely to be used directly in industry), and as much as 0.5 MBD from direct solar heat.

Transportation

Most of the energy needed for transportation in the U.S. is used in cars and trucks. The fuel economy of the average car on the road in 1978 was about 14.3 miles per gallon. It should be possible to increase this to 60 mpg or more without major sacrifices in comfort, safety, or performance. If the average car on the road in the year 2000 obtains 55 mpg, the nation would consume nearly 3 MBD less gasoline than it does today, even with significant increases in both the population and the miles driven by each person. The average cost of saving this fuel could be less than a dollar a gallon. Improved fuel efficiency can be achieved through increased emphasis on small, efficient engines and with policies for increased performance standards or taxes--for example a "gas

guzzler" tax keyed to high efficiency and a fuel or petroleum tax. Methanol and other alcohol fuels derived from biomass could deliver up to 25-45 percent of the fuel needed to operate the national transportation system by 2000. Use of these fuels can be encouraged with research on methanol production and engines that use methanol, conversion of federal fleets to use methanol, and fuel taxes. Methanol is an attractive synthetic fuel since it can also be produced from coal and natural gas.

The efficiency of national freight services has been declining recently because of a continuing shift of freight from rail to truck. This decline in freight service efficiency must be reversed. Road fees that tax trucks for their full share of road construction and maintenance and a relaxation of the regulations that prevent the rail industry from raising needed revenues are an essential beginning. It should also be possible to increase the efficiency of both aircraft and trucks at least 30 percent with adequate research. Programs to improve freight service efficiency could save 1 MBD by the year 2000.

Utilities

Uncertainties about future energy demands have already greatly complicated the investment strategies of the nation's gas and electric utilities. The problem is particularly difficult for electric utilities, since a decision to build a plant that will be under construction for a decade or more requires an accurate forecast of future electrical demand. Many companies have found themselves with expensive overcapacity-- the result of overestimating demands. As a result, it may be appropriate for the nation's utilities to support an increase in electric demand of 0.1-0.7 percent per year for the next 20 years. This is true even if no plants are brought on-line after 1985, if 80 percent of all oil- and gas-burning generating plants are retired, and if all fossil plants built before 1961 are retired. A growth rate of 0.9-1.5 percent per year could be supported if optimum use is made of the potential for cogeneration in the six major steam-consuming industries.

The analysis presented in this report suggests that the ambiguity about future electric demands is likely to increase rather than diminish. Net national demand for electricity could well grow more slowly than 1.5 percent per year even given a rapid growth in demands for electric vehicles and electrolysis-produced aluminum.

The planning dilemma presented by these circumstances has no easy resolution. The central question is whether utilities should adjust construction programs on the assumption that cost-effective, demand-reducing investments will succeed, or whether they should assume that such investments will not play an important role and run the risk of over-investment.

The thesis in this study is that the best response to such a complex problem may be to confess that public regulation is unlikely to anticipate an optimum solution, and that some move in the direction of deregulation deserves careful attention. Three approaches are examined: (1) establishing rates that send the market proper signals about the real marginal cost of supplying electricity; (2) allowing utilities to expand their investment portfolios to include some investments in energy-saving equipment on their customers' premises (with proper safeguards to prevent unfair exploitation of monopoly power); and (3) allowing nonregulated companies greater access to markets heretofore reserved to utility monopolies. (The Public Utilities Regulatory Policy Act has already deregulated many promising categories of small electric-generating technologies.) Most of these activities must occur at the state level, where utility regulation is controlled. The federal role can be limited to assistance in developing sophisticated planning tools and imaginative management of the federally regulated Federal Power Marketing Administrations.

The role of solar electric-generating devices becomes very difficult to anticipate, given the uncertainties just described. Solar technologies are examined in this study to determine whether they can be an economically preferred technique for displacing existing oil- and gas-burning capacity. The analysis indicates that there will be circumstances in which solar equipment is preferred to a coal-burning plant designed for a similar function. As expected, the largest contributions from solar electric equipment are likely to come from additions to the nation's hydro-electric capacity and from wind machines. Photovoltaic equipment, some of which will be located on buildings, will also be a preferred generating source in some circumstances.

Economic Assumptions

This analysis proceeds from a straightforward series of economic calculations based upon the price of conventional fuels. To ensure that any errors were on the side of conservatism, only modest price increases were assumed. For example, it was assumed that, by the year 2000, the price of oil (in 1978 dollars) would reach $40 per barrel, that the price of natural gas would reach $5 per million Btu, and that electricity would be priced at 7¢ to 8¢ per kilowatt-hour. Critics might charge that all of these prices are likely to be reached (or surpassed) by 1985. If so, that would only encourage even greater investments in improved efficiency and solar technologies than those called for in this report.

In order to gauge the sensitivity of this analysis to higher prices, we occasionally perform calculations based on a $60-per-barrel price for oil. This can be thought of as (1) a more realistic forecast of oil prices;

or (2) a premium we would pay to avoid the national security concerns, the balance-of-trade problems, or the environmental externalities associated with higher energy use. In any case, whenever a $60-per-barrel figure is used, it is clearly called to the reader's attention as being outside the mainstream of the report's calculations.

GOALS FOR NATIONAL GROWTH

Taken together, the goals outlined in this study suggest possibilities for an enormous growth in national income over the next two decades. Plainly national energy consumption could be even lower than the levels discussed here if some of the goals for growth are not achieved. How can so very much more work be accomplished using less energy? The answer is straightforward. The United States currently uses energy less efficiently than any other country in the industrial world. Just introducing technologies hat are already widely employed elsewhere would produce dramatic reductions in energy use without diminishing our economic productivity or detracting from the comforts of current lifestyles. If, in addition to drawing the best from abroad, we also take advantage of the full repertoire of Yankee ingenuity now sitting on our own shelves, we can, over a twenty-year period, move from last place to first. Through intelligent market choices calculated to increase the productivity of energy use, we can rebuild our nation as the world's most efficient industrial state. In an era of mounting energy prices, the country that makes the most efficient use of energy will possess a key advantage in the global marketplace. It will also have controlled a major source of inflation.

No individual factor of production--labor, capital, materials, energy, or human ingenuity--is valued for itself. Value adheres in the goods and services produced when those factors are combined. These factors are almost always somewhat substitutable. The economic health of the nation demands that they be combined in a way that produces the most goods and services for the least total cost. Left alone, market forces would determine how much should be invested in energy production and how much should be invested in other factors that improve the productivity of energy use.

This may seem to be an unremarkable truth. Yet the energy literature is full of tests that proceed from the assumption that well-being is inextricably linked to the increasing use of energy. The more energy we use, they contend, the better off we will be. A little careful thought demolishes this fallacy. A person living in a drafty house in which several rooms are closed off entirely and the thermostat is set at 55 degrees is not obviously better off than the individual who lives in a tight, well-insulated home that is kept at an even 70 degrees. Yet the person shivering at 55 degrees might consume several times as much fuel oil as the person in the more comfortable house.

A host of similar examples can be provided. Is there an advantage to an inefficient refrigerator if a better model can keep the same volume of food crisp and cold using only 40 percent as much electricity? Certainly not, as long as the incremental cost of the more efficient refrigerator remains less than the incremental cost of the generating plant and the fuel required to produce the additional power needed by the less efficient model. Is there an advantage to producing aluminum in yesterday's refineries if new plants can produce an identical product using only 70 percent as much energy? Certainly not, as long as the new process makes economic sense.

Efficiency improvements--rather than being a retreat--are actually an intrinsic step in the march of progress. Early television sets required several hundred watts of power to produce images that left much to the imagination; modern television sets need no more power than a 60-watt light bulb to produce pictures of remarkable clarity. Thirty years ago, modest computational efforts required a room full of equipment drawing large amounts of power; today, the same functions can be performed using a five-ounce device that runs for many hours on a pen-light battery.

It is also important to understand that the quality of personal life can be increased with the strategies proposed in this analysis in ways that cannot easily be shown in summaries like Table 3. As mentioned, many Americans now routinely set their thermostats at 60°F throughout the day and are forced to close off parts of their homes during the winter. The analysis conducted for this study assumes that all thermostats are set at 70°F and that no sacrifice is necessary to save energy in homes. In fact, homes will almost always be made more comfortable when they are made more efficient. The quality of the natural environment can be improved in direct proportion to a decrease in the consumption of fossil fuels. Mobility can be increased if efficient cars make it possible for people to keep driving instead of relying heavily on mass transit. Equity can also be served: the growth in services shown in Table 3 can be used to increase the amenities available to low-income families more rapidly than to high-income families. A strategy that maximizes energy-use efficiency, therefore, has the effect of improving the quality of life rather than decreasing it. Far from requiring sacrifice, it provides more alternatives and more freedom.

THE POTENTIAL FOR NEW BUSINESS

Encouraging the nation's industries and buildings to increase their energy productivity will create an enormous array of profitable new businesses. Most of this growth will be straightforward extensions of existing businesses, although there will be a need for the manufacture of a variety of new parts and materials (glass, plastics, insulation, heat exchangers, computer controls, photovoltaic cells, and many others). There will be a greater need for skilled architects, engineers, and designers familiar with the subtleties of energy efficiency. Businesses will grow around the need to install, finance, insure, maintain, and operate new energy systems as well as around opportunities to audit existing facilities and undertake retrofits.

The total incremental investment required by this strategy over the 20 years would be in the range of $750 to $800 billion. This enormous number can be put in some perspective, however, by comparing it to the national energy bill in 1980--about $360 billion. Moreover, the analysis that produced the proposed strategy was indifferent to whether energy needs were met through investments in efficiency and solar technologies or investments in conventional fuels. Economic cost was the driving criterion. Therefore, the conservation and solar investments proposed in this report are necessarily comparable to those that would otherwise be needed to produce the conventional fuels that are displaced through heightened efficiency and renewable resources. The distribution of these investments by end-use sector is given in Table 4.

None of the proposed $750 billion will be spent on imported oil. Because it will be used to purchase domestic goods and services, it will have a desirable multiplying effect upon the American economy. The total investment shown in Table 4 must be compared with the investment that would be required in the absence of the programs proposed here. One can take the case of commercial buildings for an example. Later analysis will show that the investment of $110 billion in solar and efficiency equipment in commercial buildings results in a net national savings of at least 7.5 quads of energy per year by the year 2000. If 7.5 quads of energy were generated by new power plants, the capital costs of the plants would be $114 billion (assuming $1000 per installed kilowatt). This does not count the capital needed to maintain the fuel cycle, and it does not include the costs involved in operating the facility over a number of years. The investments in efficiency, of course, have very low operating costs and no fuel costs. It is interesting to observe that the efficiency investments may actually have a lower initial cost as well.

THE POLICY FRAMEWORK

A wide array of specific policy recommendations are outlined in later sections of this report. All are governed by the following principles.

1) Distortions that prevent the marketplace from making rational decisions about the value of different energy investments should be removed. Currently, energy markets are warped by large federal subsidies to energy sources. These subsidies hide the full cost of

energy from the consumer and thus discourage cost-effective investments in increased efficiency. (Several of the more easily quantified subsidies are summarized in Table 5.) Government policy has artificially encouraged inefficient energy use in many ways. Until phased decontrol of liquid and gaseous fossil fuels was begun, the combined effect of state and federal price regulations was to keep energy costs several tens of billions of dollars per year lower than they would have been in a free market. Federal insurance and legislated liability ceilings are other important measures that have helped steer the private sector toward investments it might not otherwise have made.

2) No distinction is made between investments at the margin that reduce the use of conventional fuels and investments at the margin that increase the supply, as long as the investments in efficiency and renewables neither cost more nor cause a reduction in the production of goods and services. Whereas existing policies provide rich incentives to operate inefficient factories burning foreign oil, this report assumes that an industrialist who would rather build a more efficient factory that burns far less oil deserves even-handed treatment.

3) Policies should be designed to maximize competition for energy services. The policies suggested in this study would allow utilities greater entry into markets for energy equipment on a customer's premises; at the same time, they would allow nonregulated firms greater access to markets that had previously been barred by regulation.

For example, if rooftop photovoltaics (solar cells) can generate electricity more cheaply than a utility can deliver power for equivalent purposes, the whole basis for utility monopolies over power generation is called into question. (Utility monopolies over transmission and distribution of power, however, would be unaffected by such a development.) In such an instance, rooftop photovoltaics should not be barred by regulation. If the recommendations in this study are implemented, between 10 and 15 Quads of natural gas can be displaced from traditional uses. This surplus gas could put gas utilities in a position to compete with electric utilities for many markets--especially if efficient cogeneration equipment begins to see greater penetration in buildings and industry.

4) One of the greatest market imperfections today is a lack of general access to reliable information. Several recommendations are aimed at providing both industry and consumers with solid information about the effectiveness of various technologies, with which they can make sound investment decisions.

5) While this report is structured around conservation principles, it embodies a "conservatism with a human face." Market forces are harnessed to provide the most efficient allocation of resources. However, an unbridled market cannot, will not, and perhaps should not provide for such "externalities" as environmental quality or a safety net for the poor.

The New Prosperity

Historian Frederick Jackson Turner wrote convincingly about the effect of the disappearance of the "American Frontier" on the nation's spirit. For a period afterward, the nation's prospects seemed more constrained; there appeared to be nowhere for the nation's agricultural base to expand and grow. As later historians, such as David Potter have pointed out, however, the disappearance of the frontier coincided with advent of the modern American industrial revolution. The daring, the intelligence, and the resources that had previously been devoted to settling the geographical extremities were now harnessed to the task of building a modern economic state. The result has been a century of rapid growth.

Now, as the end of the 20th century approaches, the nation appears to be exhausting yet another frontier. This time, the limiting factor is not unsettled land but inexpensive fuel. Again, some researchers are offering disturbing predictions about the consequences for America's future.

The central conclusion of this study is that, again, the nation is focusing too little attention upon the opportunities afforded by its changed circumstances. We have today a unique opportunity to begin rebuilding the nation's economy on a more productive and sustainable base. Although we have largely squandered the last eight years, enough time still remains. The change must be rapid, but it can be deliberate.

Imaginative use of technologies available to improve the productivity of fuels in short supply, along with widespread adoption of technologies to utilize renewable resources, constitute the heart of the cheapest, fastest, and safest strategy through the next twenty years.

Adoption of this strategy, however, will require us to overcome a six-decade-old set of preconceptions about the linkage between growth and energy. It no longer makes sense to equate an increase in national wealth with a growth in energy consumption. We are today at the threshold of a wide-reaching revolution in the physical and biological sciences. It is too early to predict in detail the eventual outcome of these advances. But one fact is clear: they are teaching us how to do far more with far less. Efficiency and sustainability will be the hallmarks of the new prosperity.

Table 1. END-USE ENERGY DEMAND POTENTIALS (including no renewable contribution)
(Quads* of Oil Equivalent)

Sector	1977***			2000 Potential		
	Fuel	Electric	Total	Fuel	Electric	Total
BUILDINGS	13.2	13.4	26.6	5.5	12.3	17.8
Residential	(8.8)	(7.8)	(16.2)	(3.8)	(7.1)	(10.9)
Commercial	(4.7)	(5.6)	(10.4)	(1.7)	(5.5)	(7.2)
INDUSTRY	19.8	9.3	29.1	18.7	10.7	29.4
AGRICULTURE	1.3	0.3	1.6	1.4	.3	1.7
TRANSPORTATION	19.5	--	19.5	12.6-16.5	**	12.6-16.5
Personal	(15.1)	--	(15.1)	(6.9-10.5)	**	(6.9-10.5)
Freight	(4.3)	--	(4.3)	(5.7- 6.0)	**	(5.7- 6.0)
TOTALS****	53.8	23.0	75.1	38.3-42.2	23.7	62.0-65.9

*One quad per year is approximately equal to 500,000 barrels of oil per day.

**Aggressive rail-electrification and electric-vehicle programs could create between .75 and 1.15 Quad (primary equivalent) demand for electricity in the transportation sector, with the displacement of .46-.76 Quad of petroleum (fuel) demand.

***1979 Total Consumption was roughly 79 Quad.

****Not including about 2 Quad of fuel saving possible through cogeneration

( ) = Not additive within end-use sector.

Table 2. POTENTIAL RENEWABLES CONTRIBUTION BY SECTOR
(Oil Equivalent Displaced in Quads)

Sector	Solar Thermal	Biomass*	Wind	Photovoltaics	Hydro	Total
BUILDINGS	1.9-2.3	1.0	.8-1.1	.4-.7		4.1-5.1
Residential	(1.6-1.9)	(1.0)	(.8-1.1)	(.3-.45)	--	(3.7-4.45)
Commercial	(0.3-0.4)	--	--	(.1-.25)	--	(.4-.65)
INDUSTRY	.5-2.0	3.5-5.5	--	--	--	4.0-7.5
AGRICULTURE	--	.1- .7	--	--	--	1.-.7
TRANSPORTATION	--	.4-5.5	--	--	--	0.4-5.5
UTILITIES	--	--	.5-3.4	--	3.4-3.7	3.9-7.1
TOTAL	2.4-4.2	4.8-10.5**	1.3-4.0**	.4-0.7	3.4-3.7	12.3-22.5**

*Biomass estimates are given in terms of oil displaced, rather than primary biomass supply.

**These columns do not add; high end of penetration is limited to less than total of potential applications in end-use sectors.

( ) = not additive within end-use sector.

Table 3. GROWTH RATES USED IN SETTING GOALS
(a representative sample)

		Ratio of value in 2000 to value in 1977
1.	Population	1.17
2.	Gross National Product	1.80
3.	Floor area of average new housing unit	1.16
4.	Floor space in commercial buildings	1.59
5.	Fraction of homes with central air-conditioning	1.33
6.	Fraction of homes with dishwashers	1.56
7.	Fraction of homes with freezers	1.39
8.	Fraction of homes with swimming pools	1.59
9.	Miles traveled by car (per person)	1.3-1.7
10.	Miles traveled by air (per person)	1.6-1.9
11.	Freight carried by truck and rail (ton miles)	1.80
12.	Freight carried by air (ton miles)	4.0
13.	Industrial value added	1.48
14.	Tons of primary materials (cement, steel, aluminum, etc.)	1.32

Table 4. INCREMENTAL INVESTMENTS FOR EFFICIENCY AND RENEWABLE RESOURCES* (billions of dollars)

1. Buildings		
a. Residential		
--retrofits	180	
--new construction	100	
--appliances	100	
b. Commercial		
--retrofits	75	
--new construction	45	
2. Industry		
a. rebuilding	200+	
b. direct solar heat	5-10	
3. Transportation		
a. rebuilding the automobile industry	10-30+	
b. rebuilding the railroads	50+	
4. Total	755-790	

*Not including investments in biomass or solar electric systems.

Table 5.   ESTIMATED FEDERAL GOVERNMENT SUBSIDIES IN SUPPORT OF THE ROUTINE PROVISION OF ENERGY SUPPLIES IN 1977

Federal Activities	(million 1977 dollars)
Low interest appropriations plus tax exemptions for hydroelectric and transmission facilities[a]	290
Enrichment services[b,c]	180
NRC regulatory costs[b]	146
Privately Owned Utilities	
Liberalized Depreciation[a,b]	2000 (approx)
Publicly-Owned Utilities	
Exemption from federal income taxes[a]	
Tennessee Valley Authority	130
State power authorities and municipal utilities	80
Rural Electrification Administration (REA)	294
Tax exempt bond subsidies[a]	260
Loans and loan guarantees provided by REA	260
Oil and Gas Industries	
Percentage depletion allowance[a]	
Oil	550
Gas	1150
Expensing of intangible exploration and development costs	
Oil	740
Gas	460
TOTAL	6540

Notes:

(a)   B.W. Cone et al., (An Analysis of Federal Incentives Used to Stimulate Energy Production), second revised report prepared by Battelle Pacific Northwest Laboratory, PNL-2410 Rev. II, February 1980.

(b)   General Accounting Office, "Nuclear Power Costs and Subsidies," June 13, 1979.

(c)   The subsidy for enrichment services is the estimated difference between what the private sector would have charged and what the government actually charged.

(d)   The liberalized depreciation allowance for utilities translates into an outright tax savings (instead of just a deferral) for a utility growing sufficiently rapidly or for a utility that is not growing, as long as there is significant general inflation.

Source: Robert H. Williams, Princeton University
Testimony before the Energy Conservation and Power Subcommittee, U.S. House of Representatives, February 24, 1981.

CHAPTER 1
BUILDINGS

Section One
Summary of Findings

The energy which can be saved in buildings in the United States represents the largest and least expensive source of energy that can be supplied during the next two decades. Buildings now consume about a third of all energy used in the U.S. using 13 Quads per year (Quadrillion Btu) of oil and gas and 13 additional Quads per year of primary energy to generate electricity. (see Table 1.1) It is evident that energy used for space heating and cooling dominates the demand for energy in both residential and commercial buildings and consequently these uses will receive the greatest attention in this analysis. Refrigeration, lighting, dishwashers, and other appliances were responsible for about 28% of the energy used in residences in 1980 but the importance of appliances will grow if energy used for heating and cooling can be significantly reduced. Even with the substantial improvements in efficiency anticipated for appliances, they may consume as much as half of the energy used in residences in the year 2000. The third major category of energy use in residences is water heating which now consumes about 14% of the energy used. Its share of total building demands is expected to remain relatively constant since energy used for water heating can be saved both by reducing the amount of hot water required for a task (e.g., with an efficient clothes washer) and by heating water more efficiently.

The analysis presented here will argue that given a sensible set of program initiatives it should be possible to encourage the market to act in a way that reduces this consumption by over 8 Quads during the next 20 years without the use of any solar energy equipment. (see Table 1.2 and 1.3) The potential demand reductions resulting from this study are compared with a "baseline" energy demand developed using techniques used by the Energy Information Administration. These 'conventional' forecasts, of course, also contain an assumption that building owners will react to higher energy prices by investing in efficiency; the savings

postulated here assume the impact of programs which would take savings beyond 'business as usual'. Solar heating and hot water systems, daylighting, wood heat, onsite photovoltaic equipment and rural wind machines could further reduce consumption of conventional energy sources in buildings by another 4 to 5 Quads per year by the end of the century. (see Table 1.4)

All of these savings can be achieved while maintaining the growth in the building industry, in the size of the average home and in the use of building appliances forecast in the conventional estimates of building demand. The changes in buildings examined here would, without exception, improve the comfort of buildings by reducing drafts, ensuring the comfort of occupants even during power failures, and eliminating the need to turn thermostats to uncomfortable levels or close off rooms.

Savings of the magnitude suggested are possible because five decades of inexpensive energy have left the U.S. with a stock of buildings with low energy efficiency. It is fair to say that during the past two decades the country has probably spent more money in examining energy flows in rocket nozzles than it has in examining efficient use of energy in buildings. Buildings in West Germany, Great Britain, Sweden, and the Netherlands are typically much more efficient than those in the U.S. The average building in Sweden, for example, uses approximately 35 percent less energy than the average U.S. building in spite of the severe Swedish climate. It is therefore not surprising to learn that builders who pay careful attention to energy efficiency have been able to construct buildings which use 1/3 to 1/5 as much energy per square foot as the typical U.S. building of the same general design; with careful application of conservation and solar technologies, the energy used to heat a building can be reduced by 90 percent or more.

The increase in energy prices during the 70's caused total energy use in buildings to plateau and in some

Table 1.1. U.S. BUILDINGS ENERGY USE BY SECTOR, FUEL, & END USE - 1977
(10^{15} Btu)

	Electricity[a]	Gas	Oil	Other	Total
Residential					
Space heaters	1.25	3.64	2.26	0.54	7.69
Water heaters	1.17	0.87	0.14	0.08	2.25
Refrigerators	1.49				1.49
Freezers	0.64				0.64
Ranges/ovens	0.52	0.31			0.83
Air conditioners	1.10				1.10
Lights		0.96			0.96
Other	0.68	0.48			1.15
Total Residential	7.81	5.30	2.40	0.62	16.12
Commercial					
Space heaters	0.37	1.94	1.90	0.35	4.56
Air conditioners	2.03	0.16			2.19
Water heaters	0.04	0.09	0.10		0.23
Lights	2.23				2.23
Other	0.85	0.20			1.05
Total Commercial	5.62	2.39	2.00	0.35	10.36

[a]Electricity is reported as primary energy (11,500 Btu/kwhr).

Source: Blue, 80 R-A1

cases to decline, even though the total number of homes and offices increased every year. Suddenly insulation manufacturers found their inventories depleted, real estate agents heard questions about ceiling insulation and storm windows, and the gas- guzzling office building of the sixties, with its acres of lighting and acres of glass, found itself in the midst of re-examination.

A change in attitudes is also widely evident. Nearly 11% of all households undertook a major retrofit of their residence during 1977, and nearly an equal number undertook a major retrofit during 1978. Evidence indicates that this trend may well continue with increasing strength into the 80's. A recent survey conducted by the Opinion Research Corporation for Dow Chemical found that 44% of the households interviewed were planning to take a major energy-saving action during 1980 and 1981.

Even the new home market, which for years has been dominated by convenience and luxury items such as landscaping and wall-to-wall carpeting, has undergone a major shift. The energy integrity of the home is becoming the number one buyer criterion. Energy-saving features made a clean sweep of the top five features rated "most important" in new home purchases by respondents to the Opinion Research survey. Attic insulation, wall insulation, weather stripping, storm windows and basement insulation surpassed even the ever-popular fireplace and garage door opener.

Commercial building practices have also undergone rapid change. The standard for construction of new commercial buildings developed by the building industry in the mid-1970's (ASHRAE 90-75R) calls for buildings using about half of the energy required by buildings built before 1973; an even more efficient standard is likely to be adopted within the next few years. Both professional and business publications have carried an increasing number of articles on successful energy retrofits.

The changes that are already under way will certainly lead to a significant improvement in the energy efficiency of the nation's building stock. Few buildings being built today repeat the wasteful mistakes of the past. But it will take more than a laissez-faire ap-

Table 1-2. SUMMARY OF PRESENT AND YEAR 2000 RESIDENTIAL ENERGY CONSUMPTION

	Year 1980			Year 2000 Baseline			Year 2000 Conservation Measures to Current Cost of Oil and Electricity			Year 2000 Conservation and Solar Measures to Current Cost of Oil & Electricity		
	Fuel (Q)	Elec. (Q resource)	Total	Fuel (Q)	Elec. (Q resource)	Total	Fuel (Q)	Elec. (Q resource)	Total	Fuel (Q)	Elec. (Q resource)	Total
Existing Residences Space Heating and Cooling	7.04	2.65	9.69	5.50	1.98	7.48	1.95	0.88	2.83	1.37	0.66	2.03
New Residences Space Heating and Cooling	--	--	--	1.59	1.96	3.55	0.74	0.92	1.66	0.54	0.81	1.35
Water Heating	1.29	1.00	2.29	1.46	2.11	3.57	0.75	1.26	2.01	0.55	0.92	1.47
Appliances	0.47	4.16	4.63	0.68	6.67	7.35	0.43	4.13	4.56	0.43	4.13	4.56
Residential Total	8.80	7.81	16.61	9.23	12.72	21.95	3.87	7.19	11.06	2.89	6.52	9.41

Note: The conservation and solar technology potential is based on conservation measures with a cost less than 7.5 $/million Btu or 5.7 ¢ per kilowatt hr. The additional savings for all measures with a cost between current cost and marginal cost is 0.70 quads.

Table 1.3. SUMMARY OF PRESENT AND YEAR 2000 COMMERCIAL ENERGY CONSUMPTION (Q resource)

	Year 1980			Year 2000 Baseline			Year 2000 Conservation Measures to Current Cost of Oil and Electricity		
	Fuel (Q)	Elec. (Q resource)	Total	Fuel (Q)	Elec. (Q resource)	Total	Fuel (Q)	Elec. (Q resource)	Total
Existing buildings	4.4	5.6	10.4[a]				1.57	2.44	4.01
New buildings			10.4				.16	3.09	3.25
Commercial Total	4.4	5.6	10.4	3.2	10.1	13.3	1.73	5.53	7.26

[a]Includes 0.4 other

Table 1-4. SUMMARY OF YEAR 2000
POTENTIAL SOLAR CONTRI-
BUTION: ALL BUILDINGS
(QUADS)

Residential Buildings

Wood Usage	1.0
Small Wind*	0.8-1.1
Photovoltaics*	0.3-0.45
Water Heating	0.5-0.6
Fuel	(.2-.3)
Electric	(.3)
Space Heating	1.1-1.3
New Bldgs Fuel	(.2-.3)
New Bldgs Elec	(.1-.2)
Exist Bldgs Fuel	(.6)
Exist Bldgs Elec	(.2)
Subtotal	(3.7-4.45)

Commercial Buildings

Water Heating	0.1
Daylighting	0.2-0.3
Photovoltaics*	0.1-0.25
Subtotal	(0.4-0.65)

TOTAL RENEWABLES   4.1-5.1

( ) = Non Additive
* = Not included in Table 1-2 or 1-3.

proach to overcome the influence of decades of cheap energy.

RESIDENTIAL BUILDINGS

Figure 1.1 illustrates the use of energy in residential buildings in 1980 and the demand which can be expected in the year 2000 assuming a conventional extrapolation of past behavior. The figure also indicates the savings which could be achieved given a successful program to encourage economically efficient investments which reduce demands in these buildings.

Several features of this figure deserve attention. First, it is interesting to notice that a majority of the energy required to heat and cool residential buildings in the year 2000 will be used in buildings now standing even though about a quarter of these buildings are expected to have vanished by the year 2000. (The stand-

ard forecast used here assumes that about 60 million of the 80 million residential units now standing will last to the year 2000 and that about 40 million new units will be built between 1980 and 2000.) The energy required to provide heating and cooling to residential units constructed between 1980 and 2000 can be surprisingly small given care in building design and construction. As noted earlier, the importance of appliances will increase dramatically if goals for reducing heating and cooling demands are met. Even with significant improvements in appliance efficiency, Figure 1.1 shows that appliances may consume nearly one half of the energy used in residential buildings by 2000.

The techniques used to select optimum levels of investment in buildings are summarized in curves such as the ones shown in Figures 1.2 and 1.3. These figures were constructed from a detailed analysis of the options available for reducing the energy demands of new and existing housing units. Different sets of measures were selected for different classes of existing homes (for example 'uninsulated' and 'partially 'insulated' existing units were treated separately) and for new buildings. The demand reducing technologies treated onsite solar energy equipment simply as another "energy saving" investment. Each investment was ranked in order of the 'effective cost' of the energy saved. This effective cost was calculated by computing the annual payments needed for a loan covering the cost of the incremental investment and dividing this payment by the annual energy saved. The loans were assumed to last for the expected lifetime of the equipment purchased and the interest rates were either 3% or 10% (the 'discount rates' shown on Figures 1.2 and 1.3). The analysis was all done in "constant dollars," which ignore inflation. The 3% and 10% "constant dollar" interest rates are comparable to 13% and 20% interest rates if inflation is 10%. The figures were prepared by showing how much energy could be saved if each class of investment were taken in every housing unit judged to be eligible for such an investment. For example, 0.85 quads per year could be saved in the year 2000 if R-19 insulation were added to the attic of every fuel-heated uninsulated building now standing. The effective cost of such investments would be about 50 cents per million Btu's saved. The sequence of investments selected, and assumptions made about costs and savings of each measure, are discussed in detail in later sections.

The combined figures show how much energy could be saved on a national basis given that all investments made cost less than a specified cost of fuel (or electricity). A perfectly rational investor would invest up to the point where the cost of the last incremental investment exceeded the cost of competing energy. For reference, several different possible competing costs of energy are shown in the Figure. The current cost of home heating oil, for example, is about $7.50 per million Btu. Using the baseline assumptions made about

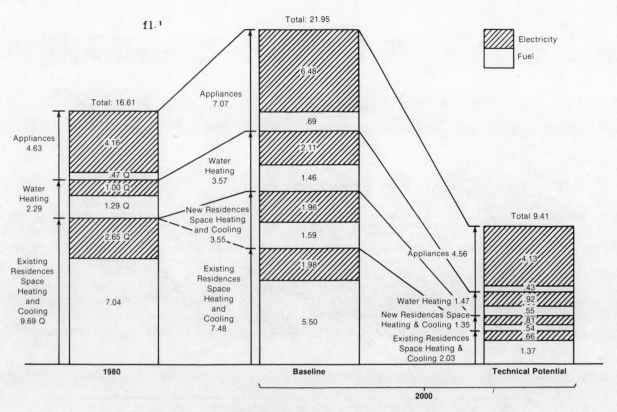

Figure 1.1. **Graphic Comparison of Annual Energy Use, in Quads (10¹⁵ Btu), 1980 and 2000**
(Baseline and Technical Potentials)

future energy prices shown in the appendix, the levelized cost of residential heating fuel (oil and gas weighted by market share) is assumed to be $7.15/MBtu in 1990 and to reach $8.70/MBtu by the year 2000. The current cost of heating oil is about $7.50/MBtu. The marginal cost of electricity is estimated to be 5.7¢/kwh in 1980 and 6.4¢/kwh in 2000 (not including transmission or distribution costs) see the section on "Baseline Assumptions" at the end of this volume. The costs actually used to select 'cost effective' conservation and solar investments are shown by solid lines in the figures.

Technical Potentials

Space Heating in Existing Residential Buildings

The potential for saving energy by retrofitting existing buildings is shown in Figures 1.4 and 1.5. These figures are constructed from a detailed analysis of retrofit conservation and solar measures which were applied to several classes of "typical" existing housing units. Techniques considered include use of simple techniques for reducing air-infiltration, using storm windows, adding ceiling insulation, efficient furnaces, heat pumps and use of active and passive retrofits.

The wide-range of measures applicable below the current cost of fuel and electricity demonstrates the inefficiency of the existing housing stock and the savings which are possible given an aggressive home retrofit program. Savings of over 5 Quads per year by the year 2000 are possible using the criteria discussed above.

Some of the major measures considered in these retrofits are shown explicitly on these figures. The rest are referenced by "measure nmbers" which can be located in a key found later in this text.

Space Heating in New Residential Buildings

Energy-saving measures can reduce the annual energy requirement of a typical new residential unit in an average U.S. climate to 14 MBtu per year. The average new house built in 1980 consumes about half as much energy per square foot and further reductions are plainly achievable using better insulation, improved furnaces and air conditioners, tighter construction, better windows and doors, and a variety of other straightforward design improvements. The savings achievable in buildings built between 1980 and 2000 could be 1.9 quads per year by the year 2000 even if no solar heating is used. These savings are about half the savings possible from retrofits. Figures 1.6 and 1.7 summarize savings potentials for new homes.

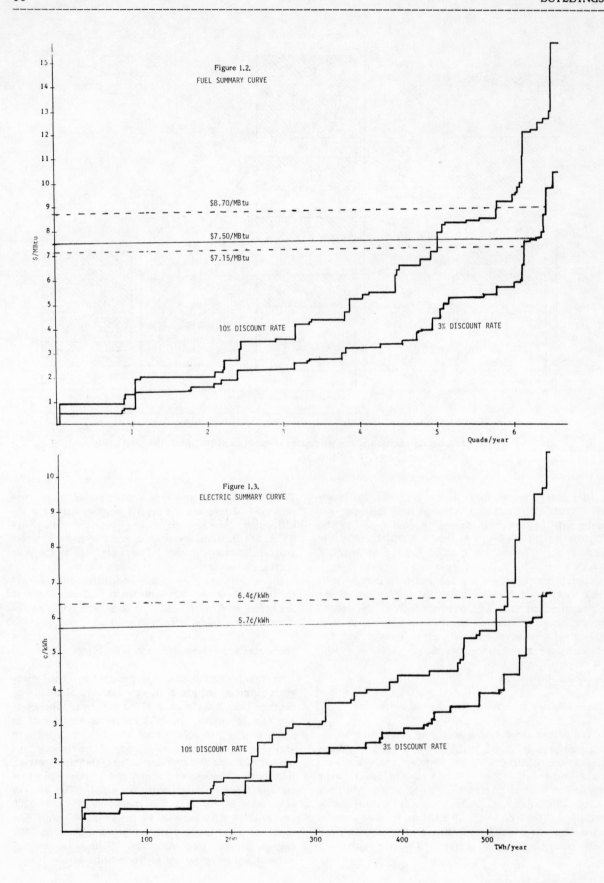

Figure 1.2.
FUEL SUMMARY CURVE

Figure 1.3.
ELECTRIC SUMMARY CURVE

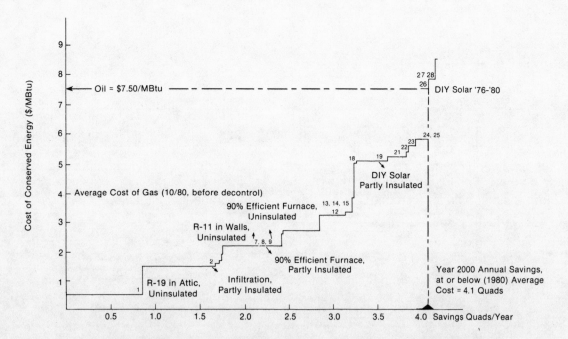

Figure 1.4. **Year 2000 Supply Curve of Conserved Energy (in quads/year) for Space Heat of Fuel Heated Dwellings Built Through 1980.**

Year 2000 baseline annual use for this sector is _5.5_ Quads, where "baseline" assumes continuation of 1980 average unit energy consumption for existing stock or new additions in that year. Unit cost of conserved energy (in constant 1980 $) assumes that all increased costs are amortized over the useful life of this measure, using a 3% (real-dollar) interest rate. Potential annual savings in 2000, at or below today's cost of oil (7.5 $/MBtu) is 4.1 quads, or 75% of the year 2000 baseline.

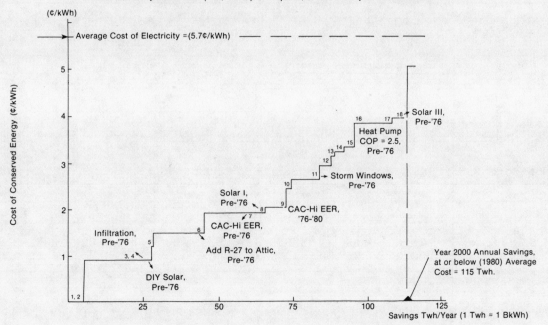

Figure 1.5. **Year 2000 Supply Curve of Conserved Energy (in Twh/year) for Space Heat in Electric-Heated Dwellings Built Through 1980 and All Central Airconditioning in Dwellings Built Through 1980.**

Year 2000 baseline annual use for this sector is _172_Twh, where "baseline" assumes continuation of 1980 average unit energy consumptions for existing stock or new additions in that year. Unit cost of conserved energy (in constant 1980 $) assumes that all increased costs are amortized over the useful life of the measure, using a 3% (real dollar) interest rate. Potential annual savings in 2000, at or below today's average cost (5.7¢/kWh) is 115 Twh, or_67_% of the year 2000 baseline.

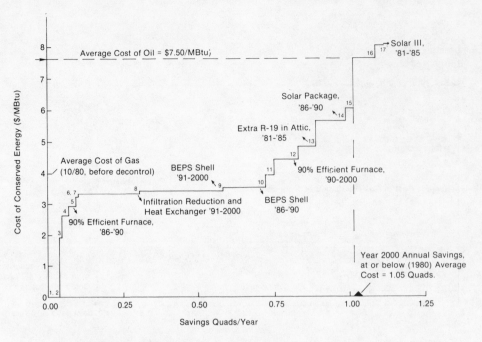

Figure 1.6. **Year 2000 Supply Curve of Conserved Energy (in quads/year) for Space Heat of New Fuel-Heated Dwellings.**

Year 2000 baseline annual use for this sector is 1.59 quads, where "baseline" assumes continuation of 1980 average unit energy consumption for existing stock or new additions in that year. Unit cost of conserved energy (in constant 1980 $) assumes that all increased costs are amortized over the useful life of the measure, using a 3% (real-dollar) interest rate. Potential annual savings in 2000, at or below today's average cost ($7.50/MBtu) is 1.05 quads, or 66% of the year 2000 baseline.

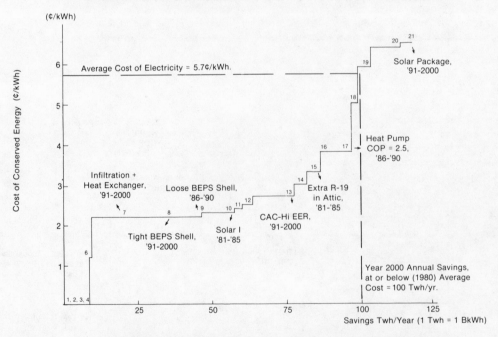

Figure 1.7. **Year 2000 Supply Curve of Conserved Energy (in Twh/year) for Space Heat in Electrically-Heated New Homes and All Central Airconditioning in New Homes.**

Year 2000 baseline annual use for this sector is 170 Twh, where "baseline" assumes continuation of 1980 average unit energy consumptions for existing stock or new additions in that year. Unit cost of conserved energy (in constant 1980 $) assumes that all increased costs are amortized over the useful life of the measure, using a 3% (real-dollar) interest rate. Potential annual savings in 2000, at or below today's average cost (5.7¢/kWh) is 100 Twh, or 59% of the year 2000 baseline.

Water Heating

The savings possible in water heating by 2000 (for all residential buildings) is only slightly less than the potential savings in space heating for new residential buildings built between 1980 and 2000. From a baseline year 2000 use of 3.57 quads, the energy needed to provide hot water to showers, dishwashers, clothes washers, and other uses could be reduced by slightly more than 1.5 quads per year. These savings are made possible by using inexpensive low-flow showerheads and faucets, more efficient water-using appliances, solar hot water heaters, and by increasing the efficiency of conventional water heaters.

Figures 1.8 and 1.9 summarize the potential savings. Once again, a wide range of savings is shown which could be achieved at costs less than the current cost of conventional fuels. The diversity of measures listed highlights the importance of not only increasing the efficiency of water heaters, but also decreasing standby losses through insulation and improving the efficiency of hot water-using appliances.

Appliances

One of the most dramatic change in the use of energy shown anywhere in this analysis was found in residential appliances.

Figures 1.10 and 1.11 show how appliance demand can be reduced by nearly 3 quads per year from the baseline case. Efficiency improvements were considered in refrigerators, freezers, clothes dryers, room air conditioners, ranges, televisions, dishwashers, swimming pools and lighting. Since most appliances are difficult to retrofit, a significant improvement in energy efficiency was assumed only by the replacement of existing models when they wear out with more efficient ones; the EIA schedule of replacements was used in the analysis.

The impact of improvements in appliance efficiency are often difficult to calculate because of their diversity. For example, the use of pilotless ignition in gas ranges can save 30 to 40% of consumption. Two studies of improvements in refrigerators concluded that improved motors and condensors, increased condensor and evaporator surfaces, addition of an antisweat heater switch, improved insulation and other improvements can reduce energy consumption by factors of 3-4 from today's models which use 1800 kwh per year. (ADL 1977; ORNL 1977) A prototype refrigerator recently already tested by Amana uses only 675 kwh per year; this is within 35% of the technical potential for refrigerators (500 kwh/year) assumed in this study for the year 2000.

Improved temperature and moisture sensors (which can automatically turn off a dryer when the clothes reach the desired dryness) can save 10% of current consumption of electric clothes dryers. Another possible improvement is an adaption of a technique used in dry cleaning; i.e., the use of a vacuum pump to permit rapid low temperature evaporation of water. Later in this century clothes dryers may use a heat exchanger to recover energy from the exhaust air stream. The consumption of electric clothes dryers is expected to improve from 900/kwh/year/unit for new 1980 models to 600 kwh/year/unit with such improvements.

Some Empirical Support

It is reasonable to ask for empirical evidence that the kinds of savings indicated in the previous discussion can actually be achieved. Data on the energy performance of real buildings is, unfortunately very sparse. Efforts now underway will provide a much improved data base in the next few years. What follows is the best currently available.

Residential Building Retrofits

Table 1.5 presents data on residential retrofits. Some of the cases identified represent one home, while others represent the averages of a large group of retrofits. This data is graphically presented in Figure 1.12. In this figure, each home (or set of homes) is plotted as a point.

The Twin Rivers House and perhaps some of the other houses shown in the Figure, almost certainly meet the 1982 proposed BEPS standard for new houses with retrofits costing less than $3000. The achievements of the Oregon houses deserve special attention because of the size of the sample (1896 single-family units), the measured fuel use reduction (20%) and the cost of conserved energy (2¢/kwh).

Three kinds of buildings are shown:

o Homes pressurized with a fan, air leaks detected with smoke sticks and/or an infrared camera, and fixed on the spot. These homes are marked with a "+." They seem to give the best return on investment, and hence the lowest cost of conserved energy. They seem to lie along or below the sketched line. They are all fuel heated. Princeton experience shown that this costs $150, but saves 9% of the space heat.

o Homes in which infiltration and attic bypasses were not found and fixed. These retrofits have mainly been sponsored by utilities, and by the Canadian CHIP program.

o Apartments which have been retrofitted by Scallop Thermal Management, a subsidiary of Shell Oil. Scallop is noteworthy, because it is willing to supply its own capital. Scallop has not tried to seal infiltration and bypasses.

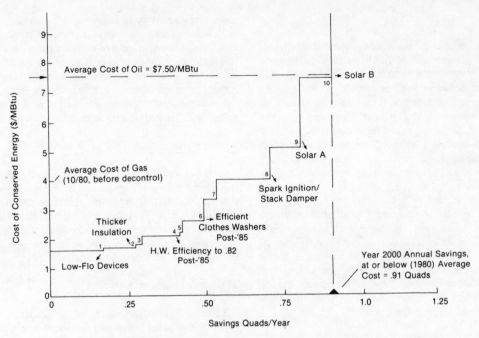

Figure 1.8. **Year 2000 Supply Curve of Conserved Energy (in quads/year) for Fuel-Heated Water Heaters.**

Year 2000 baseline annual use for this sector is 1.46 quads, where "baseline" assumes continuation of 1980 average unit energy consumption for existing stock or new additions in that year. Unit cost of conserved energy (in constant 1980 $) assumes that all increased costs are amortized over the useful life of the measure, using a 3% (real-dollar) interest rate. Potential annual savings in 2000, at or below today's cost of oil ($7.50/MBtu) is 0.91 quads, or 62% of the year 2000 baseline.

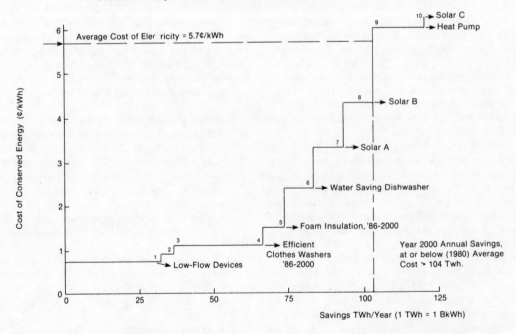

Figure 1.9. **Year 2000 Supply Curve of Conserved Energy (in Twh/year) for Electric Water Heaters.**

Year 2000 baseline annual use for this sector is 173 Twh, where "baseline" assumes continuation of 1980 average unit energy consumption for existing stock or new additions in that year. Unit cost of conserved energy (in constant 1980 $) assumes that all increased costs are amortized, over the useful life of the measure, using a 3% (real-dollar) interest rate. Potential annual savings in 2000, at or below today's average cost of 5.7¢/kWh, is 104 Twh, or 60% of the year 2000 baseline.

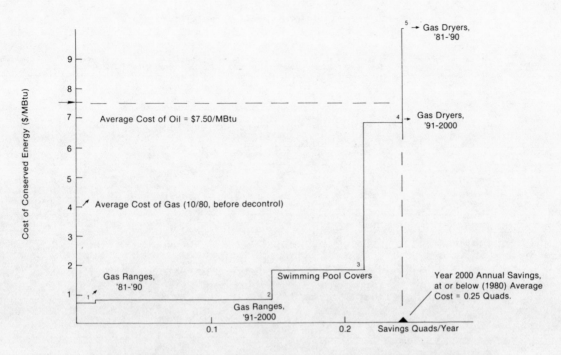

Figure 1.10. **Year 2000 Supply Curve of Conserved Energy (in quads/year) for Fuel Appliances.**
Year 2000 baseline annual use for this sector is 0.68 Quads, where "baseline" assumes continuation
of 1980 average unit energy consumptions for existing stock or new additions in that year. Unit cost
of conserved energy (in constatnt 1980 $) assumes that all increased costs are amortized over the
useful life of the measure, using a 3% (real-dollar) interest rate. Potential annual savings in 2000,
at or below today's cost of oil ($7.50/MBtu) is .25 quads, or 37% of the year 2000 baseline.

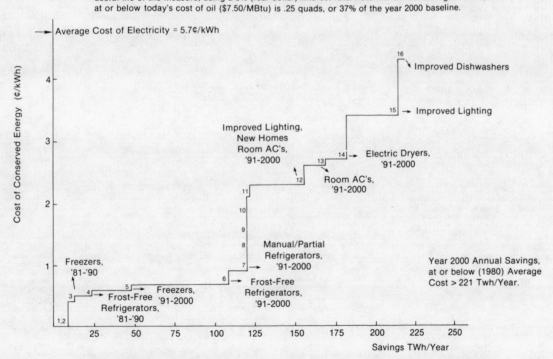

Figure 1.11. **Year 2000 Supply Curve of Conserved Energy (in Twh/year) for Electric Appliances.**
Year 2000 baseline annual use for this sector is 581Twh, where 'baseline" assumes continuation
of 1980 average unit energy consumption for existing stock or new additions in that year. Unit cost
of conserved energy (in constant 1980 $) assumes that all increased costs are amortized over the
useful life of the measure, using a 3% (real-dollar) interest rate. Potential saviangs in 2000, at or below
today's average cost of 5.7¢/kWh is 221 Twh, or 38% of the year 2000 baseline.

Table 1.5. RESIDENTIAL RETROFIT SURVEY, GAS-HEATED HOMES

Label	Number of Homes	Location/Sponsor	Annual $Saved/$Invested	Energy (MBtu) after/before		Savings (MBtu)	Cost of Conserved Energy at 6.7% Capital Recovery Rate	Heating/ Heating and Hotwater	Comments
G1	1	Bowman House, Maryland NBS total	$\frac{\$175}{\$2840}$	$\frac{52}{125}$	= 41.5%	73.	$2.59/MBtu	H	
		(a) storm windows	$\frac{\$75}{\$780}$	$\frac{93}{125}$	= 74%	30.	$1.84/MBtu	H	
		(b) insulation	$\frac{\$100}{\$1870}$	$\frac{83.0}{125}$	= 66.7%	41.8	$3.00/MBtu	H	
		(c) infiltration	$\frac{\$0}{\$190}$	$\frac{125}{125}$	= 100%	0.	0/MBtu	H	
G2	1	Twin Rivers, NJ Princeton	$\frac{\$120}{\$1750}$	$\frac{20}{60}$	= 33.3%	40.	$2.93/MBtu	H&W	
G3a	6	Freehold, NJ Princeton	$84	$\frac{96}{}$	= 80%	24.	$.61/MBtu	H	No hot water system retrofits
G3b	6		$\frac{\$1105}{\$260}$	$\frac{143}{173}$	= 83%	30	$.58/MBtu	H&W	This included the hot water retrofit
G4a	5	Tom's River, NJ Princeton	$\frac{\$31}{\$200}$	$\frac{60.5}{67.5}$	= 89.6%	7.	$1.91/MBtu	H&W	This was a partial retrofit
G4b	5		$\frac{\$95}{\$500}$	$\frac{42.5}{67.5}$	= 63%	25.	$2.14/MBtu	H&W	This was a partial retrofit
G5	1	Miller House, NJ	$1000	$\frac{103}{134}$	= 76.8%	31.	$2.16/MBtu	H&W	
G6	1	Settles House	$500	$\frac{74}{90}$	= 52%	10.	$2.09/MBtu	H&W	
G7	74	Ramsey county, Minn. Northern States Power	$789	$\frac{180.4}{192}$	= 93.9%	11.6	$1.67/MBtu	H	
G8	8	Oakland, CA CSH	$357		= 93%	10.	$.35/MBtu	H	
G9	10	Atlanta, Ga. CSH	$874		= 76%	36.	$1.62/MBtu	H	
G10	29	St. Louis, Mo. CSH	$1752		= 82%	27.	$4.35/MBtu	H	
G11	10	Tacoma, Wash. CSH	$1779		= 53%	75.	$1.59/MBtu	H	
G12	10	Chicago, Ill CSH	$2162		= 59%	120.	$1.21/MBtu	H	
G13	17	Minneapolic, MI CSH $1761	——		= 78%	54.	$11.93/MBtu	H	
G14	13	Fargo, N.D. CSH $1638	——		= 59%	90.	11.23/MBtu	H	
G15	17	Colorado Springs, CO CSH	$1852		= 50%	120.	$1.03/MBtu	H	
G16	14	Charleston, S.C. CSH $996	——		= 73%	15.	$4.45/MBtu	H	Bottled gas
01	12	Portlant, ME CSH	$2377		= 60¢	120.	$1.33/MBtu		
02	12	Easter, PA CSH	$798		= 74%	33.	$1.62/MBtu		
03	4	Washington, D.C. CSH	$2972		= 67%	94.	$2.08/MBtu		
04	7-800,000	Canada CHJP	$910	$\frac{500-550}{1000}$	= 80-85%	150-200 gal.	$.30-.40/gal.		Estimate no measurement
E1	500	Tennessee TVA	$310	$\frac{7937}{10,145}$	= 78%	2211.	.9¢/kWh		Savings only for heating
			$310	$\frac{2012}{2646}$	= 76%	632.	3.3¢/kWh		Savings only for cooling
			$310	$\frac{9949}{12795}$	= 75%	2843.	.7¢/kWh		Combined heating and cooling
E2	138	Tennessee TVA	$150-$300	$\frac{123353}{10446}$	= 75%	4111.	.24-.48¢/kWh		A Hic insulation only
E3	81	Tennessee	$440	$\frac{11367}{17488}$	= 65%	6121.	.48¢/kWh		Weatherization
E4	1896	Oregon Pacific Power and Light	$1335	$\frac{17044}{21305}$	= 80%	4261.	2¢/kWh		

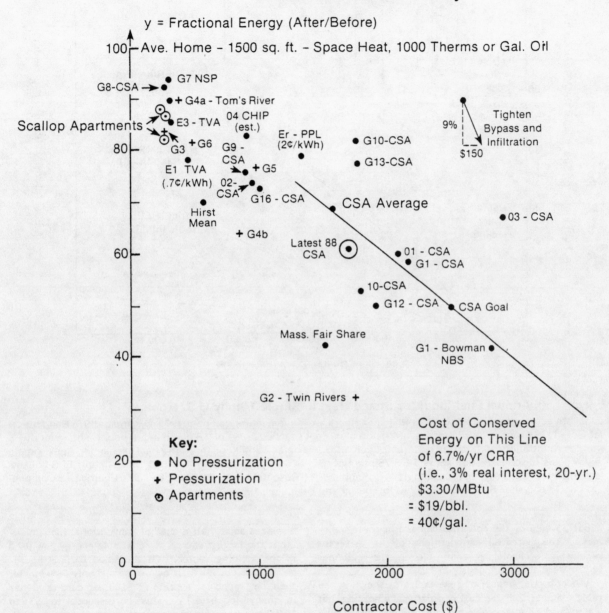

Figure 1.12.
Residential Retrofit Survey

The degree of savings projected at the costs shown by these retrofits are within 20 percent of the results of the supply curve analysis detailed below.

New Residential Buildings

There is considerable evidence indicating that homes can be built to meet or exceed the savings shown on the supply curves. Figure 1.13 illustrates the remarkable success of several different buildings in re-

ducing conventional heating requirements. These homes are well beneath even the "strict" level for the proposed Building Energy Performance Standard, and some show reductions in energy demand of more than 90% from 1975/76 practice.

Renewable Energy Technologies

Renewable energy technologies could provide up to 3.7-4.45 quads in residential buildings by 2000, or

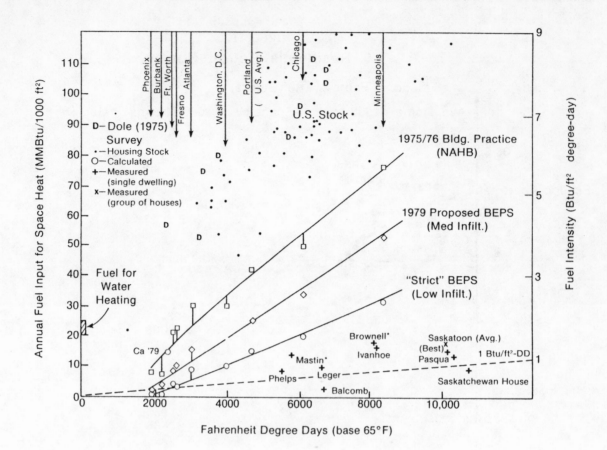

Figure 1.13. **Annual Fuel Input for Space Heat in Single-Family U.S. Homes.**
Dots are actual gas sendout to residential customers for space heat for calendar 1978. D's are the 1975 Rand report of Dole 1979 proposed BEPS leaves infiltration at current practice levels of 0.6 air changes/hour (ach); strict BEPS reduces infiltration to 0.2 ach but restores 0.4 ach with mechanical ventilation through a heat exchanger. Approximate extra costs for conservation above 1975/76 practice: BEPS, $1000; tight BEPS — $1500; better-than-BEPS houses: several hundred to several thousand dollars. Source: A. H. Rosenfeld et al., BECA-A: Building Energy Use Compilation and Analysis, LBL 8912, submitted to Energy and Buildings, 1980.

about 35-40% of the potential demand in these buildings. The potential contributions of six different renewable technologies applicable in the buildings sector were shown previously in Table 1.4.

The potential impact of renewables as shown in the table assumes that the cost of solar space heating, hot water, and wind technologies will decline by 20-40% (in constant dollars) over the next 20 years and that installed photovoltaic systems will reach the DOE cost goals for 1986-2000.

Residential Wood Usage

There are already five million homes with wood stoves and 18 million with fireplaces. These homes use about 1.0 quad of wood annually (poll conducted by Gallup for the Wood Energy Institute 1979). About half of the wood is used in stoves. We estimate that it

displaces about half a quad of conventional fuels, making it the largest source of renewable energy now used in the building sector. Of the wood burning households, 42% cut all their own wood and only 36% buy all their own wood (at a price of $100 per cord or higher in many localities). This corresponds to $3 to $6/MBtu, or nearly the present price for gas and oil. Although we expect residential woodburning to rise to 1.5 to 2.0 quad during the 1980's, a number of factors should force wood consumption back down to a stable consumption rate of approximately 1.0 quad at 2000. Among these factors are improvements in efficiencies of wood stoves, a desire to reduce the labor of cutting and splitting wood for the large segment of the population that cust its own wood, and market pressures from wood-derived methanol or other biomass-uses of wood that would drive the cost still higher. Lastly, more intensive uses of forests resulting from greater use of biomass will decrease the availability of fallen and waste wood supplies that can currently be cut "free."

Solar Heating in New Housing

Some passive heating techniques are clearly competitive with average consumer costs today. Figure 1.14 illustrates the performance of a number of recent designs, many of which are better than the proposed BEPS standards. A detailed regional analysis is necessary to understand how much of the new housing stock is to be built in regions of high insolation and/or high heating demand that make solar heating attractive. Solar space heating is more than competitive if oil prices rise to $60/BBL by 2000, and a very high penetration can be expected if consumers are induced to invest up to the social cost of conventional supply. In addition, we assume that many new homes take advantage of direct solar gain by reorienting their distri-

bution of window area towards the south. From the supply curves, it is estimated that these units could displace as much as 0.3-0.5 quads.

Space Heating in Existing Homes

A wide range of approaches and technologies may be used to realize the potential for solar heating in existing housing. Choices can be made among a variety of active and passive techniques. In many cases, the active solar retrofit is the only one that can be placed to receive enough sunlight or that can be installed in a way that is aesthetically acceptable. Passive solar retrofits may include the additions of greenhouses or sunspaces; homes can be improved by adding more south facing windows (particularly in the case of an addition), by installing glazing over an existing masonry walls to create a passive Trombe wall, or by adding thermosyphoning collectors to a south wall. Some-

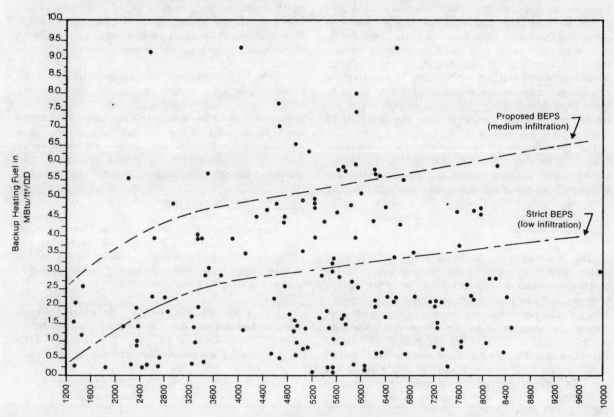

Figure 1.14. **Fahrenheit Heating Degree Days**
Estimated backup heating fuel for 150 winners in the HUD Passive Solar
Competition compared with proposed BEPS standards.

times awnings, windbreaks, shade trees, porches, or moveable insulation panels considered passive retrofits. However, they are not included in our calculations.

The number of homes with adequate sunlight exposure for a solar retrofit is not well known. An audit conducted for the RCS program in Portland, Oregon, found that three-fourths of the homes inspected had adequate access for solar water heaters. Some of the homes, however, had areas of south-facing roof that would not have been sufficient to support the greater demands of a space-heating system. Determining the potential for passive retrofits is also difficult and the range of estimates is wider. Preliminary surveys in Iowa, Oregon, and California suggest that 50 to 60% of existing single-family homes have adequate access for passive retrofits. (Hull et al. 1979; SERI audit of RCS 1980; California Energy Commission, Wornum Report 1980). On the other hand, the California Energy Commission study has determined that only about 25% of the units in large California apartments would be suitable for retrofits. Based on these sample surveys, it is assumed that up to 60% of the existing housing stock has access for a solar heating retrofit. (Since 60 million homes will survive to 2000, 36 million units represents the maximum technical potential for solar space heating retrofits.). Using the supply curves, solar heating retrofits appear to have the potential to displace up to 0.8 quad by 2000.

The combination of the markets for solar space heating in both retrofits and new housing is large by any standard. For both categories, it is estimated that active or passive solar space heating measures by 2000, could potentially displace 1.1-1.3 quads.

Solar Water Heating

Solar water heating is competitive with electric water heating in parts of the country today. Heat pump water heaters have recently entered the market however, and are expected to provide strong competition for solar water heaters in parts of the country reducing the level of solar use which that otherwise occur in the electric water heating market. Projected decreases of 2-5% per year in the real (constant dollar) cost of solar hot water systems and increases in the price of natural gas will make solar water heating competitive with gas water heating in many parts of the country before 1990.

It is estimated that aggressive programs to induce consumers to invest in water heaters up to the marginal cost of new supply could result in solar water heaters in 40-50 million dwelling units by 2000. Since hot water consumption is expected to decrease as a result of conservation measures, the energy contribution of these units would be 0.5-0.6 quad.

Residential Wind Supplies

Over 6.5 million rural households in America are located in regions windy enough to make small wind machines competitive as sources of electricity. Existing machines would produce electricity in these areas for 4¢/kwh with the 40% tax credit (6¢ without the credit) in the windiest regions where the average wind velocity at the top of a 40-foot tower is 10 mph or higher. Approximately 400,000 rural households are located in such regions. Breaking down this population by the marginal cost of electricity in 7 groups and finding the appropriate degree of penetration in each group, we find that small wind machines have the potential to displace 0.8 to 1.1 resource quad by 2000.

Photovoltaics

By 2000, the penetration of photovoltaics is expected to be substantial, almost all of it in new homes where panels can be used as part of the building shell. New Residential buildings by 2000 could potentially to displace 0.3 to 0.45 quad of resource energy with photovoltaics.

Figure 1.15 illustrates the prototype design analyzed. Table 1.6 indicates that an efficient building with a photovoltaic system could meet 100% of its energy needs even in Madison, Wisconsin. The house would be a net producer of electricity in California and Washington. (In most such designs the house would both buy and sell electricity to a utility rather than storing electricity in batteries located in the house.)

Summary of Policy Options

A combined buildings program should be developed with the clear, coherent objective of creating an aggressive, profitable business capable of saving energy in buildings. The program should be constructed around the following principles:

(a) It should address buildings as integrated entities (as the building industry must) and deal systematically with the full range of efficiency and renewable technologies

(b) It should approach the building industry in the context of their normal business practices

(c) It should provide a single point of contact for the building industry in DOE and base its programs around clear objectives endorsed and reviewed by the industry.

(d) It should ensure that federal buildings and buildings built or operated with significant

Figure 1.15.   Typical Photovoltaic Array (about 380 ft^2) for Residential Electricity Production

Table 1.6. ANNUAL ELECTRICITY USE AND GENERATION OF PROTOTYPE
HOUSE WITH 4.5 KILOWAT (380 sq. ft.) PHOTOVOLTAIC ARRAY

	Fresno, CA	Washington, D.C.	Madison, WI
Baseline house without PV	5230	5000	5800
PV output used by house	2830	2160	2210
PV output sold to utility	4990	3650	3590
Electricity purchased	2400	2840	3590
Net sales of PV output to utility	2590	810	0

federal support are models of cost-effective energy-efficient designs and that lessons learned in the research program are rapidly implemented in federal buildings.

Retrofits of Existing Residential Buildings

At present no major industry specializes in retrofitting existing residential buildings to save energy. A program designed to encourage the development of such an industry should be a critical part of a balanced national energy program. The work could be divided into three generic categories:

o   designing an applied research program to determine the best technical approaches for retrofitting buildings with current and future buildings technology and the most effective way to communicate this information;

o   working through utilities to develop a competitive supporting industry capable of offering a single package of audits, retrofits, and financing;

o   rationalizing existing federal incentives to encourage optimum building retrofit package designs selected by industry (rather than by arbitrary government choices) and to meet the special needs of low income families and renters.

Estimated impacts of these programs are summarized in Table 1.7.

Information and Applied Research. Actions explored under this program fall into three categories:

o   an applied reserach program designed to measure the effectiveness and improve the reliability of different auditing procedures and retrofit techniques carried out through regional centers supported jointly by the federal government, professional organizations, builders, utilities, state and local governments, and other interested parties;

o   programs designed to help the market to operate more effectively by providing individual and institutional decision-makers with better information about building-energy performance and energy costs. This could include, the development and testing of a building energy rating system, similar to the mpg rating used for new cars, that can be used by home buyers, lending institutions and others and; support for experiments in feedback billing and metering from "report card billing" to special meters which instantaneously display information on rates of energy usage; and the development and testing of simple, yet relatively accurate formulas for lending institutions to determine design energy use for new and existing homes

(and, if necessary, a requirement by Federal financial insurance agencies that lending institutions include energy costs within their determination of loan qualifications); and

o   a program of direct public outreach designed to increase the awareness of and response to RCS, tax credits, and other programs through state and local governments, community groups, and others, and including community based workshops on audits and "do-it-yourself" retrofit measures.

Strengthen the Residential Conservation Service Program. The Residential Conservation Service (RCS) program was designed to encourage utilities to move actively to broaden their investment portfolios with investments in building. The program could be adjusted to encourage smaller, nonregulated enterprises to benefit and to respond more immediately to the products of the applied research program just described. In addition, this program should be recognized as possibly the most effective means of providing homeowners with specific information on the possible investments in their residence, encouraging their undertaking cost-effective measures, and facilitating a more productive use of capital by providing easy access to financing.

Five major changes in the currently authorized RCS program are examined:

o   improving the skills of energy auditors through expanded support for text materials for training programs, equipment for training facilities, and increased support for auditor training at technical and vocational schools; assistance to the states in developing auditor certification; assistance in acquiring state-of-the-art auditing equipment; and providing feedback to auditors to sharpen analytical and communications skills:

o   encouraging auditors to make simple retrofits during their visit to the residence such as installing low-flow showerheads, water heater insulation, and other cost-effective measures which are inexpensive and simple to install; and offering comparable subsidies for such low-cost measures installed by other qualified audit firms through a voucher system;

o   providing incentives for private auditing firms not associated with utilities and revising the RCS regulation to require that utilities offer to share costs with private auditing firms that provide at least comparable services to their customers (the utility share of the cost being equal to the cost of the audit if conducted by the utility itself), and direct federal credits

Table 1.7. COMBINED RETROFIT PROGRAM COSTS AND IMPACTS

Program	Total Savings (MBD)[a]	Total Investment ($ billion)[b]*	Cost of Saved Fuel ($/Bbl)[c]*	Est. Federal Program Cost ($ billion)[d]*
1) Information & Applied Research				
Fuel Saving	1.1	64	18	
Electric Savings	.3	12	13	
SUBTOTAL	(1.4)	(76)	(17)	0.23
2) Enhanced RCS				
Fuel Savings	.2	11	17	
Electric Savings	.1	.2	6	
SUBTOTAL	(.3)	(13)	(14)	0.13
3) Low Income/ Rental				
Fuel Savings	.5	32	20	
Electric Savings[e]	.15	6	13	
SUBTOTAL	(.65)	(38)	(19)	11-25 low income[f] 0.5 rental
4) Finance				
Fuel Savings	.2	33	52	
Electric Savings	.1	7	22	
SUBTOTAL	(.3)	(40)	(42)	20[f]
TOTAL (may not add due to rounding)	2.5	169	21	

[a] primary energy equivalent of electricity using 11,500 Btu/kwh

[b] total investment required from all sources for equipment or services

[c] computed using financing available for investor-owned utilities (fixed charge rate = 0.116 in constant dollars)

[d] estimated cost of federal programs in each subgroup

[e] assumes that 15% of rental units react to programs (1) and (2) as if they were owner-occupied (pure speculation)

*1978 constant dollars

for homes served by oil and bottled gas companies to cover all or part of the cost of an audit;

o providing easy access to financing through the RCS process including an expanded program of support for one-stop-shopping services and encouragement for auditing firms to facilitate consumer financing as required by law; and

o increasing the DOE staffing effort for RCS at both Washington and Regional office levels.

Low-Income Households and Rental Housing. The programs described previously have assumed that households are able to respond to information made available to them about economical retrofits. Low-income families and families living in rental housing, however, may be unable to take advantage of this information or of the many forms of financial incentives (like low-interest loans or tax credits). The number of housing units occupied by renters and low-income families is surprisingly large--30 to 40 percent of all residential units. Many of these low-income and rental dwellings are extremely inefficient. When the Mellon Institute divided residential units into low-, medium- and high-energy efficiency; 66% of low-income and 51% of rental dwellings fell into the "low" category. These households are the ones most often forced to respond to rising fuel bills by sacrificing comfort and health. Thus, improvement in these units should be measured by improvements in living conditions as well as the net energy that can be realized. From either viewpoint a more effective way of approaching low-income housing must be implemented.

For low-income households, large-scale retrofitting can only be accomplished through some form of direct financial assistance such as that now available through the weatherization program. Approximately 16 million units now qualify for this program, but only a fraction of these will be served in the next ten years at the present level of funding.

Designing programs to assist renters has proven extremely difficult. While financial incentive programs available to the rental market have received poor responses, such programs have been limited since the rental market has rarely been specifically addressed by utility or government initiatives. In addition to financing programs, there are a number of actions which should be undertaken to encourage retrofitting of the rental stock. These initiatives include: (1) an experimental locally-administered program to mandate retrofits for rental property at the time of sale; (2) other local demonstration programs designed to modify rent control laws and rental contract agreements to allow building owners to pass retrofit costs through to tenants, when retrofit investments would work to both the tenants and owners advantages.

Actions examined under this program are:

o development of a program to shift the funds (available from the Windfall Profits Tax Bill) from fuel bill subsidies to permanent savings, through weatherization of low-income households;

o development of an experimental grant program for low-income households, which if successful could be expanded nationwide by 1984;

o financial support for state and local demonstration programs for conservation in rental buildings; including financial incentives, local retrofit mandates, and programs aimed at tenant information and energy-using behavior.

Financial Incentives. In the section of this report dealing with technical potentials, a case is made that virtually all of the conservation and solar measures considered are clearly cost-justified, even at today's average cost of energy. Nevertheless, there is still a justification for using well-designed and carefully

monitored financial incentives, wherever these can help to offset institutional and market barriers and speed the widespread adoption of such measure.

o the fragmentation of current incentives results in a confusing array of eligibility requirements and procedures. Some systems are eligible for several incentives while others receive none;

o rigidities in current programs lead to sub-optimal retrofit investment decisions. For example, the structure of the Solar/Conservation Bank makes it difficult to obtain support for an integrated conservation-plus-solar package.

The proposals outlined below try to address each of these problems, suggesting a few specific changes in financial incentive programs now in effect, but most importantly, arguing that all levels of government must work together, with utilities, to improve the coherence of their separate programs, increase their flexibility to respond to diverse needs and opportunities, and above all to improve feedback channels that can increase their cost-effectiveness over time.

Actions under this program include:

o establishment of programs of technical assistance, rate-case intervention, and demonstration funding to induce utilities and PUC's to adopt marginal cost pricing and to authorize or require direct utility financing of on-site investments, including the rapid development and testing of a variety of zero-interest loan programs, and programs to leverage federal funds (such as low-income weatherization funds) with utility funding;

o utilization of the full funding authority and flexibility of the Solar and Conservation Bank to supplement utility financing, for well-balanced "packages" of solar and conservation measures;

o demonstration through the Solar and Conservation Bank of a progressive incentive payment, with the amount of the incentive linked to expected energy savings and consumer investment;

o federal support of monitoring, evaluation and information exchange for conservation and solar loan and other financial assistance programs, including comparisons of the results of programs using different approaches.

New Residential Buildings

An Integrated Research Program for New Residential Buildings. A carefully designed research program should be developed for new building designs. Experiments should include a variety of shell design options, new appliances, and solar equipment. Photovoltaic demonstrations should be included when appropriate.

The experiments should allow builders, building code officials, and others to familiarize themselves with a variety of advanced building designs and convince them that the new features will not be overly complex or require specialized skills. The initiative for new designs should be firmly in the hands of private builders and designers; government help should be offered in a way that they find helpful. Tests should be conducted in a variety of regions. They should also be designed to allow prospective customers to visit model buildings and allow local utilities to observe the performance of a well-documented building operating in their service areas.

The programs outlined here build upon the applied research centers recommended in the section on Existing Residential Buildings policy (jointly supported by the federal government, professional organizations, builders, utilities and others) to improve new buildings technology and disseminate that technology. The building industry is dominated by relatively small firms; even the largest building firms cannot support the type of applied research necessary to improve new building techniques, equipment, design tools, and the like.

Information is also needed for consumers. One useful concept involves the use of a building energy rating system, providing the consumer and lending institutions with a means of determining the energy soundness of his investment. This rating system should be coordinated. If energy costs are incorporated into loan qualifying formulas, the real "first-cost" barrier to energy efficient homes could be reduced. Such homes should have larger potential markets than inefficient houses with lower first costs.

Five categories of programs are examined:

o establishment of management centers for new buildings technology demonstrations linked with existing R&D programs (see Applied Research under Existing Residential Buildings Policy);

o improvement of analytical tools used to design energy efficient buildings with and without solar devices;

o development of effective manuals of good building design useful to the building industry;

o development and demonstration of a building energy rating system for new buildings which can also be used by lending institutions to include the cost of energy in loan qualification;

o incorporation of both funding and processing priorities into HUD multi-family housing programs for energy efficient buildings.

Building Standards. Since efforts to achieve early national implementation of a Building Energy Performance Standard (BEPS) have been delayed, it may be advantageous to assist states interested in undertaking performance-based standards equal to or better than the proposed BEPS. Federal assistance could insure greater uniformity of codes and that the standards (which some states are already considering) are well designed and reasonable. In addition, the experience gained through this early "demonstration" would be valuable to other states interested in developing performance-based standards. The program can be used to explore different approaches to implementing and enforcing performance-based codes, such as requiring builder warranties of design performance, issuing building permits with a rating of the building design energy use which would be posted during time of sale, or other approaches which may well be more effective and less onerous to both builders and buyers than conventional building code enforcement techniques, which have been fairly ineffective.

Actions in this program include:

o establishing indirect inducements to states to adopt efficient, performance-based building standards through the secondary mortgage market, FUA, or the Federal Power Marketing Authorities;

o encouraging states to experiment with new enforcement techniques for performance standards in new buildings; and

o providing direct grants to states to cover the costs of energy implementation of performance based standards equal to or better than the proposed BEPS standard.

Incentives For Better Buildings. Performance-based standards as they are likely to be published may not achieve a mix of buildings designed to minimize life-cycle costs or the full social cost of energy. This is in part due to the limitations of the methods now used to compute building design performance, which rely upon national average energy prices and a limited set of technical opportunities. Repeated analysis has indicated that life-cycle costs are nearly the same for a broad range of building designs. What is often chosen for the standard, however, is in fact the front end of this broad and uncertain range. The Solar and Conservation Bank has the flexibility to induce the construction of new homes which go beyond the proposed BEPS levels which would be cost-justified by the full social cost of conventional energy sources, and be a source of innovation in building design and construction (see Table 1.8).

Actions include:

o clarify and, if necessary, revise the law regarding the Bank's ability to fund conservation investments in new buildings; and

o advertise the availability of programs offered by the Bank.

Appliances

Residential Appliances account for a major share of the residential energy use. The following discussion includes heating and air conditioning equipment under the term "appliances."

There is limited evidence about market reaction to improvements in appliance efficiency. The evidence available, however, indicates that while considerable improvements can be induced by information and incentive programs, a large portion of the appliance market will not be directly impacted by labeling, early-retirement or other consumer-oriented programs because approximately 65% of all appliances purchased are not purchased by their ultimate user but by owners of rental housing, and construction firms.

Minimum efficiency standards may therefore be the only sensible way to improve appliance performance. Such standards are currently being promulgated for

1981, with more aggressive standards taking effect in 1986. These federal standards can reduce the difficulties created by a multiplicity of state standards. Careful attention should be paid to assisting the industry re-tool to produce more efficient appliances, particularly those firms willing to re-tool early and/or produce appliances significantly more efficient than the proposed standards. The promulgation of standards should not be seen as a cure-all; it should be complemented by information and consumer incentive programs (see Table 1.9).

Specific actions include:

o information and incentive programs to help shift the distribution of new appliances that already meet minimum requirements, towards the most efficient available models (and to encourage consumer purchase of "close substitutes" such as chest rather than upright freezers, that meet the most important needs while using les energy);

o increased emphasis, through RCS and other utility, state, or local programs, on improved appliance maintenance, energy-efficient operation, and the early retirement of very inefficient models (or seldom-needed appliances, such as second refrigerators).

o provisions for manufacturers to use a "model-mix" approach (similar to the fleet-averaging for new auto efficiency standards), to meet or exceed efficiency targets while still providing for the diversity of consumer preferences; and

Table 1.8. IMPACT OF PROGRAMS TO SAVE ENERGY IN NEW RESIDENTIAL BUILDINGS

	Cumulative Saving in 1990			Cumulative Saving in 2000			Capital Requirements
	Fuel Saving	Electric Saving (Bil. kwh)	Total Saving (Bbl/day)	Fuel Saving	Electric Saving (Bil. kwh)	Total Saving (Bbl/day)	Consumer Investment ($ billions)
Information and demonstration	66	4.6	58,000	327	18.3	263,000	16.06-16.93
Incentives for States to adopt Performance Standards	96	6.7	85,000	418	24.9	340,000	22.71-23.77
Mandatory Performance A	118	8.3	105,000	514	31.9	430,000	29.5-30.55
Mandatory Performance A With Performance B Incentives	128	8.9	113,000	534	33.4	450,000	30.98-33.38

Table 1.9. POTENTIAL SAVINGS FROM APPLIANCE STANDARDS

Appliance Category	Annual Savings in year 2000 (QUADS)	Total Investment in all Appliances existing in the year 2000[c] ($ Billion)	Cost of Saved Energy
Electrical Appliances	2.54	39.4	0.4-3¢/kwh
Heat Pumps (Exist. Bldg)	.17	6.5	3.2-2.8¢/kwh
Heat Pumps (New Bldg)	.26	13.3	3.0-6.4¢/kwh
High Effic. CAC (Exist. Bldg)	.27	5.5	1.9-2.0¢/kwh
High Effic. CAC (New Bldg)	.2	5.8	2.5-2.7¢/kwh
Electric Water Heating	.85	6.9	0.7-1.5¢/kwh
Heat Pump H.W.	.2	12.8	6.0¢/kwh
Solar H.W. (Elec)	.35	15.1	2.4-6.5¢/kwh
SUBTOTAL ELECTRIC	(4.84)	(105.3)	
Fuel Furnaces (Exist. Bldg)	.29	9.7	2.2-2.6$/MBtu
Fuel Furnaces (New Bldg)	.13	7.4	2.6-4.4$/MBtu
Fuel Appliances	.25	4.4	0.7-10.2$/MBtu
Fuel Hot Water	.7	15.9	1.6-4.0$/MBtu
Solar Hot Water (Fuel)	.35	15.1	2.4-4.3$/MBtu
SUBTOTAL FUEL	(1.72)	(52.5)	
TOTAL ALL APPLIANCES	6.56	157.8	

[a]Amortized assuming a 3% discount rate for the lifetime of the appliance.

[b]Appliance savings includes some items which are also accounted for in other end-uses; discrepancy is attributable to different definition of appliances used in supply curve analysis and the proposed Appliance Efficiency Standards.

[c]The cost is calculated from the supply curves, where an existing appliance is replaced by a more efficient model when it wears out, only the incremental cost is included; where one is replaced by a different system (e.g., heat pump replacing resistance heating) the total cost is included, and some item costs are assumed to change over time (e.g., high efficiency heat pumps).

( ) = non additive

o regularly announced upward revision of these standards, adopted (as "long-range standards") well in advance of their effective date to provide long-term guidance to manufacturers;

o research, demonstrations, and manufacturer capitalization incentives to raise the efficiency level of the "best available" models.

COMMERCIAL BUILDINGS

Office buildings, schools, hospitals, warehouses, churches, hamburger stands, massage parlors, and other "commercial buildings" collectively consume approximately 2/3 as much energy as residential build-ings. In 1980 the nation provided energy to heat, cool, and light an area under one roof approximately the size of the state of Rhode Island (32.5 billion square feet of commercial space, to be exact). Additional energy was needed to operate computers, xerox machines, typewriters, drink machines, and other appliances in these buildings. Standard forecasts (See Table 1.3) indicate that commercial floor space will increase a spectacular 57% by the year 2000, making the amount of commercial floor-space per American increase by more than 36%, a number that may seem a bit high, but it will be used as the basis of all calculations conducted in this study. A forecast showing rapid increase in commercial floor space, however, can be justified by the fact that the nation's "service" industries are growing more rapidly than the nation's industrial sector (an issue examined in considerable detail in the chapter dealing with industry). Since this study assumes that personal income will increase by more than 50% by the end of the century, it is reasonable to assume that a significant fraction of this increased income will be spent in areas requiring services performed in new commercial space. The rapid increase in commercial floor-space means that about 53% of the commercial area standing in the year 2000 will be built during the next 20 years. For comparison, about 38% of the residences standing in the year 2000 will have been built after 1980.

The following discussion will demonstrate that it is possible to build commercial buildings that use about a quarter as much energy per square foot as the average commercial building standing today. The energy of the existing stock of buildings can also be cut in half using relatively straight-forward retrofit techniques. This means that expected growth in commercial buildings can occur while the total demand for energy in this sector is actually reduced by about 30%. Savings of this magnitude may seem incredible but seem more plausible given a clearer understanding of the way heating and cooling systems were designed for commercial buildings in the past: until recently, engineers calculated the amount of cooling that would be required on the hottest day expected in 10 years, installing a chilling unit that was able to meet this peak requirement. Once installed, the chiller was operated at full capacity for most of the year and the space temperature was controlled by "terminal reheat," i.e., by reheating the chilled air. As a result the buildings often were operating both an air-conditioner and a furnace when the outside temperature was perfectly comfortable. This is perhaps the most egregious example of waste, but it is not atypical of many practices which made sense in an era when energy was very cheap and reliable controls were expensive. It does not take much imagination to achieve significant improvements but saving 75% will require some ingenuity.

Technologies

Assessing the potential for saving energy in commercial buildings is complicated because these buildings house an enormous range of functions and come in all shapes and sizes, while residential buildings can be treated in a much simpler way since they all are called on to provide roughly similar functions. The energy needs of a church, which may have an enormous sanctuary used only a few times a week, are plainly very different from those of a laundry or office building. There may be a number of different businesses in a single building, each with a unique energy problem. One of the buildings examined in this study, for example, used 50% of its energy to support office space,

20% of its energy to supply a retail card store, and 30% of its energy for a pizza parlor. Figure 1.16 illustrates some of this diversity by showing consumption of fuel and electricity per square foot of building. (Electric consumption is shown in "resource energy"-- that is the energy required to generate enough electricity to meet the building's demand, 11,500 Btu per kwh.) Hospitals and clinics need more energy per square foot than any other major category largely because of their heavy use of large quantities of fresh air (which is not recirculated as in conventional buildings), laundry equipment, sterilizing devices, and other energy consuming appliances. As could be expected, the nation's 2.7 billion square feet of warehouses use the least energy per square foot. The wide range in energy consumption for commercial buildings results from enormous differences in the number of people using the buildings, ventilation requirements, use of appliances, required lighting levels and basic architectural strategies. The importance of each type of commercial building is shown in Figure 1.17, which shows how commercial space is divided by function.

The techniques used to save energy in commercial buildings may differ significantly from the repertory of tricks used to save energy in residences. In general the shell of large commercial buildings tends to be less important than the walls and ceilings of a residence simply because commercial buildings tend to be larger, having less surface area per unit of floor area. It is, of course, important to ensure that the walls and ceilings of commercial buildings are adequately insulated, that double or triple glazing is used where appropriate, and that vestibule entrances are added wherever feasible. It is interesting to notice, however, that a study which examined a well built commercial office building in Denver that used R-11 walls indicated that the building's use of energy would decrease only 1% if the insulation thickness was doubled. (AIA/RC 1980). A similar study of a building in Minnesota, however, indicated that triple glazing, R-30 walls, and R-40 ceilings were cost effective.

The energy used by the heating, cooling and ventilating equipment of commercial buildings can also be reduced by proper care in the selection and installation of equipment. For example, the fans used in many ventilating systems are driven by relatively inefficient motors and are frequently too large for the function they are designed to serve. Ducts and pipes may not be adequately insulated and the boilers and chillers may be much less efficient than state-of-the-art equipment. Clearly it is easiest to design a good system for a new building but much can be accomplished in a retrofit.

Many commercial buildings require more outside air per square foot of building than a typical residential structure; minimum levels of ventilation are frequently prescribed by local building codes. The outside air can be provided at relatively low cost even during per-

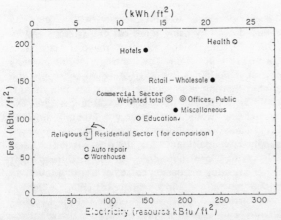

Figure 1.16.

ENERGY USE BY BUILDING TYPE FOR THE COMMERCIAL STOCK

Energy Intensity: The square for the residences, plotted for
comparison, was calculated from the total energy use by residences,
divided by the number of homes and an assumed average floor space
of 1350 feet.

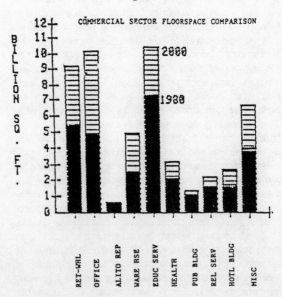

Figure 1.17.

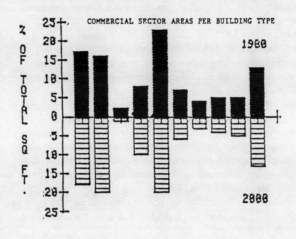

SOURCE: ORNL Commercial Energy Use Simulation, 1970-2000

iods of peak heating or cooling by using "recuperators"
or heat exchangers which bring outside air into a building by passing it over a stream of interior air that is
being flushed out of the building; the two air streams
are separated by a thin layer of metal or plastic. Such
devices allow the heat in the building air to transfer to
the incoming air during the heating season, and the
heat from the outside air to be removed during the
cooling season. In regions where day and night temperatures differ significantly, automatic ventilating
systems can be designed to flush the building with cool
air during the night (perhaps operating chillers during
periods of cheap electric rates), so that cooling demands can be reduced during the day.

Another major difference between commercial and
residential buildings is that commercial buildings tend
to have a number of zones with widely divergent heating and cooling demands. In large office buildings, for
example, the heat generated by people and equipment
creates a need in the central core of the building for
air conditioning all year long. Rooms on the outer
walls of buildings may require heating during the
winter while the core of the building is being cooled.
In other cases, the southern half of a large building
may require cooling on a sunny day, while shaded
north-facing rooms require heat. This means that
there can be significant advantages in systems which
move heat from one part of the building to another.
Modern designs can combine such systems with a storage system acting as a buffer. For example, it is possible to have small heat-pumps in each building zone
connected through a water pipe with each other and
with a large tank of water in the basement. Each heat
pump unit either takes heat from the circulating water
to heat a room or puts heat into the water when it
cools a room. It is also important that each zone be
equipped with its own thermostat so that temperatures
can be adjusted to reflect actual needs.

Another innovation which leads to significant energy
savings in commercial buildings is the "dead-band"
thermostat. This device has the effect of shutting
down the heating and cooling system during periods
when the temperature of the room is within some

range of acceptable temperatures (e.g. from 68°F to 78°F). Many of the buildings built during the past decades have controls which maintain a nearly constant interior temperature; thus either the furnace or the chillers must be operating at all times.

It is estimated that if "dead-band" thermostats are retrofitted into each zone of an existing building, heating and cooling demands can be reduced by 10-30%. Additional savings can be achieved with centralized, computer controls which automatically adjust ventilation, boiler temperatures, room temperatures, and other heating and cooling operations.

A significant amount of energy can also be saved by reducing the energy needed to light commercial buildings. Lighting requires nearly a quarter of the energy demands of commercial buildings; the relative importance of lighting can be expected to increase as the heating and cooling systems become more efficient. In a recent study conducted by the American Institute of Architects, lighting was responsible for nearly half of the energy needed by three buildings designed for optimum energy efficiency even though considerable care was taken to minimize buildings' lighting requirements.

Adjustment in standards for lighting levels can result in energy savings, for lighting levels in modern commercial buildings are often far in excess of actual requirements. Lighting requirements increased spectacularly during the 1960's and early 1970's. For example, in 1971 the New York City Board of Education Manual of School Planning called for a level of lighting that was three times the level specified in their 1952 manual. Buildings typical of the early 1970's required about four watts to light each square foot. Typical buildings constructed in 1980 required only about 2 watts per square foot and it is estimated that requirements could be reduced to 0.5-1.0 watts per square foot by 1990. (Berman 1979). The impact of the potential savings is enormous. Reducing the lighting of commercial buildings by an average of 2 watts per square foot, for example, would reduce the demand for electricity by the equivalent of 150 large (1000 MW) nuclear or coal-fired electric generating plants by the year 2000.

Too much lighting is especially wasteful in buildings with large cooling loads since lights can add significantly to cooling requirements. Even efficient fluorescent lights convert only about 19% of their energy into visible light; all of the energy sent to the lamps, however, must eventually be removed by a cooling system during hot weather. Lighting can, of course, contribute to the heating of a building during the winter season but often provides too much heat to the building core and too little to the periphery. The combined impact of lighting, and the added cooling demands it imposes, can be costly if utilities charge a higher rate for electricity during periods of peak utility demand. It is mainly the demands of commercial buildings that

have caused U.S. peak loads to switch from winter to summer.

The energy needed to light commerical buildings can be reduced both by using more efficient lamps and by matching lighting to the needs of the different tasks performed in a building. The lighting systems in many commerical buildings are still designed to supply the maximum lighting requirements of any possible tenant. It would be preferable to design a building so that the fixed lighting would supply only minimum ambient lighting needs; the building tenants could be expected to supply any additional task lighting needed. Additional savings can be achieved by installing light switches allowing occupants to adjust lighting levels to individual needs (and allow a single office to be lighted without turning on lights for an entire floor). Lighting efficiency can also be improved with high-frequency ballasts and parabolic reflectors for fluorescent bulbs. High pressure sodium vapor lamps are extremely efficient and should be used whenever possible.

Another way to reduce lighting demands is to use natural daylighting. Table 1.10 shows how a careful redesign of a building in Raleigh, North Carolina reduced the building's lighting demands by nearly a factor of two. The net heating demand increased somewhat but this increase was more than offset by the savings in lighting. Use of daylighting requires careful attention to the use and placement of windows. Performance can be improved by using lighting fixtures equipped with sensors that automatically dim lighting fixtures when daylight provides adequate levels of illumination. Such sensors have the additional advantage of adjusting light levels to compensate automatically for lighting fixtures which dim with age or the accumulation of dirt. In current practice lighting levelse are frequently set too high so that they can provide adequate light even if they are left for years without cleaning.

The energy demands of commercial buildings can be reduced by devices to prevent unusable heat gains that would necessitate increased air-conditioning: Reflective "solar control" coatings can be used on windows to block heat gains, overhangs can shade southern, eastern, and western walls during the summer, and projecting fins on east or west walls can provide similar effects early and late in the day.

The previous discussion has covered only some of the more common approaches to saving energy in commercial buildings; it does not begin to deal with the full range of possibilities. The major opportunities are summarized in Table 1.11. Because optimum solutions will vary widely in individual cases, energy efficiency in commercial buildings will depend on the imagination and skill of the architects and engineers challenged with the problem.

Table 1.10. SELECTED ENERGY RESULTS FOR 3 BUILDINGS IN LIFE-CYCLE COST ANALYSIS

Different Design Solutions to Same Building Design Problem	Buildings											
	Denver				Minneapolis				Raleigh			
	A*	B*	C*	D*	A	B	C	D	A	B	C	D
Original 1976-1976 Design												
Heating and Cooling	20.2	(41)			40.6	(59)			41.3	(58)		
Lighting	25.9	(52)			18.7	(27)			18.8	(27)		
Total	49.4	(100)			70.1	(100)			70.7	(100)		
ASHRAE 90-75R Exact Applicable Requirements												
Heating and Cooling	12.6	(39)	(38)		31.4	(53)	(22)		26.1	(42)	(37)	
Lighting	16.1	(50)	(38)		20.5	(34)	(0)		13.1	(26)	(30)	
Total	31.9	(100)	(35)	(-1)	59.8	(100)	(15)	(0)	50.2	(100)	(29)	(-1)
ASHRAE 90-75R Requirements Per Component, or 1975-1976 Value, whichever is Better												
Heating and Cooling	8.4	(30)	(58)		28.5	(55)	(30)		9.5	(29)	(77)	
Lighting	16.1	(58)	(38)		15.8	(30)	(16)		13.1	(40)	(30)	
Total	27.7	(100)	(43)	(-1)	51.9	(100)	(26)	(-1)	32.6	(100)	(46)	(-1)
Phase-2 Redesign												
Heating and Cooling	8.0	(30)	(61)		18.1	(45)	(55)		8.5	(28)	(79)	
Lighting	14.8	(56)	(42)		14.1	(35)	(25)		15.4	(51)	(18)	
Total	26.5	(100)	(46)	(11)	40.1	(100)	(43)	(7)	30.0	(100)	(58)	(7)
Phase-3 Redesign												
Heating and Cooling	6.0	(29)	(70)		5.2	(21)	(87)		9.8	(40)	(76)	
Lighting	11.2	(55)	(57)		12.7	(50)	(32)		9.2	(37)	(51)	
Total	20.4	(100)	(59)	(16)	25.3	(100)	(64)	(11)	24.6	(100)	(65)	(9)

*A = Energy in KBtu/Sq. ft./year.

*B = Percent of total building energy.

*C = Percent reduction from original design.

*D = Percent of first cost increase.

NOTES:

1. Energy results for HVAC fans, domestic hot water, elevators, escalators, and general exhaust fans not listed here.

2. "Process" energy is not included in this analysis.

Table 1.11. TECHNOLOGIES FOR IMPROVING THE PERFORMANCE OF COMMERCIAL BUILDINGS

1. New Buildings

    —good wall and ceiling insulation
    —multiple pane windows
    —reflective materials on windows
    —careful attention to the use of daylighting
    —lighting levels adjusted to actual needs
    —maximum use of efficient flourescent and sodium vapor lamps
    —computer control systems and "dead-band" thermostats in each zone
    —variable air-volume systems or water-loop heat pump system
    —thermal storage in building mass or in a water tank
    —maximum use of recuperators and ventilation control strategies

2. Building Retrofits

    —installation of deadband control systems and centralized computer management system
    —insulated piping and ductwork where appropriate
    —replacement of inefficient fan motors and matching of fan-size to actual requirements
    —lower boiler temperatures
    —installation of economizers and recuperators where appropriate
    —improved sealing and caulking
    —lower ambient lighting levels and use task lighting
    —use of efficient fluorescent lights wherever possible
    —replacement of heating and cooling equipment when it is obsolete or very inefficient
    —use of waste heat from chillers and refrigeration units.

Source: Sant, 1980; Scott, 1980; Derringer, 1980

Forecasting

Measuring the impact of the technologies just described requires data on building energy consumption, the cost of the systems developed, and the attitudes taken by the variety of individuals and organizations in position to influence decisions made about the design and operation of commercial buildings. Existing information is scarce on all of these issues. However, it is clear that the owners of commercial buildings are reacting quickly to increased energy prices. Figure 1.18 shows the effect dramatically. The energy used per square foot of building increased by nearly 40% between 1952 and 1973 (when the real price of energy was decreasing) but the situation was reversed dramatically when energy prices began to climb. The average building designed in 1976 was more than twice as efficient as a buildings built in 1972. Where will the decline end? One indication is the performance of Swedish buildings also shown on Figure 1.18. These buildings have used less than half of the energy required for a U.S. office building in spite of the fact that the Swedish climate is significantly more frigid and gloomy than that of the U.S., and that in Sweden (as in most of Europe) every office must have a window.

The analysis conducted for this study has been based both on a theoretical determination of new building potentials and a collection of empirical information about the performance of real buildings. The results of this work are summarized in Table 1.12. The most dramatic changes occur in the use of fuel. Fuel use per square foot of commercial buildings built after 1993 goes practically to zero. This is due in part to a

Figure 1.18.

Office Building Resource Energy Intensity, 40 year trends

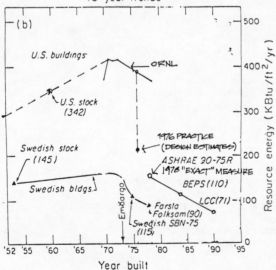

shift toward electric heating. Overall energy efficiency is expected to increase by a factor of nearly four.

New Buildings

The potential for reducing energy use in new buildings presented here is based on analysis performed by the Research Corporation of the American Institute of Architects (AIA/RC) working in connection with the American Society of Heating, Refrigeration, and Air-conditioning Engineers (ASHRAE). The work began in 1976 with an analysis of 1661 building designs developed by commercial architects in that year. A sample of 125 buildings was selected to test the industry association's voluntary energy performance standard--called ASHRAE Standard-90. The analysis showed that if the 125 buildings had been built to ASHRAE Standard-90 overall energy consumption would have been decreased by an average of 22%. The savings were greatest for warehouses, which saved an average of 41%, and smallest for hospitals, which saved an average of 3%. This standard has been adopted by approximately 40 states, but its impact is difficult to judge.

The Massachusetts Building Code Commission has recently examined the impact of the Massachusetts state code, which is very similar to the ASHRAE Standard-90. They found that the average commercial building shell constructed in the state was nearly 20% better than called for in the code, but that 90% of the planned buildings had heating, cooling or ventilating equipment which did not meet the standard. The impact of these findings is difficult to judge. For present purposes, however, it has been assumed that the average energy consumption of commercial buildings built in the U.S. from 1980 to 1982 will be midway between 1976 practice and the requirements of the industry consensus standard (see Table 1.17). In most cases the projected cost of the new buildings did not increase when the designs were changed to meet the standard; in fact the average cost actually decreased!

In the second phase of their study, the AIA/RC selected 168 buildings from their original sample and challenged the building designers to improve the energy efficiency of their designs without violating any of the requirements specified by their original clients and without changing the overall cost of the building significantly. Only "off-the-shelf" technologies could be used. Since few of the original design teams were expert in energy efficiency, the AIA/RC assisted them by providing a three-day training session, and a workbook with practical design ideas and simplified analytical tools. The proposed designs were carefully reviewed twice by experts in the field.

The results of this "Phase-2" redesign study are summarized in Figure 1.19. The overall savings are astonishing given the level of training available to the architects: overall savings averaged 40%. The study

Table 1.12. ENERGY CONSUMPTION OF COMMERCIAL BUILDINGS
(1976-2000)

	Resource Energy in KBtu/Ft^2		
	Electric	Fuel	Total
A. Retrofits			
1. Average commercial building stock (1978)	207	135	342
2. 1978 Building stock after phase-1 Retrofit [a]	160	100	260
3. 1978 Building stock after phase-2 Retrofit [a]	105	67	170
B. New Construction			
1. 1976 Standard Practice	190	15	205
2. 1980 - 1982[b]	160	14	174
3. 1983 - 1992[c]	120	6	126
4. 1993 - 2000[d]	84	4	88

[a]Hirst 1980, EBASCO 1980, Ohio State 1980.

[b]Average of ASHRAE 90-75R (1980 interpretation) and 1976 practice.

[c]Based on AIA/RC Phase-II redesign of 1975-1976 buildings (AIA/RC-1980).

[d]Based on AIA/RC Life cycle cost minimum designs (AIA/RC-1980).

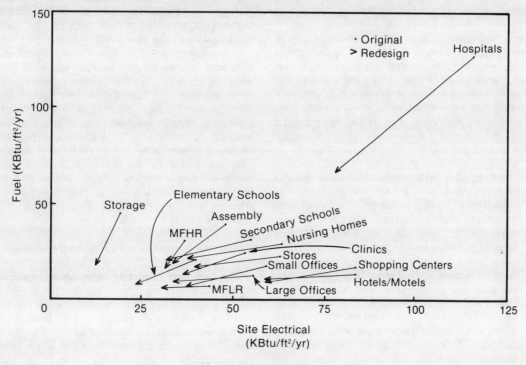

Figure 1.19.

was, however, as much a test of the state-of-the-art of energy analysis as it was of building technology. In fact 7% of the building redesigns ended up consuming more energy than the original design! ASHRAE is examining the results of the redesign project to determine whether it should be used as the basis for a revision of their Standard-90. A new standard close to the redesign levels has been proposed, but it is likely to be 1-2 years before it receives clearance from the elaborate series of reviews which an ASHRAE consensus standard must obtain. The redesign study has also been used as the basis for proposed state and federal commercial building standards. In the analysis presented here it will be assumed that a national consensus standard based on the Phase-2 redesigns will be implemented in 1983.

Three of the buildings selected for redesigns were chosen by AIA/RC for a much more elaborate examination. In this third phase of their project, architects were not selected at random but chosen because of demonstrated skill in designing efficient commercial buildings. They were asked to take one of the original building designs and alter it to save energy in a way that could give the owner at least a 10% return on his investment in constant dollars. The results of this work are summarized in Table 1.10. These buildings all use careful lighting techniques, state-of-the-art controls, and hydronic heat pump systems, or the equivalent. In Table 1.12 it is assumed that all commercial buildings constructed after 1993 will be built to these technical standards.

Retrofits

There is also significant potential for saving energy by retrofitting commercial buildings. Again data is not readily available, but enough experience has been gathered over the past few years to make a convincing case that the energy used in an average U.S. commercial building can be reduced by 50% using techniques which save energy at costs well below the marginal cost of producing new conventional fuels. Evidence includes:

o A survey of 14 experienced architects and engineers asked to estimate the potential for retrofitting existing buildings. Their responses indicated that reductions of 25% seemed plausible by 1990 and an additional 25% seemed plausible by the year 2000.

o Articles appearing in Energy User News during the past five years which dealt with commercial retrofits were reviewed. Out of 190 articles found, only 20 contained enough information for a quantitative evaluation of costs and savings. Most of the investments reported seemed to be "cream skimming." Nearly 90%

of the investments cost less than 50 cents per square foot and achieved savings of 20-35%. Three of the retrofits, however, saved more than 50%.

o Several buildings retrofit by Ohio State University between 1974 and 1978 that achieved 40-50% reductions in energy consumption (starting with relatively inefficient structures).

o Audits performed for state office buildings in the state of Minnesota that indicated that even though these buildings were, on the average, already more efficient than the U.S. average, fuel use could be cut in half and electric consumption reduced by 27%. The cost was estimated to be approximately $1.22 per million Btu saved--well below the marginal cost of all current energy supplies.

o Proposals submitted by a private auditing company (EBASCO) which guaranteed savings averaging 20% of electric consumption and 45% of fuel consumption for seven commercial buildings. EBASCO estimated that the average cost of energy saved with such retrofits would be approximately $1 per million Btu saved.

The results of these surveys are summarized in Figure 1.20. Using this information, it is estimated that it should be possible to reduce the energy consumed by commercial buildings at least 25% by 1990 and an additional 25% by the year 2000.

It is also necessary to report on an unsuccessful experiment in commercial building retrofits conducted between 1975-1977 by the Lawrence Berkeley Laboratory working in cooperation with the American Association of School Administrators. Retrofits of 10 elementry schools resulted in an average of only 20% savings. Of course in 1975 there was still little interest in conservation. The relatively small savings were attributed to the primitive state of knowledge about building retrofits at the time and to the fact that schools were allowed to deal directly with contractors without the advice and supervision of a qualified engineering firm or a utility staff.

Renewable Resources in Commercial Buildings

The two most important renewable contributions to commercial-building energy use are the use of natural lighting and photovoltaics. Although commercial buildings typically do not have large demands for hot water, we have estimated that as much as 0.1 quad of energy can be contributed by solar water heaters on commercial buildings.

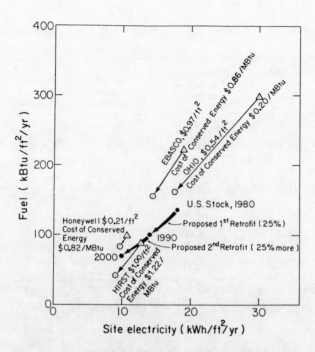

Fig. 1.20 Comparison of our two-stage retrofit assumptions (heavy arrows) with four sets of measurements, guaranteed savings, or audits.

Improved use of daylighting techniques could reduce the large artificial lighting loads associated with most commercial buildings. We are only beginning to recognize the significance of this "passive" solar techology but recent advances in building modeling have permitted careful evaluations of a number of daylighting strategies. The analysis is complex since a reduction in artificial lighting increases the demand placed on the heating system and reduces the demand placed on the cooling system. Based on a careful analysis of a commercial office building in Raleigh, North Carolina where proper attention to daylighting reduced the overall energy demands of the building by approximately 5600 Btu/ft^2/year), we have estimated that daylighting in commercial buildings can reduce U.S. energy demands by at least 0.2-0.3 Q by the year 2000.

The analysis discussed earlier indicates that is should be possible to virtually eliminate the thermal requirements of commercial buildings and that the remaining demands will be dominated by demands for electricity. As a result, commercial buildings are likely to provide excellent markets for photovoltaic systems attached to racks on building roofs or, if costs are reduced sufficiently, to the walls of the buildings. Using costs of capital typical for owners of commercial buildings, we have estimated that enough photovoltaic systems could be sold in this sector by the year 2000 to contribute approximately 0.1 to 0.25 Quads of energy.

The Bottom Line

The potential energy intensities summarized in Table 1.12 can be combined with standard forecasts of growth in commercial buildings to establish goals for energy use in the year 2000. The results of this assessment are summarized in Table 1.13. Total savings from efficiency improvements can be on the order of 5.5 Quads from a baseline which assumes a continuation of 1980-82 practice. Standard forecasts show that energy use in commercial buildings will increase from 10.3 Quads in 1977 to 13.3 Quads in 2000. The analysis presented here, however, leads to an estimate of 7.26 Quads in the year 2000.

Capturing the Potential

Designing a uniform policy for dealing with commercial buildings is no simpler than designing a uniform technical approach for saving energy in commercial building designs. Commercial buildings are owned by a variety of owners with a variety of expectations. Some are owned by federal, state or local governments. Some are owned by speculative investors and rented to commercial clients who may face the same difficulties as individuals in rented residences. Some are owned or rented by small businesses lacking any sophistication in energy or financial analysis. And some are owned by large enterprises with design teams

Table 1.13.   SAVINGS FROM IMPROVEMENTS IN COMMERCIAL BUILDINGS ACHIEVABLE BY THE YEAR 2000 AND THE COST OF THOSE SAVINGS

	Energy Savings in Resource Quads[c]			Total Cost in Billions of Dollars	Cost of Saved[d] Electricity[c] (¢/kwh)	Cost of Saved Energy[e] $/MBtu
	Electricity	Fuel	Total			
1. Improved designs[a]	0.88	0.13	1.01	38		
a. 1983-1992 improvements	(0.54)	(0.11)	(0.65)	(13.7)	3.1	2.2
b. 1993-2000 improvements	(0.34)	(0.02)	(0.36)	(14.3)	5.1	4.2
2. Retrofits[b]	2.4	1.6	4.0	71		
a. Phase-1 (1980-1990)	(1.2)	(0.8)	(2.0)	(23.7)	2.4	1.3
b. Phase-2 (1991-2000)	(1.2)	(0.8)	(2.0)	(47.4)	4.8	2.6
3. Solar	0.4-0.6	--	0.4-0.6	19-26	6.1	
a. Daylighting	(0.2-0.3)	--	0.2-0.3	(9)	5.7	
b. Photovoltaics	(0.1-0.3)	--	0.1-0.2	(8-14)	8-11	
c. Hot Water	(0.1)	--	0.1	(2-3)	2-7	
4. Total Savings	3.7-3.9	1.7	54-5.6	118-125		

[a] 27.3 billion square feet of commercial buildings are built between 1980 and 2000 (Oak Ridge, 1980).

[b] 23.7 billion square feet of commercial buildings standing in 1980 will survive to the year 2000 (Oak Ridge, 1980). Buildings retrofit but not surviving to the year 2000 are not counted.

[c] Electricity converted to resource quads at 11,500 Btu/kwh.

[d] All costs are charged to electricity savings; the effective cost would be lower if credit were taken for savings in fuel.

[e] Costs assume an annual fixed charge rate of 0.116 (typical of utility financing or a 10% real rate of returned over 30 years).

which have carefully tailored the building to meet specialized needs. If no action is taken, rapid increases in energy prices--possibly accelerated by rapid deregulation of gas--could leave the owners of commercial buildings in a desperate situation without adequate information on which to base a response, or without capital to invest adequately even if information were available.

The programs here are based on a structure similar to the residential policy analysis. Policy proposals are presented for information, incentive and regulatory programs, reflecting the preference for market-based information and incentive programs, unless market-failure makes government regulation unavoidable.

New Commercial Buildings

Information and Demonstration Programs.   Actions examined in this area are:

o   Analysis and Applied Research programs to improve the quality of new commercial buildings should be developed along the same general lines as the programs designed to improve national understanding of energy use in residential buildings discussed earlier. A series of experiments should be conducted in close cooperation with ASHRAE and BOMA as well as the organizations involved in the residential applied research. The research effort should also expand the currently limited efforts to monitor buildings in the demonstration pro-

gram. Another practical product of the applied research program should be the development of a standard method for rating the performance of commercial buildings in a way that can be of direct assistance to potential renters and purchasers.

o   Training and Education programs should be expanded from their currently limited levels. The two seminars per year which DOE now provides for faculty at schools of architecture and engineering should be expanded and professional architectural and engineering associations should be supported in the development and implementation of training courses for practicing professionals, possibly in conjunction with the schools of architecture and engineering established as analytical centers.

Careful attention should be paid in this effort to provide analytical tools to both architects and engineers for all phases of the design of commercial buildings. Particular emphasis should be given to insure the availability of simple, accessible tools especially for small and medium sized A&E firms, while continuing work on complex tools such as computer programs as well as education on their use.

o   Professional Business Organizations should be tapped as a channel to their memberships. These organizations offer the advantage of representing a relatively uniform segment of the commercial buildings spectrum, allowing the development of specific programs for each group. While hundreds of such organizations exist, five or six could be approached initially to undertake cooperative information programs, and the program expanded based upon a careful evaluation of its success. This effort should be complemented by an on-going effort to analyze the commercial building market by both ownership and building type. This analysis should seek to identify where the marketplace is not working and why.

Standards. The existing ASHRAE 90 Standards are currently being revised, and should produce a more energy efficient base-line for commercial building design. In addition, the Building Energy Performance Standards (BEPS) being promulgated by the Department of Energy are intended to apply to commercial buildings as well as residential structures. However, without the life-cycle cost analysis now being conducted by Oak Ridge National Laboratory, implementation of BEPS could be haphazard.

The proposed incentives for states to adopt BEPS in the residential sector should also require that they include the design energy budgets proposed for BEPS commercial buildings. The continuing programs of DOE, ORNL, and ASHRAE should be coordinated more closely to produce coherent and consistent documentation on life-cycle cost minimum designs for commercial buildings by both prescriptive and performance methods. This process will most probably take two years, and in the interim the effectiveness of information and demonstration programs and the experiences of states adopting the proposed energy budgets should be evaluated. Careful attention should be paid to actual compliance with voluntary standards, such as ASHRAE; the actual energy use of buildings designed to that standard should be compared with their design requirements. A program within the Department of Energy should be established to undertake this analysis and evaluation, and provide a recommendation to the Secretary within two years on the role of a federally mandated BEPS in the commercial buildings sector.

Existing Commercial Buildings

Information and Demonstration Programs. Actions recommended under this program are:

o Applied Research program should follow the same general format as the applied research program directed at residential buildings. The program should be conducted in cooperation with private firms interested in working in the area of providing commercial building audits;

provision should be made to allow some of these participants to maintain proprietary interests in specialized retrofit techniques.

o Cooperative Programs with Business and Professional associations should also be extended to existing commercial buildings. In addition to building retrofit information, information should be provided and exchanged on efficient building/business energy management. Also, DOE support for organizations to collect data on the costs and savings of retrofits in a consistent format would provide a very cost-effective method of developing a commercial building retrofit data base.

o Feedback Programs, such as feedback billing and metering, should be extended to commercial buildings. At present, little federal support for feedback programs in commercial buildings exists, while the potential for their effectiveness should be equal to or greater than those in the residential sector.

o Restrictive Lease Provisions are a major impediment to energy efficiency improvements in commercial buildings. The Department of Energy could undertake a program to develop model lease forms, and support state and local legislation to eliminate restrictive lease provisions.

o Federal buildings should be retrofit in a way that supports the applied research program.

Improvement of the Building Audit Program. Actions recommended under this program are:

o increase allocation of DOE staff and resources to the expansion of RCS to commercial buildings

o initiate an aggressive audit development and auditor training/certification program in conjunction with the applied research program

o provide a tax credit for the costs of audits performed on buildings not qualifying for RCS, and cost-sharing grants for audits of non-profit buildings not covered by other programs

The Base Case and Limitations of the Analysis

The projections made by the Energy Information Agency in its 1979 Annual Report to the Congress were used as a point of comparison of our potentials. This EIA forecast is based on the use of an adaptation of the Oak Ridge National Laboratory (ORNL) Residential Model.

To facilitate comparisons between the EIA projections and the "potentials" developed in this study, we have adopted most of the basic assumptions about housing stock and appliance saturations used by EIA in preparing the 1979 Annual Report to the Congress (EIA, 1980). Conversations with EIA personnel indicated that a number of modifications made it difficult to state a set of input assumptions for the ORNL Residential Model which would reproduce exactly the EIA projections, since modifications were made to both inputs and outputs, but a set of input assumptions was developed in consultation with EIA and ORNL personnel which very nearly reproduced the EIA projections when used in the ORNL Residential Model and the ORNL model was run using these assumptions. Whenever the assumptions or results of these runs are discussed, they will be identified as EIA/OR.

A comparison of 1975 residential energy usage with the baseline case and the EIA/OR projection yields the following energy use levels in quads of primary equivalent (Table 1.14).

Table 1.14 RESIDENTIAL FUEL USAGE IN RESOURCE QUADS

	1980 Usage	Year 2000 Baseline Case	EIA/OR Year 2000 Projection*
Fuel	8.8	9.2	7.5
Electric	7.8	12.7	12.4
Total	16.6	21.9	19.9

*Includes EIA/OR projected savings due to BEPS, RCS, AEPS, and other Federal initiatives.

The technologies and approaches considered in this analysis are far from exhaustive. Many potentially significant energy efficiency and renewable options have not been included. This does not imply any judgment on their merits, but rather indicates that due to the limits of time and funding for this study somewhat subjective choices had to be made. Community Energy Systems, small-scale cogenerators, experimental heat pump cycles, solar cooling, and many other possibly significant technologies were not able to be included in this analysis.

Section Two
Residential Technical Potential

METHODOLOGY AND ASSUMPTIONS

General Approach

Our methodology for estimating residential conservation and solar potential had to confront two major problems: (1) the fragmented, incomplete, and poorly documented data base, and (2) the size and diversity of the nation's housing stock. Even for the 80 million or so existing homes, it is extremely difficult to establish how energy is used (the specific of energy end-use), given the range in physical characteristics of buildings, climate conditions, and the variations in occupant behavior that can greatly affect energy consumption in houses that are physically alike. Physical and occupant diversity is greater in the case of commercial buildings, as discussed below, making them even more difficult to analyze than residential buildings.

Since our intent was to look forward twenty years and estimate conservation and solar technical potentials for the total U.S. residential stock as of the year 2000 even more uncertainty was introduced. Some questions that needed to be considered were: what will be the energy-using characteristics of the roughly 40 million new housing units expected to be added to the stock; where will they be built; which parts of the existing stock are likely to be removed or replaced over the next two decades; and how will energy use and efficiency change within those existing units that remain in use by 2000?

Both the scope of these questions and the limitations of available data inevitably required compromises in our methodology. The choices we made were designed to meet several objectives:

o to let us look in the greatest possible detail at end-use measures within any given (new, or existing/retrofitted) structure;

o to take explicitly into account the physical interactions among certain measures (especially those involving space heating, cooling, and water heating); and

o to rank and compare the cost-effectiveness of individual measures in a systematic and consistent way (using "conserved energy supply curves"), in order to identify where the most significant energy-saving opportunities exist.

Finally, it was our intent that the methodology be as explicit and easily replicated as possible and that it neither disguise nor exceed the limitations of available data--while remaining open to future revisions as improved data become available.

In turn, these methodological choices forced us to give up, at least for the current exercise, some possible desirable facets of analysis. We did not disaggregate our estimates by region (climate) or by type of stock (single-family, large and small, multi-family and mobile homes). Also, we have taken a detailed look at conservation and solar potentials only in a year 2000 "snapshot." It is possible to use the same approach to look at potentials for intermediate years, but this involves a great deal of detailed accounting. It would also require a much more thorough analysis of the rates at which new technologies could be installed between now and 2000. In future updates of these technical potential estimates, we hope to examine intermediate years in detail, and to include more disaggregation.

For the current study, we developed a "prototype" single-family house to represent the average characteristics of the entire U.S. housing stock. For the analysis, we located that house in the Washington D.C. climate zone (which is close to the U.S. weighted average in terms of heating degree-days and cooling load). We corrected our estimates of cost, performance, and penetration rates of individual conservation and solar measures for any obvious biases introduced by this "U.S. average" approach. An important advantage of this technique is that it enabled us to use, with only a few corrections factors (described later in the text) the DOE-2 computer simulations of houses, prepared as part of the least-cost optimization analyses for the proposed federal Building Energy Performance Standards (BEPS). We were also able to use statistical data from the DOE/EIA "Residential Energy Consumption Survey (RECS) to guide us in allocating pre-1975 housing stock into "uninsulated" and "partially insulated" categories.

To estimate potential energy savings in the year 2000, we had to track changes in the stock and average energy use of several separate cohorts of appliance and housing stocks. Rather than set up a completely new model, we decided to use all of the basic assumptions about existing stocks, net new additions, fuel types, and other characteristics as defined in the Residential Building Model developed at Oak Ridge National Laboratory (called the "ORNL model").* This model is routinely used by the Department of Energy in preparing annual forecasts of future energy demand. Using the same stock and fuel-use assumptions allowed us to focus exclusively on improvements in the technical efficiency of buildings and appliances, rather than on "lifestyle" or consumption practices that also affect residential energy use. The differences between our estimates of "least-cost technical potentials" for conservation and solar use in homes, and the "baseline" case that we define for the year 2000, are therefore attributable entirely to improvements in end-use efficiency and appropriate use of solar heat, with other factors held constant.

In the next sections we will present an overview of the methodology used to estimate technical potentials for residential energy efficiency, including defubutuibs if "technical potential", and the baseline case; the development and correct interpretation of "conservation supply curves"; the basis for estimating energy savings and incremental costs for each conservation measure; and some overall cautions to be kept in mind in using our bottom-line results.

Defining Least-Cost Technical Potential

When we refer to "technical potentials" in this study, either for an individual house (or an end-use of energy), or larger aggregates, we mean the <u>maximum</u> level of efficiency improvement that is technically possible and economically justified, without initial regard to how likely it is that this level of energy savings will actually be achieved, through some combination of market forces and public policies and programs. We did not consider the potential savings from "exotic" technologies; all of the energy-saving measures considered are either now commercially available, or very likely to be available soon. (In several cases, our calculations take into account the delayed commercial availability of measures.)

Except as noted, our technical potential estimates assume virtually complete saturation of energy saving measures in newly constructed residences and in new or replacement appliances and heating/cooling equipment. Likewise, our technical potentials reflect a complete retrofitting of the existing housing stock by the year 2000.** The only exceptions to this are where retrofitting is clearly not now physically possible, or economic, for example, in inaccessible attic crawl-spaces, sites with poor solar access, or leaky furnace ducting that is sealed inside walls.

Almost all of the measures that we examined are economically justified (considering life-cycle energy savings vs. increases in initial cost) compared with fuel and electricity at today's average energy prices. We evaluated a few measures that cost more than today's average cost of energy, but less than the cost of new supply. Only a small fraction of the measures we looked at proved more expensive than today's marginal cost of energy. This is an area deserving more analytical attention, as we discuss later. As a rule, there is only a few percent difference in the estimates of total potential savings, if the cut-off point is at current marginal costs rather than average costs.

On the whole, we tried to be conservative in our estimates of energy savings from each measure and of fraction of the stock eligible for that measure--and generous in estimating costs, especially where the available data is scant or inconsistent. For example, we used contractor-installed costs in almost every case (except the one measure labeled "do-it-yourself solar") even though many of the measures could be wholly or partially done by the home-owner. We also tended to use retail sales prices listed in places like the Sears Catalogue for comparing the costs of currently available appliance models with varying levels of efficiency. (Often, these list prices include the cost

---

*In some cases, these parameters from the ORNL model were used for purposes of this analysis even though we do not necessarily agree that they are either plausible or desirable in terms of resource-energy efficiency. For example, the tendency of most ORNL model runs is to predict substantial substitutions of electricity for fuel use in space heating and water heating for new homes.

**In fact, even some new houses built between 1980 and 1985 were assumed to be retrofitted before the year 2000, since BEPS-levels of efficiency in new construction is not assumed to begin until 1985.

of additional features that are provided in the top-of-the-line, high-efficiency models). Current prices for high-efficiency models also may not reflect future economies of large-scale manufacture and marketing, and in some cases retail prices for efficient models may partly reflect marketing strategies rather than actual differences in production costs.

There may be some individual cases where our estimates of energy savings and cost-effectiveness of measures are too optimistic and will need to be revised downward in future iterations. In general, though, we have probably underestimated the cost-justified technical potential for saving energy in new and existing homes, both because of the conservative bias in our estimates, and because we have overlooked (or decided to eliminate) higher-cost measures, advanced concepts that are not yet widely known or well-documented, and technical advances that remain to be developed.

EIA/OR Comparisons

The forecasts of future energy demand made by the Energy Information Agency (EIA) are based on the use of an adaptation of the Oak Ridge National Laboratory (ORNL) Residential Model.

To facilitate comparisons between the EIA projections and the "potentials" developed in this study, we have adopted most of the basic assumptions about housing stock and appliance saturations used by EIA in preparing the 1979 Annual Report to the Congress (EIA, 1980). Conversations with EIA personnel indicated that a number of modifications made it difficult to state precisely a set of input assumptions for the ORNL Residential Model that would reproduce exactly the EIA projections since modifications were made to both inputs and outputs, but a set of input assumptions was developed in consultation with EIA and ORNL personnel that very nearly reproduced the EIA projections when used in the ORNL Residential Model. Whenever the assumptions or results of these runs are discussed, they will be identified as EIA/OR.

The Baseline Case

Quantification of energy savings from conservation and solar measures is meaningless unless the baseline case, from which "savings" are to be subtracted, has been clearly defined. For our analysis, baseline energy use in the year 2000 is defined as follows:

o for existing (1980) stock surviving in the year 2000, baseline usage (per household or per appliance) is equal to the average usage for all stock in 1980.

o for new additions to the appliance and building stock, 1980 to 2000, baseline energy consumption per unit equals the average usage for all new units built or purchased in 1980.

In other words, our hypothetical baseline case for the year 2000 assumes that current thermal and equipment efficiencies continue unchanged for the next twenty years. The only changes are those due to new construction, attrition of the existing housing stock, and normal turnover of appliances and equipment. A few basic demographic and housing stock trends built into the EIA/OR projections are also accounted for in the baseline case, the most significant of which are an increase in average square footage of new homes, a gradual decrease in the average number of people per household, and growing saturation of some appliances like clotheswashers and dishwashers.

It is important to note that the year 2000 baseline case is not at all the same as the conventional forecasts of future energy demand made by the Energy Information Agency, using the ORNL model. First, these EIA/ORNL demand forecasts already take into account a significant amount of energy conservation (both technical improvements in efficiency and changes in usage practices) resulting mainly from increased energy prices, but also including the anticipated direct impact of building standards, appliance standards, and limited retrofit programs. We have included in the supply curves a few end-uses which EIA/OR treats in the "miscellaneous" category. Finally, for a few end-uses, we have chosen more current and, we believe, more accurate figures for today's average unit energy consumption than those available when the EIA/OR model inputs were initially chosen. All of these differences are important only for a few end-uses. Further, we have scaled our 1975-80 residential consumption total to agree as closely as possible with the EIA/OR model runs (making the adjustment in the "miscellaneous" category).

Conservation Supply Curves

We make use of a special analytical device, called the "supply curve of conserved energy," to display and summarize the technical potentials for conservation and solar energy for each major end-use, as well as for overall residential consumption of fuel and electricity. These supply curves, illustrated in the following sections comprise many individual conservation measures, ordered in terms of increasing costs per unit of energy saved (kilowatt-hours of electricity or millions of Btus of fuel). For any given conservation supply curve, the unit cost of conserved energy (CCE) is measured on the vertical axis, while energy savings (per measure and cumulative) are measured along the horizontal axis. Because all of our conservation supply curves are constructed for the year 2000, the savings are expressed as annual energy savings due to the measures in place in that year.

The concepts underlying a "conservation supply curve" are the same as for the more familiar supply curves for other, conventional forms of fuel or electrical energy (or for most other commodities). Based on economic principles, at any given unit cost, some amount of energy is likely to be made available, but increased supplies can only be obtained at higher unit costs. Thus, supply curves slope upward to the right, and the slope becomes steeper (unit costs rise more and more sharply) at increased levels of production.

As shown in in Section 1, our conservation supply curves share this characteristic shape. Each curve begins, at the left end, with a number of relatively low-cost conservation measures. The unit cost of conserved energy begins to rise steeply at the right side of the curve.

Arranging data on individual conservation measures in this form helps to highlight the similarities between conserved energy and other sources of energy supply: as conservation frees energy from wasteful or inefficient use, that energy becomes available for productive use elsewhere in the economy. The conservation supply curves also make it easy to identify the most important opportunities for additional energy savings, both by type of stock (existing homes, new homes, appliances) and by end-use. Those measures that are low-cost are on the lower part of the curve, and those representing a relatively large fraction of potential savings (often these are the same measures) can be identified as the longer horizontal segments on a given supply curve.

Finally, the supply curve format allows the analyst or policy-maker to make an explicit comparison between the cost of conserved energy from various measures, and the (average or marginal) cost of conventional energy sources. Shown on the vertical axis of each conservation supply curve are the prices for oil and gas (or electricity). By locating the points where these horizontal lines intersect the supply curve, one can answer the question: approximately how much energy is potentially available from conservation and solar measures at a unit cost at or below the cost of other supply sources?

Constructing these supply curves required us to estimate the average energy savings, useful lifetime, and incremental purchase (and operating) costs associated with each measure. The sources for the numbers we use (or the basis for our assumptions, where these had to be made) are discussed in detail in the following sections on each residential end-use. Also required to construct the supply curves are the number of units (households or individual appliances) to which each measure was potentially applicable. As noted above, we base these estimates on the year 2000 housing stock and appliance saturation projections from the ORNL model. The text below discusses our estimates for each measure of eligible fractions of that stock.

Estimating Energy Savings and Incremental Costs

The next step is to convert the estimates of unit energy savings and increased first-cost for each measure into the average cost-of-conserved-energy (CCE) for that measure. This requires further assumptions about how the increased first costs are paid for, so that annual energy savings can be consistently compared with added annual costs. While it is admittedly a simplification of actual consumer purchasing behavior, for purposes of the analysis we assumed that the increased costs of a conservation measure are repaid (with interest) in equal annual amounts over the useful life of the measure. In other words, we acted as if these increased costs were simply added to the amount of a loan for the purchase of a new home, new appliance, or the retrofit of an existing home. Since we have kept all dollar figures in constant (1980) dollars, we use a "real" interest rate (net of inflation) of 3% per year, the same rate that was used in the cost-effectiveness analysis for proposed federal building and appliance standards.*

In developing the conservation supply curves for space heating and cooling and for water heating, we paid attention to the physical interactions among measures installed in a given structure, including the cost and performance interactions between conservation measures and active or passive solar measures for space heating and water heating. In general, our approach was to assume that measures would be implemented in an economically optimal order. In other words, those with the lowest cost-of-conserved-energy (CCE) would be done first, with the cost and energy savings from subsequent measures adjusted to take into account those already in place. For example, annual energy savings from a high-efficiency replacement furnace would be lower in a house that had already been insulated. Similarly, the size and therefore the initial cost of an active solar water heating system would be lower for a house that had already

---

*A DOE final rule for calculating cost-effectiveness for solar demonstration programs (Federal Register, 45:16, 1/23/80, p. 5620) calls for the use of a 10% real discount rate, to "reflect the opportunity cost of capital in the private sector." This might suggest that our choice of 3% for a real-dollar interest rate is too low. However, home mortgage rates have historically averaged about 3% above the rate of inflation, which suggests that 3% real interest is what most homeowners actually pay. To investigate the magnitude of changes in discount rate, we calculated the summary fuel and summary electric supply curves at a discount rate of 10%, as shown earlier. The comparison of these curves indicates that the total savings will not be significantly changed if a higher discount rate is chosen.

been fitted out with low-flow plumbing devicess and with a replacement dishwasher and clotheswasher with lower hot water requirements. This approach again oversimplifies real decision-making; in many cases homeowners may undertake conservation retrofits in a very different order, for reasons of convenience, limited budgets, aesthetic preferences, or do-it-yourself inclinations.

The advantage of our approach is that it helps to avoid double-counting of energy savings or costs, and probably yields a fair estimate of the aggregate energy savings and costs from a combination of many measures. (However, energy savings and CCE values for some individual measures would vary if a different order of implementation were assumed.) This whole approach required a careful tracking of the space heating, cooling, and water heating measures implemented, at various times, for different segments of the housing stock, especially since different age-cohorts of that stock were steadily decaying between 1980 and 2000. We explain these calculations and show detailed diagrams in the sections on each end-use.

One final point is worth mentioning: we did not include in the supply curves the estimated technical potential for on-site (residential-scale) generation of electricity from wind and solar photovoltaic systems. These potentials are estimated separately as discussed in Chapter IV. We adopted the conventional practice of considering solar and wind-generated electricity as a means of increasing supply, not of reducing demand through end-use efficiency. Arguably, our analysis of the potentials for site-generated electricity from wind and photovoltaics needs to be closely integrated with considerations of electricity "quality control," and the size and timing of loads at the same residential sites.

Limitations of the Data and Estimates

Some final words of caution are in order to help the reader interpret our estimates of technical potential for conservation and solar energy in residences. First, considerable averaging--across regions, building types, and individual households--was required to make the accounting problem manageable. A direct corollary of this is that the supply curve for each end-use can be correctly viewed only as a proxy for representative measures that might be undertaken in the building stock as a whole. The supply curve is not an accurate list of the specific items that would achieve precisely the same unit savings, at the costs shown, in any one actual house.

Second, there are several cases where the cost and savings numbers we arrived at were, quite simply, our best guesses. These are cases where reliable, measured data is either absent, or different sources disagreed widely--and we had to reconcile these differences using an educated judgment. The process of updating and extending the data base for conservation

supply curves and substituting good data for best guesses, is a crucial one. Both the aggregate numbers and the estimates of individual measures should become increasingly reliable with each future iteration.

A third caveat: the entries on the supply curve are the least complete, and have the widest uncertainty bands, for higher cost measures. Relatively little data are available on measures at or above today's average unit cost of energy, simply because fewer of them have been installed commercially, and until recently (with rapidly increasing energy prices), there has been little reason for either private industry or government-funded research to address these relatively expensive energy-saving options. (Expensive, that is, in comparison to the bulk of measures toward the left end of the supply curves, with low costs-of-conserved-energy.)

Among the technical areas in need of more careful analysis of costs, energy savings, and potential applicability in new and existing residences are: cooling-load-reduction measures (building analysis in the past has tended to concentrate on reducing energy requirements for heating); "smart" heating and cooling control systems (including those that provide real-time occupant feedback); gas-fired heat pumps for space and water heating; high-thermal-performance window systems; and "integrated" appliances that provide for heat recovery (combined refrigerator/water heaters, or air conditioner/water heaters, for example).

The following sections, covering each major residential end-use, explain our calculations in more detail. They are accompanied by conservation supply curves for each end use, which are in turn aggregated in the summary supply-curves for residential fuel and electricity conservation. The end-use categories, in order, are:

o fuel space heating - existing houses

o fuel space heating - new houses

o electric space heating - existing houses

o electric space heating - new houses

o central air conditioning

o water heating (fuel and electric)

o other appliances (fuel and electric)

SPACE CONDITIONING

Space conditioning (heating and air conditioning) is the largest single end-use of energy in residential buildings, accounting for over 50% of current residential energy consumption. By the year 2000, 80% of the currently existing houses will still be in use, com-

---

prising over 50% of the total year 2000 housing stock. Thus, the reduction of heating and cooling requirements in existing homes is one of the most important, and most difficult, aspects of achieving energy savings. Space heating in existing homes also presented the most difficulty in analysis. A tour of any urban area will quickly demonstrate the complexity of housing types, sizes, siting conditions, etc., which compound the calculations and perplex the analyst.

Some Notes on Method

In order to stay within the limitations of this report, a method was developed to simplify the housing stock into a series of "model" houses. Using data from the Residential Energy Consumption Survey (RECS) the "energy" characteristics of the housing stock were divided into three categories entitled "uninsulated," "partially insulated" and "1976-80 practice." An average home for each category was then developed using the base home for the BEPS analysis as a model that allowed the use of computer simulation. One BEPS home was originally a Washington, D.C. residence of approximately 1176 square feet. While the Washington D.C. location seemed appropriate, since it represents a rough national average for climate conditions (4230 heating degree days) and typical cooling load, the size of the house had to be adjusted to reflect more accurately the average size of the homes we were categorizing. Further, fuel heated homes had to be separated from electrically heated homes since each has unique retrofit applications.

Therefore, the analysis of existing residential dwellings is divided into six parts, based first upon the type of heating system (fuel or electric) and then on the degree of existing insulation. For each fuel type, retrofit options were identified, costs estimated, and limitations upon the application of each option quantified.

The use of the model homes allowed the construction of "supply curves of conserved energy" once costs were established for the applicable retrofit options. These supply curves are exhibited below for each of the six categories. Certain options can apply to only a fraction of the homes within a category, so application of every item to each house in order of its cost-effectiveness was impossible. For some of the options, exact data on the percentage of houses to which each retrofit option could actually be performed were unavailable; in other cases approximations could be made.

Where data were difficult to obtain, expert opinion was relied upon to develop a percentage of the stock to which the retrofit would apply, termed the "eligible fraction."

The use of eligible fractions resulted in a complex branching of the supply curves. For each category of homes the sequence of the retrofits is displayed graphically. In each, the representative home is defined, and the retrofit options are applied to it in order of cost-effectiveness. Wherever an option is encountered that can only apply to a percentage of the category, there is a branching of the diagram and the percentage of the category following each path is indicated. The basic groupings used for both electric and fuel heated homes are shown in Table 1.15.

Normalization to 1975 ORNL Fuel Use

The fuel consumed by the stock for space heating in 1975 is adjusted to agree with ORNL to facilitate later comparisons. The adjustment is a reduction in fuel usage of 5%, which may be taken as an overestimation by having all the stock as single-family homes. The effects of all the shell retrofits are also multiplied by this factor, .95.

Fuel use: a) uninsulated*: 174 MBtu/yr + 6 MBtu/yr pilot light becomes 165+6 = 171 MBtu/yr.

b) partially insulated*: 83 MBtu/yr + 6/MBTU/yr pilot light becomes 79+6 = 85 MBtu/yr.

Average fuel use is 113 MBtu/yr, as in EIA/OR (EIA/OR gives the total fuel use for space heat as 6.97 quads, for a fuel heated stock of 61.8×10^6).

Electricity Use

All electrically heated homes are considered to be partially insulated; therefore equivalent fuel use would be 79 MBtu (since there are no pilot losses), as above. With a furnace system efficiency of 60% delivered heat is 47.4 MBtu or 13,888 kwh electricity, or 160 MBtu resource electricity.

Average electricity use is 160 MBtu/yr resource energy, but ORNL gives the total electricity use for space heat as 1.39 quads for a stock of 9.05×10^6 units, for a consumption of 154 MBtu/yr.

Load Calculation Method

The house-heating and cooling loads are derived from the DOE-2 calculations made for the BEPS homes, series V. However, not all the states of the house are included in the BEPS runs, and to obtain them, interpolations between the various known states

---

*RECS Data for Insulation Levels.

Table 1.15. BASIC GROUPINGS OF HOMES

| | Houses Surviving in 2000 (10^6) | | |
Cohort	Electric	Fuel	Total
Existing through 1975	6.7	45.5	52.2
76-80	3.4	6.7	10.1
81-85	3.9	7.6	11.5
86-90	3.8	7.5	11.3
91-2000	7.1	13.8	20.9
TOTAL	24.9	81.1	106.0

must be made. Interpolations are made differently for insulation level changes, glazing changes and infiltration changes since these items have such different properties.

a) Infiltration changes are scaled directly with the volume of air. BEPS runs exist for both 0.6 air changes per hour (ach) and 0.2 ach. The infiltration part of the load is then calculated as follows:

$$\text{Load at 1 ach} = \frac{\text{Load at 0.6 ach (BEPS high)} - \text{Load at 0.3 ach (BEPS low)}}{0.6 \text{ ach} - 0.3 \text{ ach}}$$

b) Window loads are scaled directly with the U value of the window; BEPS runs exist for double and triple but not single glazed windows.

Load single-glazed - Load double-glazed

$$\frac{(U_1 - U_2)}{(U_2 - U_3)} \text{ (Load double-load triple)}$$

c) Insulation changes are scaled by effective degree-days (DD) and change in U-value:

$$\text{Change in load} = (\text{Change in U}) \times \text{Area} \times \text{DD} \times 24 \text{ hours/day}$$

The value for DD is found by inserting known values for load and U changes, i.e., from known BEPS states. For this particular house in Washington, D.C., 5000 DD gives results within 20% of values calculated using DOE-2, and so this value was used.

Fuel-Heated Homes

Existing Housing

Fuel heating currently accounts for nearly four times more energy use in residential dwellings than electric heating. It also requires the most complex analysis. The process and results of the analysis will be discussed for each "model home" with the "uninsulated" category of fuel heated homes first, which represents the most complex supply curve, followed by discussions of the partially insulated" and "1976-1980 practice" homes. The basic characteristics of these homes are shown in Table 1.16.

To each of these, a series of retrofit options was applied in their order of cost-effectiveness. Table 1.17 lists the measures and the costs assumed for all fuel heating retrofits.

Since each group of homes has different initial shells, different retrofits are appropriate; the following Table 1.18 shows the retrofit measures applicable to each group.

Table 1.16. BASIC CHARACTERISTICS OF CATEGORIES OF HOMES

| Fuel Heated Homes | Pre-1976 | | 1976-80 Practice |
	Uninsulated	Partially Insulated	
Number of homes surviving to 2000 (millions)	15.2	30.3	6.7
Total Area (sq. ft.)	1350	1350	1560
Window Area (sq. ft.)	202	202	234
Door (sq. ft.)	20	20	20
Ceiling Insulation	0	R-11	R-19
Wall Insulation	0	R-7	R-11
Window Glazings	1	60%-2 & 40%-1	2
Infiltration (ach)	1	1	.6
Furnace Efficiency	75%	75%	75%
Duct Efficiency	80%	80%	90%
Heating System Efficiency	60%	60%	67.5%
Annual Fuel Use (MBtu)	171	85	49

Table 1.17 RETROFIT OPTIONS AND COSTS; SPACE HEAT[a]

Measure	Unit Cost	Area (ft^2)	Source	Total Cost ($)
Add: R-19 to to ceiling	35¢/ft^2 (20¢ for materials)	1350	PG&E; Means cost data	470
R-11 to walls	70¢/ft^2	1232	Wide Range; Means cost data	860
Infiltration Reduction	$20/manhour	12 manhours	B. O'Regan, LBL - House Doctor	240
Double pane storms	$7/ft^2	200	See Note 2	1400
Add: R-27 ceiling	43¢/ft^2	1350	Means cost data	580
Seal Ducts			Wright et al	300
Retrofit Spark Ignition			PG&E SF Project	170
90% eff. furnace				400
Re-siding + R-5	Cost of R-5 only		See note (1)	500
Double pane storms to 40% of windows	$7/ft^2	80	See note (2)	560
Single pane storms	$4.5/ft^2	200	See note (2)	900
DIY Solar	(Occur after		See note (3)	35/MBtu/year delivered
Solar I	Phase I)		See note (3)	86/MBtu/year delivered
Solar II	(Occur after Phase II,		See note (3)	122/MBtu/year delivered
Solar III	after 90% furnace)		See note (3)	170/MBtu/year delivered

[a]See Chapter IV for further discussion of cost assumption and references.

Table 1.18

Measure	Uninsulated	Partially Insulated	76-80 Practice
R-19 ceiling insulation	x		
Reduce infiltration to 0.6 ach	x		
R-11 wall insulation	x		
Extra R-19 ceiling insulation	x		x
Extra R-27 ceiling insulation		x	
Double storm windows	x	x	
Spark ingnition to furnace	x	x	
90% efficient furnace	x	x	x
Solar Energy	x	x	x
Seal Heating ducts	x	x	x
Add insulation on re-siding		x	

Notes to Retrofit Costs

1) Re-siding

In metropolitan Denver in one year, 10,700 of 334,000 owner-occupied houses were resided; 1/4 to 1/3 were insulated at the same time. The typical cost for a 1500 ft^2 house was $2600 for aluminum siding, plus $450 for 3/4" styrofoam, at R-4 or R-5 per inch. As confirmation, we note that the 1980 Means, "Building Construction Cost Data," (p 118) gives the price of installing 1" thick polystyrene, R = 5. 4, as 49¢/ft^2, or $603 for 1232 ft^2.

2) Storm Windows

The price of installing a window is given in 1980 Means "Building Construction Cost Data" (p. 148) as $2.18/ft^2. We add $2/ft^2 to account for retrofit diffi culties, and round up to $4.50/ft^2 for single-pane storm windows. For double storm windows we took two windows, i.e., $4.36/ft^2, plus the $2 for retrofit, and round up to $7/ft^2.

3) Infiltration Reduction

Some infiltration reduction measures can be accom- plished extremely cheaply. Reducing infiltration from its present levels of about 1.0 air change per hour (ach) to the level obtained by new homes (0.6 ach) requires the identification of the major air leaks, and sealing them with caulk or fiberglass. Many of the leaks involve air loss to the attic, which occurs around ducts, vents, recessed light fixtures, partition walls, and dropped ceilings. These leaks, along with the more widely recognized leaks around windows, can be spotted by smoke tests in a house which has been pressurized by a window-mounted or door-mounted fan. Pressure and flow gauges on the fan can inform the retrofitter of the success of the leak plugging.

The materials needed to seal leaks are exteremly cheap, so the cost of infiltration reduction is largely the cost of the labor needed to find and seal holes. If we assume that lobor costs $25/hour, then a two-man crew working 1/2 day could seal a house for $200. This 1/2-day period seems reasonable based on present research houses.

4) Attic Insulation

It is economic to retrofit ceiling insulation up to the BEPS levels (R-38 or 12 inches of fiberglass equivalent) in most parts of the country. For example, since the marginal cost of an extra R-1 of insulation is about the same for a new house as for a retrofit, the optimal insulation of R-38 for Washington D.C. attics in BEPS houses is also optimal as a retrofit. It is, of course, cheaper to install R-38 at one shot rather than two batches of R-19 at different times, since labor costs would be twice in the second case.

5) Wall Insulation

Uninsulated walls can be retrofitted at relatively low cost and with a high benefit/cost ratio by filling the stud cavities with a blown-in insulation material, such as fire-resistant cellulose. If the wall cavity is partially insulated, as is commonly the case in the colder parts of the country, it is often cost-effective to reinsulate, either by removing the siding, adding fiberglass and insulating sheathing, and re-siding, or by adding insulating material (e.g., foam) and new siding over the existing walls. If a house is to be refitted with new siding anyway, it is very cost-effective to install insulation at the same time.

6) Furnace Retrofits

Heating system efficiencies are presently in the range of 50-65%. There has not been much attention devoted to measuring _in situ_ furnace efficiencies on a systematic basis; spot measurements have ranged from less than 40% to over 75%. The lower end of the range is undoubtedly a result of large duct losses, in which improvements can be retrofitted. Using inside air for combustion increases the infiltration rate, reducing system efficiency; in many cases it will be possible to install a duct to bring in outside air for combustion as

a retrofit. Another large source of waste is the pilot light. Retrofit spark ignition systems are beginning to be available for a relatively low cost--as little as $150.

7) Solar Energy

Since there are a number of solar systems able to provide space heat, a composite of prices was made, using estimates from Chapter IV. Cost estimates for different passive retrofits are shown below.

	Capital Cost Passive Solar Retrofits $/MBTU/yr delivered
Vertical heater	15 - 215
Trombe Wall	22 - 215
South Windows	22 - 145
Solar Greenhouse	21 - 277

Active solar retrofits are estimated to cost $82-159/MBtu/yr. These ranges were reduced to discrete values as described in Chapter IV, and the active and passive costs were averaged to give the "combined" values shown below.

	Passive	Active $/MBTu/yr
DIY	35	--
Solar I	78	95
Solar II	123	120
Solar III	195	146

	Combined $/MBtu/yr	% of Market
DIY	35	10
Solar I	86	30
Solar II	122	30
Solar III	170	30

The applicable market shares were reduced to account for limitations on solar access. It was assumed that 60% of existing homes had adequate access for an active or passive system. (See complete discussion in Chapter IV.)

Figure 1.21 and Table 1.19 respectively show the supply curve for all existing fuel heated homes and the supply Table which details the cost and Savings for each measure shown on the supply curve. Note carefully that not all retrofits apply to all houses. Thus, it is not possible to simply add the "savings" column in the cost of conserved energy table, since no home receives for example, Solar I plus extra R-19 insulation in the ceiling. The retrofit sequence diagrams which follow make this clear and provide a final column

Table 1.19
****
SUPPLY TABLE TO THE
FUEL SPACE HEAT EXISTING HOUSES
****

TIME HORIZON=10 YEARS
DISCOUNT RATE(S)
RESIDENTIAL= 3.0 PERCENT

	CONSERVATION MEASURE	A. ENERGY TYPE	B. SECTOR	C. MARGINAL COST OF CONSERVED ENERGY ($/MBTU)	D. AVERAGE COST OF CONSERVED ENERGY ($/MBTU)	E. ENERGY SUPPLIED PER MEASURE (TERABTU/ YEAR)	F. TOTAL ENERGY SUPPLIED (TERABTU/ YEAR)	G. TOTAL DOLLARS INVESTED (M$)	MEASURE NUMBER
1	R=19 IN ATTIC UNINSULATED.	GAS	R	.5	.5	851.2	851.	7144.	1
2	INFILTRATION PART. INS.	GAS	R	1.5	.9	545.4	1397.	14416.	13
3	INFILT. REDUCTION UNINS.	GAS	R	1.5	1.0	273.6	1670.	18064.	2
4	DIY SOLAR PART. INS.	GAS	R	1.6	1.0	45.5	1716.	19209.	20
5	DIY SOLAR UNINSULATED	GAS	R	1.7	1.0	27.4	1743.	19902.	9
6	DIY SOLAR 76-80	GAS	R	1.9	1.1	7.6	1751.	20124.	27
7	R=11 IN WALLS UNINSULATED.	GAS	R	2.2	1.3	395.2	2146.	33196.	3
8	.90 EFF. FURN. PART. INS.	GAS	R	2.2	1.3	181.8	2328.	39256.	18
9	.90 EFF. FURNACE UNINS.	GAS	R	2.2	1.4	91.2	2419.	42296.	7
10	.90 EFF. FURN. 76-80	GAS	R	2.6	1.4	14.7	2434.	42885.	25
11	R=27 IN ATTIC PART. INS.	GAS	R	2.7	1.6	424.2	2858.	60459.	14
12	SPARK IGN. RETR. 76-80	GAS	R	3.2	1.6	15.7	2873.	60903.	26
13	RETR. SPARK IGN.PART. INS.	GAS	R	3.2	1.6	90.9	2964.	63479.	19
14	SPARK IGN. RETR. UNINS.	GAS	R	3.2	1.6	45.6	3010.	64771.	8
15	SOLAR I. UNINSULATED	GAS	R	3.3	1.7	82.1	3092.	68957.	10
16	SOLAR I. PART. INS.	GAS	R	3.3	1.8	136.4	3228.	75938.	21
17	SOLAR I. 76-80	GAS	R	3.8	1.8	22.9	3251.	77277.	28
18	EX. R=19 IN ATTIC 76-80	GAS	R	5.0	1.8	46.4	3298.	80895.	24
19	STORM WINDOWS UNINSULATED.	GAS	R	5.1	2.1	273.6	3572.	102175.	4
20	EXTRA R=19 IN ATTIC UNINS.	GAS	R	5.1	2.1	69.3	3641.	107604.	5
21	STORM WINDOWS PART. INS.	GAS	R	5.2	2.3	212.1	3853.	124572.	15
22	RESIDE +R=5 PART. INS.	GAS	R	5.4	2.3	20.0	3873.	126239.	16
23	SOLAR II.PART. INS.	GAS	R	5.6	2.4	92.7	3966.	134256.	22
24	SOLAR II. UNINSULATED	GAS	R	5.8	2.4	46.5	4012.	138415.	11
25	SOLAR II. 76-80	GAS	R	5.8	2.5	22.9	4035.	140465.	29
26	SOLAR III. PART. INS.	GAS	R	7.5	2.6	92.7	4128.	151100.	23
27	SOLAR III. UNINSULATED	GAS	R	7.8	2.6	46.5	4175.	156627.	12
28	SOLAR III. 76-80	GAS	R	7.8	2.7	22.9	4198.	159353.	30
29	DUCT SEALING PART. INS.	GAS	R	8.5	2.7	9.7	4207.	160080.	17
30	DUCT SEALING UNINSULATED.	GAS	R	8.5	2.7	9.1	4216.	160764.	6

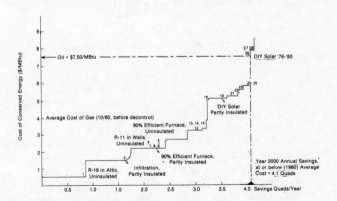

Figure 1.21. **Year 2000 Supply Curve of Conserved Energy (in quads/year) for Space Heat of Fuel Heated Dwellings Built Through 1980.**

Year 2000 baseline annual use for this sector is 5.5 Quads, where "baseline" assumes continuation of 1980 average unit energy consumption for existing stock or new additions in that year. Unit cost of conserved energy (in constant 1980 $) assumes that all increased costs are amortized over the useful life of this measure, using a 3% (real-dollar) interest rate. Potential annual savings in 2000, at or below today's cost of oil (7.5 $/MBtu) is 4.1 quads, or 75% of the year 2000 baseline.

which can be added to obtain the total savings without double-counting.

Uninsulated Pre-1976 Homes. There are over 15.2 million existing homes which will still be in use in the year 2000 that can be presently categorized as uninsulated. These homes represent the most wasteful segment of our housing stock and offer the greatest potential for energy savings. These homes now on the average use 171 MBtu of energy per year, which could be reduced to an average of 25.2 MBtu/yr through the application of conservation and renewable technologies, for an aggregate savings of nearly 2 Quads over the year 2000 baseline case or the equivalent of 1,000,000 barrels of oil per day.

Figure 1.22 shows the retrofit sequence for this group of "uninsulated" fuel heated homes.

There are four reasons for the branching of the retrofit sequence: (1) the application of active and passive solar retrofits is limited to 60% of the housing stock due to siting problems, such as obstruction by trees, other buildings, etc.; (2) it is assumed that only 10% of all homeowners would be willing and able to undertake a so-it-yourself solar retrofit (which is reduced to 6% eligibility due to siting problems); (3) only 20% of the houses could have their heating system's ducts sealed, since in most cases they will be inaccessible behind walls, etc.; and (4) since the cost-effectiveness of 90% efficient furnace replacements and spark ignition retrofits is roughly equal, it is assumed that 50% of the homes do each one.

It should also be noted that the actual sequencing chosen effects the cost-effectiveness of the retrofit options. While in most cases, the initial order of cost-effectiveness shown in the supply curve is followed, in some cases retrofit options are not undertaken because they are not cost-effective due to the sequencing on a particular branch of the diagram. For example, the homes which receive Solar I do not receive an additional R-19 in the ceiling. Because of the method used to calculate shell savings, it does not, in general, make any difference which independent shell retrofit comes first. However, the fuel savings (and thus cost-effectiveness) change when the heating system is retrofitted by adding solar or a more efficient furnace.

Partially Insulated Pre-1976 Homes. Of the homes categorized as partially insulated, 30.3 million survive to the year 2000. These homes now have an average use of 85 MBtu per year, which could be reduced to 23.7 MBtu per year through the application of solar and conservation technologies, for a savings of nearly 1.9 quad over the baseline case or nearly one million barrels of oil per day (equivalent).

Figure 1.23 shows the flow diagram for this group. The four reasons discussed previously under uninsulated homes which contribute to the branching are also applied to this category. They concerned limitation of

solar to 60%, limitation of solar do-it-yourself to 6%, limitation of duct seals to 20%, and equal penetrations of 90% efficient furnaces and spark-ignition retrofits. In addition, it is assumed that adding insulation to walls which already have R-7 insulation is not cost effective; however, it is cost-effective to put in extra insulation if the house is resided. As explained in the notes to the retrofit cost table, it is assumed that 15% of these houses will be resided during the next 20 years.

1976-80 Practice. Of the homes built between 1976-80, 6.7 million fuel heated units survive until the year 2000. There are few retrofits which can be carried out on these homes. For example, adding single-glazed storm windows (base home already has double-pane windows) only saves 5MBtu at a cost of $1050, for a cost of conserved energy of $14/MBtu. Ceiling insulation, furnace replacements and retrofits and solar installations are carried out on these homes, reducing the average units usage from 49MBtu/yr to 29.8 MBtu/yr, for a savings of .13 quads over the base case.

Figure 1.24 shows the retrofit sequence for these homes.

It should be noted that it is assumed that 50% of these homes do not have pilot lights, and that the average units area has increased from 1350 ft^2 to 1560 ft^2.

New Homes 1981-2000

The discussion of the basic methodology noted that the baseline case assumes an increase in the average size of new housing units size between 1975 and 2000. To account for this increase in home size, the addition of new housing to the stock is taken in three groups:

Year Built	Number Surviving (millions)	Square Feet
81-85	7.6	1690
86-90	7.5	1770
91-2000	13.8	1810

The baseline new home is already fairly energy-efficient compared to the existing stock in 1980. For the baseline case, each of these groups starts with the same characteristics except for square footage, and the assumptions that 50% of the new homes built in each group have pilotless furnaces. The other shared characteristics are:

R-19 Ceiling insulation
R-11 Wall insulation
100% Double glazing
.6 air changes per hour
90% efficient ducts
75% efficient furnace

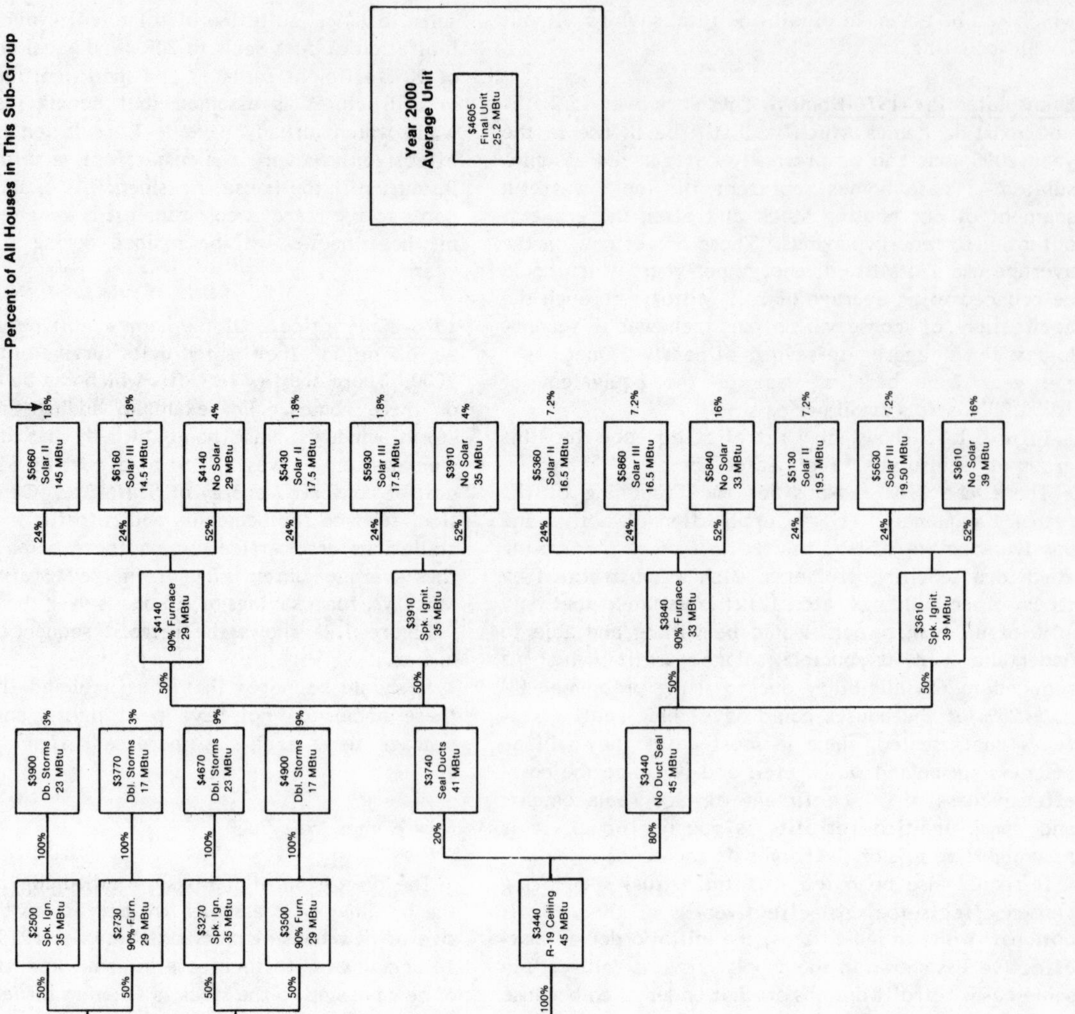

Legend
$000
QXYZ
000 MBtu
Cumulative Investment
Conservation Measure
Annual Fuel Use
Percent of All Houses in This Sub-Group
$000
Orig. House
171 MBtu
100%
$470
R-19 Ceiling
115 MBtu
100%
$710
0.6 ach Infil.
97 MBtu
100%
$1570
R-11 Walls
71 MBtu
6%
$2330
DIY Solar
41 MBtu
50%
$2500
Spk. Ign.
35 MBtu
100%
$3900
Db. Storms
23 MBtu
3%
50%
$2730
90% Furn.
29 MBtu
100%
$3770
Dbl. Storms
17 MBtu
3%
18%
$3100
Solar I
41 MBtu
50%
$3270
Spk. Ign.
35 MBtu
100%
$4670
Dbl. Storms
23 MBtu
9%
50%
$3500
90% Furn.
29 MBtu
100%
$4900
Dbl. Storms
17 MBtu
9%
76%
$2970
Dbl. Storms
51 MBtu
100%
$3440
R-19 Ceiling
45 MBtu
20%
$3740
Seal Ducts
41 MBtu
50%
$4140
90% Furnace
29 MBtu
24%
$5660
Solar II
145 MBtu
1.8%
24%
$6160
Solar III
14.5 MBtu
1.8%
52%
$4140
No Solar
29 MBtu
4%
50%
$3910
Spk. Ignit.
35 MBtu
24%
$5430
Solar II
17.5 MBtu
1.8%
24%
$5930
Solar III
17.5 MBtu
1.8%
52%
$3910
No Solar
35 MBtu
4%
80%
$3440
No Duct Seal
45 MBtu
50%
$3840
90% Furnace
33 MBtu
24%
$5360
Solar II
16.5 MBtu
7.2%
24%
$5860
Solar III
16.5 MBtu
7.2%
52%
$5840
No Solar
33 MBtu
16%
50%
$3610
Spk. Ignit.
39 MBtu
24%
$5130
Solar II
19.5 MBtu
7.2%
24%
$5630
Solar III
19.50 MBtu
7.2%
52%
$3610
No Solar
39 MBtu
16%
Year 2000
Average Unit
$4605
Final Unit
25.2 MBtu
Original House
No Ceiling Insulation
No Wall Insulation
Infiltration 1 Ach
60% Efficient Heating System
75% Efficient Furnace
80% Efficient Ducts
Single Glazed
171 MBtu/yr Fuel Use
15.2 Million Houses Surviving to 2000

Figure 1.22. **Logic Diagram — Space Heat Fuel: Uninsulated Pre-1976**

Figure 1.23.
Space Heat Fuel, Insulated Pre-1976

Original House

R-11 Ceilings
R-7 Walls
Infiltration 1 ach
60% Efficient Heating System { 75% Furnace
60% Double Glazed   80% Ducts
85 MBtu/yr

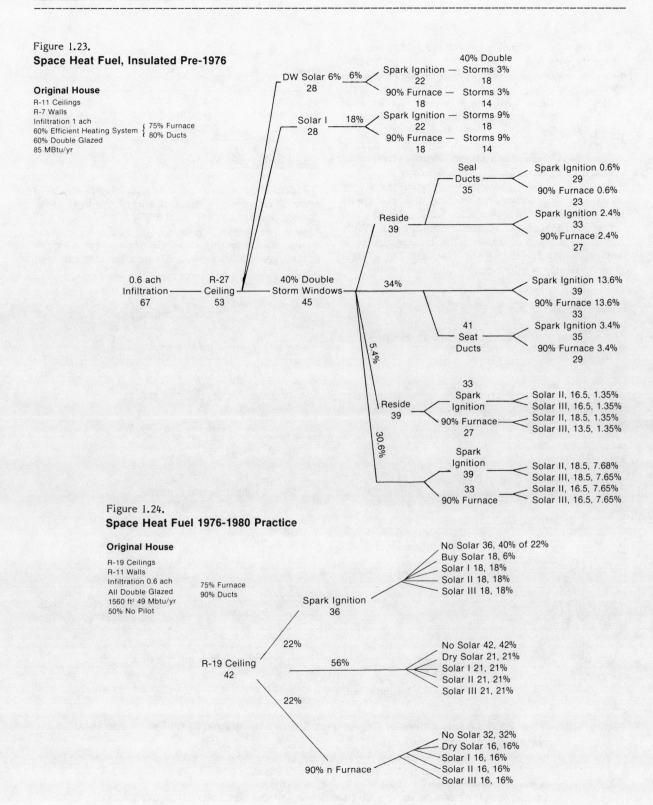

Figure 1.24.
Space Heat Fuel 1976-1980 Practice

Original House

R-19 Ceilings
R-11 Walls
Infiltration 0.6 ach
All Double Glazed   75% Furnace
1560 ft² 49 Mbtu/yr   90% Ducts
50% No Pilot

The measures applicable to each group of homes significantly overlap. Table 1.19 shows the efficiency measures applicable to each group of homes and the costs assumed.

The 28.9 million fuel-heated homes built between 1981 and 2000 have a baseline energy usage of 1.59 quads. Through the application of additional of conservation and renewable measures this could be reduced to .44 quad, for a savings of the equivalent of over one-half million barrels of oil per day.

Figure 1.25 and 1.24 respectively show the supply curve for new fuel-heated housing and the supply table, Table 1.20. Figures 1.26, 1.27, and 1.28 show the application of the measures identified on Table 1.19 to each group of houses. The average change in unit efficiency for each group (not including central air-conditioning requirements) is calculated to be:

Group	Average Unit Baseline Usage	Average Unit Potential Usage
81-85	53MBtu	28 MBtu
86-90	55MBtu	16 MBtu
91-2000	56MBtu	8 MBtu

Electrically Heatd Homes

Electric space heating and air conditioning currently account for nearly 3 quads of energy use in the residential sector; by the year 2000, the baseline case assumption is that this usage will reach to nearly 4 quads. Space heating alone will account for 2.7 quads of this year 2000 demand. Through the application of

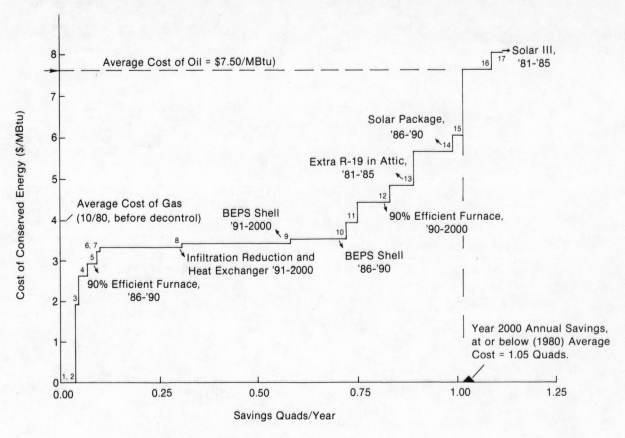

Figure 1.25. **Year 2000 Supply Curve of Conserved Energy (in quads/year) for Space Heat of New Fuel-Heated Dwellings.**
Year 2000 baseline annual use for this sector is 1.59 quads, where "baseline" assumes continuation of 1980 average unit energy consumption for existing stock or new additions in that year. Unit cost of conserved energy (in constant 1980 $) assumes that all increased costs are amortized over the useful life of the measure, using a 3% (real-dollar) interest rate. Potential annual savings in 2000, at or below today's average cost ($7.50/MBtu) is 1.05 quads, or 66% of the year 2000 baseline.

Table 1.20

```
*************************************************
FUEL SPACE HEAT NEW HOUSES    SUPPLY TABLE TO THE
*************************************************
```

TIME HORIZON=10 YEARS
DISCOUNT RATE(S)
RESIDENTIAL= 5.0 PERCENT

	CONSERVATION MEASURE	A. ENERGY TYPE	B. SECTOR	C. MARGINAL COST OF CONSERVED ENERGY ($/MMBTU)	D. AVERAGE COST OF CONSERVED ENERGY ($/MMBTU)	E. ENERGY SUPPLIED PER MEASURE (TERABTU/YEAR)	F. TOTAL ENERGY SUPPLIED (TERABTU/YEAR)	G. TOTAL DOLLARS INVESTED (M$)	MEASURE NUMBER
1	WINDOW TO SOUTH 91-2000	GAS	R	0.	0.	24.8	25.	0.	46
2	WINDOWS TO SOUTH 86-90	GAS	R	0.	0.	13.5	38.	0.	41
3	DIY SOLAR 81-85	GAS	R	1.9	.4	9.1	47.	269.	34
4	.90 EFF. FURNACE 81-85	GAS	R	2.6	1.0	20.5	68.	1090.	32
5	.90 EFF. FURNACE 86-90	GAS	R	2.9	1.5	23.6	92.	2140.	39
6	SPARK IGN. RETR. 86-90	GAS	R	5.2	1.8	14.9	106.	2561.	40
7	SPARK IGN. RETR. 81-85	GAS	R	5.2	2.0	16.9	123.	3039.	33
8	INF.+HEAT EXCH. 91-2000	GAS	R	3.3	2.8	207.0	330.	13665.	44
9	BEPS SHELL 91-2000	GAS	R	3.4	3.1	276.0	606.	32433.	43
10	BEPS SHELL 86-90	GAS	R	5.5	3.2	142.5	749.	42483.	38
11	SOLAR T. 81-85	GAS	R	5.9	3.2	27.4	776.	44111.	35
12	.90 EFF. FURNACE 91-2000	GAS	R	4.4	3.3	82.8	859.	49631.	45
13	EXTRA R-19 IN ATTIC 81-85	GAS	R	4.8	3.4	60.8	920.	54115.	31
14	SOLAR PACKAGE 86-90	GAS	R	5.6	3.6	101.3	1021.	62822.	42
15	SOLAR TI. 81-85	GAS	R	6.0	3.7	27.4	1048.	65325.	36
16	SOLAR PACKAGE 91-2000	GAS	R	7.6	3.9	74.5	1123.	74019.	47
17	SOLAR TI. 81-85	GAS	R	8.0	4.0	27.4	1150.	77357.	37

Figure 1.26. **Fuel Space Heat, 1981-1985**

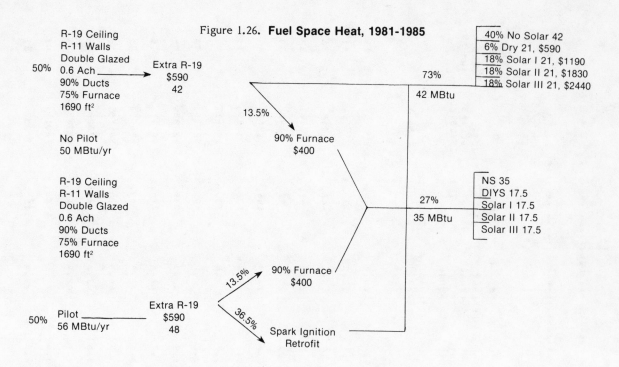

Figure 1.27.
Fuel Space Heat, 1986-1990

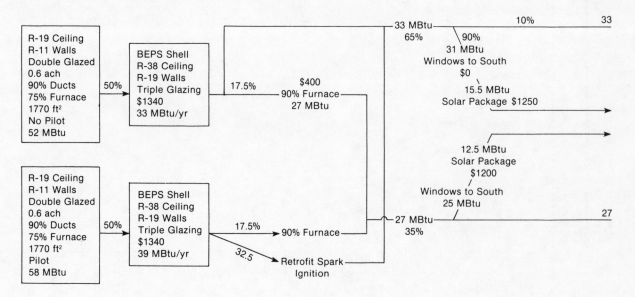

Figure 1.28.
Fuel Space Heat, 1991-2000

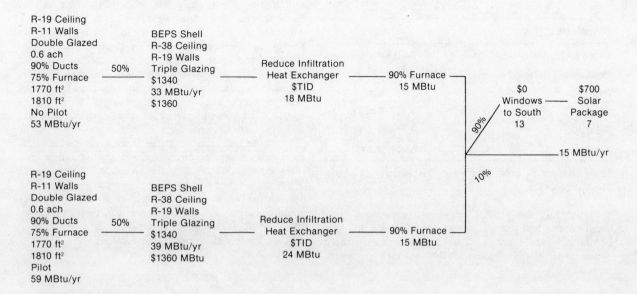

conservation and renewable measures in both existing and new homes with electric space heating, this requirement could be reduced from 2.7 quads to 0.7 quads, for a savings of the equivalent of 1,000,000 barrels of oil per day.

The assumptions used to construct the model homes for both new and retrofit electric heating are:

ASSUMPTIONS FOR BASELINE CASE

	Pre-76	1976-80	1981-86	1986-90	1990-2000
Number of homes surviving to 2000	6.7	3.4	3.9	3.8	7.1
Total Area ft^2	1350	1560	1690	1770	1810
Ceiling Insulation	R-11	R-19	R-19	R-19	R-19
Wall Insulation	R-7	R-11	R-11	R-11	R-11
Window Glazings	60%-2	2	2	2	2
Infiltration (ach)	1	0.6	0.6	0.6	0.6
Annual Space Heating Use Use Per Dwelling (kwh)	13,900	7,030	7630	8000	8180
Total Heating Use (TWh)	93.1	23.9	29.8	30.4	58.1

All pre-1976 homes are assumed to have electric resistance heating, and all post-1976 new homes are assumed to be split equally between heat pumps and resistance heating. The average Coefficient of Performance (COP) for heat pumps between 1976 and 1990 is assumed to be 1.8, and after 1990 heat pumps with a COP of 2.5 enter the market.

Existing Homes

Of the electrically heated homes now existing, 10.1 million will survive to the year 2000. These homes

currently use approximately 2.7 quads (primary equivalent) for all space conditioning, of which 1.8 Quad is for heating. Through the cost effective application of conservation and renewable retrofits, this could be reduced from 2.7 to 0.7 quad in the year 2000.

The basic stock of existing electrically heated homes are divided into pre-1976 construction and 1976-80 practice. The basic characteristics of each group were presented in the previous discussion. The retrofit measures applicable to each group, and the assumed cost of the retrofits are:

RETROFIT MEASURES: EXISTING ELECTRICALLY HEATED HOMES

Pre-1976	
Measure	Cost
Infiltration Reduction	$240
R-27 to attic	$580
Storm Windows (40%)	$560
Residing + R-5	$500
Heat Pump COP 2.5	$1900
DIY Solar	14.3¢/kwh/Yr
Solar I	29.0¢/kwh/Yr
Solar II	44.7¢/kwh/Yr
Solar III	59.4¢/kwh/Yr

RETROFIT MEASURES: EXISTING ELECTRICALLY
HEATED HOMES (Concluded)

1976-80 Practice	
Measure	Cost
Extra R-19 Attic	$540
Heat Pump COP 2.5	$1900
DIY Solar	14.3¢/kWh/Yr
Solar I	29.0¢/kWh/Yr
Solar II	44.7¢/kWh/Yr
Solar III	5.4¢/kWh/Yr

See Chapter IV for complete discussion of cost
assumption.

Figure 1.29 and Table 1.21 present respectively the
supply curve and supply table for existing electrically
heated homes. Figures 1.30 and 1.31 show the retrofit
sequence for the pre-76 and 1976-80 practice groups of
homes.

New Homes

It is assumed here that between 1980 and 2000, 14.8
million new electrically heated housing units will be
constructed. Since the average size of these units is
expected to increase by over 100 square feet, the
energy use in these homes also increases even though
the shell and heating equipment characteristics remain
identical for the model unit used in the analysis.
These homes were divided into three groups by year of
construction to account for this change. These groups
are 1981-85 construction, 1986-1990 construction, and
1990-2000 construction.

The base line case for these three groups of electrically heated new homes would require 2.0 quads (primary equivalent) annually for all space conditioning in
2000, of which 1.4 quad is required for heating. The
application of conservation and renewable measures
could reduce space heating from 1.36 to 0.29 quad by
the year 2000, for an annual savings of 1.07 quad or
500,000 barrels of oil per day equivalent.

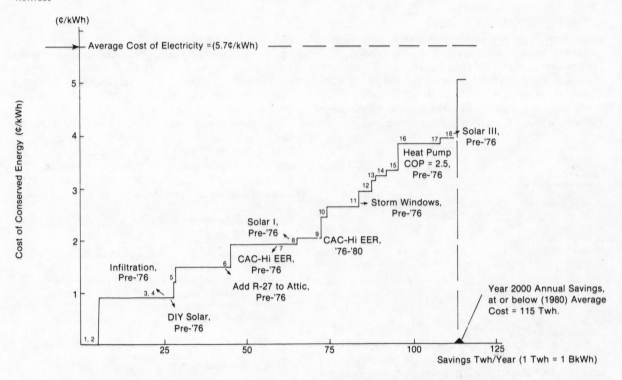

Figure 1.29. **Year 2000 Supply Curve of Conserved Energy (in Twh/year) for Space Heat in Electric-
Heated Dwellings Built Through 1980 and All Central Airconditioning in Dwellings Built Through 1980.**
Year 2000 baseline annual use for this sector is 172 Twh, where "baseline" assumes continuation of
1980 average unit energy consumptions for existing stock or new additions in that year. Unit cost
of conserved energy (in constant 1980 $) assumes that all increased costs are amortized over the
useful life of the measure, using a 3% (real dollar) interest rate. Potential annual savings in 2000, at
or below today's average cost (5.7¢/kWh) is 115 Twh, or 67% of the year 2000 baseline.

Table 1.21.

SPACE HEAT ELECTRIC AND CAC EXISTING HOUSES

SUPPLY TABLE TO THE

TIME HORIZON-10 YEARS
DISCOUNT RATE(S)
RESIDENTIAL- 3.0 PERCENT

	CONSERVATION MEASURE	A. ENERGY TYPE	B. SECTOR	C. MARGINAL COST OF CONSERVED ENERGY (CT/KWH)	D. AVERAGE COST OF CONSERVED ENERGY (CT/KWH)	E. ENERGY SUPPLIED PER MEASURE (TERAWH/YEAR)	F. TOTAL ENERGY SUPPLIED (TERAWH/YEAR)	G. TOTAL DOLLARS INVESTED (M$)	MEASURE NUMBER
1	CAC-LOAD REDUCTION 76-80	ELECTRIC	R	0.	0.	.6	1.	0.	102
2	LOAD REDUCTION PRE-1976	ELECTRIC	R	0.	0.	5.1	6.	0.	100
3	INFILTRATION PRE-1976	ELECTRIC	R	.9	.7	20.8	27.	1608.	48
4	DIY SOLAR PRE-76	ELECTRIC	R	.9	.7	1.3	28.	1801.	53
5	DIY SOLAR 76-80	ELECTRIC	R	1.2	.7	.6	29.	1913.	59
6	ADD R-27 TO ATTIC PRE-76	ELECTRIC	R	1.5	1.0	16.8	45.	5799.	49
7	CAC-HIGH EEK PRE-1976	ELECTRIC	R	1.9	1.2	16.3	62.	9543.	99
8	SOLAR I. PRE-76	ELECTRIC	R	1.9	1.3	4.0	66.	10713.	54
9	CAC-HIGH EEK 1976-80	ELECTRIC	R	2.0	1.3	7.3	73.	12495.	101
10	SOLAR I. 76-80	ELECTRIC	R	2.4	1.4	1.8	75.	13180.	60
11	STORM WINDOWS PRE-76	ELECTRIC	R	2.6	1.5	9.5	84.	16932.	50
12	SOLAR II. PRE-76	ELECTRIC	R	2.9	1.6	4.0	88.	18741.	55
13	RESIDE +R-5 PRE-75	ELECTRIC	R	3.1	1.6	1.1	89.	19244.	51
14	HEAT PUMP COP=2.5 76-80	ELECTRIC	R	3.2	1.7	3.5	93.	20615.	58
15	EXTRA R-19 IN ATTIC 76-80	ELECTRIC	R	3.3	1.7	3.6	97.	22451.	57
16	SOLAR II. 76-80	ELECTRIC	R	3.8	1.8	1.8	98.	23510.	61
17	HEAT PUMP,COP=2.5 PRE-76	ELECTRIC	R	3.8	2.0	10.8	109.	28802.	52
18	SOLAR III. PRE-76	ELECTRIC	R	3.9	2.0	4.0	113.	31002.	56
19	SOLAR III. 76-80	ELECTRIC	R	5.0	2.1	1.8	115.	32409.	62

Figure 1.30.

Electric Space Heat, Pre-1976

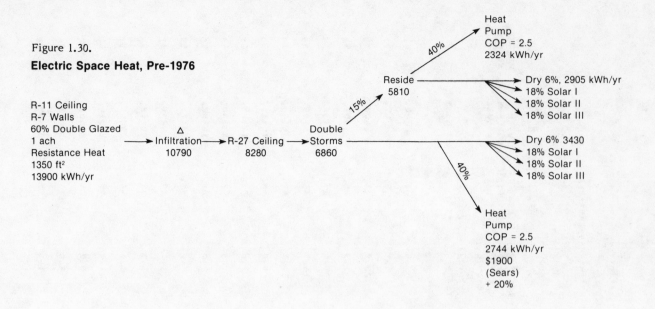

R-11 Ceiling
R-7 Walls
60% Double Glazed
1 ach
Resistance Heat
1350 ft²
13900 kWh/yr

Δ
Infiltration
10790

R-27 Ceiling
8280

Double Storms
6860

15% → Reside 5810

40% → Heat Pump COP = 2.5 2324 kWh/yr

Dry 6%, 2905 kWh/yr
18% Solar I
18% Solar II
18% Solar III

Dry 6% 3430
18% Solar I
18% Solar II
18% Solar III

40% → Heat Pump COP = 2.5 2744 kWh/yr $1900 (Sears) + 20%

Figure 1.31.

Electric Space Heat, 1976-80

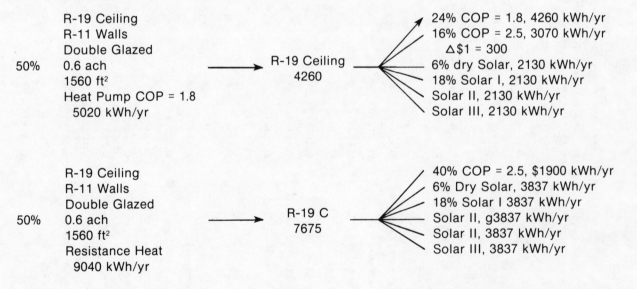

50%
R-19 Ceiling
R-11 Walls
Double Glazed
0.6 ach
1560 ft²
Heat Pump COP = 1.8
5020 kWh/yr

R-19 Ceiling
4260

24% COP = 1.8, 4260 kWh/yr
16% COP = 2.5, 3070 kWh/yr
Δ$1 = 300
6% dry Solar, 2130 kWh/yr
18% Solar I, 2130 kWh/yr
Solar II, 2130 kWh/yr
Solar III, 2130 kWh/yr

50%
R-19 Ceiling
R-11 Walls
Double Glazed
0.6 ach
1560 ft²
Resistance Heat
9040 kWh/yr

R-19 C
7675

40% COP = 2.5, $1900 kWh/yr
6% Dry Solar, 3837 kWh/yr
18% Solar I 3837 kWh/yr
Solar II, g3837 kWh/yr
Solar II, 3837 kWh/yr
Solar III, 3837 kWh/yr

The characteristics of the basic unit are uniform, except for total area and heating equipment. The 1981-1986 unit is assumed to be 1690 ft², the 1986-1990 unit 1770 ft², and the 1991-2000 unit 1810 ft². For each group, 50% of the units are assumed to have resistance heating and 50% are equipped with a heat pump having a COP of 1.8. The other characteristics of these units are:

Common Characteristics
of Electrically Heated New Homes

Ceiling Insulation	R-19
Wall Insulation	R-11
Glazings	2
Infiltration	0.6ach

The solar and conservation measures applicable to these homes vary among the three groupings described above. The basic measures employed and the assumed costs are shown in Table 1.22.

Figure 1.32 and Table 1.23 present the supply curve and supply table for these homes. Figures 1.33, 1.34, and 1.35 detail the application of efficiency measures to the three groups of homes noted above.

The following figures and tables present, respectively, the retrofit sequence and the supply curve tables for these three groups of homes.

Central Air Conditioning

Base Case 107 (Twh/yr) Potential 66 (Twh/yr)

Central air-conditioning load is reduced in two ways, by Energy Efficient Ratio (EER) improvements and by load reductions related to shell improvements. The shell improvements costs are wholly assigned to the reductions in heating loads, and the cooling load reductions are considered a bonus. However, since the improvements in EER are more dependable than the load reduction, EER improvements are considered to occur first. Load reductions are calculated by comparing BEPS cooling load results to BEPS heating load results, a slightly uncertain method.

Costs for improving EER were taken from the Appliance Energy Performance Standards (AEPS) data, increased almost 75% to 20¢ per watt reduction for a standard central air conditioner of 30,000 Btu/hr capacity. The path taken was:

Cohort	Through 1980	81-85	88-90	91-2000
EER	6.5	8	20	13
Change in cost from EER = 6.5 Unit $	0	173	323	461
Change in cost from EER = 8 Unit $	0	0	150	288

We note that there is a wide range of EERs currently, from 13 down to rather low values for winter-optimized heat pumps. Swamp, or evaporation, coolers, of course, have much higher effective EER.

The stock of units was calculated using a 94% one year survival rate, and EIA/OR saturations (0.20 in 1975 increasing to 0.37 in 2000). This calculation gave the vintage of 2000 stocks and thus their EER.

Load calculations were made by inspection of the BEPS data. For example, in 1981 to 1985 houses, the final state has R-19 enter in the ceiling compared to the home. For this house a 10% reduction in heating load compounds to a 4% reduction in cooling load. The final heating load is reduced 15%, so cooling is reduced by 6%. For other houses the reduction is as follows:

Cohort	Through 1980	76-80	81-85	86-90	91-2000
Reduction in cooling load as compared with appropriate base case house (5)	24	6*	6*	20	26

*These houses have no increase in wall insulation compared to the base.

WATER HEATING

After space heating, water heating is the largest single end-use in the residential sector, and therfore merits more attention that it has received to date. Indeed, if retrofit programs attain their potential, water heating may become the largest end-use.

In assessing potentials it is important to be clear about the base-case. For water heat the base is a little complicated because of the important components of total energy use. These are: a) the standby loss, the heat lost to the surroundings by the tank; b) water use, the amount of energy contained in the hot water used in showers etc, in the clothes washer use, and in the dishwasher use. In the base case, business as usual: standby losses remain constant; water use in showers etc is decreased by 2.4/2.85, the reduction in household size; clothes washer saturation increases from 70% to 82%, but equipment usage drops by 2.4/2.85, and diswasher saturation increases from 37% to 63%, but drops by the usage factor. In the potentials, all of these terms are reduced by improvements to equipment. The effect is shown on the table:

	Base Case	Potential
Fuel Use (Quads/yr)	1.46	0.55
Electricity use (TWh/yr)	183	79

To obtain some perspective on these figures, it may be noted that current standby losses in electric water heaters, about 1000 kwh/yr per unit, consume the output of four large (1000 Mw(e)) power plants.

Fuel Water Heaters

In the supply curves we calculate the energy saved by equipment in place in 2000. Since the efficiency of new equipment is assumed to increase, it is important to know the age of all equipment in place in 2000. We make the standard reliability theory assumption that

Table 1.22 C/S MEASURES APPLICABLE TO NEW ELECTRICALLY HEATED HOMES*

1981-85		1986-90		1990-2000	
Measure	Cost	Measure	Cost	Measure	Cost
R-19 Attic	$590	BEPS Shell	$1340	BEPS Shell	$1360
Heat Pump (COP 2.5)	$1900	Heat Pump (COP 2.5)	$1900	Infiltration Reduc. & heat exchanger	$770
DIY Solar	14.3¢/kwh/yr	Windows to South	0	Heat Pump (COP) 2.5	$1900
Solar I	29.0¢/kwh/yr	Solar Package	$117/MBtu Delivered/yr	Solar Package	$117/MBtu del./yr
Solar II	44.7¢/kwh/yr				
Solar III	59.4¢/kwh/yr				
		BEPS Shell -	R-38 attic R-19 walls triple glazed		

*See Chapter IV for discussion of costs

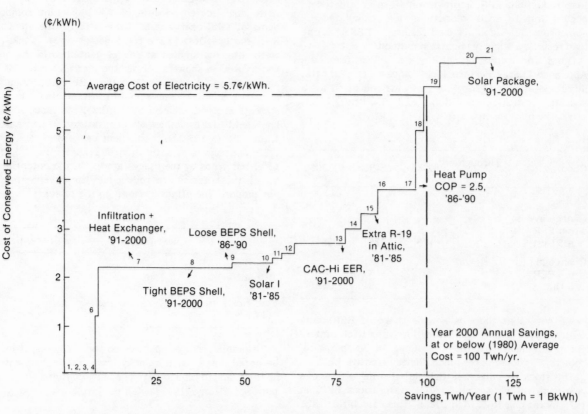

Figure 1.32. **Year 2000 Supply Curve of Conserved Energy (in Twh/year) for Space Heat in Electrically-Heated New Homes and All Central Airconditioning in New Homes.**

Year 2000 baseline annual use for this sector is 170 Twh, where "baseline" assumes continuation of 1980 average unit energy consumptions for existing stock or new additions in that year. Unit cost of conserved energy (in constant 1980 $) assumes that all increased costs are amortized over the useful life of the measure, using a 3% (real-dollar) interest rate. Potential annual savings in 2000, at or below today's average cost (5.7¢/kWh) is 100 Twh, or 59% of the year 2000 baseline.

Table 1.23

****
SUPPLY TABLE TO THE
ELECTRIC SPACE HEAT AND CAC NEW HOUSES
****

TIME HORIZON=10 YEARS
DISCOUNT RATE($)
RESIDENTIAL= 5.0 PERCENT

	CONSERVATION MEASURE	A. ENERGY TYPE	B. SECTOR	C. MARGINAL COST OF CONSERVED ENERGY (CT/KWH)	D. AVERAGE COST OF CONSERVED ENERGY (CT/KWH)	E. ENERGY SUPPLIED PER MEASURE (TERAWH/YEAR)	F. TOTAL ENERGY SUPPLIED (TERAWH/YEAR)	G. TOTAL DOLLARS INVESTED (M$)	MEASURE NUMBER
1	CAC-LOAD REDUCTION 91-2000	ELECTRIC	R	0.	0.	4.0	4.	0.	108
2	CAC-LOAD REDUCTION 86-90	ELECTRIC	R	0.	0.	1.8	6.	0.	106
3	CAC-LOAD REDUCTION 1981-85	ELECTRIC	R	0.	0.	.6	6.	0.	104
4	WINDOWS TO SOUTH 91-2000	ELECTRIC	R	0.	0.	1.3	8.	0.	76
5	WINDOWS TO SOUTH 86-90	ELECTRIC	R	0.	0.	.8	8.	0.	71
6	DIY SOLAR 81-85	ELECTRIC	R	1.2	.1	.8	9.	138.	65
7	INF.+HEAT EXCH. 91-2000	ELECTRIC	R	2.2	1.4	16.3	26.	5605.	74
8	THIGHT BEPS SHELL 91-2000	ELECTRIC	R	2.2	1.8	21.3	47.	15261.	73
9	LOOSE REPS SHELL 86-90	ELECTRIC	R	2.3	1.9	11.1	58.	20353.	69
10	SOLAR I. 81-85	ELECTRIC	R	2.4	1.9	2.3	60.	21202.	66
11	CAC-HIGH EER 1981-85	ELECTRIC	R	2.5	1.9	3.8	64.	22380.	103
12	CAC-HIGH EER 1986-90	ELECTRIC	R	2.7	2.0	4.3	68.	23783.	105
13	CAC-HIGH EER 91-2000	ELECTRIC	R	2.7	2.1	9.7	78.	27031.	107
14	HEAT PUMP,COP=2.5 81-85	ELECTRIC	R	3.0	2.1	4.4	82.	28626.	64
15	EXTRA R-19 IN ATTIC 81-85	ELECTRIC	R	3.3	2.2	4.5	87.	30927.	63
16	SOLAR II. 81-85	ELECTRIC	R	3.8	2.2	2.3	89.	32240.	67
17	HEAT PUMP COP=2.5 86-90	ELECTRIC	R	3.8	2.4	8.4	98.	36144.	70
18	SOLAR III. 81-85	ELECTRIC	R	5.0	2.4	2.3	100.	37865.	68
19	SOLAR PACKAGE 86-90	ELECTRIC	R	5.9	2.6	4.4	104.	41921.	72
20	HEAT PUMP COP=2.5 91-2000	ELECTRIC	R	6.4	2.9	9.9	114.	49731.	75
21	SOLAR PACKAGE 91-2000	ELECTRIC	R	6.5	3.0	3.8	118.	53565.	77

Figure 1.33.
Electric Space Heat, 1981-1985

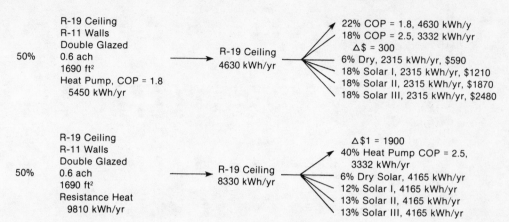

50% R-19 Ceiling
     R-11 Walls
     Double Glazed
     0.6 ach
     1690 ft²
     Heat Pump, COP = 1.8
       5450 kWh/yr

→ R-19 Ceiling
  4630 kWh/yr

22% COP = 1.8, 4630 kWh/y
18% COP = 2.5, 3332 kWh/yr
△$ = 300
6% Dry, 2315 kWh/yr, $590
18% Solar I, 2315 kWh/yr, $1210
18% Solar II, 2315 kWh/yr, $1870
18% Solar III, 2315 kWh/yr, $2480

50% R-19 Ceiling
     R-11 Walls
     Double Glazed
     0.6 ach
     1690 ft²
     Resistance Heat
       9810 kWh/yr

→ R-19 Ceiling
  8330 kWh/yr

△$1 = 1900
40% Heat Pump COP = 2.5,
    3332 kWh/yr
6% Dry Solar, 4165 kWh/yr
12% Solar I, 4165 kWh/yr
13% Solar II, 4165 kWh/yr
13% Solar III, 4165 kWh/yr

Figure 1.34.
Electric Space Heat, 1986-1990

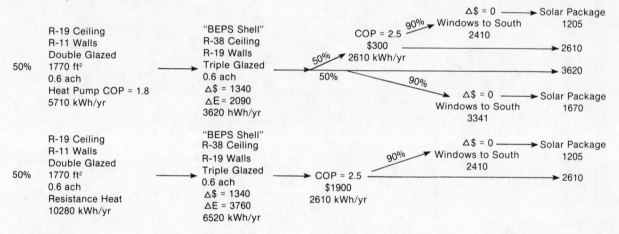

50% R-19 Ceiling
     R-11 Walls
     Double Glazed
     1770 ft²
     0.6 ach
     Heat Pump COP = 1.8
     5710 kWh/yr

→ "BEPS Shell"
  R-38 Ceiling
  R-19 Walls
  Triple Glazed
  0.6 ach
  △$ = 1340
  △E = 2090
  3620 hWh/yr

50% 2610 kWh/yr
COP = 2.5
$300   90% Windows to South
              2410   △$ = 0 → Solar Package
                                         1205
                            → 2610
50%
         90%   → 3620
              Windows to South   △$ = 0 → Solar Package
              3341   1670

50% R-19 Ceiling
     R-11 Walls
     Double Glazed
     1770 ft²
     0.6 ach
     Resistance Heat
     10280 kWh/yr

→ "BEPS Shell"
  R-38 Ceiling
  R-19 Walls
  Triple Glazed
  0.6 ach
  △$ = 1340
  △E = 3760
  6520 kWh/yr

→ COP = 2.5
  $1900
  2610 kWh/yr

90% Windows to South   △$ = 0 → Solar Package
        2410   1205
                          → 2610

Figure 1.35.
Electric Space Heat, New Houses 1991-2000

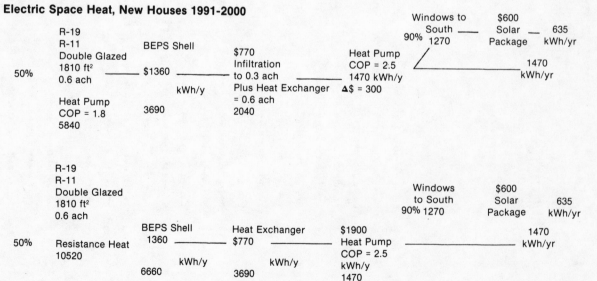

50% R-19
     R-11
     Double Glazed
     1810 ft²
     0.6 ach
     Heat Pump
     COP = 1.8
     5840

— $1360 ——
       kWh/y
  3690

BEPS Shell $770
               Infiltration
               to 0.3 ach
           —— Plus Heat Exchanger
               = 0.6 ach
               2040

Heat Pump
COP = 2.5
1470 kWh/y
△$ = 300

90% Windows to $600
    South ——   Solar —— 635
    1270   Package   kWh/yr
                          1470
                          kWh/yr

50% R-19
     R-11
     Double Glazed
     1810 ft²
     0.6 ach

     Resistance Heat
     10520

BEPS Shell   1360 ——
                  kWh/y
             6660

Heat Exchanger $770 ——
                    kWh/y
               3690

$1900
Heat Pump
COP = 2.5
kWh/y
1470

Windows
to South $600
90% 1270   Solar   635
            Package   kWh/yr
                      1470
                      kWh/yr

equipment remaining after t years is reduced by a fraction exp-t/c, where t is the lifetime. For fuel water heaters this is ten years. The average purchase date for present water heaters is assumed to be 1975, to a first approximation, and the average purchase date for 1980-85 heaters is taken as 1983. With these assumptions, it is found that the 2000 population is as follows:

Year of Sale	Number of Units
pre 1981	5.31
1981-86	11.34
86-2000	40.58

For details of the calculation, see the appendix. A similar calculation was made for dishwashers (t = 13.6 years) and clotheswashers (t = 12.3 years).

Year of Sale	Number of Clothswashers	Number of Dishwashers
pre 1981	7.48 10^6	8.78
1981-85	10.14	
1986-2000	69.68	58.30

The units are assumed to each have the water usage and extra cost noted in Table 3.8. The one retrofit measure, low-flow devices, is only assumed to apply to 90% of houses because goose-neck showers and un-threaded faucets prevent universal installation of these devices. Low-flow devices are assumed to reduce total water use by 55%, made up of a 50% reduction in shower use (two thirds of the total), plus a 33% reduction in all other use.

Figure 1.36 and Table 1.24 present the supply curve and supply table for fuel water heaters. Figure 1.37 shows the application of the measure shown in Table 1.25.

Electric Water Heaters

The lifetime of electric heaters is 12 years. For simplicity, the same age structure is taken for them in 2000 as for the fuel heater 2000 stock. Since the saturation of electric heaters increases towards the turn of the century, this assumption overestimates the number of older units; as a result, our savings estimates are slightly low. Only one efficiency improvement applies exclusively to electric heaters; foam insulation which is used from 1986 on. The electric heat pump with a COP of 2 is assumed to form half of sales from 1986 on.

Figure 1.38 and Table 1.26 present the supply curve and supply table for electric water heaters. Figure 1.39 details the application of the measures shown in Table 3.8.

Solar Water Heaters

Whether or not a house can have a solar water heater depends on the year in which it was built. 60% of houses before 1980 are assumed to have adequate solar access, and 80% for those built after 1980.

Type of Water Heat	Fuel	Electric
Houses Existing in 1980 (10^6)	48.24	26.83
Houses Existing in 2000 (10^6)	57.17	45.68
1980 Houses Existing in 2000 (10^6)	38.00	20.36
1980 Post 1980 houses in 2000 (10^6)	19.17	25.32
1980 Homes in 2000 with solar access (10^6)	38.14	32.48

All of the fuel heaters can be replaced with solar heaters. Since one half of all "electric heaters" are heat pumps post-1985, only 21.1 million solar water heaters are installed. A certain number of the solar systems are combined with space heat systems; for details see the appendix. The costs of the solar systems are taken from Chapter IV and are from $46 to $176 per delivered MBtu per year. This range is reduced to discrete values of $63, $96, and $130/MBtu/yr as described in Chapter IV.

Appliances

The energy consumed by household appliances is substantial. For example, refrigerators alone consumed over 90 TWh of electricity in 1980, equivalent to the output of fifteen large power plants. Although most appliances cannot be retrofitted, new units can generally be made much more efficient for a modest increase in cost. (Televisions are anomalous in that reduction in electricity consumption can come at the same time as reductions in first-cost). Appliances' comparatively short lifetimes permit rapid introduction of new stock. The supply curves show, as always, the energy supplied by more efficient units in place in 2000 by comparison with the energy that would have been consumed had 1980 buying practices continued unchanged until 2000. Miscellaneous usage is assumed constant at 0.09 Quads per year fuel and 16 TWh of electricity.

	Base Case	Potential
Total Fuel Use (Quads/yr)	0.67	0.43
Total Electricity Use (TWh/yr)	581	361

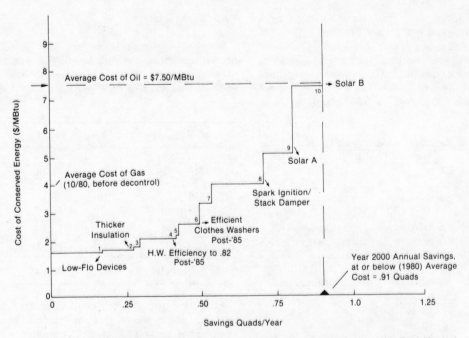

Figure 1.36. **Year 2000 Supply Curve of Conserved Energy (in quads/year) for Fuel-Heated Water Heaters.**

Year 2000 baseline annual use for this sector is 1.46 quads, where "baseline" assumes continuation of 1980 average unit energy consumption for existing stock or new additions in that year. Unit cost of conserved energy (in constant 1980 $) assumes that all increased costs are amortized over the useful life of the measure, using a 3% (real-dollar) interest rate. Potential annual savings in 2000, at or below today's cost of oil ($7.50/MBtu) is 0.91 quads, or 62% of the year 2000 baseline.

Figure 1.37.

Fuel Water Heat

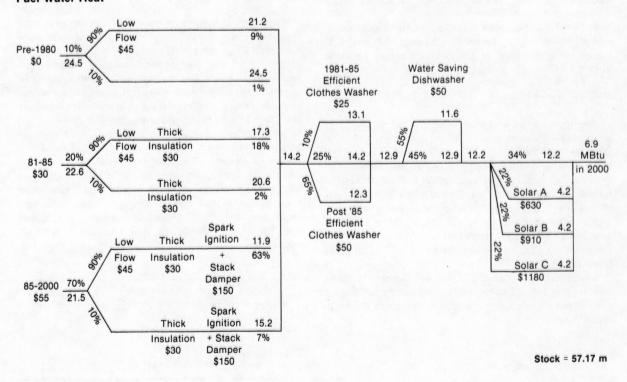

Table 1.24.

****
SUPPLY TABLE TO THE
FUEL WATER HEATING
****

TIME HORIZON=10 YEARS
DISCOUNT RATE(S)
RESIDENTIAL= 3.0 PERCENT

	CONSERVATION MEASURE	A. ENERGY TYPE	B. SECTOR	C. MARGINAL COST OF CONSERVED ENERGY ($/MBTU)	D. AVERAGE COST OF CONSERVED ENERGY ($/MBTU)	E. ENERGY SUPPLIED PER MEASURE (TERABTU/YEAR)	F. TOTAL ENERGY SUPPLIED (TERABTU/YEAR)	G. TOTAL DOLLARS INVESTED (M$)	MEASURE NUMBER
1	LOW FLOW DEVICES	GAS	R	1.6	1.6	169.8	170.	2315.	78
2	THICKER INSULATION	GAS	R	1.7	1.6	102.9	273.	3859.	79
3	EFFICIENCY TO .78 81-85	GAS	R	1.8	1.6	21.7	294.	4202.	81
4	EFFICIENCY TO .82 POST-85	GAS	R	2.1	1.8	120.1	414.	6403.	82
5	EFFICIENT CWLS 1981-85	GAS	R	2.2	1.8	6.3	421.	6546.	83
6	EFFICIENT CWLS POST-85	GAS	R	2.6	1.9	70.6	491.	8404.	84
7	DISHWASHER WATER SAVING	GAS	R	3.3	2.0	40.9	532.	9976.	85
8	SPARK IGN & STACK DAMP.	GAS	R	4.0	2.5	172.1	704.	15979.	80
9	SOLAR A	GAS	R	5.1	2.8	100.6	805.	23903.	86
10	SOLAR B	GAS	R	7.4	3.3	100.6	906.	55348.	87
11	SOLAR C	GAS	R	9.5	4.0	100.6	1006.	50190.	88

Table 1.25 RETROFIT* AND IMPROVED FUEL AND ELECTRIC WATER HEATERS

Measures	Cost $
Low-flow devices for showers and faucets	45
Thicker insulation on new fuel water heater (1981-2000)	30
Thicker insulation on new electric water heater (81-85)	20
Foam insulation on new electric water heater (86-2000)	60
Spark ignition and stack damper (1986-2000)	150
Efficiency increased from 72% to 78% (1981-85)	30
Efficiency increased from, 72% to 82% (1986-2000)	55
Efficiency of 1981-85 clothe washer increased; water use falls 30%	25
Efficiency of 86-2000 clothes washer; water use falls 50% from 1980	50
Efficiency of 86-2000 diswashers; water use falls 50% from 1980	50
Electric heat pump (50% of sales 1986-2000)	800

*The only retrofit measure is low-flow devices. So few water heaters from pre-1980 survive until 2000 that retrofitting on insulating blanket, a common and cost effective measure, would only make a very minor contribution and so is not included here.

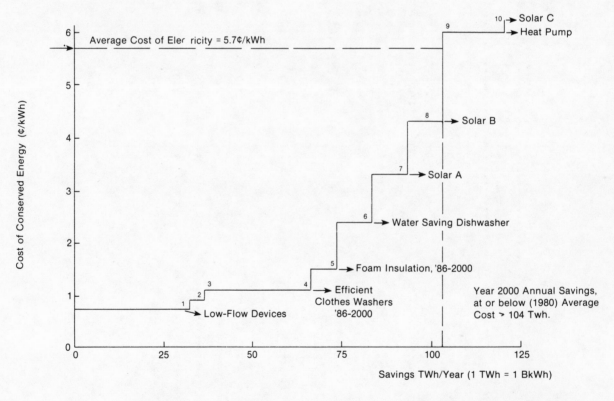

Figure 1.38. **Year 2000 Supply Curve of Conserved Energy (in Twh/year) for Electric Water Heaters.**
Year 2000 baseline annual use for this sector is 173 Twh, where "baseline" assumes continuation of 1980 average unit energy consumption for existing stock or new additions in that year. Unit cost of conserved energy (in constant 1980 $) assumes that all increased costs are amortized, over the useful life of the measure, using a 3% (real-dollar) interest rate. Potential annual savings in 2000, at or below today's average cost of 5.7¢/kWh, is 104 Twh, or 60% of the year 2000 baseline.

Table 1.26.

```
***********************************
    ELECTRIC WATER HEATING   SUPPLY TABLE TO THE
  ***********************************
```

TIME HORIZON=10 YEARS
DISCOUNT RATE(S)
RESIDENTIAL= 3.0 PERCENT

CONSERVATION MEASURE	A. ENERGY TYPE	B. SECTOR	C. MARGINAL COST OF CONSERVED ENERGY (CT/KWH)	D. AVERAGE COST OF CONSERVED ENERGY (CT/KWH)	E. ENERGY SUPPLIED PER MEASURE (TERAWH/YEAR)	F. TOTAL ENERGY SUPPLIED (TERAWH/YEAR)	G. TOTAL DOLLARS INVESTED (M$)	MEASURE NUMBER
1 LOW FLOW DEVICES	ELECTRIC	R	.7	.7	32.1	32.	1850.	89
2 THICK INSULATION 1931-85	ELECTRIC	R	.9	.7	3.1	35.	2124.	90
3 EFFICIENT CKTS 1941-85	ELECTRIC	R	.9	.7	1.2	36.	2238.	92
4 FOAM INSULATION 85-2000	ELECTRIC	R	1.1	.8	17.1	54.	4157.	91
5 EFFICIENT CKTS 86-2000	ELECTRIC	R	1.1	.9	13.1	67.	5641.	93
6 DISHWASHER WATER SAVING	ELECTRIC	R	1.5	.9	7.3	74.	6898.	94
7 SOLAR A	ELECTRIC	R	2.4	1.1	9.9	84.	10509.	95
8 SOLAR B	ELECTRIC	R	3.3	1.3	9.9	94.	15524.	96
9 SOLAR C	ELECTRIC	R	4.3	1.6	9.9	104.	22006.	97
10 HEAT PUMP	ELECTRIC	R	6.0	2.2	17.4	121.	34797.	98

Figure 1.39.

Electric Water Heat

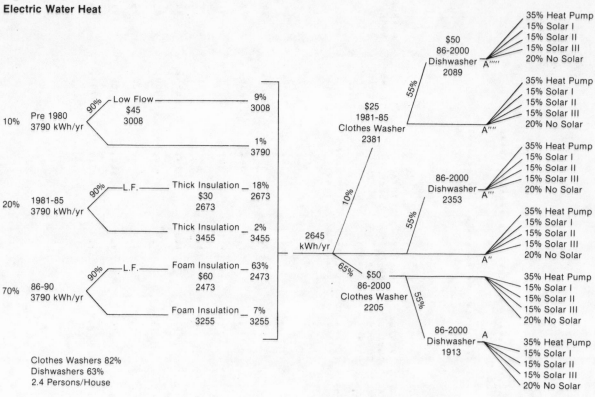

Clothes Washers 82%
Dishwashers 63%
2.4 Persons/House

Refrigerators

	Base Case	Potential
Frost Free	151	80
Partials/manuals	21	16

The current stock of refrigerators consumes 120 kwh/month on the average, according to EIA/OR. Refrigerators form two fairly distinct groups, the frost-free (which never need to be defrosted), and the partially frost-free (with separate frozen food compartments that are insulated from refrigerator compartment; the freezer must be defrosted but not the refrigerator); and manual (which have no insulation between freezer and refrigerator compartment and no defrost mechanism at all). There are efficiency improvements that only apply to frost-free units, which use more energy on the average, and though they only form 53% of the current stock, they from 80% of current sales. We assume this sales ratio will continue, so they form 76% of the stock in 2000. The saturation of refrigerators continues to increase, as in the ORNL model.

To estimate possible improvements in future refrigerators, we first note that some efficient refrigerators already exist. The stock of frost-free units

consumes 1700 kwh/yr/unit, but Amana currently makes an 18-ft^3 top-freezer unit that uses only 1344 kwh/yr. Further, an Amana prototype, also 18-ft^3 top freezer, has been built and tested to use only 785 kwh/yr. The reductions are obtained by improved insulation and defrost systems. Much of the information on refrigerators comes from an A.D. Little 1977 report, which suggested the following projections for a 16-ft^3 frost-free top-freezer unit:

	Base	1972	1976	1980	Beyond 1980	Ultimate
kwh/yr	1668	1314	934	621		402
$ above base			33	33	48	NA

Using this data, and similar reasoning for the manual and partially automatic models, we arrived at the following projections: (Intermediate years have intermediate costs and electricity use.)

	Stock 1980	New 1980(Base)	New 1990	2000
Frost free kwh/yr	1700	1700	750	500
Cost above base ($)			75	150
Manual & Partial kwh/yr	635*	800*	450	400
Cost above base ($)			45	55

*Larger manuals and partials, and a greater proportion of partials in sales account for the difference between stock and sales.

Freezers

	Base Case	Potential
Electricity Use (TWh/yr)	70	42

The stock of freezers is said to consume 1643 kwh/yr/unit (ORNL, as quoted in LBL 11338), or 1285 kwh/yr/unit (LBL 9110), or 1170 kwh/yr (1978 sales weighted data, ADL, quoted in LBL 11338). The middle figure was taken, and by means of the 1977 ADL report, the costs and effects of various improvements were calculated. The report did not go into great detail on freezers, and costs had to be deduced by adding the costs of a number of individual measures. This may result in an overestimation of costs. Efficient freezers consume ____ kwh/yr/unit, according to California Energy Commission data. Units bought in intermediate years are assumed to have intermediate costs.

	Stock 1980 new 1980(Base)	New 1985	1995
kwh/yr	1285	1000	500
$ above 1980 base		20	90

Clothes Dryers

	Base Case	Potential
Fuel dryers (Quads/yr)	0.11	0.08
Electric dryers (TWh/yr)	53	40

Electric

The energy consumption of electric clothes dryers is given variously as 1032 kwh/yr/unit by an Electric Power Research Institute (EPRI) Report, 993 kwh/yr/unit by the Edison Electric Institute, 1100 kwh/yr/unit by ADL, and 950 kwh/yr/unit by LBL. We assumed 1000 kwh/yr/unit. It is estimated that improved temperature and moisture sensors reduce use by 10%. Another possible innovation is an adaption of a technique used in dry cleaning, i.e., use of a vacuum pump to permit rapid low temperature evaporation of water. A heat exchanger could be used to recover energy from the exhaust airstream. This speculative technique is assumed to be introduced in 1990.

Gas

Figures for gas consumption in clothes dryers are more scattered than for electricity, perhaps because of the almost 50% difference that the presence of a pilot light can make. The range we found for 1975 was 4.2-7 MBtu/yr. We assume the mid-point to be the stock of 1975, the low-point sales value. Conservation measures that have occurred between 1975 and 1980 lowers these figures slightly. The same measures are carried out as for electric dryers, except that sensors, already in electric dryers, are introduced post-1980. See table at top of next page.

Ranges

	Base Case	Potential
Fuel (Quads/yr)	0.38	0.24
Electricity (TWh/yr)	50	44

Electric

There is some disagreement as to the energy consumption of electric ranges and some confusion as to whether a figure should represent all cooking devices

	1980 Stock	1980 New (Base)	1981-90 New	1991-2000 New
Fuel(MBTU/yr)	5	4	3.8	2.5
$ above 1980 base			25	125
Electric(kwh/yr)	970	900	900	600
$ above 1980 base				100

associated with the range, (including toasters, skillets, coffee makers, etc.) or just the range. Since this study looked at engineering improvement in the range, it was not appropriate to use a "total cooking equipment" figure. Small cooking devices are part of this study's "Miscellaneous" category.

The EIA/OR model uses 1200 kwh/yr/unit, the CEC uses 775 ADL (PG&E) uses 778, EPRI EA-682 uses 782, EEI (which formerly used 1200) gives 700 kwh/yr/unit for non-self-cleaning ranges. Improvements came from better insulation and reduced losses in the area of the door seal, as indicated (for example) by ADL (for PG&E 1980 and NBS 1977.

	Stock and New 1980 (Base)	New 1990	New 2000
kwh/yr	780	702	624
$ above 1980 base		10	10

These projections are slightly more conservative than ADL (for PG&E), which gives the following reduction for single-family homes: 768 to 674 to 613 kwh/yr/unit. Costs are taken from the NBS study, increased by almost 50% to account for inflation.

Fuel

All fuel ranges were taken to be gas ranges, since the ORNL model gives no oil ranges and "other fuels" are less than 20% of gas usage for cooking. For gas stoves, the prime means of improving efficiency is pilotless ignition, since pilots consume 4 MBtu/yr (CPS) or 30 to 47% of consumption (ORNL Blue Book). Total consumption is given as 9.6 MBtu/yr/unit (ORNL in LBL 11448), 4.4 (ADL 1978 sales, in LBL 11338), or 8.8 (ADL 1980 for PG&E). The ADL sales figure presumably contains a large volume of pilotless stock. The cost of an ignition system was simply deduced from the Sears Catalogue.

A rather high value for current stock is assumed to ensure that no double counting of pilotless ignition occurs. The pilotless range is then assumed to undergo the same insulation and door seal changes as the electric stove. These occur later than in the electric case

because pilot savings occur first, since they are very lucrative. Venting of the gas stove means that its efficiency is still far below that of the electric stove (4 MBtu/yr vs. 624 kwh/yr).

	Stock and New 1980	New 1990	New 2000
MBtu/yr	9.0	5	4
Cost above 1980 base ($)		40	60

Televisions

	Base Case	Potential
Electric use (TWh/yr)	28	19

Televisions present a rather special case for conservation, since their energy use is dropping rapidly. It is almost impossible now to buy a new television with vacuum tubes; solid state technology has reduced prices, increased reliability, and reduced energy consumption. Therefore, no cost is assigned to the energy savings. It is difficult to keep track of the reduction in consumption. UCID gives 450 kwh/yr/household, while PG&G data for 1978 gives 354 kwh/yr/household. Following UCID, we assume 1900 hours of TV watched in every house per year, with two thirds of the time being for color TV's. To calculate new 1980 stock, we take the TV's from the Sears 1979 Fall/Winter Catalogue with the highest energy consumption, and assume these to be typical; these are a 25" diagonal color television at 1700 watts, and a 12" diagonal black and white at 45 watts. Japan already has 18" diagonal color at 71 watts, or 135 kwh/yr at 1900 viewing hours/year; we confidently assume that continued improvements in solid-state technology will reduce the larger color televisions to a lower level by 2000, while black and white units continue to decline in use.

	1980 Stock	1980 New	2000
kwh/yr/household*	350	250	100

*No cost assigned.

Dishwashers

	Base Case	Potential
Electric use (TWh/yr)	17	12

A 1977 LBL report discussed various data sources for energy consumption of dishwashers, and concluded that 250 kwh/yr is the stock average. We take this also to be the new sales. Some of the electricity goes to booster heaters, but most is for the motor and pump. We assume that there are improvements to motor and pump which require annual usage to 150 kwh. The cost is assumed to be $150.

	Stock and New 1980	New 1990
kwh/yr	250	150
Cost above 1980 base ($)		50

The saturation of dishwashers rises from 43% in 1980 to 63% in 2000. The Statistical Abstract gives 42% saturation for 1980.

Swimming Pools

	Base Case	Potential
Fuel use (Quads/yr)	0.09	0.02

Swimming pools do not consume a large proportion of total U.S. energy, but there is a dramatic retrofit that can reduce energy consumption by up to 80%. Swimming pools are included for illustrative purposes only, since so many assumptions must be made.

There were 1.6 million residential pools in 1978. We assume 1/3 are heated, and use only 88 MBtu/yr each. The saturation rises to 3% (from about 2% in 1978) by 2000, with 1/3 heated. The heated stock is thus one million (1/3 x .03 x 106.47). The use of a pool cover, costing $250 with a two-year lifetime, saves 70 MBtu/yr (CPS) per pool. The electricity saved by fewer hours of filter and pump operation is not included in these calculations.

Lighting

	Base Case	Potential
TWh/yr	122	54

To calculate the savings from lighting improvements we modelled the fixtures in the 1350-ft^2 home and calculated which fixtures could be replaced by new improved units. Electric consumption was 1000 kwh/yr, with twenty-five light bulbs; consumption increases directly with floor area. The fixtures were two (2) three-way bulbs, three (3) fluorescent tubes, and twenty (20) other incandescent bulbs. Replacement bulbs are either Halarc, or other types of high-efficiency bulbs, or three-way fluorescents. Both three-way bulbs are replaced, as are the twelve incandescents which are used four hundred hours or more per year. The replacement bulbs are assumed to have a lifetime of ten years, though twelve years is expected at the usage rates we assume. The costs are calculated on the basis of $10 per Halarc-type bulb, and $34 per three-way fluorescent (since a new fixture must be bought), for an average of $160/house.

Room Air Conditioners

	Base Case	Potential
Twh	53	38

The ORNL model treats air conditioners differently from many other appliances. One "unit" is one house with room air conditioners, although the ORNL buildings data book shows that 35% of such homes in 1974 had two or more units. We assume 1.5 room air conditioners per house, and EIA/or saturations, i.e., 0.29 in 1975 and 0.42 in 2000. The building data book gives sales weighted size as 12,500 Btu/hr, which we take as constant over the period up to 2000. Taking current EER as 6.5 and usage as 500 hr/yr, electricity per unit is 960 kwh/ur, or 1440 kwh/yr/house. EIA/OR give 1708.7 kwh/yr (LBL 1138). The costs of increasing this EER we calculated using AEPS data, adding 10% to be conservative. The projection for room air conditioners is:

	Stock 1980	New 1980 (Base)	New 1990	1991-2000
EER	6.5	8	10	13
kwh/y/house	1440	1170	936	720
$ above base/house		0	50	120

Figure 1.40 and Table 1.27 present the supply curve and supply table for fuel appliances, and Figure 1.41 and Table 1.28 show the supply curve and supply table for electric appliances.

INDOOR AIR QUALITY

The two main methods of decreasing heat loss from a building are to increase its insulation level and to decrease its infiltration rate. These two methods are quite independent, and are measured in quite different units. Insulation is resistant to heat flow, and is measured in U-value, Btu/ft^2 per hour.* Infiltration is air leakage, and is measured in ach, air changes per hour. It is quite possible to have a house with very good insulation (high R-value), but high infiltration rate. Confusion arises between these two terms because a balanced approach involves attacking energy loss by both methods at once.

Houses act as traps for a number of pollutants that are released into indoor air. Water vapor from showering and cooking, chemicals released by glues and paints, cooking odors, moisture and radon gas released by the soil, cigarette smoke--all these are trapped inside a house to some degree. Air leaking into and out of a house flushes out these indoor air pollutants. The rate of leakage is measured in air changes per hour, or ach; for example, a $1350 ft^2$ house with 8 foot ceilings and an infiltration rate of 1 ach would permit $10,800 ft^3/hr$ of air to leak in, on average. Infiltration is caused by temperature and pressure differences between the inside and the outside. In winter, warm house air rises and escapes via cracks, drawing in cold outside air, while wind forces air in through cracks in the surface against which it is blowing. These effects are usually smaller in summer, when temperature differences between inside and out are smaller. Thus infiltration tends to be highest in winter, lowest in summer.

Air leaking out of the house takes with it the undesired pollutants; the indoor air pollutant concentration depends both on how much pollutant is being released and on how much air moves through the house. Measurements of radioactive radon gas inside houses in Maryland, Illinois, New Mexico, and California show that for houses with about the same ventilation rate massive variations in radon levels are possible, apparently caused by different levels of radon in the soil and ground water. The radon release rates may differ by a factor of 1,000. The same effect occurs with less exotic pollutants; indoor air might be acceptable if one member of a household smoked occasionally but very stale if all members chain-smoked. It is therefore impossible to set a single rate that will lead to acceptable air for all places at all times; the implications of this problem differ in new and existing houses, which will, therefore, be discussed separately.

In cold areas, such as the northern U.S., Canada, and northern Europe, small drafts can make a house very uncomfortable. Special techniques are often adopted for new construction in these and other areas to reduce air leakage to a minimum. One such technique is placing a continuous plastic sheet in the walls and ceiling of the house. Infiltration rates can be reduced to 0.1 to 0.2 ach by this and other methods. A house is very likely to have air pollution problems at 0.1 ach; moisture and all smells will linger. However, it is possible to increase the ventilation rate to any desired value, and to recover heat from exhaust air by means of an air-to-air heat exchanger.

An air-to-air heat exchanger is a simple device that pre-heats or pre-cools ventilation air. It does this by bringing the exhaust and supply air streams close together so that heat can be transferred from the hot to the cold stream. When combined with an air distribution system, such an exchanger can increase ventilation to a house while avoiding much of the cost of heating and cooling the extra air. Heat exchangers are inexpensive to run; the only power needed during normal operation, typically 150 watts, is electricity for the fans.

A cost-effective conservation measure in new homes is to reduce natural infiltration to perhaps 0.2 ach and to provide the necessary outside air via a heat exchanger. The table below shows the heating and cooling loads for the Hastings Ranch house in Washington, D.C., calculated for the Building Energy Performance Standards (BEPS) in two situations, one in which the house has a natural infiltration rate of 0.6 ach; the other in which the house has a natural infiltration rate of 0.2 ach, and receives 0.4 ach through a 75% efficient heat exchanger (see Table 1.29).

Table 1.29. HEATING AND COOLING LOADS FOR BEPS HOUSE IN WASHINGTON, D.C.

	Heating Load (MBtu/yr)	Cooling Load (MBtu/yr)
0.6 ach natural infiltration	15.29	9.42
0.2 ach nat. infil; 0.4 ach yeat ex.	8.17	8.76

NOTES:
House has R-38 ceiling, R-19 walls, and triple glazing.
The figure given is the load, i.e., energy supplied by the heating/cooling system.
To obtain fuel use, divide by total system efficiency.

Constructing a low infiltration house and putting in a heat exchanger to provide necessary ventilation thus appears to be cost effective.

---

*Or, R-value, which is the inverse of U-value.

Table 1.27.

****

SUPPLY TABLE TO THE

FUEL APPLIANCES

****

TIME HORIZON=10 YEARS
DISCOUNT RATE(S)
 RESIDENTIAL= 5.0 PERCENT

	CONSERVATION MEASURE	A. ENERGY TYPE	B. SECTOR	C. MARGINAL COST OF CONSERVED ENERGY ($/MBTU)	D. AVERAGE COST OF CONSERVED ENERGY ($/MBTU)	E. ENERGY SUPPLIED PER MEASURE (TERABTU/ YEAR)	F. TOTAL ENERGY SUPPLIED (TERABTU/ YEAR)	G. TOTAL DOLLARS INVESTED (M$)	MEASURE NUMBER
1	1981-90 GAS RANGES	GAS	R	.7	.7	13.6	14.	136.	115
2	91-2000 GAS RANGES	GAS	R	.8	.8	132.3	146.	1606.	116
3	SWIMMING POOL COVERS	GAS	R	1.8	1.1	70.0	216.	1856.	129
4	91-2000 GAS DRYERS	GAS	R	6.8	1.8	29.9	246.	4350.	126
5	1981-90 GAS DRYERS	GAS	R	10.2	1.8	.7	247.	4443.	125

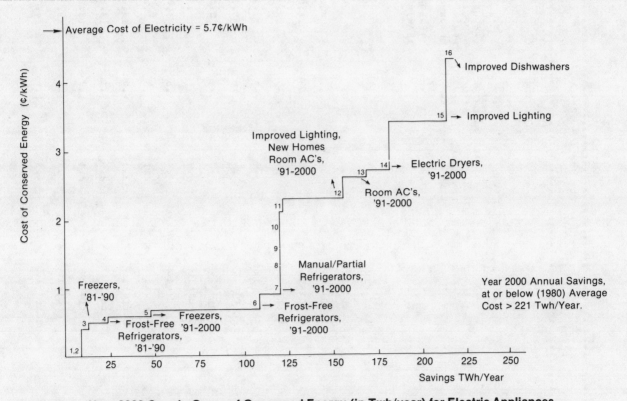

Figure 1.40. Year 2000 Supply Curve of Conserved Energy (in Twh/year) for Electric Appliances.
Year 2000 baseline annual use for this sector is 581 Twh, where 'baseline" assumes continuation
of 1980 average unit energy consumption for existing stock or new additions in that year. Unit cost
of conserved energy (in constant 1980 $) assumes that all increased costs are amortized over the
useful life of the measure, using a 3% (real-dollar) interest rate. Potential savings in 2000, at or below
today's average cost of 5.7¢/kWh is 221 Twh, or 38% of the year 2000 baseline.

Table 1.28.

```
****************************************************
  ELECTRIC APPLIANCES   SUPPLY TABLE TO THE
****************************************************
```

TIME HORIZON=10 YEARS
DISCOUNT RATE(3)
RESIDENTIAL= 5.0 PERCENT

	CONSERVATION MEASURE	A. ENERGY TYPE	B. SECTOR	C. MARGINAL COST OF CONSERVED ENERGY (CT/KWH)	D. AVERAGE COST OF CONSERVED ENERGY (CT/KWH)	E. ENERGY SUPPLIED PER MEASURE (TERAWH/YEAR)	F. TOTAL ENERGY SUPPLIED (TERAWH/YEAR)	G. TOTAL DOLLARS INVESTED (M$)	MEASURE NUMBER
1	1991-2000 TELEVISIONS	ELECTRIC	R	0.	0.	7.8	8.	0.	122
2	1981-90 TELEVISIONS	ELECTRIC	R	0.	0.	.9	9.	0.	121
3	1981-90 FREEZERS	ELECTRIC	R	.4	.1	4.0	13.	282.	113
4	FROST FREE REFR. 1981-90	ELECTRIC	R	.5	.3	11.0	24.	1208.	109
5	1991-2000 FREEZERS	ELECTRIC	R	.6	.5	24.5	48.	3994.	114
6	FROST FREE REFR. 1991-90	ELECTRIC	R	.7	.6	60.3	108.	10161.	110
7	MAN-PARTIAL REFR 1991-90	ELECTRIC	R	.9	.6	5.3	114.	10864.	112
8	1991-2000 ELECTRIC RANGES	ELECTRIC	R	.9	.6	5.5	119.	11554.	118
9	1981-90 ELECTRIC RANGE	ELECTRIC	R	.9	.6	.5	120.	11611.	117
10	MAN-PARTIAL REFR 1981-90	ELECTRIC	R	.9	.6	1.0	121.	11759.	111
11	1986-90 ROOM ACIS	ELECTRIC	R	2.1	.6	1.6	122.	12114.	119
12	IMPROVED LIGHTING NEW HOMES	ELECTRIC	R	2.3	1.0	34.2	157.	19138.	124
13	1991-2000 ROOM ACS	ELECTRIC	R	2.6	1.1	15.2	170.	22655.	120
14	1991-2000 ELECTR.DRYERS	ELECTRIC	R	2.7	1.2	15.2	183.	27050.	127
15	IMPROVED LIGHTING	ELECTRIC	R	3.4	1.6	35.2	216.	37066.	123
16	IMPROVED DISHWASHERS	ELECTRIC	R	4.5	1.6	4.7	221.	39414.	128

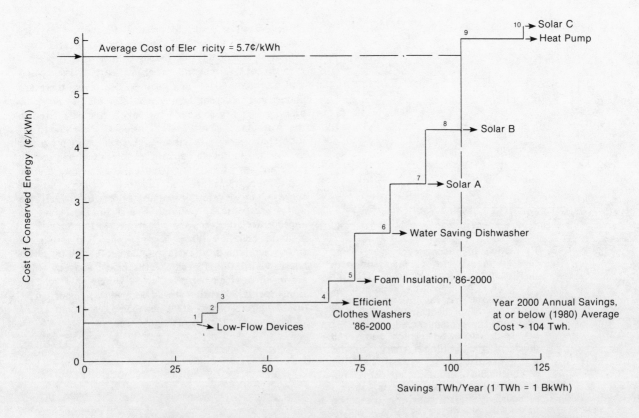

Figure 1.41. **Year 2000 Supply Curve of Conserved Energy (in Twh/year) for Electric Water Heaters.**

Year 2000 baseline annual use for this sector is 173 Twh, where "baseline" assumes continuation of 1980 average unit energy consumption for existing stock or new additions in that year. Unit cost of conserved energy (in constant 1980 $) assumes that all increased costs are amortized, over the useful life of the measure, using a 3% (real-dollar) interest rate. Potential annual savings in 2000, at or below today's average cost of 5.7¢/kWh, is 104 Twh, or 60% of the year 2000 baseline.

To calculate the potentials, we assume that this strategy is followed in all houses built after 1991, by which time experience with heat exchangers should permit their widespread use.

An infiltration rate of 0.6 ach was used in our calculations for new houses per 1991 for uniformity with the BEPS calculations. It may also be noted that this is close to the value for recently built houses. 0.6 ach is not proposed as a standard infiltration rate, since insufficient data exist at present to make such a judgment.

In existing houses, it is not possible to retrofit a continuous plastic sheet in the walls and ceilings, and conventional weatherstripping and caulking are not expected to be able to reduce infiltration rates below 0.6 ach, except in unusual cases. (However, there are already in existence well-built houses whose infiltration rates are below 0.6 ach.) On a simple model, pollutant concentration is inversely proportional to air change rate, so going from 1 ach (our assumed base) to 0.6 ach increases pollutant concentrations 1.7 times, whereas going from 1 ach to 0.1 ach increases it ten-fold. We assume this decrease will not convert acceptable air quality into unacceptable air quality. Owners of houses which now have high levels of formaldehyde, radon, or other pollutants are assumed to have to take remedial action anyway, perhaps including the use of a heat exchanger. Figure 1.42 shows reduction in radon levels by use of mechanical ventilation with a heat exchanger in a house with very high radon levels.

The importance of indoor air quality for the health of the American public is unknown, for several reasons. First, exposure to indoor air pollutants takes place in millions of buildings whose pollutant and occupancy levels are simply not known. A valid sample of the pollutants at such a large number of sites has not yet been collected. Second, indoor air often contains a mixture of chemicals which could interact to produce an effect far greater than the sum of the expected effects of the individual components. This interaction has not been studied, perhaps because the components have not all been identified. Third, the pollutants are most often present at levels far below those at which acute health effects have been

82 BUILDINGS

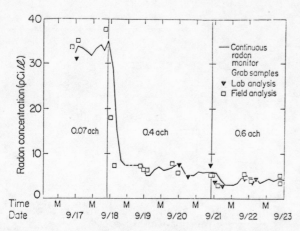

Figure 1.42. Reduction in radon level in a home located in an area of Maryland with high natural radon levels. The air exchange rate was increased above 0.07 ach, the natural infiltration rate, by means of a mechanical ventilation system. The system included in air-to-air heat exchanger which recovered much of the available energy in the warm exhaust air. The health effects of continuously breathing air containing radon (and its decay products) at one picocurie per litre has been compared to smoking one cigarette daily.

observed. The result of prolonged exposure to very low levels of a known toxin is uncertain at best.

Given present data on indoor pollution, it is premature to set up remedial programs on the assumption that a problem exists in all homes. However, there are undoubtedly current homes whose levels of radon, or other pollutants approach the levels not permitted in the workplace. Near term action to find and treat such "hot spots" is advisable. Longer term programs are needed to develop strategies to control and reduce pollutant levels in current homes and in new, tighter homes. Programs are also required to plan pollutant surveys, to develop instrumentation, to investigate human response to pollutants, and finally to set up appropriate standards for indoor air. In the meantime, a little common sense is required. Objectionable smells and humidity levels may be dealt with by whole house ventilation (perhaps with heat recovery using air-to-air heat exchangers), or by simple spot ventilation. Special attention should be given to indoor pollutant sources in very tight new homes, or in old homes which are very thoroughly tightened.

Section Three
Renewable Energy Technologies in Buildings

The analysis of renewable energy technologies serving buildings will concentrate on the following technologies: passive solar space heating, combined active solar heating and hot water systems, solar water heating, wood stoves, small wind machines, and photovoltaic systems designed for residential and commercial buildings. The first part of this section will review the techniques available and provide background information about costs and performance. A final section will motivate the assumptions actually selected to construct the "supply curves" exhibited in Section 3. Photovoltaics, wind, and wood stoves were not examined as a part of the "supply curve" analysis; markets for these technologies were treated separately. Solar cooling using thermal processes was not examined.

PASSIVE SOLAR HEATING

Passive Systems for the New Construction

Passive solar heating systems use the structure of a building for solar collection and storage. Natural conduction, convection and radiation are used to move heat within the building, although in some systems fans are used to augment natural heat flows when this would improve temperature control or increase system performance. In some cases, movable insulation is used to reduce heat loss at night or on cloudy days. While a pure definition of "passive" structures would exclude any use of mechanical equipment, systems which use fans and movable insulation will be treated here as "passive."

It is extremely difficult to establish the cost of a passive system since these systems are such an integral part of the design of the building. The design requires a sensitive analysis of both performance and aesthetics. Many architectural features of passive buildings, such as large expanses of glass, clerestory windows, massive fireplace sections, shading overhangs and shutters, and greenhouses can be skillfully used as elements in an efficient passive house but may also be attractive to homeowners for reasons having nothing to do with energy. The elements of a passive design can add variety and interest to a building design. A greenhouse, for example, can be a splendid living space where plants and changing light offer beauty and variety. Plainly it is impossible to account for such features in a strict cost analysis. Recent surveys of consumer attitudes, however, indicate that these intangible benefits do add directly to the value of a building. Consumers have indicated that they are attracted to the appearance and feeling of passive buildings as well as to the feeling of independence that such buildings offer (Cook, 1979; Towle, 1979; Skurka, 1976).

Another unmeasured benefit of passive buildings results from the fact that the systems offer direct radiant heat from large surface areas such as wall-sections, floors, or large containers full of rocks or water. Surfaces which are near room temperature can provide uniform heat without drafts or cold-spots. The air circulation is imperceptible and quiet.

Systems Cost and Performance

Without taking credit from any of these intangibles just described, the following analysis indicates that the cost of heat delivered from a passive solar system can range from 0 to $20 per million BTUs. The higher end of this cost range applies to the more elaborate systems where considerable costs have been incurred to create attractive new building spaces. The majority of the systems built provide heat for 0 to $10 per million Btu.

Costs. The cost of a passive solar system can be difficult to establish even if no credit is taken for intangibles, since the equipment is integrally part of the building structure. Reported costs are almost impossible to compare since they are often based on very different assumptions about what is counted as part of the passive system. Table 1.30 summarizes the range of costs reported for a variety of different solar designs by Nichols (Nichols 1978). An informal telephone survey of designers, builders, and occupants of existing passive buildings produced costs consistent with those in Table 1.30. Apart from the differences in costing techniques, the range of prices results from differences in construction technique, materials, and workmanship. The designer's choice about whether to use night insulation accounts for a substantial amount of the cost range since insulation costs can range from 4-12 dollars per square foot. Costs can also vary with region. For example, brick construction is much less expensive in the southeast because both material and labor is cheaper. System prices can be reduced in any region if attention is paid to the use of local materials.

Information obtained directly from builders can be difficult to interpret but an informal survey for this study indicates that the use of passive techniques can add from 0-12% to the construction cost of a house, but 5-6% is typical. Some of the information gathered is shown in Table 1.31

A somewhat more systematic evaluation of passive system costs is available from TVA's "Solar Homes for the Valley" program. This program has built 34 passive homes in a seven-state region. The passive homes cost from 4 to 6 percent more than conventional homes of similar design (Born, 1980).

Performance. The performance of passive systems has also been reported in ways that are difficult to interpret. For example, poor passive designs often allow interior temperatures to vary through ranges that most occupants would find uncomfortable but which dedicated residents may willingly suffer in order to minimize use of backup heat. Published reports of such systems often do not take the care to indicate whether temperatures have been kept within reasonable bounds. Systems with unusual temperature excursions have been eliminated from the statistics presented in this study.

The largest compendium of passive system performance data now available in a roughly consistent format is the calculated performance of 162 building designs that received awards in HUD's 1978 Passive Solar Design Competition (HUD 1979). (Recent experience seems to indicate that passive buildings have frequently been able to meet or exceed the predicted performance.) The backup energy required by these buildings was calculated in a way that allows consistent comparisons with other estimates of building performance. It has therefore been assumed that all buildings use furnaces which are 70% efficient. It is interesting to note that nearly 90% of

Table 1.30.   INSTALLED COST OF TYPICAL PASSIVE HEATING SYSTEMS

System Type	Cost ($/ft^2 of receiving area)
Direct Gain	2 - 11
Thermal Storage Wall	8 - 18
Attached Sunspace	5 - 20

Nichols 1978. An informal telephone survey of designers, builders, and occupants of existing passive buildings and a review of published sources below produced costs consistent with those shown in the Table.

Solar Age Magazine, 1976-1980
Proceedings of the Second, Third, and Fourth National Passive Solar Conferences, 1978-1979
AIA Solar Building Survey, 1978
AIA Survey, to be published
AIA Monitored Building Survey, 1978
Stromberg, 1979
Shurcliff, 1978
Franklin Research Center, 1979

the designs are expected to exceed the performance requirements of the proposed BEPS standard and that more than half would exceed the requirements of the low-infiltration standard proposed in the previous discussion. It must also be recognized, however, that a significant fraction of the "award winning" passive homes illustrated on this figure are either unable to meet the proposed BEPS standard or are only modestly better than the BEPS standard. Since the BEPS standard can be met inexpensively without the use of any passive technology, it is plain that there is a considerable room for improvement in the state-of-the-art of passive technology. The sophistication of designers appears to vary enormously.

Case Studies. Table 1.31 indicates the cost and performance of nine successful systems selected because they illustrate applications in different parts of the country and because information is available on both cost and performance. The systems were selected to represent a range of passive system techniques; frequently more than one technique is employed. The results are impressive. Several of the systems required no backup heat and in others the backup heat required was minimal. The system in Bozeman, Montana for example, maintained an interior temperature above 68°F during the Montana winter of 1978-1979 (a climate requiring approximately twice as much heat as the U.S. average). Backup requirements were provided entirely by a wood stove that was lighted only 16 times during the winter. The systems shown in the table will be discussed in greater detail in the following section.

A Survey of Passive Building Design Techniques

Direct Gain.   A direct-gain passive system relies simply on a relatively large area of south-facing windows. At some increased cost the performance of such systems can be improved by adding insulation that can be moved into place at night and on cloudy days to

Table 1.31. COST AND PERFORMANCE FOR SEVERAL PASSIVE SOLAR HEATING INSTALLATIONS

Installation Location and Date Completed	Backup Heat Provided (Btu/ft^2/DD)	Incremental Cost Of Passive Solar System Total $ ($/ft^2-glazing)		Heat Supplied to House in MBTUs		Cost of Solar Heat[d] ($/MBtu)	System Description/Notes
				Solar	Backup		
1. Sommersworth, NH. (1979)	1.5	1500	5.08	37	19 (27)	2.55	295 ft^2 Direct Gain, 2388 ft^2 living area, woodstove backup 7252 DD, 1 1/2 cords backup for 1979.
2. Ames, IA. (1979)	a. 2.9 b. 0.6	6000	12.00	a. 14.4 b. 35.4	31 (43) 9	26.30 10.70	500 ft^2 Direct Gain, concrete cored slab floor, 220 ft^2 living area, 31MBtu gas backup 1979, 6800 DD. Range numbers are for present performance without night insulation, and for predicted performance with night insulation.
3. Prescott, AZ. (1978)	2.6	3440	9.83	58	14 (20)	4.23	350 ft^2 Trombe wall, greenhouse, with rock-bed storage bermed north side. Amortization period of 8 yrs monitored data, 1400 ft^2 living area, 5500 DD.
4. Royal Oak, MD. (1976)	3.3 – 5.8	2500	5.56	55 – 58	15 – (21 – 27 39)	3.30 – 3.49	460 ft^2 Direct Gain, 1300 ft^2 living area, woodstove backup and electric. Range numbers for 3 yrs period.
5. Atascadero, CA. (1976)	0	1815	1.65	32	0	4.53	1100 ft^2 Roof pond area on concrete slab ceiling, 1140 ft^2 living area, 2970 DD. 100% solar.
6. Santa Fe, NM. (1976)	0.3	3450 – 6900	8.40 – 16.75	66	3 (4)	4.30 – 8.54	409 ft^2 Direct Gain, greenhouse, rock-bed storage, 2300 ft^2 living area, 6000 DD, close to 100% solar conditioning, monitored data.
7. Bozeman, MT. (1977)	a. 0.9 b. 1.3	7920	16.85	a. 68 b. 76	a. 12 (17) b. 18 (26)	a. 8.94 b. 8.00	470 ft^2 water wall, 2200 ft^2 living area, 9000 DD. woodstove backup, extremely low backup requirements.
8. Yakima, WA. (1977)	1.7	1000	2.65	20 – 25	12 (17)	3.83 – 3.07	378 ft^2 Direct Gain, triple glazing, 1728 ft^2 living area, 5941 DD., 55-65% annual sunshine, 1 cord of wood backup used for 1979.
9. East Pepperell, MA. (1979)	1.9	0	0	6	11 (16)	0	100 ft^2 Direct Gain, "super insulated" house, 1200 ft^2 living area, 6766 DD. 11M3tus backup gas in 1979. No additional cost for house construction over conventional construction because of elimination of large backup heating system.

[a]Backup assuming 70% efficient fuel-fired furnace; electric use converted at 3413 Btu/kWh 0.7, woodstoves assumed to be 50% efficient (Btu/ft^2/DD wood x 0.5/0.7) except Bozeman, Montana, which was measured at 59%.

[b]Backup heat supplied, i.e., furnace output.

[c]Values in parentheses are fuel used if furnace 70% efficient.

[d]Assumes 30 year system life, 3% real discount rate and 1% annual operating cost. Annual fixed charge rate is therefore 0.71.

Sources:
1. Franklin Research Center, 1979;
   Fike, 1980
2. Third Passive Conference, 1979, p. 771
   Franklin Research Center, 1979;
   Block, 1980
3. Second Passive Conference, 1978, p. 117;
   Frerking, 1980.
4. Solar Age Magazine, 1977;
   Crosley, 1980;
   AIA Survey, to be published

5. AIA Passive Solar Buildings Survey, 1979, p. 111;
   Third Passive Conference, 1979, p. 173
6. Sandia, 1979, p. 19;
   Third Passive Conference, 1979;
   AIA Survey, to be published, Tape 28;
   Fourth Passive Conference, 1979, p. 704;
   Second Passive Conference, Tape 114.

7. AIA Survey, to be published, Tape 114.
   Second Passive Conference, 1978, p. 8.
8. Fowlkes, 1980;
   SERI, to be published.
9. Fourth Passive Conference, 1979, p. 317;
   Leger, 1980.

reduce heat losses. Performance (and costs) are further increased if storage is added to the interior of the building in the form of large planters, masonry walls on a building interior etc. A direct-gain system should be able to provide 30-40% of the space heating for a building using only the thermal mass associated with typical building interiors, but additional mass is typically required if solar heat is expected to supply much more than 40% of the total heating demand (Balcomb, 1978). Costs of direct-gain systems vary greatly depending on the equipment used and the quality of the glazing and insulating systems.

Several of the houses shown in Table 1.31 employ direct-gain systems. The Leger House in Massachusets employs no passive feature other than having more windows than normal on the south side. In fact the designer does not consider the building to be a "solar" home. The estimated net solar contribution is only about 6 million BTU per year, but the total demand of this "super insulated" house is so low that the solar heat provides approximately a third of the total demand.

The house in Yakima, Washington is also a simple direct-gain building with most of its windows on the south side. This building uses thermal mass in the form of a concrete slab floor. The only costs associated with the passive system result from the use of approximately 100 more square feet of windows than would have been used on a typical house (and placing most of the windows on the south wall). The direct-gain house in Somersworth, New Hampshire is very similar to the Yakima house, relying on large amounts of south facing glass and a concrete slab floor.

The direct-gain house in Ames, Iowa is somewhat more elaborate. It uses an imaginative contemporary design with 545 ft^2 of south facing glass and stores heat in a cored concrete slab. Air is continuously circulated between the glazing and storage slab; insulation is moved over the entire glazed area at night.

All of the direct-gain systems, with the exception of the Ames house (in which some costs were clearly attributable to the aesthetics of the contemporary design), provided heat for less than 4 $/MBTU.

A detailed analysis of the cost of a direct-gain system has been performed by Total Environmental Action (Sullivan 1979) using professional cost estimators to examine working drawings. The results of their study are summarized in Table 1.32. It can be seen that this estimate is close to the high end of the range of costs for direct-gain systems indicated in Table 1.30. This estimate assumed that all of the south-facing glazing would be high quality windows which could be opened. Most direct-gain installations, however, use only a standard amount of openable window area and use fixed glazing for the remainder of the area required. This lowers costs substantially.

Table 1.32.

CSI	Item	$Glazing
3.1	Foundation	
4.0	Masonry and Stone	
6.1	Carpentry - rough	(-277)
6.2	Carpentry - finish	300
7.2	Thermal insulation	599
7.9	Caulk and Sealant	40
8.5	Glass and Glazing	1483
9.1	Lath and Plaster	(-793)
9.2	Gyp Wall Board	(-56)
9.8	Painting	24
Subtotal		1370
1.10	General conditions	106
17.10	Contingency	137
Subtotal		1613
	Contractors Fee	242
TOTAL		1855
$/sf added aperture		14.27

Source: Sullivan, Paul W. and Katz, Malcolm R., Total Environmental Action, Inc., Harrisville, New Hampshire, the Proceedings of the 4th National Passive Solar Conference, October 1979, p. 673.

This study also considered the addition of a large interior storage wall which increased costs to $34/ft^2. However, many direct-gain systems use concrete slab floors for thermal mass, and these often add nothing to the cost.

Thermal Storage Walls. Thermal storage walls tend to be expensive because they require additional materials beyond those used in conventional building construction. Thermal storage walls are normally made either of masonry or water in containers. Masonry is expensive, especially if the wall is very thick, but it has excellent reliability and permanence. The principal advantages of the use of water-filled containers are the availability and low cost of water, and water's ability to achieve large thermal storage capacity in a small volume since the volumetric heat capacity of water is about twice that of concrete.

Thermal storage wall costs will exhibit significant regional variability due to the significant cost variance in structural elements (footing depth, earthquake requirements, etc.). The choice of concrete or water may depend on local building practices and labor costs in an area. For example, in most parts of the country frame construction is used almost exclusively for residential buildings. Masonry construction is expensive partly because it is unusual. Water storage is easily compatible with frame construction.

Two of the houses listed incorporate a Trombe wall to provide heat storage. The house in Royal Oak, Maryland uses a concrete wall 8 inches thick and double glazing. It is interesting to note that while this house employs rather minimal insulation levels by 1980 standards, overall performance is quite acceptable. The house was originally built with electric resistance backup and used 5.8 MBtu/ft^2/dd, but a wood stove was installed later; during the winter of '79-'80k it used only 3.3 MBtu even assuming a 50% efficiency for the wood stove (presumably because the wood stove was used only to supply heat to areas where it was needed when it was needed). The Trombe wall is a concrete blockwall with the cells filled with compacted sand. The house in Bozeman, Montana also uses a storage wall. In this case storage is provided by fiberglass tubes 16 inches in diameter and 17 feet long, each filled with 1500 lbs of water. These tubes stand immediately behind the large glazed area. The house also uses a thick movable insulating curtain. The Bozeman house has provided outstanding performance in one of the most severe climates in the country. Both the Royal Oak house and the Bozeman house provide heat for approximately $5/MBtu.

A recent paper reports the incremental costs of a passive house built in New Mexico and compares them with a conventional house and with a house with special "energy conserving" features (Taylor 1979). The house studied was built by Communico, Inc., a Santa Fe, New Mexico firm which has been building energy conserving passive solar homes for several years. Communico has been monitoring building costs closely. Their "Model 1," La Vereda house is a 1370 ft^2 frame house, with 128 sq. ft. of 16" thick poured concrete mass wall and a mass-backed greenhouse with 122 sq. ft. of glazing.

Table 1.33 shows the construction cost comparison between the passive home and an energy conserving version of the same home. The cost of the passive system shown in Figure 1.43 was $5.66 per square foot more than the conventional building primarily because of the cost of glazing (costs are reported here in dollars per square foot of solar collecting area). It is interesting to note that the incremental cost of passive solar heating is $1414 or about $1/sq.ft. of floor area more than an "energy-conserving home" and the cost of an "energy conserving home" is estimated to be about $2/sq. ft. of floor area more than an "inefficient conventional home." The cost of the conservation and passive systems reported may be overstated since the block stemwalls could well be considered an expense due to an architectural decision to lower the profile of the house instead of an energy-related cost. Using this accounting would lower the incremental costs of the passive and "energy conserving house" by $1358.

The yearly heating loads and heating bills for each house were estimated and annual costs determined using the 1980 cost of a KWH in New Mexico (currently $.05). Table 1.51 shows the yearly heating load and heating bill for each house. Comparing this data to the costs presented in Table 1.33 it can be seen that an investment of $3,512 yields a yearly savings of $510 on the heating bill. Another investment of $1414 yields an additional savings of $192 per year.

Attached Sunspaces. Calculating costs for an attached sunspace is particularly difficult. The solar green-

Table 1.33. CONSTRUCTION COST COMPARISON

Building Category	Passive Home	Energy Cons.	Conventional
Block Stemwalls	$ 1,935	$ 1,935	$ 577
16-in. Trombe Wall	818	-0-	-0-
Rigid Insul.	577	577	-0-
Exterior Framing	1,868	2,278	2,100
Exterior Sheathing	685	819	757
Interior Sheathing & Thincoat	2,104	2,317	2,317
Roof Insul.	909	909	587
Wall Insul.	496	872	582
Exterior Stucco	3,320	3,658	3,658
Interior Plaster	599	311	311
Venting Windows	1,194	1,870	1,692
Fixed Glass	2,029	-0-	-0-
	$16,734	$15,546	$12,581
20% Labor Fringe	645	604	515
	$17,379	$16,150	$13,096
15% Overhead & Profit	2,697	2,422	1,964
	$19,986	$18,572	$15,060
Total House Costs	$62,151	$60,737	$57,225
Cost per sq. ft. floor area:	$45	$44	$42

Source: Taylor 1979.

house has much in common with the thermal storage wall approach but the space created serves both as solar collector and as additional useful building space; its value for each function should be considered. The solar greenhouse is capable of providing significant heating but can also provide space for growing plants and humidification. Although an attractive, well constructed greenhouse may cost $20/ft^2 or more, such high costs typically result from providing the attractive designs featured in magazines such as Better Homes and Gardens; in such cases style is more important than heating.

An economic analysis of greenhouse systems conducted by Scott Noll of Los Alamos Scientific Laboratory, based on a recent survey of 19 New Mexico greenhouses, found the cost per sq. ft. of greenhouse floor area ranged from $3 to $29/sq. ft. (Noll 1979). Only four greenhouses were estimated to have a negative net present value of dollar fuel savings over their expected lifetime.

Simple payback periods for the cost effective designs ranged from four to eight years when escalating fuel costs were considered. The greenhouses had a capital cost for annual energy savings in the range of $30 to $110/MBTU/yr., which compares favorably with

Figure 1.43.   An Architect Sketch of the La Vereda 1 Home Costed in Table 1.33.

wood, electric resistance, and liquid petroleum gas heating.

The Santa Fe house relies primarily on a large sunspace or greenhouse for passive heating. It has worked well and provided comfortable interior temperatures using only small amounts of backup heat. In estimating the incremental cost of the structure, it has been assumed that half of the value of the greenhouse can be creditied to the additional value of the building space provided.

Passive Solar Retrofits

A wide range of passive systems are available for retrofits. The addition of an attached greenhouse or sunspace is rapidly gaining popularity. Passive heating can also be provided by adding more south facing windows (particularly in the case of an addition), by installing glazing over an existing masonry wall to create a passive Trombe (thermal storage) wall, or by adding vertical thermosyphon collectors to a south wall. Other measures sometimes considered to be passive retrofits include installing awnings, windbreaks, porches or moveable insulation panels.

Table 1.34 shows the projected range of savings that can be accomplished by different passive retrofit heating techniques (TVA, 1980). The cost of the heat provided ranges from less than $1.00/MBtu to about $40.00/MBtu. The range of costs typically reflects the fact that some systems are self-built or while others are designed and installed professionally. For example, moveable insulation for windows can be very cheap polystyrene or Thermax sheets cut to fit a window and put in place by hand, or it can be an automated insulating curtain. Some greenhouse retrofits

are very cost inexpensive if built only for the heat provided while others, such as those featured in a recent issue of House Beautiful (May 1980) must be viewed as pleasant additions to living space, which also provide heating benefits.

ACTIVE HEATING AND HOT WATER SYSTEMS

Combined Systems

Active solar systems that provide heating and hot water come in a tremendous variety of sizes and forms. They range from simple systems entirely built and installed by the homeowner to those that employ sophisticated technologies, such as evacuated tube collectors combined with complex control circuitry. Most installations use factory-built flat plate collectors with one or two glazings to heat air, water, or an anti-freeze solution. Collected heat is either sent directly into the building, or to storage, typically some type of rock bed (for air systems) or water tank (for liquid systems). When liquid systems are used in buildings equipped with forced air heating, a heat exchanger similar to an auto radiator is used to transfer heat from the liquid to the air.

Data for cost-effectiveness and performance of heating and hot water systems has been elusive for two reasons. Most carefully-monitored systems were installed as part of a HUD demonstration program, and those program specifications resulted in inordinately high prices. On the other hand, privately installed systems reflect have realistic costs, but are seldom monitored accurately. Although backup heating requirements have been reported for numerous systems, this information has been of limited value since heating

Table 1.34. PASSIVE SOLAR RETROFIT: COSTS AND ANNUAL ENERGY SAVINGS

Measure	Cost ($) per Unit Size[a]	Annual kWh Saved per Unit Size[a]	Fuel Displaced (MBtu per Unit Size)[b]	Cost ($) per Annual Fuel Displaced (MBtu)
Moveable Insulation	0.50 - 5.00	10 - 50	.053 - 0.26	1.90 - 94.00
Vertical Heater	1.50 - 9.50	13 - 29	.068 - 0.15	10.00 - 140.00
Awnings	3.50 - 6.50	5 - 15	.026 - .079	44.00 - 250.00
Trombe Wall	3.00 - 15.00	20 - 40	0.11 - 0.21	14.00 - 140.00
South Windows	5.00 - 25.00	25 - 35	0.13 - 0.18	28.00 - 190.00
Porch or Overhang	4.00 - 15.00	5 - 15	.026 - 0.79	5.00 - 580.00
Solar Greenhouse	4.00 - 30.00	16 - 28	.084 - 0.15	27.00 - 360.00
Windbreak	1.50 - 2.50	2 - 26	.011 - 0.14	11.00 - 230.00

[a] A unit size is equal to a one square foot aperture for the following: moveable insulation, Trombe wall, south windows, vertical heater, and solar greenhouse. A unit size equals one square foot of window shaded for the following: awnings, porch or overhang, and windbreaks.

[b] Assumes 65% furnace efficiency.

demands of houses can vary by a factor of ten due to characteristics and operation of the house.

Heating and hot water systems generally fall into three main categories. (1) Relatively high temperature systems use sophisticated technology in their collectors, storage, and controls. An example would be a system designed with evacuated tube collectors, coupled to a heat pump for source input offering direct or stored heat use, and with heatpump and resistance backup. (2) A second type of system uses modular collectors with single or double glazing (and perhaps a selective absorber surface) coupled to water or rock storage and a conventional heating system. A third group of systems use simple collectors designed to produce heat at 70°-100°F for use in low velocity distribution.

As might be expected, more sophistication in a system results in a higher cost per square foot of collector. However, cost per square foot is not a particularly good measure of system merit since there is a significant difference in collector efficiencies. In a recent study done for TVA (Reed, 1979) the cost and size of ten different systems capable of providing 51% of the heating and hot water demand for a 2000-square-foot house in Knoxville, Tenn. (i.e., 33 MBtu/yr), was investigated. As shown in Table 1.35, there is virtually no correlation between cost per square foot and estimated total installed system cost.

Some of the least expensive systems in cost per square foot have been built onsite, for around $10 per square foot (Site-Built Conf., 1978). Reports have suggested that these installations are very cost effective; we have not located any careful measurements to verify those claims. Indeed, some reports have cited examples of systems that have had numerous problems to suggest that site built systems will never be installed with sufficiently consistent quality to become significant in the marketplace. The performance of several private installations is shown in Table 1.36. Note that most of the systems sold at the given incremental costs (assuming replacement of a $2500 conventional system) also include hot water preheat, which could provide an additional 10MBtu/yr of output.

Locating enough well-monitored installations with good cost data to use them as the basis for cost-effectiveness of systems proved impossible; the results of simulation programs were used to compute the performance of a system typical of many being sold today. Table 1.37 compares the actual and predicted performance of a number of different types of systems

Table 1.35.   COLLECTOR AREA NECESSARY FOR ANNUAL SYSTEM HEAT
DELIVERY OF 33 x 10^6 Btu/yr AND ESTIMATED SYSTEM COST FOR
SEVERAL COLLECTORS LOCATION - KNOXVILLE, TENNESSEE

Brand	Type	Collector Area	Estimated Installed System Cost/ft^2	Estimated Installed System Cost
Grumman*	Liquid, 1-Cover Black	400 ft^2	$27	$10,800
Grumman	Liquid, 2-Cover Black	300	28	8,400
Sunworks	Liquid, 1-Cover Selective	350	28	9,800
Sunworks	Liquid, 2-Cover Selective	305	29	8,845
Lennox	Liquid, 2-Cover Selective	225	32	7,200
Owens-Illinois	Liquid, Evacuted Tubular	290	47	13,630
Scientific-Atlanta	Liquid, 1-Cover Open-Flow	480	22	10,560
Sepco*	Air, 1-Cover Black	505	22	11,110
Solaron	Air, 1-Cover Selective	495	24	11,880
Solaron**	Air, 2-Cover Black	625	24	15,000

Source: Robert Reed, John Tombuson and David Chaffin.   "Solar Energy in the TVA System; a Proposed Strategy," Volume IV, Paper 2, p.7, Sept. 1979, TVA.

*"Standard" models used for Tables

**Obsolete model

monitored by the National Solar Data Network (Operational Results Conference, 1979). These systems show satisfactory agreement between the predicted fraction of the heat supplied by the solar systems (f_p) and the measured fraction of the heat supplied by the solar system (f_m). Since the simulation program (F-Chart) uses "typical" weather data and the systems operate with real weather, differences are to be expected. We note some cases where the program over predicts and some cases where it underpredicts. These systems suggest that predictions are relatively good for well-maintained and installed systems; a number of other systems on the National Solar Data Network showed very poor agreement with predictions due to a variety of installation errors and equipment problems. Since lessons learned from these earlier problems have been and are being incorporated in new systems, it seems reasonable to use these model predictions for system performance over the period of this study.

As noted earlier, it is necessary to know both the cost and performance of a system before its cost effectiveness can be determined. The Solar Energy Industries Association (SEIA Financial Workshops, 1980) estimates installed costs for typical heating and hot water systems (see footnote to Table 1.38 for performance specifications) to be $30-40 per square foot of collector in early 1980. This corresponds to $25-34 per square foot in 1978 dollars and can be seen to be consistent with the costs shown in Table 1.35. Estimates for the future cost of systems also vary. Some industry sources offer private estimates that system costs will increase at a rate 2-3% below general inflation, or a real decrease of 2-3% per year for systems of comparable performance. The active heating and cooling program plan released by DOE earlier this year has a goal of 5% annual decrease in system costs. It is difficult to know how long to extrapolate such estimates, but 10 years of 5% annual reductions would

Table 1.36.   MEASURED HEATING PERFORMANCE OF SEVERAL HOMES WITH ACTIVE SOLAR HEATING SYSTEMS

Location	Energy Supplied by Solar System (MBtu/yr)	Incremental Cost of System	Backup Heating Requirements (Btu/ft^2-DD)	System Type
DeKalb, Ill.[a]	56	$6000	1.1	Active air w/green-house & berms
Quechee, Vt.[b]		7500	3.1	Liquid collectors, heat pump
Escanaba, Mich.[c]	61	4300	3.0	Active air and direct gain
Santa Fe, N. Mex.[d]	87	6500	2.1	Active air and direct gain
Acton, Mass.[e]	27	5000	4	Liquid system, heating only
Madison, Wis.[f]	20	12,000	2.4	Active air and heat pump
New Hampshire[g]		8000	0.9	Active and passive; cost also includes conservation
Jamestown, R.I.[h]		5000-9000	0.0	Active liquid; at House kept to 70-80°F.
Wilder, Vt.		7500	3.0	Vertical air collectors

[a] Solar Age, Oct. 1978; p. 29.

[b]

[c] Solar Age, Jan. 1979; p. 21.

[d] Passive Solar Buildings, Sandia p. 39.

[e] W. Hapgood and John R. Bemis, "Report on the Performance of the Acorn Solar Heated House for Winter of 1975-76," Acorn Structures, Inc., Concord, Mass.; Solar Age, Oct. 1979, p. 65.

[f] Proceedings of the 1978 Annual Meeting, AS/ISES; Vol. 2.1, p. 299; and Private Communication, Robert E. Terrell, Designer/Occupant, July 17, 1980.

[g] Bruce Anderson, Testimony at BEPS Hearing, March 27, 1980, Washington, D.C.; and William A. Shurcliff, Solar Heated Buildings of North America, p. 144, Brick House Publishing Co., Harrisville, N.H., 1978.

[h] William Shurcliff op. cit., p. 226; and Private Communication, Mike Smith, President, Solar Homes Inc., Providence, R.I., 1979.

Table 1.37. OPERATIONAL RESULTS COMPARED TO PREDICTIONS

Project Name	Dwelling Size ft^2	Collector Size ft^2	Measured Solar (MBtu)	Measured Aux (MBtu)	Measured Solar Fraction f_m	Predicted Aux MBtu	Predicted Solar Fraction f_p	Monitored Period
Saddle Hill Massachusetts	1696	360	7.75	17.99	.30	15.31	.34	March–April 79
Ortiz & Reill California	1620	196	8.09	13.38	.38	26.11	.24	August 78–April 79
Design Construction Montana	2255	750	13.60	20.50	.40	18.24	.43	Feb.–April 79
Sir Galahad Virginia	1510	624	2.73	2.32	.54	3.32	.45	Feb.–June 78 Sep.–Nov. 78
Chester West Alabama	2398	208	2.96	2.59	.60	3.29	.55	July–Nov. 78
J. D. Evans Maryland	2252	374	9.97	8.26	.55	10.45	.49	Oct.–Dec. 78 Mar.–Apr. 79
USAF Academy Colorado	1900	546	30.0	20.3	.60	11.1	.73	Jan.–Dec. 78

Source: Conference Proceedings - Solar Heating and Cooling Systems Operational Results, 1979

Table 1.38. ESTIMATED COST AND PERFORMANCE OF TYPICAL SOLAR HEATING AND HOT WATER INSTALLATIONS

City	Collector Size (sq. ft.)	Solar Fraction	Heat Delivered by Solar System		Cost of Delivered Heat** (1978$/MBtu) Present Systems		1990-2000 Systems	
			MBtu/yr	Btu/Sq ft/yr	w/40% ITC	w/o ITC	w/40% ITC	w/o ITC
Hartford, Conn.	180	.27	26.2	145,000	10.1-13.5	15.4-20.5	6.1-8.1	9.2-12.3
	252	.34	32.9	130,000	11.3-15.0	17.2-22.9	6.8-9.0	10.3-13.7
Washington, D.C.	180	.39	27.2	151,000	9.7-12.9	14.8-19.7	5.8-7.8	8.9-11.8
	216	.43	30.2	140,000	10.5-13.9	15.9-21.3	6.3-8.4	9.6-12.8
Atlanta, GA	144	.44	24.5	170,000	8.6-11.5	13.1-17.5	5.2-6.9	7.9-10.5
	216	.56	30.9	143,000	10.2-13.6	15.6-20.8	6.1-8.2	9.4-12.5
Chicago, IL	180	.32	30.3	168,000	8.7-11.6	13.3-17.7	5.2-7.0	8.0-10.6
	216	.35	33.8	157,000	9.3-12.4	14.2-19.0	5.6-7.5	8.5-11.4
Kansas City, MO	180	.39	32.4	180,000	8.1-10.8	12.4-16.5	4.9-6.5	7.4-9.9
	216	.44	36.2	167,000	8.8-11.7	13.3-17.8	5.3-7.0	8.0-10.7
Denver, CO	252	.64	60.4	240,000	6.1-8.1	9.3-12.4	3.7-4.9	5.6-7.4
	396	.81	75.6	191,000	7.7-10.2	11.7-15.6	4.6-6.1	7.0-9.3
Los Angeles, CA	144	.74	30.0	208,000	7.0-9.4	10.7-14.3	4.2-5.6	6.4-8.6
	216	.88	35.6	165,000	8.9-11.8	13.5-18.0	5.3-7.1	8.1-10.8
Seattle, WA	198	.38	28.3	143,000	10.2-13.6	15.6-20.8	6.1-8.2	9.4-12.5
	234	.41	31.3	134,000	10.9-14.6	16.7-22.2	6.6-8.7	10.0-13.3

*The solar system assumed a single-glazed flat-plate collector sloped at latitude plus 15° with $F_R( ) = 0.72$, $F_R U_L = 0.82$ and a storage capacity of 15 Btu/^{o}F-ft^2. The house uses 80 gallons of HW daily and has heating loads of 12,000 Btu per degree-day.

**The cost of delivered heat was calculated using FCR = 0.0878 including maintenance: present systems were assumed to cost $30-$40/ft^2 installed (1980 $) and 1990-2000 systems $18-$24/ft^2 (1980 $). Cost of delivered heat was converted to 1978 dollars using a multiplier of 0.846.

result in 60% of present cost and 10 year of 2.5% annual reduction would result in 78% of present system cost. These estimates are consistent with the cost and performance goals of the Low Cost Collector Task Force at SERI which are shown in Table 1.39.

The task force has examined a number of low-cost collectors and concepts and has recently purchased several collectors costing $5/ft^2 or less for testing to determine performance and lifetime. Most of the low-cost systems identified to date substitute plastics, wood products, and/or treated papers for some of the more traditional materials.

Table 1.38 combines these costs and performance estimates for systems in several different cities, both with and without the 40% Federal tax credit. The system's installed after 1990 are expected to provide heating and hot water for $6-14/MBtu. It should be noted that these systems do not provide high solar fractions in unfavorable climates and delivered energy costs can increase substantially if the systems are designed to provide, say, 80% of the demand in a city like Seattle.

Solar Water Heating

Most solar hot water systems presently installed circulate either water or an antifreeze medium through collectors when the sun is shining. "Thermosyphon" systems normally heat water directly and do not require a pump since the storage tank is above the collectors and the buoyancy of the heated water causes adequate circulation. "Bread box" systems are very simple, consisting of an uninsulated tank which is painted black and mounted in an insulated box with a clear plate cover to admit sunlight. However, these systems cool off overnight unless some kind of insulating cover is placed on them every evening.

Collectors are protected from freezing by several methods. The collector may be drained whenever the pump is not operating, or an antifreeze may adequately prevent damage. Warm water from the storage tanks may be circulated through the collectors whenever temperatures drop. Protection of thermosyphon systems from freezing is more difficult, although a thermosyphon system that provides automatic freeze protection has been proposed (Farrington, et al., 1980, Appendix B) and experimental systems using a refrigerant as the thermosyphoning fluid have been built (Morrison 1980).

Many builders resist the idea of placing a heavy hot water tank in the attic or on the roof, whether it is a thermosyphon or bread box system, because of possible structural inadequacies. This has slowed the acceptance of thermosyphon and bread box heaters in the United States, although thermosyphon systems are used more than any other water heating method in both Israel and Japan (Farrington, et al., 1980, p. 3).

The performance of solar hot water systems varies widely. Some provide nearly all of the user's hot water needs, with the backup shut off for 6-9 months per year, while others provide very little useful output. The industry is confident that it now knows how to manufacture and install high quality systems which perform according to model predictions. Several studies of system performance are currently ongoing, but very little information is available on the actual performance of systems installed in the last three years.

SERI is aware of only two experiments where residential hot water systems were monitored carefully enough to be used for computer model validation. The National Bureau of Standards measured the performance of six different generic types of hot water systems. These results were compared with the widely

Table 1.39. COST AND PERFORMANCE GOALS OF SERI LOW COST COLLECTOR TASK FORCE

	Present Demonstration Unit		Better than 50% probability that this goal can be reached by the year 2000		Possible given success of ongoing research programs	
	Cost	Relative Performance	Cost	Relative Performance	Cost	Relative Performance
Components	$20/ft^2	60	$9/ft^2	50	$7/ft^2	50
Systems	$40/ft^2	50	$18/ft^2	40	$14/ft^2	40

Source: Barry Butler, memo to Henry Kelly, April 22, 1980.

used F-Chart simulation for four system configurations (Buckles, et al., 1979), and it was found that F-Chart 3.0 consistently over-predicted the output of the solar hot water system as shown in Table 1.40. However, when the F-Chart was modified to consider more accurately the losses from the storage tanks, the model predictions were much more closer to the actual measured performance (as shown in Table 1.40 for the modified F-Chart 3.0).

Table 1.40.   MEASURED FRACTION OF HOT WATER SUPPLIED BY SOLAR COMPARED WITH MODEL PREDICTIONS

(eight month average)

System	NBS Measurement	F-Chart 3.0 Prediction	Modified F-Chart 3.0 Prediction
1-Tank, Ext. HX	0.34	0.46	0.35
2-Tank, Ext. HX	0.35	0.61	0.41
1-Tank, Int. HX	0.42	0.53	0.45
2-Tank, Int. HX	0.32	0.53	0.32

Source: W. E. Buckles, S. A. Klein, and J. A. Duffie, "Analysis of Solar Water Heating Systems," ISES Annual Meeting, Atlanta, GA; 1979.

A model developed at California State University at Sacramento was compared with the performance of four different commercially available solar hot water systems. The model was compared with experimental performance under a variety of climatic conditions for two different water use profiles. The agreement between the model and experimental results was generally excellent (Young, 1980; Bergquam et al., 1979) as illustrated in Figure 1.44 for two different systems with a 100 gallon/day load.

Table 1.41 shows results of this model (Bergquam, et al., 1979) comparing the yearly output in Btu/sq ft in five cities on the West Coast. A 40-square-foot system is assumed in each case, with a 75-gallon/day load, and each system uses a single glazed collector with a flat black absorber. Ranges are shown to reflect the difference in performance of two pumped systems (using a 1/20-HP pump) and two thermosyphon systems. Most of the area in the U.S. averages 1100-1900 $Btu/ft^2/day$, which is quite close to the range for the cities shown. However, there is little population in areas at the extreme high insolation end, so 125-300 kBtu/sq ft/yr is probably representative of hot water system output for most U.S. locations.

The cost of solar hot water systems also varies widely. Table 1.42 gives the average costs encountered in two separate surveys and gives the range stated recently by the Solar Energy Industries Association. Aside from obvious differences in system

quality, several other factors are shown. For example, in Florida and California, simpler freeze protection systems can be used, resulting in lower system costs, but the increased cost of retrofit installations is clearly apparent in the two entries for Florida. Lower costs per square foot in Boston and Washington as compared with Denver are possibly due to the larger system sizes in those two cities.

The cost of owner-built or installed systems can be much lower than shown in Table 1.42. Bread box systems appear to be somewhat lower in cost. Twenty thermosyphon systems in Arizona had costs comparable to active systems, (Solar Energy Research Associates, 1979), while another study found thermosyphon systems could cost about half to two-thirds that of pumped systems (Farrington, et al., 1980, p. 44).

The cost of solar hot water is compared with the cost of several conventional alternatives in Table 1.43. The table assumes that electric hot water heaters are 82% efficient and that gas and oil water heaters are 45-55% efficient (DOE, 1978). The installed cost of a conventional water heater is assumed to be $290 (Sears Roebuck, 1980). The solar estimates use the costs from Table 1.42 ($32-47/ft^2) and assume 40-60 square foot systems with output of 140,000-300,000 $Btu/ft^2/yr$ (Table 1.41). The table assumes that the conventional portion of the solar system is replaced after 10 years. The heat pump estimate is based on an incremental installed cost of $720 (E-Tech, 1980), replacement after 10 years, a COP of 1.8-2.3 (based on measurements by ORNL and Duke Power Co.), and daily use of 74 gallons of water.

It should be noted that some states have substantial tax credits (in addition to the Federal credit shown) which lower the solar cost still further. The cost shown for the heat pump would be lower if similar tax credits were available for heat pumps.

The projected 1990 cost of solar water heating is compared with conventional alternatives in Table 1.44, where it is assumed that conventional fuels are priced at the "social" cost to reflect externalities. (see Baseline Assumptions Appendix). The estimates assume that oil and gas water heaters have efficiencies of 55-75%. The ranges for gas water heating reflect prices without distribution cost included (low) and with them included (high). Estimates for solar systems are shown for system costs 20% and 40% below those used in Table 1.43. These estimates are discussed in the section on solar heating. The heat pump estimates assume that hot water heat pumps are available at an incremental cost of $286 (Hirst, 1980) and have a COP of 1.8-2.0 as measured by ORNL for this unit. It is plausible that the range shown should be extended in both directions to reflect potential technical improvements and higher costs suggested by recent experience, especially for some cold climate applications.

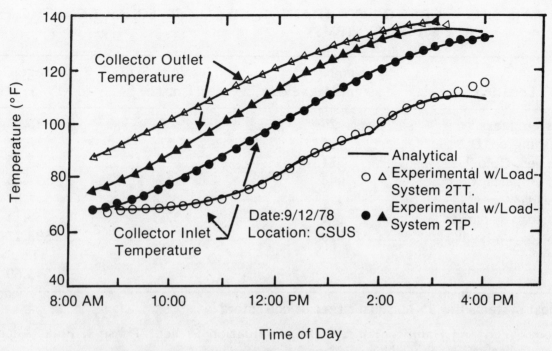

Figure 1.44. **Comparisons of Analytical and Experimental Collector Temperatures for Two Solar Hot Water Systems with a 100 Gallon per Day Load.**
SOURCE: M. F. Young and J. W. Baughn, "Economics of Solar Domestic Hot Water Heaters in California," Systems Simulation and Economic Analysis Conference Proceedings, January 23-25, 1980, Bahai Hotel, San Diego, California.

Table 1.41. ANNUAL OUTPUT OF SOLAR HOT WATER SYSTEMS IN FIVE CITIES[1]

| Location | Annual Output[2] | | Average Daily Radiation on Horiz. Surface[3] (Btu/sq. ft.) |
	Thousands of Btu/sq. ft. Collector	Millions of Btu for 40 sq. ft. System	
El Centro, CA	300–380	12–15	1880
San Diego, CA	250–325	10–13	1730
Sacramento, CA	220–290	9–12	1590
Medford, OR	170–230	7–09	1360
Eureka, CA	140–180	5–07	1230

[1]Output predicted by the model developed by Bergquam, et. al. as discussed in the text.

[2]Annual output ranges based on Table 6 in James Bergquam, et. al., "A Comparative Study of SDHW Systems in California," California Energy Commission, June 1979. Assumes 40 sq. ft. system with 75 gallon/day load single glazed collector with flat black absorber. Ranges represent performance of two pumped systems (with 1/20 HP pump) and two thermosyphon systems.

[3]Average daily radiation on a horizontal surface for most United States locations is in the range 1100-1900 Btu/sq. ft./day.

Table 1.42.   COST OF CONTRACTOR-INSTALLED SOLAR DOMESTIC HOT WATER SYSTEMS*

Location	Number of Systems Surveyed	Average Size (sq. ft.)	Average Cost (1978 $/ft^2
Boston, Mass.[a]	26	58	$38.97
Washington, DC[a]	28	57	$37.77
Denver, Colo.[a]	71	41	$45.15
Los Angeles, Ca.[a]	045	45	$31.65
Florida - new[b]	556	36.8	$31.89
Florida - retrofit[b]	805	37.3	$39.14
Florida - mixed[b]	919		$36.27
SEIA[c]			$33.90 - $46.60

*Most systems use 35-60 square feet of collector.

[a]Booz-Allen and Hamilton, Inc. SHAC Evaluation Project Phase I, Final Report, December 1979.

[b]Energy Information Administration, US/DOE, "New and Retrofit Solar Hot Water Installations in Florida, January Through June, 1977," HCP/15663-01, April 1978, p. 12.

[c]Solar Energy Industries Association, "Solar Workshops Financial Incentives," under DOE Contract #DE-FG-01-79cs30293, March 1980.

Table 1.43.   CURRENT PRICE OF HOT WATER FROM VARIOUS SOURCES*

	1978 $/kwh
Electric HW	0.03 - 0.07
Solar (includes 40% ITC)**	.016 - .059
Solar (without ITC)	.027 - .094
Heat Pump	.034 - .045

	1978 $/MBtu
Gas HW	$ 3.50 - 10.00
Oil HW	$11.00 - 13.50
Solar HW (includes 40% ITC)	$ 4.75 - 17.30
Solar HW (without ITC)	$ 8.00 - 27.50

*Price of delivered hot water = residential fuel price/heater efficiency for conventional systems. Heat pump entry is the levelized cost of the energy saved by the heat pump.

**ITC is "investment tax credit"

Table 1.44.   PROJECTED 1990 COST OF SOLAR AND HEAT-PUMP HOT WATER COMPARED WITH CONVENTIONAL SHADOW COSTS

	1978 $/kwh
Electric	
Heat Pump	.018 - .020
Solar (40% cost reduction)	.014 - .054
Solar (20% cost reduction)	.021 - .074

	1978 $/MBtu
Oil (.75 - .55 efficiency)	15.60 - 21.30
Gas (.55 efficiency)	14.80 - 18.70
(.75 efficiency)	10.90 - 13.70
Solar (40% cost reduction)	4.20 - 15.70
(20% cost reduction)	6.10 - 21.60

Table 1.43 shows that with the present 40% tax credit, solar hot water is substantially less expensive than oil water heating and should be competitive with gas in some instances. By 1990 (Table 1.44) it should be less than the social cost of gas or oil in most locations.

The solar penetration in the electric market is difficult to predict. With the 40% tax credit, it is clearly competitive today with electric water heaters in many parts of the country. Heat pump water heaters have recently appeared on the market, and Table 1.43 indicates that they are now competitive; if the decreased cost shown in Table 1.44 for 1990 occurs, then they will have an economic advantage over solar in most areas. The present generation of heat pump water heaters must be operated in a "tempered" space where the temperature seldom goes below 45°F. However, there would seem to be no technical reason why heat pump water heaters cannot be built to operate in lower temperatures or backed up by demand (resistance) heaters, although it could raise the cost of the hot water provided.

RESIDENTIAL WOOD USAGE

There are already five million homes with wood stoves and 18 million with fireplaces, that now consume about 1.0 quad of wood annually (Wood Energy Institute 1979). About half the wood is used in stoves, the remainder in fireplaces. Of the wood burning households, 42% cut all their own wood and only 36% buy all their wood (at a price of $100 per cord (a stack 4x4x8 feet) or higher in many localities). This corresponds to $3 to $6/MBtu, or nearly the present price for gas and oil. Thus, it is cost-effective to retrofit a wood-heated house. An additional benefit to retrofitting these houses would be apparent in the lowering of pollution levels since less wood would then need to be burned to heat those units. Although we expect residential woodburning to rise to 1.5 to 2.0 quad during the 1980s, based on the present rate of stove sales, a number of factors should force wood consumption back down to a stable consumption rate of approximately 1.0 quad at 2000. Among these factors are improvements in efficiencies of wood stoves, a desire to reduce the labor of cutting and splitting wood for the large segment of the population that cuts its own wood, conservation retrofits and market pressures from wood-derived methanol or other biomass-uses of wood that would drive the cost still higher. Lastly, more intensive uses of forest due to an overall increase in biomass usage will decrease the availability of fallen and waste wood supplies that can currently be cut "free."

RESIDENTIAL WIND SUPPLIES

Over 6.4 million American rural households are located in regions where small wind energy conversion systems (SWECS) and large generators should become competitive as sources of electricity (see Figure 1.45 and Table 1.45). The feasibility of generating mechanical power and electricity from these systems is well-established; technical problems of reducing costs, improving performance, and ensuring reliability remain to be solved. However, recent 5-year operating exper-

iments funded by DOE have shown no technical problems that would be a major barrier to the widespread use of wind generating equipment. There are over 400,000 rural households living where commercially available machines could produce electricity for 4¢/kwh or less, using the 40% tax credit (or 6¢/kwh without the credit), wherever SWECS are privately owned. Assuming utilities were to own the systems, electricity could be bought for 5.4¢/kwh. These costs assume reliable operation for 20 years; some minor improvements are needed to meet this assumption. Figure 1.45 shows that there are 18.4 million people in rural areas where the average annual velocity is at least 10.6 mph. This determination is based on county rural population estimates (Bureau of the Census, 1980), county estimates of annual wind velocity from an enlarged version of a national wind source map (DOE 1978), and on a county survey of average rural household sizes (Bureau of the Census, 1978) determining the average household size to be 2.88 persons. Wind energy conversion systems at a residential scale are assumed to be a significant energy source only in rural regions in the U.S. because large ground areas must be available for their efficient operation. Since these machines must be separated to prevent the wake from one operating system from interfering with the performance of adjacent systems, a wind machine must use 25-50 units of land area for each vertical unit of receiving area.

Wind machines can typically convert the energy available from wind to electricity more efficiently than photovoltaic or solar thermal devices can convert sunlight. An ideal wind machine could operate at about 60% efficiency, but considering mechanical and other real design limitations, actual machines are about 40-50% efficient. In comparison, photovoltaics and solar thermal electric devices provide net efficiencies of 10-20%. Wind energy is also more reliable than direct solar energy in many regions. In most parts of the country, for example, sites can be found where a wind device operates at full rated capacity for the equivalent of 3000 hours annually. Even in the most sunny climate, a direct solar electric system with a peak output of 1 MW_e could not be expected to produce more than about 2500 MWh and in most regions the direct solar electric systems would produce less than 1700-1800 MWh. The greater load factor of the wind resource means that more energy can be derived from a given piece of equipment.

SWECS will operate economically competitively in rural areas where the average wind velocity at 40' is at least 10 mph. Estimates of the future cost of wind-generated electricity from improved versions of three currently available machines are shown in Table 1.45. Separate estimates of installed system cost are based on the use of 95% learning curves and an examination of specific changes in manufacturing procedures. The unit price shown for "manuf. scale up" is based on an

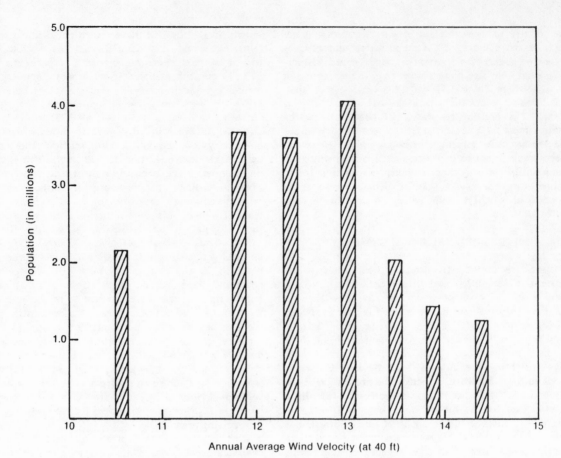

Figure 1.45.   Rural Population Distribution Wind Systems.

Table 1.45.   1985-90 PRICE OF WIND-GENERATED ELECTRICITY, PRIVATE RURAL OWNERSHIP

10 kW Wind Machine Type	Ititial Cost	Average wind speed at 40 feet, mph COST, $/kWh[a] (kWh generated annually)[b]						
		10.6	11.8	12.4	13.0	13.5	13.9	14.4
Millville (Manuf. scale-up)	$9432	.072 (11,500)	.0571 (14,500)	.0499 (16,600)	.0436 (19,000)	.0394 (21,000)	.0366 (22,600)	.0348 (23,800)
Millville (Manuf. scale-up & innovative installation)	$7686	.0587	.0465	.0407	.0355	.0321	.0299	.0284
Millville (95% Learning curve)	$7568	.0511	.0391	.0343	.0299	.0282	.0272	.0262
Astral Wilkon (95% Learning Curve)	$8619	.0455 (15,500)	.0348 (18,500)	.0302 (21,200)	.0262 (24,500)	.0248 (26,250)	.0244 (29,450)	.0238 (30,600)
Wind Power Systems (95% Learning Curve)	$10,852	.0469 (18,720)	.0389 (21,960)	.0354 (24,120)	.0320 (26,550)	.0303 (28,575)	.0293 (30,195)	.0282 (31,680)
Average cost	$8831							
average electricity cost		.0546	.0433	.0381	.0334	.0310	.0295	.0283
median electricity cost		.0583	.0460	.0401	.0349	.0321	.0305	.0293

[a]The cost of electricity produced is calculated assuming an FCR=.0878 which includes operations and maintenance costs of 1% annually. The fixed charge rate is based on a 20-year useful life and 3% discount rate.

[b]The annual output of each machine (as shown in parentheses) was calculated at each wind velocity by

estimate of specific manufacturing process changes which could be implemented at an annual production rate of 10,000 units, and a continuation of present installation practices. "Manuf. scale-up and innovative installation" assumes that installation costs are reduced by redesign of towers and pads as the manufacturing scale-up proceeds, "95% Learning Curve" prices apply a 95% learning curve to present installed costs to estimate the cost of systems when cumulative production totals 100,000 units. The cost per kilowatthour is calculated for each of seven different average wind velocities, assuming a 20-year useful life of each machine. Also, the average and median prices of electricity generated at each velocity are shown. Table 1.46 shows the cost of purchasing electricity from SWECS in identical wind conditions, if utilities were to own them (based on a FCR of 0.13). Based on the assumed reductions in future costs, all household-owned SWECS shown would be able to produce electricity below the marginal cost of new central station generating plants. In regions where costs are only about half the marginal costs, it is reasonable to assume that 60-80% penetration of that market will occur by 2000, while in areas where wind-powered generation is closer to the marginal cost, a likely penetration would occur from about 20-40%. These estimates result in a potential displacement of 0.8 to 1.1 quad of primary fuel by 2000.

In order to produce such significant amounts of energy, wind generators will have to be placed at a large number of sites. These sites must meet several demands: wind speeds and constancy adequate to be reliable; be relatively close to transmission lines (or production of enough energy to make a new transmission facility or long-distance tie-in affordable); placement in a location which would not create a nuisance to abutting residences for reasons of safety, noise, aesthetics, TV interference, etc. And, as mentioned, there must be enough room around a wind generator to enable efficient operation.

The primary engineering challenge of wind machines is the need to construct mechanical systems with high reliabilities and moderate costs. Blades would be subjected to 450 million cycles during a 30-year lifetime (28.5 rpm) presenting a challenging problem in design and material selection. Since they are typically sited in exposed, isolated locations, machines can also be subjected to the stresses of severe weather or environmental conditions such as icing or salt spray. None of these problems appears to present an insurmountable barrier, however.

Another potential difficulty of large rotating equipment is safety. A broken blade could be thrown 700 feet or more. While the probability of such an occurrence is very low for a well-designed machine using contemporary materials, the risk could result in a further limiting of possible sites. Icing might also cause safety hazards in some areas; it may be necessary to stop machines when ice has formed on the blades to prevent ice from being thrown.

RESIDENTIAL PHOTOVOLTAIC SYSTEMS

Photovoltaic (PV) applications in the residential sector can furnish a significant contribution to U.S. energy needs if technical questions relating to reliability, performance, and maintenance are solved, and costs are reduced. The Department of Energy has established goals for installed system costs and module prices. By 1986 enough research, development and mass production of photovoltaics should be completed to enable a module selling price of 70¢/Watt-peak (W_p) and an installed system price of $1.60 to $2.20/$W_p$. These prices and others in this photovoltaic discussion are expressed in 1980 dollars for consistency with DOE price goals. From 1990 to 2000, prices should be much lower, such that modules would cost 15¢/ to 50¢/W_p and an installed system would be $1.10 to $1.30/$W_p$ (DOE, 1980). Cost reduction and technical progress during the past 5 years provides considerable confidence that these goals can be reached.

Table 1.46. 1985-90 PRICE OF WIND-GENERATED ELECTRICITY, UTILITY OWNERSHIP

10kW Wind Maching Type		Average wind speed at 40 feet, mph Electricity cost ($/kWh)						
		10.6	11.8	12.4	13.0	13.5	13.9	14.4
Millville	(Manuf. scale-up)	.107	.0845	.0739	.0645	.0583	.0542	.0515
Millville	(Manuf-scale up and installation)	.0869	.0688	.0602	.0525	.0475	.0443	.0420
Millville	(95% Learning curve)	.0756	.0579	.0508	.0443	.0418	.0403	.0388
Astral Wilkon (95% Learning curve)		.0658	.0515	.0447	.0388	.0367	.0361	.0352
Wind Power Systems (95% Learning curve)		.0694	.0576	.0524	.0474	.0448	.0444	.0418
	average electricity cost	.081	.0641	.0564	.0494	.0459	.0437	.0419
	median electricity cost	.085	.0681	.0593	.0517	.0475	.0451	.0434

One breakdown of the installed system costs has been estimated (Cox, 1980) as follows:

Array purchase	$0.70-$1.08/W_p
Installation	$0.32-$0.64
Installed power conditioning	$0.23-$0.36
Design	$0.05-$0.05
Total	$1.30-$2.13/W_p

The above range assumes 100,000 residential installations/year.

The ability of photovoltaics to displace the use of other energy sources was calculated for the system cost assumptions of Table 1.47 for each of the ten FEA regions, based on average insolation for that region. Average insolation on photovoltaic arrays was estimated by detemining the typical latitude for each region and finding the appropriate midpoint of the insolation range, as determined from a national map of mean daily horizontal insolation. Horizontal radiation was converted to the proper latitude slant for every region (ASHRAE 1978). Photovoltaic output is based on annual simulation runs made for Madison, Wis.; Fresno, Calif.; and Washington, D.C. Radiation striking the slanted panels was then corrected in this model by a factor of 0.76 to account for power conditioner efficiency, dirt on the collecting face, and operation under elevated temperatures during the summer months (Table 1.48). Penetration estimates were made using the METREK (MITRE 1977) penetration function:

$$Y = \frac{1}{1 + 1/\sqrt{F} \tanh \left[(T_n) \, e^{(1 - 1/F)T_n} \right]}$$

Table 1.47. INSTALLED SYSTEM COSTS OF PHOTO-VOLTAIC SYSTEM PER PEAK WATT (1980 dollars)

1986	1.60 - 2.20
1990	1.30 - 1.60[a]
1995	1.20 - 1.45[a]
2000	1.10 - 1.30

Source: U.S. Department of Energy, "Solar Energy Program Summary Document FY1981," p. III-29
[a]The 1990 and 1995 estimates have represented interpolation between the 1986 DOE goal and the 1990-2000 goal (which is shown here for year 2000).

$$\text{where } T_n = \frac{T - T_o}{t^2}^2$$

Y = solar share

F = figure of merit = $\dfrac{\text{Levelized conventional cost}}{\text{Levelized solar cost}}$

T = time

T_o= year in which technology first becomes generally available (1986)

t = time scale factor representing number of years required for the technology to mature (10 years)

The photovoltaic cost was compared to $0.061/kwh (1978), which is the levelized conventional cost with T&D cost excluded. Note that use of t = 10 implicitly assumes aggressive policies promoting rapid market penetration of PV. The penetration for each 5-year period was assumed to be the average of the entire

Table 1.48. REGIONAL RESIDENTIAL DISPLACEMENT OF CONVENTIONAL FUEL USE BY PHOTOVOLTAICS BY 2000

FEA Region	Annual Output- 1 kW array (kWh)	Insolation kWh/m²yr	Year 2000 Energy Displaced (trillions of Btu) 30 Year Life & Lower System Prices	Year 2000 Energy Displaced (trillions of Btu) 20 Year Life & Higher System Prices	Mean Daily Horizontal Radiation	Average Latitude
1	1260	1655	16.4	11.5	325	42
2	1270	1675	15.0	10.6	330	41
3	1340	1760	34.9	25.0	350	40
4	1450	1910	112.7	80.4	400	33
5	1260	1655	47.5	33.4	325	42
6	1530	2015	70.2	51.0	425	32
7	1440	1890	16.4	11.9	375	40
8	1550	2040	22.5	16.8	400	42
9	1760	2310	113.2	88.1	475	35
10	1350	1780	15.1	10.8	340	45
Total Energy Displaced			464 = .46Q	340 = .34Q		

period and it was assumed that each single family home with photovoltaics had a 5 kWp system, each mobile home with PV had 4 kWp, and each multifamily unit with PV had a 2 kWp system. The construction rate for each type of housing in each region was based on the ORNL Residential Model. By 2000, the penetration of PVs is expected to be substantial, mostly in new homes where the panels can be used as a part of the building shell. By 2000, these new residential buildings by are projected to displace 0.34 quad to 0.46 quad of energy with photovoltaics as shown in Table 1.48.

Photovoltaic penetration into the market for the most favorable case as calculated assumes residential ownership with an annual fixed charge rate (FCR) in 1978 dollars of 0.0613, and an additional 1% for operation and maintenance expenses. This corresponds to 30-year system life and the lower prices shown in Table 1.47 This fixed charge rate and the others in this section follow the methodology developed by the Office of Technology Assessment (OTA, 1978, Vol. II) and consider the impact of insurance, property taxes, income taxes, etc.

The lower penetration assumes the higher prices shown in Table 1.47 and a 20-year system lifetime corresponding to an annual FCR of 0.0878 (including 1% O&M).

The same regionalized penetration calculation shown in Table 1.48 was repeated for utility ownership of the photovoltaic systems. It was assumed that the utilities received a 10% investment tax credit for photovoltaics as they do for conventional equipment. This results in fixed charge rates of 0.116 for 20-year lifetime and 0.106 for 30 year lifetime when 1% annual O&M expense is assumed. These fixed charge rates were combined with the high and low price assumptions as before. The potential residential photovoltaic displacement with utility ownership is 0.21-0.37Q annually by 2000.

To explore the impact of incentives, a final case was run assuming that utilities received a 40% investment tax credit for photovoltaic investments. For this case, the fixed charge rates were 0.0759 and 0.0736 and the potential displacement is 0.43 to 0.58 quads.

PURPA Title II requires utilities to buy back, at a fair and reasonable rate, excess electricity produced by their customers. An efficiently designed house with a modestly sized photovoltaic array on the roof could be a net producer of electricity in most parts of the country. Table 1.49 shows that a highly insulated 1176 sq ft house with modest direct gain passive heating could produce as much excess electricity for sale to the utility as it must purchase, even in Madison, Wisconsin. In Fresno, Calif. and Washington, D.C. it can be a substantial net producer. One should note that in every case, the PV array assumed was only 380 square feet—just over half of the roof area readily available even if the roof is constructed with a steep pitch on the south face as illustrated in Figure 1.46.

The prototypical house modeled is well insulated and uses highly efficient appliances as shown in Table 1.50, but both of these conditions are expected to be commonplace in the 1990's.

Clearly the house could be made to be a substantial net producer of electricity in virtually any part of the country if all of the south-facing roof area was used for a PV array. The houses modeled were not fully optimized, and the Fresno house has a much larger cooling requirement than the Washington, D.C. house. In an optimized case, we would expect that electricity usage in the baseline Fresno house would be lower than for the Washington house.

COST/PERFORMANCE ASSUMPTIONS USED IN ESTIMATING ENERGY DISPLACEMENT POTENTIAL OF SOLAR HEATING AND HOT WATER

The preceding discussion of the different renewable technologies reveals wide ranges in the cost of actual active and passive installations and in the projections of costs for future installations. These ranges reflect variations in the solar resource in different parts of the country and in the cost of different types of equipment. It does not seem likely that this variation will disappear. However, the supply curve methodology used in Chapter III to estimate potential penetration of the various technologies requires fixed numbers for the cost/performance of a technology (e.g. in $/MBtu/year) rather than ranges.

The remainder of this section shows how the cost/performance ranges given previously have been broken into specific cost categories for use in the analysis of potential energy displacement.

Solar Hot Water Assumptions

The projected 1990 cost of solar hot water shown in Table R-19b can be expressed as $46-$176/MBtu/year if the 40% reduction in installed system costs is achieved. The cost of retrofits was about 23% greater than for new installations for the Florida study shown in Table 1.42. This was rounded to 20% and it was assumed that the low end of the range above is for new installations, so $55-$176/MBtu/yr was used as the range of costs for retrofit hot water installations. We arbitrarily split this into three ranges ($55-$95, $95-$135 and $135-175) and took the mid-point of each range. Thus we assumed that one-third of potential retrofit hot water installations cost $75/MBtu/year, one-third cost $115/MBtu/year and one-third cost $155/MBtu/year.

The same assumption that retrofit installations cost 20% more than new installations suggests that $46-$147/MBtu/year be used as the range for new installations. Splitting this into three ranges and taking the mid-point of each range as above yields the assumption that one-third of the potential new hot water installations cost $63/MBtu/year, one third cost $96/MBtu/

Table 1.49   ANNUAL ELECTRICITY USE AND GENERATION OF PROTOTYPE
               HOUSE WITH 4.5 KILOWATT (380 sq. ft.) PHOTOVOLTAIC ARRAY

	Fresno, CA	Washington, D.C.	Madison, WI
Baseline house without PV	5230	5000	5800
PV output used by house	2830	2160	2210
PV output sold to utility	4990	3650	3590
Electricity purchased	2400	2840	3590
Net sales of PV output to utility	2590	810	0

Figure 1.46.   Typical Photovoltaic Array (about 380 ft^2) for Residential Electricity Production

year and one third cost $130/MBtu/year. These values must also be multiplied by the appropriate water heater efficiency before being used on supply curves.

For electric backup, one should divide by 293 to obtain $/kwh/year for the values given, then multiply by the appropriate heater efficiency or COP.

Table 1.50.   BASIC ENERGY-RELATED FEATURES OF
              PROTOTYPICAL PHOTOVOLTAIC HOUSE

Size	1176 sq. ft.
Attic Insulation	R-38
Wall Insulation	R-25
Glazing	Double on south
	Triple on other sides
South-facing Windows	100-175 sq. ft.
Night Insulation	R-9 on south
4 inch Slab Floor	
Infiltration	0.3 ach effective
Internal loads	41,000 Btu/day
(uses 1990 appliances)	
Minimum Temperatures	65°F days
	60°F nights
Heat Pump COP	2.6 at 40 F heating
	2.9 at 90 F cooling
	1.8 water heating
Hot Water Use	20 gallons/day

Retrofit Solar Heating Costs

In Table 1.38 the 1990 cost of active heating is estimated to be 6-14 $/MBtu delivered to the house. We do not have separate estimates for the difference between the cost of new and retrofit active heating and hot water installations so we assume that retrofit installations typically cost about 20% more than new installations on the same basis as for hot water. The range on Table 1.38 included new installations and retrofits, so we drop the bottom of the range (6 - 6 x 1.2) and use $7.2-14/MBtu delivered, or $82-159/MBtu/ year when divided by the FCR of 0.0878. Somewhat arbitrarily splitting this into three separate costs we get: $95, $120, $146/MBtu/year.

Passive retrofit costs are given in Table 1.34 as:

	$/MBtu/yr.
Vertical heater	10-140
Trombe wall	14-140
South windows	28-190
Solar Greenhouses	27-360

One should note that these costs are also expressed in terms of heat displaced if a 65 percent efficient furnace is used. Greenhouses and south window retrofits are generally done at least partially for the added space and light. We will arbitrarily assume that half of the value of greenhouses and south windows is not energy related and convert the ranges to represent heat delivered to be consistent with the active heating and hot water costs given above to obtain:

	$MBtu/yr delivered
Vertical heater	15-215
Trombe wall	22-215
South windows	22-145
Greenhouses	21-277

The lowest costs clearly apply to do-it-yourself in-stallations (DIY). We assume systems which cost $50/MBtu/year are DIY and that these installations represent 10% of the market. If we assume equal numbers of each system type and a uniform distribu-tion, the average DIY cost is $35/MBtu/year. Again assuming equal numbers of each system type and uni-form distribution within each cost range for systems that cost $50/MBtu/year or more, three different "average" passive systems would cost $78 (30%), $123 (30%), and $195 (30%) per MBtu delivered annually.

Considering a similar level of owner-installed active systems and combining both categories, one obtains the values shown in Table 1.51.

Table 1.51.   SOLAR HEATING RETROFIT
              ASSUMPTIONS

	Solar Heating Percent of Installations	Cost of Delivered Energy ($/MBtu/yr)
DIY	10	35
Solar I	30	86
Solar II	30	122
Solar III	30	170

These assumptions will be multiplied by appropriate furnace efficiencies before use in the supply curves.

Retrofit Potential

Determining the potential for active or passive solar heating retrofits is difficult and the range of estimates is wide. Preliminary surveys in Iowa, Oregon, and California suggest that 50 to 60% of existing single-family homes have adequate solar access for at least some passive retrofits. (Hull et al. 1979; SERI RCS 1980; Wornum Report 1980). On the other hand, the Wornum Report has estimated that only about 25% of the units in large California apartments would be suit-able for retrofits. An audit conducted for the Resi-dential Conservation Service program in Portland, Oregon (SERI RCS 1980) found that three-fourths of the homes inspected had adequate access for solar water heaters. Some of the homes, however, had areas of south-facing roof that would not have been suffi-cient to support the greater collector area required for a space-heating system. Based on these sample surveys, it is assumed that up to 60 percent of the existing housing stock has access for a solar heating retrofit. This number, 60% of 70 million homes (present stock excluding units which will be demolished within 10 years) or 42 million units represents the maximum technical potential for space heating retro-fit.

The degree to which the potential will be implemented is another question. The data is sparse. Two areas already have solar heating in 4-6% of the houses (San Luis Valley Solar Energy Association and the Taos Solar Energy Association report these numbers for their areas in Colorado and New Mexico.) (SLV, 1980; TSEA, 1980). Over half of the installations in these two areas are retrofits, and the San Luis Valley group has a goal of 20% solar by 1985 (mostly from retrofits). Because of aesthetic considerations it may be optimistic to assume that half the homes with solar access will ever be retrofitted. On the other hand, if a sunny but low income area like the San Luis Valley can plausibly project 20% implementation, the nation as a whole should do at least as well.

To determine what sort of energy savings this degree of retrofitting represents, it is necessary to estimate the average energy displacement of the solar retrofits. Some of the passive retrofits are cost competitive with certain conservation options and will likely be substituted for conservation measures. In other cases, solar retrofits will no doubt be applied to buildings where some conservation measures are difficult to apply. Hence we have assumed that the average solar retrofit is applied to a house which uses 35 MBtu/year (Phase 1 and Phase 2 levels for a "partially insulated" house). An active system which provided 60% of the heating would displace 21 MBtu/yr. Sample calculations (RCS Program Announcement, 1980) indicate that passive retrofits will provide 10 to 15 MBtu/yr, and two other studies are consistent with this result (Hull, et al., 1979; Yanda, 1980). The energy displaced will depend on whether fuel or electricity is used as backup and on the utilization efficiency. We assume 20 MBtu displaced per retrofit for the mix of active and passive systems installed.

Section Four
Residential Buildings Policy

A decentralized approach to achieving the energy-efficiency and solar goals is essential to achieve national energy goals and preserving local flexibility, innovation, and accountability. This approach has a number of advantages over an extremely centralized, federally managed effort:

o many of the barriers to effective conservation and solar applications originate within the jurisdiction of state and local governments, and conversely . . .,

o many of the existing levers on the building sector are traditionally within the jurisdiction of state and local governments (e.g. building codes),

o state and local governmments have the unique ability to adapt energy programs to local conditions and to make optimum use of institutional and human resources,

o state Public Utility Commissions can insure full utilization of and coordination with the utility industry,

o large, centralized institutions tend toward large, centralized, and uniform solutions, often perceiving the problem as the same everywhere; the scale of the effort needed is simply too large for a centralized federal effort to be effective.

Numerous communities throughout the nation and in many states have already taken or are now actively considering a wide range of actions to promote energy efficiency and renewable energy use. Some examples of the innovative and varied actions at these levels of government are:

o Santa Clara, California, has had a Municipal Solar Utility in operation since 1976.

o The state of Connecticut requires life-cycle costing for all state financed public housing.

o San Diego, California requires solar hot water systems for all new homes using electric heat.

o Portland, Oregon is implementing a comprehensive energy efficiency program, based on a plan initially developed by a broad based panel of citizens which held extensive public hearings.

o The Center for Neighborhood Development in Chicago has established a private neighborhood development corporation for energy efficiency and solar applications, in conjunction with Northwestern University.

o Six New England states have developed a regional energy plan intended to shift the region to alternative (non-oil) fuels.

o California has adopted and the Washington legislature is considering legislation allowing utilities an increased rate of return on electricity generated from renewable resources.

o Fitchburg, Massachusetts organized local community groups and volunteers to retrofit the community's housing.

o Champaign, Illinois has enacted an energy conservation code for new housing.

o Port Arthur, Texas has enacted a local ordinance requiring proper building orientation for solar use.

105

These are but a few examples of on-going activities at the state and community level. While the array of state and local approaches to the energy problem is broad, it also presents a picture of somewhat scattered and unplanned efforts, many of which have been unable to take full advantage of experiences gained earlier, in other communities. This is an area in which the federal government can and should play a role. The federal government can:

o provide for the exchange of information among state and local programs, and make state-of-the-art technical information and assistance more accesible to states, local governments and community groups;

o assist in the evaluation of programs and facilitate the dissemination of information on both successes and failures;

o provide funding for state and local programs, to both initiate projects which require only one-time financial assistance and provide more sustained assistance for programs which need to develop ongoing technical and policy planning capabilities;

o eliminate or reduce barriers to local action created by the federal government. For example, the pre-emption of state or local actions (tougher appliance efficiency), locally mandated should not occur expect where clearly justified. Unfulfilled (or partially, or slowly fulfilled) federal promises, such as the delay in enacting the Solar and Conservation Bank or EMPA, may lead state and local governments to stall actions while waiting for promised federal support. In the absence of changes to the federal tax code, state and local tax incentives for conservation and renewable sources are partly "taxed away" in the form of reduced tax deductions at the federal level.

In addition to the above areas, there seems to be a clear need for federal support for programs to deal with the lower-income groups. The states have too often been unable or unwilling to commit adequate resources to meet the energy needs of lower income households.

The federal role should be careful to recognize both the potentials and limitations of state and local governments, particularly:

o State and local governments often lack the resources needed for planning and coordination, and only a sustained program of federal funding will allow development of necessary staffing. Conversely short-term or indefinite commitments will most probably result in exten-

sive contracting with consultant groups for work that has more positive impact on local printers than on local institutional capabilities, public or private.

o The federal efforts must be designed to allow sufficient flexibility in meeting unique local needs and conditions, developing local resources, and implementing innovative approaches.

o The federal program should focus on creating and integrating state and local plans and activities which move towards well-defined, achievable energy efficiency and solar policy goals--instead of concentrating on programmatic requirements and the issuance of extensive, detailus on creating and integrating state and local plans and activities which move towards well-defined, achievable energy efficiency and solar policy goals--instead of concentrating on programmatic requirements and the issuance of extensive, detailed, and inflexible federal regulations.

The prior Congress considered a proposed Energy Management Partnership Act, which sought to create a framework for an effectively decentralized energy program. The general ideas embodied in EMPA are laudable. However, it is often not enough to pass a law, or even to appropriate funds. The regulations which are issued, the administration of the program, and the development of a staff sensitive to local conditions and needs is the most critical element of a decentralized approach. Whether the federal government can turn the ideals of a program like EMPA into a workable program remains to be seen.

While one of the advantages of state and local governments is their ability to adapt to meet unique local needs, the diversity of state and local programs that results made it impractical to conduct an analysis of state and local programs. The policy analysis in this study was directed towards the federal government. However, the general themes and approaches in the following sections can provide a framework for nearly any level of government, with some exceptions.

EXISTING RESIDENTIAL PROGRAMS

Over the next twenty years, by far the largest potential for saving energy in buildings, as outlined in the preceding chapters, lies the tens of millions of structures already in place. Designed to use relatively cheap energy, most of these existing homes and commercial buildings will still be in use well into the next century, when energy costs are likely to be many times higher.

A considerable amount of private investment in energy efficiency and renewable sources is already underway, at least in owner-occupied buildings. Legislation enacted over the past few years has now put into place a number of promising programs at all levels of government, but in many cases further changes are needed to make these programs more effective, broader in scope, and more coherently tied together. In a few cases, entirely new approaches appear to have merit; these should be implemented first on a demonstration basis to determine what works best, in varying local circumstances. Finally, the consequences of changing energy costs and availability need to be taken into account more systematically by both government programs and private institutions outside of the energy field.

The proposals outlined below include some new approaches, some changes in existing programs, and suggestions for introducing energy concerns into non-energy institutions. They fall into three major categories: information, financing and incentives, and regulations and standards. The order of the sections that follow indicates our general preference for first seeking ways to help the market system work better (improved information, incentives to offset past subsidies for conventional energy supplies, and widespread access to loan capital) before turning to government's regulatory powers in cases where the market approach is insufficient. Some of these proposals are applicable to both homes and commercial buildings.

Disseminate Information and Stimulate Applied Research

Recent analysis and experimental work have demonstrated that it should be possible to reduce the heating energy requirements of existing residential structures by as much as a factor of four at costs which would be acceptable to typical owners of residential buildings. A major barrier to such investments is poor information. First, the current state of knowledge about the technical performance and cost-effectiveness of a large number of products, given the variety of buildings now standing is extremely primitive. Second, there are too few channels for reliably communicating to building owners the information which exists about techniques for retrofitting buildings. The measures described here have the objective of filling both gaps through applied research and through information programs designed to supplement the Residential Conservation Service (RCS) program.

Actions recommended under this program fall into three categories:

o   an applied research program designed to measure the effectiveness and improve the reliability of different auditing procedures and retrofit techniques;

o   programs designed to help the market operate more effectively by providing individual and institutional decision makers with better in-

formation about building energy performance and energy costs (specific actions are including energy in considerations of mortgage qualification, feedback billing, and a building rating system); and

o   a program of direct public outreach designed to increase the awareness of and response to RCS, tax credits, and other programs.

Applied Research

The applied research program would be designed with the following objectives:

o   improving our understanding of energy flows in existing residential buildings (particularly the interaction of solar equipment with building designs and occupant behavior);

o   developing optimum techiques for improving the energy efficiency of buildings by reducing conductive, convective, direct air-flow heat losses (or unwanted gains), improving the efficiency of direct or indirect solar gains, and documenting actual costs and savings associated with retrofit options;

o   developing better analytical techniques for prioritizing retrofit investments, both as a basis for conservation and solar "supply curves" and to support a Building Performance Rating System (see next section);

o   developing diagnostic instruments that can allow simple and inexpensive identification of leaks, along with simplified computer programs capable of evaluating field measurement data and accurately estimating costs and savings;

o   developing information about consumer responses to different retrofit packages and marketing approaches;

o   refining the available techniques for optimal selection and sizing of passive solar equipment and active solar devices, in conjunction with conservation measures in existing buildings;

o   training energy auditors in the use of more sophisticated on-site measurement, analysis, and partial retrofit techniques;

o   developing new products for improving the thermal integrity of buildings (better foam insulation, caulking, window treatments, etc.); and

o   testing indoor air quality in well-sealed houses.

It will be necessary to develop information about a variety of building designs in a number of different climates. Moreover, it will be important to continue the program for a number of years so that the accumulated experience can continuously improve the cost-effectiveness of the retrofit programs.

It is recommended that an initial program of applied research examine 10 building designs in 10 climatic regions with 50 sites for each building type in each region. This would result in 5,000 thoroughly documented residential retrofits per year. Several retrofit options should be examined for each building type and careful attention given to the proper selection, installation, and use of solar equipment particularly in cases where it may prove difficult to achieve significant savings with improvements to the building shell. These residences should also provide local examples of the latest, optimum techniques for its locale.

It is extremely important that adequate attention be paid to the initial research design and long-term management of the program. Each research site should be carefully monitored, the results evaluated systematically, and successes communicated in a timely and routine way to utilities, state agencies, and commercial retrofit firms. In order to accomplish this, management of the program should be delegated to roughly 10 centers around the nation, preferably the same centers used to monitor the proposed applied research program in new buildings (see below). As in the case of new buildings, the government should coordinate all other applied building research retrofit programs with these regional research centers.

These applied research centers should be staffed by a core group of personnel supported by the federal government, professional organizations, builders, utilities, state and local governments, and other interested parties. Only through a broad base of support will these centers succeed in achieving their goals. Approximately 15 personnel should receive direct federal support for a lead center, and at least 3 for each regional center. Additional staff should be provided by financial contributions from other supporting institutions. The demonstration, monitoring, information, analytical and other activities at each center should be proportionally supported by the federal government and other participants.

Other features of the program should include:

o Close cooperation and preferably financial commitment by utilities, private energy service companies, and others involved in audits and energy efficiency and renewable applications in existing buildings. At a minimum, cost sharing and joint sponsorship of research and data-gathering should be encouraged.

o Energy use should be measured for a year before and at least a year after retrofits and corrected for weather; control houses in the same climate zone should be measured similarly.

o All construction work and costs should be carefully documented using a consistent, nationwide format (with supplemental data gathered locally, where appropriate.

o The performance of all buildings in the program should be monitored through utility bills and physical measurements. At a minimum, arrangements should be made to receive utility bills from each building for five years. A low cost monitoring system (less than $1500 per unit) should be developed and installed in as many buildings (retrofitted under this program) as possible, at least 5% (250 sites/year) should be carefully monitored in this way.

o Indoor air quality research should be integrated with the demonstration program so that a statistically valid sample of indoor air quality measurements can be developed and correlated to specific building shell and heating system retrofits.

o All major categories of residential buildings should be examined, including multi-family buildings, duplexes, single-family detached, and mobile homes.

Information About Energy Use, Efficiency, and Energy Costs

While building owners are clearly responding to increasing energy prices, their responses have not been as rapid as they would be if better information were available about the energy costs associated with buildings. A national program should be developed to provide building owners and occupants, real-estate agents, and staffs of financial institutions with better information about the energy performance and resultant life-cycle costs of existing buildings. Three specific steps are recommended:

o develop one or more "building rating" systems;

o require that all banks issuing loans insured by FDIC or FSLIC and all institutions handling VA and FHA loans use a "PITIE" (principal, interest, taxes, insurance and energy) criterion to establish the eligibility of a loan applicant; and encourage such banks to offer other energy related services; and,

o encourage "feedback billing" and metering demonstrations.

Develop a Building Energy Rating System. If credible estimates of the energy consumption of a residential building are not available at the time of sale, prospective purchasers are likely to base their decisions more on the building's initial cost than its future operating costs for energy. With a simple and reliable performance indicator, however, customers can make more informed decisions, builders and real estate agencies can use energy efficiency as a selling point, and mortgage lenders can consider energy bills as a part of the total costs of owning the building when they estimate a customer's ability to qualify for a loan. Such a rating system would also clearly encourage the home-seller to undertake improvements in the house before it is resold.

The Department of Energy should, therefore, move immediately to develop a rating system that would provide:

o a simple performance index of the building, similar to the MPG rating used for automobiles;

o an estimate of the total energy bills of the building for each month of a typical year (making standard assumptions about occupant behavior); and

o a brief list and explanation of the good and bad energy features of the building.

These rating systems should be developed in close cooperation with the Federal Home Loan Bank Board (see below) and carefully coordinated with the development of Building Energy Performance Standards for new buildings and the Residential Conservation Service for existing homes. There are several possible models to build upon, including the performance index used by the PG&E Company in California to calculate each incentive for builders on the basis of estimated performance.

The program should include the following elements:

o Tests of the impact of alternative performance rating systems working with about 5-10 state or local agencies, utilities, or building industry groups. Initially, the ratings could be based on actual utility bills, with the seller required to provide either 12 months of past billing information or show that the building had been audited by a qualified building auditor.

o Improved calculation techniques for determining the performance of existing buildings should also be developed and empirically validated, in connection with the applied research program discussed previously. These could supplement or replace billing records, as soon as they become available.

o The accuracy of the building performance rating systems should be carefully evaluated in

several communities during a period of 2-3 years. If an acceptable system is developed, states or local jurisdictions should be encouraged to require a building energy rating and disclosure requirement as a condition of sale.

Home Mortgages. The Department of Energy should significantly expand its direct and cooperative efforts to fully incorporate energy concerns into lending practices. This effort should include:

o the development and testing of various systems for providing lending institutions with a simple but reasonably accurate formula to determine design energy use for new and existing homes;

o transfer of information gained from the applied research program to lending institutions, real estate agencies, and others in a format cooperatively designed to meet their information needs.

By definition, if an energy-efficiency investment is cost-effective, a homeowner or buyer making such an investment will have more net disposable income than another mortgage applicant with an identical house who does not. However, since energy costs typically are not calculated into conventional loan formulas directly, such loan decisions are actually biased against the person making the financially wise investment to the detriment of both the lender and the borrower.

In passing the National Energy Act, Congress found that the lack of consideration of energy costs in mortgage loans was a serious problem. In some areas, energy costs are reaching major proportions, up to 20% of a family's income. A recent survey by the FHLBB noted that some lending institutions (12%) had begun to identify high energy costs as a cause for delinquencies or foreclosures.

While many lending institutions have made major strides in the last few years to incorporate energy considerations into lending practices, the ability of these institutions to directly include energy costs into calculating a borrower's ability to repay has been difficult because no simple method exists to determine an unbiased energy cost estimate for homes.

A recent Federal Home Loan Bank Board survey indicated that 59% of respondents consider energy in their first mortgage evaluation. However, many different levels of "consideration" were demonstrated. Some indicated that they held the debt-to-income ratio lower to allow for higher energy costs, while others claimed to thoroughly examine the structure involved.

However, without an unbiased method of determining a buildings design energy use, changing this situation from being biased against energy efficiency to spurring energy efficiency will be constrained. Many lending institutions argue that using existing energy bills would be biased against the prospective loan applicant, since individual life-style accounts for variances of over 100% in energy use in identical houses. The existing methods for determining design energy use are too difficult and time consuming for consistent use with every loan application.

A simple, relatively accurate, and unbiased test would allow lending institutions to directly add energy costs into their calculations for every home sale. The inclusion of monthly energy costs in this formula could have wide-ranging effects on energy efficiency possibly well beyond the housing market. The framework should be adaptable to various regions, housing types and energy costs, and could start with federally assisted housing and FHA, VA and FmHA mortgages. The formula should attempt to include all energy costs associated with living in a house including transportation to work and shopping. While starting with residential buildings, eventually commercial buildings could be included as well.

While the Department of Energy has a limited program to assist financial institutions, it has not received adequate staffing and support. A program should be established in cooperation with the effort to develop a building rating system to assist and encourage financial institutions to incorporate energy concerns into lending procedures, and specifically to develop simple, yet relatively accurate, analytical techniques tailored to lending institutions needs. At least five full-time staff should be allotted to this program.

In addition, lending institutions need to be kept up-to-date on the information about actual performance of energy efficiency techniques and renewable energy systems which would be gained through the applied research program outlined previously. Not only will this aid in their consideration of home mortgages, but will enhance their ability to consider loans in tandem with mortgages to finance energy improvements in dwellings.

It is important that the Department of Energy conduct these programs in full cooperation with Federal institutions already working with the lending community, including the Federal Home Loan Bank Board, the Federal Deposit Insurance Corporation, The Federal Savings and Loan Insurance Corporation, the Federal National Mortgage Association, and the Solar and Conservation Bank. These organizations should encourage member lending institutions to go beyond including energy costs into loan considerations. Lending institutions should encourage loan applicants to undertake energy efficiency improvements which are cost-effective, and offer to include such costs in the mortgage or in a tandem home improvement loan. In addition, they should inform loan applicants of special programs, such as tax credits or loan eligibility through the Solar and Conservation Bank. DOE and these Federal institutions should encourage lenders to offer special loan consideration for the most energy efficient homes.

The recent movement of lending institutions to consider energy cost lends some credence to believing that with such assistance and encouragement the lending market will adequately take these important steps. However, the Federal government should consider taking direct action to require its present financial insurance agencies, FDIC & FSLIC, to require that all federally insured lending institutions fully include energy costs within their loan calculations. Once a method of determining building design energy use is developed and demonstrated, such a requirement would be in the insuring agencies interest in securing the solvency of their member financial institutions.

Consumer Feedback, Billing, and Metering. DOE should substantially increase its present level of funding for individual utilities and state regulatory agencies and others, to test innovative methods of providing consumers with prompt feedback on their levels of energy use or energy expenditures.

The inclusion of well-designed feedback information as part of utility bills can improve customer awareness of usage and costs, and thus encourage energy efficient practices. The programs to date have shown that 10 to 15 percent energy usage reductions are achievable through feedback In addition, increased awareness of the costs of conventional energy will begin to put the higher first-cost investments in conservation and renewable sources on a more equal footing with purchases of conventional fuels and electricity.

A number of utilities are already conveying such information to their customers through what is generically termed "report-card billing." The intent is to redesign the utility bill format so that individual customers can readily compare their current level of energy use (or the amount of their bills) with those from a previous period--or in some cases with those of "similar" customers served by the same utility. The experience in California, in response to an order by the Public Utilities Commission, shows that the costs of such a program are minor if the utility already has some form of automated billing and is given some leeway to convert it over two or three years.

Some utilities, however, have been reluctant to include dollar costs as well as energy-usage comparisons in their "report-card" bills. The experience of the Atlantic City Electric Company in New Jersey, for example, was that such information can highlight the rapid increase in utility rates, thus intensifying customer resentment against the utility in general (and on occasion, against conservation itself). However, other utilities have found ways to handle increased customer complaints, and learned that emphasizing the increases in utility bills over time can be a particularly effective selling-point for their other conservation program.

A related effort, also being introduced by California utilities in response to a PUC order, is the clear display of rate structures on each month's bill, for residential customers. The aim is to increase customer-awareness of how much energy is being purchased at lower "lifeline" rates (which apply to all energy used up to a fixed "lifeline" allowance), and how much is being purchased at the significantly higher intermediate and "tail block" rates (the latter are set roughly equal to replacement costs of energy).

Other forms of consumer feedback may also be effective tools for reducing energy use, especially if they provide consumers with instantaneous information on their rates of energy usage (i.e., information available at the moment the energy-using behavior is occurring). Special meters which display usage and cost data are being tested in a number of areas. The Tennessee Valley Authority, CON ED, and Pacific Gas and Electric (PG&E) are preparing a program that will equip test homes with a meter that gives dollar and cents display of the accumulated electric bill. DOE is supporting a number of additional pilot programs in the residential sector, and could further encourage states to adopt consumer feedback mechanisms for utility customers through report-card billing or other methods, in conjunction with the RCS program. In addition, the current DOE program should extend its support to commercial sector demonstrations. Presently, only one such program is being conducted, again by PG&E. After this demonstration phase, a review of program results should be conducted to determine if "feedback" mechanisms should be a required component of utility programs, under RCS and/or PURPA, for both residential and commercial buildings.

Direct Information Dissemination and Outreach

One of the barriers to residential retrofits is a lack of clear, credible technical information. Many more building owners can be expected to respond to conservation opportunities if they have access to reliable data about the potential savings and how to achieve them. An expanded information program should consist of the following elements:

o A program designed to encourage building owners to undertake efficiency retrofits. This program should attempt to reach the largest possible number of building owners and persuade them of the wisdom of accepting an audit through the RCS program or some other retrofit program, and acting on the advice given. Outreach should be through advertisements, public speeches by government officials, information from utilities and other means.

Special effort should be made to reach the opinion leaders in each local community, who may be the ones most likely to effect large numbers of other consumers by their example. Similarly, information programs should be specifically targeted toward individuals contemplating major improvements to their homes, such as re-siding, reroofing, building an addition to an existing home, or other major

remodeling. A further emphasis of this program should be to support state, local, utility, community group and other outreach programs tailored to local conditions. Finally, the federal role should include behavioral research on consumer decision-making, which can be used to properly structure, target, and reinforce local efforts with a limited direct federal campaign of information and advertising.

o Establishment of an integrated information program which combines solar and energy efficiency. An integrated conservation and solar information network should be built on the existing rather fragmented pattern of agencies and programs. The federal role in such a network, other than providing a share of the funding, should be primarily information wholesaling, that is, improving the quantity, quality, consistency and accessibility of technical information on the cost and performance of available conservation and solar measures in buildings. The Applied Research Program discussed earlier should be one of the fundamental parts of this effort. In addition, the Federal government can help to support information linkage among local groups dealing directly with consumers, builders and lenders. One promising approach is the documentation and dissemination grants envisioned under the Energy Management and Partnership Act, to facilitate sharing of information and experiences among diverse regions, localities, and community groups. Information on the cost and performance of all relevant technologies can be steadily improved as a result of the applied research program. This technical information should be accompanied by up-to-date information on the private firms or local government organizations capable of providing needed services, information on loans and tax incentives, specific data on product reliability, etc.. All forms of communication have a potential role to play, including local information centers, toll-free phone numbers, do-it-yourself training courses, printed and electronic media, community education programs, and so forth.

o Develop information about product lifetime and performance. One of the major barriers to the purchase of residential solar and conservation equipment is the absence of reliable information about the durability and performance of equipment offered for sale. Data on all products offered for sale should be developed in a uniform way, with industry cooperation, and this data made widely available through the networks discussed above.

o Development of community workshops. Community-based workshops on building retrofits should be encouraged and partially financed by the federal government. These workshops should provide information about "do-it-yourself" energy audits, as well as instruction about how to correctly apply insulation, caulking, and other simple conservation measures. Workshops should also be held to provide instruction in installing basic passive and active solar systems, including those constructed by homeowners from readily available materials. Training manuals and instructional materials could be developed on a national basis, with workshops themselves conducted through state and local governments and community groups, with assistance from the regional conservation and solar energy centers described above.

Strengthen the Residential Conservation Service Program

The programs described thus far are designed to reinforce private market mechanisms by generating better information on energy use and efficiency, and by ensuring that the information is adequately communicated. The Residential Conservation Service (RCS) program is designed to encourage utilities to move actively to broaden their investment portfolios with investments in building. The program should be adjusted to encourage smaller, nonregulated enterprises to benefit and to respond more immediately to the products of applied research program just described. In addition, this program should be recognized as possibly the most effective means of providing homeowners with specific information on the possible investments in their residence, encouraging their undertaking cost-effective measures, and facilitating a more productive use of capital by providing easy access to financing.

Five major changes in the currently authorized RCS program are recommended:

o improving the skills of energy auditors;

o allowing or requiring auditors to make simple retrofits during their visit to the residence;

o providing incentives for private auditing firms not associated with utilities, and direct federal credits for homes served by oil and bottled gas companies;

o providing easy access to financing through the RCS process; and

o increasing the DOE staffing effort for RCS at both Washington and Regional office levels.

Improving the Skills of Auditors

The key to a successful nationwide home-energy-audit program is the development of a sufficient number of auditors skilled in determining the most economical way of saving energy in homes, and in persuading the owners of the buildings to take action on the recommendations. This requires both technical skills and a talent for dealing with people. A program capable of meeting the conservation and solar potentials described in this analysis would require approximately 10,000-20,000 auditors working for 15-20 years.

Plainly, recruiting, training, and continuing to refine the skills and knowledge of such a large number of auditors will be a major challenge. The program for training these individuals should be carefully integrated with the applied research program discussed earlier. In that program, it was suggested that 5000 carefully measured and monitored retrofits be undertaken each year over a 5 year period. If auditor training (and re-training) were made an integral part of this program, it would allow the thorough training of 20,000 auditors, each of whom had participated in 30-40 well-documented retrofits (assuming that two auditor-trainees participate at each retrofit site).

A program for improving the knowledge and abilities of RCS auditors should consist of the following:

o An Intensive Training Program. The federal government should subsidize the training of auditors by providing text materials, equipment for training facilities, and scholarship funding for individuals interested in receiving training in these skills. It may be most attractive to undertake these training programs in technical colleges and schools which are located in most parts of the nation. It seems possible to give an auditor rudimentary skills in auditing in 6-8 weeks; a one-year program, including periodic classroom sessions and supervised field experience, should be adequate to train a skilled professional.

o Auditor Certification. While each state RCS plan must provide some form of qualification procedure for RCS auditors, states should be encouraged to more formally certify or license skilled auditors. If necessary, the federal government may need to require that states have such certification procedures to be in compliance with the RCS statute. The federal government could provide assistance by developing a model certification program. Certification should require completion of an approved training program, as well as at least 6 months of supervised apprentice work. Examinations could be used to test the applicant's understanding of energy flows in buildings, practical skills in identifying and correcting defects in buildings, use of diagnostic equipment, familiarity with all efficiency and solar energy techniques which are likely to be employed in that region's climate and building stock, and communication skills needed to persuade homeowners to follow through on recommended actions.

o Subsidizing the Purchase of Diagnostic Equipment for Auditors. Several readily available measurement devices can be used to improve the quality of audits. For example, pressurization systems and hand-held infrared scanners have proven useful in identifying sources of heat leaks in buildings, and flue-gas analyzers can be used to time furnaces and water heaters for improved combustion efficiency. Calculation of an appropriate set of recommendations can be speeded and simplified with the use of hand-held programmable calculators at remote computer terminals. The federal government could encourage the use of this equipment by providing direct grants to utilities and other qualified auditing organizations for the purchase of such equipment.

o Feedback to individual auditors. A final means of improving the quality of residential audits is to provide for regular feedback to individual auditors, on both the accuracy and completeness of their recommendations, and their personal effectiveness in persuading building owners to take action. Under the federal regualtions for state RCS plans, each utility must make some provision for monitoring the results of audits, through follow-up inspections and other means. The results of all such follow-up visits, even if done only on a sample basis, should be routinely made available to the individual auditor, in order to sharpen his or her analytical skills and ability to communicate recommendations.

DOE is already committed to providing assistance to auditor training; $25 million was provided for FY's 1981 and 82. It is important that this program continually advance the capabilities of state auditor training programs, and minimally be maintained at that budget level through the next 3 to 5 years.

Partial Retrofits in RCS Audits

The RCS regulations should be changed to encourage or require that all residential and small-commercial audits include at least some immediate partial retrofits with measures that are quick and economical.* Utilities should also provide those customers who don't receive utility audits with vouchers to obtain comparable (free or comparably subsidized) materials to do their own low-cost partial retrofits, or obtain such services from private home-improvement contractors.

A number of retrofit items are so inexpensive and simple to install that once a trained utility auditor is already visiting the residence or commercial building, it makes no sense to leave these measures for possible later follow-up by the customer--which requires additional effort, as well as increased expense. In general, those measures which can be installed in about an hour or less, require little training or specialized equipment, and create few concerns for safety or other side-effects, are good candidates for immediate on-site retrofits. Examples of such measures include water heater insulation in some commercial buildings as well as residences, low-flow showerheads and faucet fittings, adjustment of refrigeration doors and seals, and in many cases demonstration of caulking and weather-stripping products and techniques. A number of these measures have paybacks of a year or less, and in many service areas it would probably pay for the utility to simply provide the materials free (or at a very nominal, subsidized charge) as part of the audit package--based on the utility system savings (replacement energy costs minus average customer costs). Incorporating immediate, partial retrofits as part of the RCS audit would serve as an incentive for more customers to sign up for audits, would be equally applicable to rentals as well as owner-occupied buildings (unlike many of the more expensive retrofit options), and would provide at least some guaranteed savings as a result of the very first visit of an RCS auditor.

The choice of measures to be included as "immediate partial retrofits" could be left largely to state regulatory agencies, within some rather broad federal guidelines concerning auditor time, materials cost, and cost-effectiveness.

To help alleviate concerns over anti-competitive practices, and to reach customers who do not choose to participate in the RCS program (including those who have previously been audited and/or have retrofitted on their own), utilities should also be required to offer comparable subsidies for low-cost partial retrofit measures in the form of vouchers. Customers could use such vouchers to purchase the materials from retail outlets and install them on a do-it-yourself basis or through contractors. Likewise, utility vouchers should be applied to energy audits by private firms, including low-cost partial retrofits. Not only will this help alleviate anti-competitive concerns over partial retrofits by utility auditor, will also encourage healthy competition in the audit field.

---

*In many cases, the inclusion of partial retrofits will be of advantage to the utility, the homeowner and the public.

Specific federal actions include:

o   Changing the RCS regulations to require that utilities offer their customers an appropriate package of immediate partial retrofits, either at no cost to the customer or at a reduced cost. Also, requiring utilities to develop an optimal system of vouchers for those customers who prefer to obtain the materials or audit elsewhere.

o   Monitoring the utilities' choices of retrofit measures, customer response to each element of the program, and the resultant costs and energy savings; and revising the regulations as needed.

Provide Encouragement for Private Auditing Firms

While utilities represent a major resource for providing audits for many customers within their service territories, there is also a clear need to encourage the growth of qualified private auditing firms. Such firms can provide needed competition for utility audit programs thereby improving the quality of RCS audits. Private firms may provide the only acceptable source of audits for homes not heated with electricity or pipeline gas, or for customers whose servicing utilities have not embraced the RCS program with enthusiasm because of overcapacity or some other reason.

Finally, the creation of a competent, well-trained private sector energy audit "industry" offers an important buffer against the possibility of a sudden upsurge in demand for audit and retrofit services--in the event of an emergency fuel curtailment or sudden jump in oil prices, for example. DOE should extend to these firms the same support services it provides to utilities. Man of the program initiatives described ealier would also be of direct benefit to the private auditing firms, including: auditor training, grants for purchasing diagnostic equipment, and applied research programs.

In addition, the RCS regulations should be amended to require that utilities offer to share costs with private auditing firms that provide at least comparable services to their customers and meet RCS requirements. The utility share of costs should be at least equal to the cost of audit conducted by the utility itself. Thiscould be done either through physical voucher or by the utility making payment directly to state-certified auditing firms. Of course, provisions whould be made to spot-check the audits conducted under this program to ensure that audits by private firms meet the federal and state standards. Customers should have effective means of recourse if audits are not adequate, just as with utility company audits.

Private auditing firms, moreover, will be especially useful in regions where a significant fraction of the homes use oil or bottled gas for heat. In these cases, the RCS program currently relies upon the oil jobbers to provide audit services. While it is reasonable to expect that at least some utilities would be willing to partially fund a private firm audit for its customers whose primary energy is oil or bottled gas (since the utility is fulfilling its legal requirement under RCS without the need to hire and train additional personnel) there will be little enthusiasm to make major inroads into these markets due to a lack of any clear economic incentive. The oil jobbers may be willing to undertake audits, but will either be faced with "hiding" the audit cost by increasing retrofit costs or be placed in a poor competitive position with utility audits which are largely rate-based or expensed. While the oil heating market should be the immediate focus of RCS, it seems to have been side-stepped by the legislative structure of the program.

The federal government should provide funding for a voucher program involving audits by qualified energy audit firms for all owners of residential units heated by oil or bottled gas at least on a demonstration basis. Including administrative costs, these vouchers might cost $150-$200 each. In some cases, the oil jobbers may find it attractive to extend their line of work and offer such audits themselves if they can hire or affiliate with trained RCS auditors. This could offer them the further advantage of a more stable income, during periods of widely fluctuating oil prices and supplies.

Since approximately 31% of all housing units use a fuel other than electricity or utility supplied gas, the total cost of supporting such audits will be high. Some additional source of revenues must be found to pay for the audit if this program is extended nation-wide. One possibility would be a 2.5 cents/gallon tax, which would cover the full cost of such audits, or a proportionally lower tax covering a portion of the audit cost. Whatever system is used, even considering general revenues, the extension of RCS to oil and bottled gas heated residences would correct a flaw in the current structure of the program and could have a significant impact on our current oil import problem.

One-Stop-Shopping Loans

Recent analysis has shown that few individuals are taking the extra steps or are willing to pay today's high interest rates to finance retrofits with the available home improvement loans (Sandra Rennie, Aceee Summer Study). Instead they are relying on savings and out-of-pocket funds. In order to extend conservation actions beyond a certain segment of the market (middle- and upper-income homeowners), and to increase access to the more expensive measures which have an even higher savings potential, it is important that loans be easily arranged at the same time that the auditor is explaining his recommendations to the building owner. If the owner finds it necessary to both shop for products and installers and to arrange financing himself, it is likely that many potential customers will find the process to be too troublesome, time consuming, or otherwise risky, and many promising retrofit investments will not be made. If the auditor is able to offer the client a single form which describes the recommended options, the contractors who can do the work, the financial institutions able to provide full financing, all available tax credits or other incentives (like grants from the Conservation/Solar bank), and offers to handle the contracting and financial arrangements for the homeowner it is reasonable to expect a much higher response rate, and significant amount of financing of more expensive but still cost-effective retrofits will result.

The RCS program should proceed rapidly with its currently planned demonstrations of a "one-stop-shopping" approach to loans, and if these demonstrations prove successful, require such a program nationwide by 1983. In the interim, DOE should pay careful attention to the compliance with existing requirements for utilities to facilitate financing, and insure that the intent of the law is being met.

Need for Expanded DOE Staffing of RCS

The current staffing of the RCS program by the Department of Energy is completely inadequate. For a major program of its nature and scope, particularly one with its experimental nature, a staff of 20 or 30 in Washington and representatives in the DOE regional offices would seem minimal to support the required state and utility programs. For the past year efforts have been underway to increase the staffing of the RCS program, however, they have proceeded slowly. This should be corrected immediately, and the RCS program should be allotted the necessary qualified staff.

Programs for Low-Income Households and Rental Housing

Actions recommended under this program are:

o   development of a program to shift the funds (available from the Windfall Profits Tax Bill) from fuel-bill subsidies to permanent savings, through weatherization of low-income households;

o   development of an experimental grant program for low-income households, which, if successful, should be expanded nationwide by 1984;

o   financial support for state and local demonstration programs for conservation in rental buildings; including financial incentives, local retrofit mandates, and programs aimed at tenant information and energy-using behavior; and

o   inclusion in the Energy Management Partnership Act of a requirement that states develop specific plans to address rental buildings.

The number of housing units occupied by renters and low-income families is large--representing 30 to 40 percent of all residential units. Many of these low-income and rental dwellings are among the worst energy wasters. When categorized into low-, medium-, and high-energy efficiency; nearly twice as many low-income and rental dwellings fall into the "low" energy efficiency rating (66 and 51 percent respectively).

The programs described previously have assumed that households are able to respond to information made available to them about economical retrofits. Low-income families and families living in rental housing, however, may be unable to take advantage of this information or of the many forms of financial incentives (like low-interest loans or tax credits).

For low-income households, large-scale retrofitting can only be accomplished thru some form of direct financial assistance such as that now available through the weatherization program. Approximately 16 million units now qualify for this program, but only a fraction of these will be served in the next ten years at the present level of funding. These households are the ones most often forced to respond to rising fuel bills by sacrificing comfort and health. Thus, improvement in these units should be measured by improvements in living conditions as well as the net energy that can be realized. From either viewpoint a more effective way of approaching low-income housing must be implemented.

Designing programs to assist renters has proven extremely difficult. While financial incentive programs that have been available to the rental market have received poor responses, such programs have been limited since the rental market has rarely been specifically addressed by utility or government initiatives. In addition to financing programs, there are a number of actions that should be undertaken to encourage retrofitting of the rental stock. These initiatives include: (1) an experimental locally-administered program to mandate retrofits for rental property at the time of sale; (2) other local demonstration programs designed to modify rent control laws and rental contract agreements to allow building owners to pass retrofit costs through to tenants, when retrofit investments would work to both the tenants and owners advantages. These demonstration programs could be funded and in fact emphasized through the proposed Energy Management Partnership Act (EMPA) if this or similar legislation becomes law. We recommend that EMPA specifically require each state energy plan to address the rental market.

Low-Income Households

o   A major share of the funds designated for subsidizing the energy costs of low-income families through receipts from the Windfall profits Tax Act should be shifted to the weatherization of these same homes--to focus effort on the cause of high heating bills and uncomfortable (or unlivable) homes, rather than just alleviating the symptoms through subsidization of energy costs.

_____

The Congressional Conference Report on this Act suggested that $36 billion in expected receipts should be directed to assist low-income families' energy needs. If the current programs are continued, most of this funding will be used to simply subsidize the purchases of fuel, unless a major program is initiated to use it more effectively as a supplement to the present Weatherization program. Use of the full $36 billion to improve building energy efficiency would allow an investment of over $2000 per housing unit. If this funding were used for retrofits, it could finance approximately 50-70% of the costs required for the full package of conservation and solar retrofits examined in this study. Moreover, if more effective techniques can be found for making use of the CETA program, the labor requirements for weatherizing those homes can make major contributions to reducing unemployment and improving job skills and job opportunities for low-income or disadvantaged groups. The funding available is reduced, of course, if significant amounts of funds are used during the next few years as direct subsidies for fuel bills.

There is another very important effect that would substantially reduce the net federal cost of such a program. The federal government currently subsidizes the rent of approximately 3.3 million low-income residential units at an annual cost of roughly $6 billion. Rising energy costs are contributing to the rapid increase in budget requirements for these programs. Investments of the kind recommended in this analysis could save the federal government more than $1.3 billion annually in fuel bills, if all of these units are retrofitted fully. Nearly the amount of funding necessary to support a program of retrofitting all low-income housing!

    o   An alternative approach of direct grants or vouchers, similar to the Canadian Housing Improvement program should be demonstrated, and if successful should be adopted nationwide as an option to the current Weatherization program.

The Weatherization program as now formulated should be simplified and freed from many procedural entanglements. An alternative approach of direct grants or vouchers might reduce admininstrative costs, and by involving private contractors as well as CETA trainees, allow a significant expansion and acceleration of the weatherization program. Grants or vouchers along the lines of the Canadian Housing Improvement Program (CHIP) should be initiated and, if successful, the nationwide Weatherization program should allow such an option beginning in 1984, if not earlier.

This program could operate through qualified auditors and contractors thus building upon the same infrastructure as RCS. The government would simply pay for an audit and recommended cost-effective retrofits conducted by certified private, non-profit, utility, or local government organizations, including CETA programs. There would be advantages in working with a

utility in the region, particularly if the utility were able to provide financing that could complement or partly replace the direct federal grant. Due to the many structural problems in low-income housing, up to one-third of the grant should be allowed for necessary structural improvements.

Rented Buildings

Approximately 26 million units, or 35% of all residential units in the United States, are rented. It has proven difficult to design programs which will adequately motivate the owners of these buildings to make necessary retrofits. Often other kinds of investments appear more profitable to the building's owner--even if access to capital is not a problem. A national apartment association survey found that 72% of apartment owners would undertake energy improvements only if the payback period were 3 years or less. Because owners of rental housing often expect a 2- to 3-year return on investment, they are reluctant to make major investments in conservation or solar energy, even though they may be extremely cost-effective over the building's remaining useful life. Only measures such as caulking or weatherstriping, that have short paybacks and minimal initial costs, would be undertaken in this sector, given such an investment criterion.

Whether or not the landlord pays the fuel bills, there is little incentive for either owner or tenant to make energy-efficiency improvements in rental buildings. Landlords can usually pass on fuel cost increases to their tenants, while the tight housing market in many areas makes energy costs a relatively minor consideration for most prospective tenants. Also, in a number of communities, restrictions in rent-control ordinances may make it difficult for landlords to pass costs and savings through to their tenants in an equitable fashion.

Information programs, such as including the cost of utilities along with monthly rental payment in apartment advertisements, or rating the energy efficiency of rental buildings with rights to inform prospective tenants of a building's rating, may be effective in those limited rental market areas where available rental units exceed demand. However, most rental markets are extremely "tight," and such information programs may make little impact on rental decisions or, therefore, on owner decisions to retrofit.

Basically, most conservation programs to date have had little or no impact on the rental stock. It appears that a carefully designed package of low-interest, deferred-payment loans and financial incentives may offer some opportunities for encouraging retrofits in rental units. The early TVA loan program included rental units but received only a 0.02% response rate from rental sector. However, recent loan programs by TVA and Portland General Electric have received higher response rates from the rental sector, the PGE program, in particular, had a rental response rate equivalent to that from owner-occupied housing in the case of rental units with electric heating. Other utilities could be encouraged to try similar programs, offering zero-interest deferred payment loans to the rental market. The investment criteria of rental property owners, cited earlier may limit the effectiveness of loan programs alone.

The only completely effective approach may be to require by law conservation retrofits of existing buildings. This is, in fact, beginning to happen in many local communities around the country. Portland, Oregon and the state of Minnesota have recently enacted mandatory retrofit laws for rental housing. Davis, California, has a local ordinance applying to all existing residences. For political acceptance and equity, such retrofit requirements should be coupled with a well-designed system of loans and other financial incentives.

Until more is learned about how to make mandatory retrofit programs effective and equitable, the appropriate federal role is to provide financial support for a variety of promising local and state-level demonstrations, along with assistance in monitoring the results. Once the most effective approaches are determined, it may be desirable to impose some form of requirements for improved efficiency in existing buildings, if market pressures and non-mandatory programs have not made substantial headway in the meantime.

Another useful step would be for the federal government to channel HUD funds through the existing state housing agencies to help provide subsidized loans wherever a mandatory retrofit standard has been adopted at the local or statewide level. Either approach to a retrofit mandate/loan program should be carefully coordinated with state and local governments, especially where local rent control ordinances exist.

The final design of an appropriate loan and retrofit program could be left up to each state, with the federal role limited to financial and technical help, for at least a five-year period. At that time, it would be reasonable to reconsider the need for a nationwide requirement for improved efficiency in existing (rental) buildings, possibly implemented through provisions of the proposed Energy Management Partnership Act as a condition for states to receive federal funds. Initially, though, EMPA legislation should at least require each states to develop a specific plan to deal with the rental market.

Retrofit requirements imposed without financial assistance and incentives could increase the risk of building abandonment or condominium conversion. Abandonment may indeed be the only solution where the property is only a marginal income-producer and retrofit is so difficult that the costs of meeting the standard are prohibitive.

Abandonment is already becoming a serious problem in many communities, and skyrocketing energy costs bear a large part of the blame. For some of these older, inefficient buildings, monthly energy costs are substantially higher than monthly mortgage payments. In Massachusetts, the state government estimates that as many as 25% of the buildings in the low-income area of Springfield risk abandonment because of rising energy costs. Ironically, some of these buildings may in fact be saved from abandonment by a combined incentive/mandate program, since cost-effective conservation improvements to basic components such as leaks, roofs, missing or broken windows, and worn-out heating systems would greatly reduce the energy expenses of these structures while improving their habitability. A conservation loan programs could provide the necessary capital for such investments, in a form that is attractive to the building owner. For those buildings that are incapable of being substantially retrofitted in a cost-effective manner, case-by-case exemptions may be necessary, in a combined incentive/mandate program.

For those multi-family buildings already converted to condominium ownership, energy-efficiency improvements can be encouraged by the programs discussed earlier, for owner-occupied units. However, mandatory local retrofit laws could also be extended to require substantial improvements at time of condominium conversion, as many communities already require, for fire safety, parking, and other measures.

In addition to the programs to encourage the physical retrofitting of rental units, the Department of Energy should encourage and support programs designed to induce changes in tenant behavior that affects energy use and efficiency. There is growing evidence that improved information, feedback, and other behavioral inducements can achieve significant savings in the rental market. Demonstration programs to highlight energy costs as a separate item in the monthly rent bill--even where these costs are centrally metered and paid by the building owners--have shown 15 to 30% reductions in energy use by tenants which is in the same range as (or slightly higher) than the percentage savings by owner-occupants in response to utility bill "feedback" (see above). Additional efforts should be undertaken to test and demonstrate such feedback methods to reduce energy usage, and state and local governments should be encouraged to assist or require owners of rental units to implement these measures once they have been proven effective.

Financial Incentives

In the section of this report dealing with technical potentials, we argue that virtually all of the conservation and solar measures considered are clearly cost-justified, even at today's average cost of energy. Nevertheless, there is still a justification for using well-designed and carefully monitored financial incentives, wherever these can help to offset institutional and market barriers and speed the widespread adoption of such measure.

Of course, there are already in effect, at the federal, state, and even local levels, a variety of such financial incentives for improved efficiency in existing residences. New incentives are constantly being introduced, among them the recently-authorized loan and grant programs of the Federal Solar and Conservation Banks, and the increased federal tax credits for conservation and solar measures. Even the limited evidence available shows that millions of homeowners have already taken advantage of one or more of these incentive programs (the federal tax credits for conservation, in particular). What is far less clear is what difference such incentives made in the investment decisions that might have been made, anyway--or in the resultant energy savings and net cost to society for each unit of energy saved. Nor is it clear why some financial incentive programs seem to have elicited surprisingly little response from consumers. Still other programs remain largely unexamined.

In general, there are at least four main problem areas with the current patchwork system of financial incentives for conservation and solar measures in existing residences:*

o   the very fragmentation of the current incentives, which not only result in a confusing array of eligibility requirements and procedures, but duplicate coverage of some technologies or categories of buildings, while leaving others uncovered;

o   despite the multiplicity of incentive programs, overall lack of flexibility to meet the diversity of technical opportunities, ownership patterns, and economic conditions within the existing residences sector;

o   several specific rigidities that lead to seriously sub-optimized retrofit investment decisions, and in some cases effectively prevent any such decisions (examples are the barriers to use of Solar/Conservation Bank programs to support an integrated conservation-plus-solar retrofit package, and the federal prohibitions against combining conservation loans with tax incentives); and

o   There has been little or no emphasis on good monitoring, evaluation, and feedback to improve the cost effectiveness of financial incentive programs, and thus reduce total government outlays as well as increase the impact of public funds that are spent.

The proposals outlined below try to address each of these problems, suggesting a few specific changes in financial incentive programs now in effect, but most importantly, arguing that all levels of government must work together, with utilities, to improve the coherence of their separate programs, increase their flexibility to respond to diverse needs and opportunities, and above all to improve feedback channels that can increase their cost-effectiveness over time.

Actions recommended under this program element are:

o   establishment of programs of technical assistance, rate-case intervention, and demonstration funding to induce utilities and PUC's to adopt marginal-cost pricing and to authorize or require direct utility financing of on-site investments;

o   utilization of the full funding authority of the Solar and Conservation Bank to supplement utility financing, for well-balanced "packages" of solar and conservation measures; and

o   demonstration through the Solar and Conservation Bank of a progressive incentive payment, with the amount of the incentive linked to expected energy savings.

Background: Choosing an Appropriate Level of Incentives

Designing and evaluating financial incentives for building retrofits is difficult because the criteria now used by homeowners to make purchasing decisions are complex and frequently do not reflect carefully reasoned economic analysis. It appears that under existing circumstances, most homeowners make decisions on the basis of a preconceived notion about the amount they are willing to pay for an energy saving home improvement; behavior does not indicate a clear link between patterns of investment and the economic merits of the investment. Statistics from EIA and others discussed in the next section, indicate that typical investments are on the order of $700 per housing unit.

If financing is easily available from a commercial lending institution through the "one-stop-shopping" program discussed earlier, it can be argued that a reasonable fraction of the population would be willing to accept a combined retrofit and loan program which would leave the owner with total monthly bills no higher than the bills which would have been paid only to cover energy costs. Table 1.52 indicates the amount which such a person would be able to invest in a retrofit measure given current average delivered energy prices and the cost of a typical commercial 10-year loan for a house improvement.

Examining the information about the cost and savings possible from different kinds of retrofits, it is interesting to notice that many consumers using this cri-

---

*Similar problems exist with financial incentives for new residences and for commercial buildings, areas to be discussed below.

Table 1.52   ALLOWED RETROFIT INVESTMENTS
USING DIFFERENT FINANCIAL CRITERIA[a]

	Fuel ($/MBtu/yr)		Elec. ($/MBtu/yr)	
A. No change in monthly bill (retrofit)[b]	28	42	280	380
B. No change in level-ized 20 year costs[c]	86	105	707	938
C. Utility finance[d] (no dist.)	34	48	510	535
D. Social cost[e] (20 years - 3% 20$/bbl Oil Costs)	164	185	882	925

[a]All costs in 1978 dollars.

[b]13% loan (current dollars) covering 100% of investment. Loan period = 10 years.

[c]20 year loan 3% in constant dollars, 2% property tax 0.0025% insurance, investor marginal tax rate = 35%, cost levelized over 20 years assuming 3% discount.

[d]Utility financing assuming typical costs of private utility capital the allowed costs in this case compare the cost of the saved energy with the cost of generating and transmitting electricity from new plants--these costs do not include distribution costs (see chapter on assumptions).

[e]Capital costs assume a social discount rate of 3% in constant dollars. The fuel costs are levelized over 20 years. These costs are the marginal costs of fuel plus a "social cost" of $20 per barrel.

teria would be likely to invest only in the first phase of building retrofits and would be unlikely to undertake investments in equipment such as heat pumps and solar units. It is possible, however, that with skillful marketing, purchasers using this criteria could be persuaded to purchase a package of retrofits whose average cost was low even though the incremental cost of the most expensive measure in the package would exceed his economic criteria if considered in isolation. For example, in an uninsulated house burning oil or gas, the average cost of a complete retrofit package that reduced the consumption of the average uninsulated unit by 135 million Btus per year would have an average cost of 32 dollars per million Btu per year. This would be well within the 42 $/MMBtu/year allowed as an investment in 1995 (See Table 1.52). The package would however, include measures that, taken alone, would cost more than $60 MMBtu/year.

If a 20-year loan can be made available, and the homeowner can be persuaded to use total life cycle costs as an investment criterion, a larger investment will be allowed per unit. Table 1.52 indicates the impact of using such a criterion. It must be recognized that investments of this kind will result in higher monthly bills during the years immediately following the investment which are returned after fuel prices have risen. In the calculations leading to Table 1.52, it has been assumed that this sophisticated investor will take the trouble to compute the savings which can be achieved through deducting interest from income taxes and the costs of added property taxes and insurance which would result from the investment.

Finally, Table 1.52 indicates the level of investment that would be attractive to a privately owned utility, assuming that the utility would be willing to invest up

to the point where the cost of saved energy was equal to the marginal cost of generating and transmitting energy using new facilities. It is interesting that investments justified by using this procedure can be less than the investments that would be made by sophisticated homeowners comparing costs to the delivered cost of energy (which includes distribution costs). These utility investments can be made through zero interest utility loan or other mechanisms.

Table 1.52 also indicates the investment which could be justified using a "national" investment criteria in which all costs are evaluated using a 3% discount rate in constant dollars and the full social costs of fuels (in this case, assumed to be marginal fuel costs plus a penalty of 20 dollars per barrel). This is the level of investment which could be used as a target for incentives.

Specific Measures

The object of programs in this section is to bring private investment as close as possible to the optimum national investment. This can be done by:

(1) encouraging utilities to use rates as close as possible to marginal rates (with adequate provision for low income groups using the weatherization subsidies). This can be accomplished through information provided to utilities and PUCs and through a federal intervention authority.

(2) Encouraging utility investment in residential solar and efficiency retrofits using the authority now available under Section 216 of NECPA (as amended by the Energy Security Act). This encouragement would come in the form of direct grants to local public utility commissions for the purpose of designing regulations. Several experimental programs can be financed using funds authorized through the existing ESA statute.

(3) Using the authority available under the solar and conservation banks to provide grants and interest subsidies, which could supplement utility financing programs where such programs are not available, and amend Section 509 of the Energy Security Act to allow a rational coupling of incentives and subsidies where justified by social cost benefits. This bank can also be used to provide subsidies to encourage investments that go beyond the allowed level of utility financing and to take investments to levels closer to the national optimum.

Utility Financing. DOE should support the rapid development and testing of a variety of zero-interest commercial and residential conservation loan programs, by utilities and other lenders, and help with comparative evaluations of their effectiveness.

The creation of zero-interest loan programs (or "ZIP" programs), particularly those allowing deferred repayments of principal, may be one of the most important single steps to overcome the first-cost barrier to conservation and solar investments in existing buildings. In addition to accelerating the rate of residential and commercial building retrofits, such zero-interest loans may also have the effect of bringing entire new sectors into the market—notably low and moderate income households (which are either unwilling or unable to incur a market-rate loan) and owners of commercial and residential rental property (who can gain cash-flow benefits by combining a deferred-payment ZIP loan with tax benefits from depreciation and conservation/solar tax credits). Encouraging the spread of zero-interest loan programs tied to well-designed technical information and audit programs must be a central feature of the near term national retrofit strategy. It must be recognized, however, that zip programs do not result in large utility equity investments in onsite units even when larger investments are warranted.

A number of utilities now have programs underway* (or in advanced stages of planning) to provide their customers with zero-interest loans for conservation measures. At least some of these programs also allow the customer to defer any repayment of principal on the loan for several years, or until the building is sold. In other cases, repayment of principal begins immediately, but payments are stretched out over a long enough time to assure that monthly costs will be lower than the customer's monthly savings on utility bills.

Most utilities have at least some restrictions on the measures that can be financed under "ZIP," all require some test of cost-effectiveness (often through a utility energy audit), and several of the electric-only utilities limit their financing to electricity-saving measures or to all-electric buildings. With two exceptions (Puget Power and TVA) all utility ZIP programs are currently confined to the residential sector.

To the best of our knowledge, there are no ZIP loan programs now being offered by commercial lenders or state and local agencies; this is an important target for future federal assistance.

The recent passage of the Energy Security Act authorized federal loan subsidies through a Solar and conservation Bank, removed previous prohibitions against direct utility financing of customer conservation, and expanded RCS-type services to small commercial customers. All three provisions can encourage the spread of ZIP-type loan programs and closer links between such loans and the utility energy audits—in effect moving at least some utilities closer to a "full energy services" role in dealing with their customers.

Customer response to existing utility ZIP programs has generally been quite good—and growing rapidly over the past year or so. A number of utilities report a near doubling of customer participation in audit and zero-interest loan programs during 1980, compared with the previous year. In many cases, very little promotional advertising is needed to sustain as many audit and loan requests as the utilities' administrative machinery and technical staff can handle. Of particular note: several utilities with deferred payment ZIP programs preliminarily report good response rates from owners of rental property, although such programs are of fairly recent origin, and expected to expand even more rapidly in the months ahead.

Recommended Federal Action. The expansion of utility-sponsored zero-interest loan programs will depend mainly on the actions of innovative utility companies and state utility regulatory commissions. But, there are several specific actions that the federal government can take, to help this process along, as well as to encourage similar efforts by private lenders, and state or local agencies:

o Start-up Funds and Technical Reports. DOE should make available funds for start-up planning of zero-interest loan programs, to state-PUCs, utilities, local government agencies**, and private lenders (or associations of lenders). Similarly, DOE could fund several regional centers to provide on-site technical assistance to public and private organizations that are developing loan programs.

Typically, non-utility organizations will need additional technical support to establish guidelines for energy-saving measures, and perhaps help in reviewing individual proposals. This training and technical support could be provided either by local utility companies, federal

*Programs now underway include those at Pacific Power and Light (Oregon), TVA (and its local distributor-utilities), Portland General Electric, and Puget Power (Washington). In California, Pacific Gas & Electric, Southern California Edison, and Southern California Gas Co. all have programs in advanced stages of planning. Since we have not made an exhaustive survey, it is likely that other utilities have similar efforts underway or in planning stages.

**For example, a local housing or redevelopment agency could offer zero-interest, deferred payment financing on the extra costs of adding conservation and solar measures to a residential or commercial rehabilitation project.

funding, or by one of the DOE-funded regional centers.

o Support for Monitoring, Evaluation, and Information Exchange. DOE should also provide funding for the thorough evaluation of conservation and solar loan programs, including comparisons of the results (costs, energy savings, participation rates, and special target group response) of programs using different approaches. Federal support could also provide a forum for the ready exchange of information among utilities and other organizations providing conservation and solar loans.

Solar and Conservation Bank. A Solar Energy and Conservation Bank has been established in the Department of Housing and Urban Development (HUD) to supplement the availability of conventional financing. It will provide subsidized loans to persons who make energy conservation improvements or install solar applications in both residential and commercial buildings. The new bank will operate until September 30, 1987, with the same corporate powers as the Government National Mortgage Association (GNMA). The existing financial and high-interst rate probelms increase the need for such an intermediary even if conventional financing can be found. The bank is therefore authorized to make payments to local financial institutions willing to provide below-market rate loans, or a principal reduction on loans to borrowers for solar and conservation improvements.

The solar/conservation bank has been selected as a preferred mechanism for delivering the incentives for retrofits, because its enabling legislation permits considerable flexibility, and because it is arguably preferable to use an existing institution rather than to establish new programs. The bank can, for example, be used to fill holes in existing incentive programs by supplementing tax credits and other programs which are attached to specific technologies, if Sec. 509 of E.S.A. and Title II, Sec. 203 of P.L. 96-223 are properly amended. If properly managed, it should be able to encourage the most efficient retrofit investments. An attachment to this paper evaluates the impact of several different techniques for delivering incentives, including tax credits, interest subsidies, and grants. It indicates that major difference between these incentives is the equity with which they treat different income groups. The federal costs incurred in delivering equivalent levels of subsidies to sophisticated investors does not depend on the details of how the subsidy is made.

Assistance for energy-conservation improvements is provided to owners and tenants of single-family dwellings whose household income does not exceed 150% of the area median income levels. Subsidies for multi-family and commercial buildings up to 20% of cost are provided for conservation. Assistance for solar energy projects is provided to owners or builders of residential or commercial buildings ranging from 40 to 60%.

The bank should allow joint loans for projects combining aspects of both solar and energy efficiency techniques. There should be no artificial division between energy efficiency and solar energy, and loans should be available for combined techniques that are often the most cost-effective. In addition, it should be allowed to repurchase construction loans made to builders and loans made to and by intermediaries such as energy service companies.

Solar Energy and Conservation Bank authorizations include $2.5 billion for fiscal years 1981-1984 for conservation purposes; and $525 million in fiscal year 1981 and $7.5 million in fiscal year 1982 and thereafter would be available annually out of each appropriation to promote the solar and conservation loan programs. However, the appropriation by Congress for the Bank has been well below these levels.

The influence of the secondary mortgage market of this bank will be extremely limited compared to the total investment needed to meet the projected potentials in energy savings. The total loan authority and appropriation need to be dramatically increased, at least in the near term, to stimulate investment markets. Otherwise, the bank may find in a no-win situation, where the funding is limited because it has not demonstrated it's effectiveness, yet private markets arenot responding because the banks funding is so limited.

A matching grant program for energy conservation expenditures is set up for persons whose income is at or below 80% of the area median annual income. The banks are authorized to set up a maximum subsidy level of 50% of the cost of conservation measures up to about $1,000 for each unit in a single- or multi-family dwelling. No grant will be provided unless it applies to a total expense of at least $250, and assurance of financial resources is given. These provisions seem to leave the lower-income households with a no-win situation. For the grant program, they must provide 50% of the funding up front, and for the loan they must make an investment for a larger dollar amount, neither of which they may be able to afford.

While matching grant provisions is a beginning towards addressing the needs of lower-income households, the level of funding appears inadequate, and the criteria for the grants seem overly restrictive. The needs of the poor may best be met by a total overhaul of the weatherization program, which was discussed earlier.

Estimate of Financial Incentives. The Department of Energy in cooperation with the Solar and Conservation Bank should initiate a demonstration program in one or more utility company service areas to test innovative incentive programs which will serve to rationalize financial incentive programs. The Bank has the authority to undertake such programs under existing statute: "In order to encourage individuals to increase such investments, the Bank may structure the incentives so that within these limits, the subsidy percentage will increase as the amount of the investment increases." A program linked to an audit, using the basic incentive logic outlined below has the potential to increase the leverage of every federal subsidy dollar and increase the level of retrofit investments undertaken. However, such a program has numerous operational difficulties which only a demonstration program can determine whether or not they can be overcome in a cost-effective manner. The potential for significantly higher leverage of subsidy funding and greatly increased retrofit investments justifies the limited risk in a series of demonstrations. Both the rate of retrofit investment by homeowners and the level of such investments should be carefully monitored and compared with a control group in these demonstrations to accurately determine its effectiveness. In addition, the basic investment required of the homeowner and the degree of subsidy could be varied by individual income, allowing a wider range of income groups to both utilize the program and optimize their retrofit actions.

The total subsidy required of the bank can be calculated using Figure 1.47. This figure represents a typical conservation/solar supply curve of the sort illustrated in the appendix. If the customer can be convinced to invest only to the point where his payment in the first month will equal the energy savings, he can be expected to invest up to the point marked A on the curve, which implies an annual saving of A' as marked on the horizontal axis. Using social costs, however, the nation would like to encourage savings of B'. In order to encourage such savings, the government would need to pay the consumer an amount equal to the area of the triangle A-B-C. This undoubtedly overstates the extent of federal investment required since a significant portion of investors may be willing to invest above the line market consumer economics on the table if they receive incentives from utilities or if they

are willing to use life-cycle costing economics. If utilities were willing to invest up to the point justified by their marginal costs, the federal contribution would need only to be equal to the area B-D-E on the figure with the remaining part of the area of A-B-C representing the utility's contribution.

The total expected cost of the financial incentive program is estimated using the method illustrated in Figure 1.47 and the program impacts estimated in Appendix A of this discussion. The total federal costs are estimated as follows:

o   First, the total amount of energy saved by each measure is estimated.

o   Second, the incremental number of investors attracted to the measure by the financing incentive is estimated using the methods discussed in Appendix A.

o   The total incremental energy saved by the financial incentive is estimated by multiplying (1) and (2) (this corresponds to the area on Figure 1.47).

o   The increment in the cost of the measure financed is estimated by subtracting the cost of the measure (in dollars per million Btus per year) from the cost which would attract a consumer investment. This consumer cost is assumed to be the first row in Table 1.52. For simplicity, an average of 1985 and 1995 costs were used. (The difference between the consumer cost and the cost of the measure corresponds to the distance A"-B" on Figure 1.47.).

o   The federal cost is then computed by calculating the area in the triangle on Figure 1.47. (For simplicity, this is assumed to be 1/2 (A'-B') x (A"-B").).

If utility financing is available for all measures that are cost-effective to a utility, the same process is repeated, only utility investment criteria are used instead of a consumer's investment criteria.

The results of the calculation were illustrated in the retrofit summary table. It is important to notice that the subsidies are most important in leveraging investments in buildings now using fuel for heat, since current fuel prices are typically below the marginal price. The incentives translate into an average of 25-30% of the total value of the incremental investments.

As in all of the estimates of cost and savings exhibited here, considerable caution must be exercised in interpreting the results. First, it may be necessary to implement a higher incentive rate during the early stages of the program for technologies whose price can be expected to decrease as markets increase. In this case, the incentives are useful to develop immature

$/MBtu/year

industries as well as to benefit consumers. Such incremental incentives would need to be of finite duration. The cost of the financial incentives computed here may therefore have been underestimated.

On the other hand, the technique used to compute consumer behavior in the absence of incentives may be conservative since it assumed that none of the consumers affected would have been willing to use life-cycle costing. Table 1.52 indicates that if consumers use a sophisticated life-cycle costing technique, they may be willing to invest considerably more than the amounts assumed in this analysis, and therefore the necessary federal subsidies would be lower than those estimated in the Summary Table. Some account has been taken of this problem since the analysis of the impact of RCS has included an assessment of the number of consumers willing to take measures recommended by RCS which can be justified only on a life-cycle cost basis. It can be assumed, therefore, that the only persons affected by the financing incentives exhibited here will be those remaining unpersuaded by the arguments presented in RCS.

It should also be recognized that the bulk of the extra investments added as a result of the financial incentives will result from the addition of heat pumps and solar heating equipment.

Table 1.53   RETROFIT ENERGY EXPENDITURES FOR MAJOR EQUIPMENT/INSULATION

Source	Years	Average Expenditure (dollars/unit/year)	Percentage of Dwelling Units (annual)
EIA[a]	77-78	798.75	11(+ 1.7)
IRS[b]	77-78	732.82	4.7
DOW[c]	80-81	547.00	22
	80-81	1400.00	7 (not additive)

[a]From Residential Energy Consumption Survey: Conservation. February 1980. U.S. Department of Energy, Energy Information Administration, DOE/EIA-0207/2.

[b]From Internal Revenue Service data on credits claimed for conservation and renewable energy tax credits April 1977-1978. Average expenditure and percentage of dwelling units combines both the conservation and renewable credits.

[c]From Perspective on Energy: "America's Homeowners Speaks Out", a study conducted for the DOW Chemical Company by the Opinion Research Corporation, January 1980 Final Report. Survey results showed 44% of responding households plan to undertake action in the next one or two years, with an average expenditure of $547. However, fourteen percent of the repondents indicated plans to spend an average of $1400 on energy retrofits in the next one or two years.

Techniques for Estimating Program Costs: Basic Assumptions

(1) Given an effective national program, the average homeowner will invest $700 on energy related home improvements after purchasing the unit (or an average of $88 per month during the average 8-year tenure in a house).

o The data presented in Table 1.53 indicate a continuing trend of homeowner investment in energy efficiency improvements. An average homeowner expenditure, as assumed, is lower than the average expenditure in either the EIA or IRS examples. In addition, a significant fraction of the ceiling, crawl space or basement insulation, and installation of other energy-efficient items in these surveys was undertaken by the homeowner, while the costs assumed in the study's supply curves are all contractor installed costs. Notably, this level of expenditure would be equivalent to less than 10% of the average annual energy bill and approximately 15% of the average annual expenditure on home maintenance, improvement and repair.

(2) Effective outreach, information and other programs can increase the rate of homeowner investment to 12.5% per year and be sustained at that level.

o The rate of investment indicated by Table 1.54 would show a lower bound of 4.7% from the IRS tax credit data, and a higher bound of 12.7% from the EIA survey for 1977 and 1978. The Dow Survey indictes a prospective level of investment well above this level. The IRS tax data, however, should be corrected for the percentage of individuals not aware of the tax credit (25% according to the Dow Survey) and for those individuals not filing for the credit for various other reasons. The lower bound of the EIA data would more probably represent a conservative estimate of the current rate of such investments, or 9.3%. It should be possible to increase this rate to 12.5% per year through aggressive programs. Sustaining the rate of investment begs the question, "How much will consumers spend on conservation improvement?" This question is addressed by assumption #3. The conservative expenditure level assumed in #1 can be assumed to increase in response to increasing fuel prices. The percentage of homeowners who have already planned investments in a variety of conservation retrofits is summarized in Table 1.54.

Table 1.54   PERCENTAGE OF HOMEOWNERS
              PLANNING TO UNDERTAKE CAPITAL
              IMPROVEMENT DURING 1980-1981,
              ACCORDING TO DOW CHEMICAL
              COMPANY SURVEY, BY ITEM

Type of Future Action Planned	Percent of U.S. Homeowners
Add Attic Insulation	33
Lower Winter Thermostat	31
Add Weather Stripping and Caulking	29
Lower Water Heater Thermost or Insulate it	27
Install Storm doors and Windows	25
Add Installation to Basement or Crawl Space	19
Add Wall Insulation	19
Replace Major Appliance With More Efficent One	18
Raise Summer Thermostat or Air Conditioner	17
All Others	2

(3) The homeowner will easily invest up to the point
     at which the cost of their investment (including an
     opportunity cost for funds from savings) plus their
     reduced utility bill is equal to or less than their
     "prior to retrofit" utility bill immediately
     following the retrofit.

     o   Life-cycle costing analysis or a strict
         payback analysis would lead to much higher
         levels of investment than this assumption
         would allow. However, few consumers seem
         to utilize such sophisticated accounting
         techniques in their investment decisions.
         Since the homeowners primary motivation for
         investing in solar or conservation retrofits is
         to reduce his or her utility payment, it was
         assumed that few homeowners will undertake
         investments which will not reduce their total
         payment (utility bill plus investment) nearly
         immediately, and up to that point the market
         should saturate over twenty years. Most of
         the investments required by the supply curve
         are under $500, with the largest single
         investment being just over $3,000. (The
         investments required do not equal the cost of
         the points used in Table 1.52 because these
         points are "packages" of retrofit items.)

(4) No more than 90% of residences will be retrofit
     with packages costing $1000 or less even if all
     recommended measures are adopted, no more than

80% of residences will be retrofit with packages
costing more than $1000.

o   Note that the assumption that 90% of homes
    receive a retrofit measure does not imply a
    90% participation level by homeowners--since
    it has been assumed that each home has an
    average 2.5 owner-occupants over 20 years.
    In addition, the older home stock statistically
    which has a lower owner turn-over rate has
    demonstrated a higher participation rate in
    major energy efficiency improvements (see
    Dow Study).

(5) Six percent of the houses surviving to the year
     2000 could have do-it-yourself solar retrofits.

     o   It would be dubious to assume that many
         homeowners will undertake more than one do-
         it-yourself solar retrofit, even though the
         average homeowner will have 2.5 residences
         in this period. However, it is assumed that
         10% of the homeowners are willing and able
         to undertake such retrofits, and that 60% of
         their dwellings have adequate access, with no
         multiplication for home turnover.

(6) The extension of financing through RCS will en-
     courage 20% of all homeowners to finance an
     additional retrofit investment costing an average
     of $600 (without sibsidy).

     o   Recent studies by the Mellon Institute
         (Sandra Rennie, et al.) have shown that most
         home retrofits are not being financed, instead
         they are being paid for out-of-pocket or
         through savings. The RCS program requires
         utilities to inform and facilitate the financing
         of home retrofits. In addition, inclusion of
         energy costs in the mortgage-qualifying
         formula will spur investments at time of sale
         by the home-seller, which will be paid for in
         the mortgage of the home-buyer. The exten-
         sion of RCS, however, is the major factor
         used to justify this additional financing.
         There is no methodology to accurately
         predict levels and rates of consumer
         response, but historical market data
         demonstrates that the introduction of financ-
         ing increasing both the rate and level of
         expenditure by consumers, e.g., the
         introduction of on-site financing such as
         GMAC to auto sales dealerships.

(7) Investments by homeowners will be made follow-
     ing the postulated supply curve.

     o   This is the value of aggressive information
         programs coupled with an applied research
         program: to inform homeowners of the

relative cost-effectiveness of retrofit measures. The Residential Conservation Service is the most effective means to accurately inform homeowners of the exact retrofit measures they should undertake in their home, and to formulate a home-specific set of paybacks for each measure. While carefully developed and easily available information should be able to bring consumer investments close to the optimum, RCS should improve consumer investment patterns significantly.

NEW RESIDENTIAL PROGRAMS

With careful design, it should be possible to build new houses that consume one-fourth the energy for heating now required by new buildings for heating at an effective cost significantly below the current delivered cost of energy. If builders, financial institutions, and building purchasers are convinced of this fact, it should be possible to achieve a significant fraction of this potential energy savings with minimal market disruption and a relatively modest federal investment.

Information and Applied Research

Actions recommended include

o establishment of management centers for new buildings technology demonstrations linked with existing R&D programs;

o improvement of analytical tools used to design energy efficient buildings with and without solar devices;

o development of effective manuals of good building design useful to the building industry;

o development and demonstration of a building energy rating system for new buildings;

o incorporation of both funding and processing priorities into HUD multi-family housing programs for energy efficient buildings.

An Integrated Building Research Program

A carefully designed research program should be developed for new building designs. Experiments should include a variety of shell design options, new appliances, and solar equipment. Photovoltaic demonstrations should be included when appropriate.

The experiments should allow builders, building code officials, and others to familiarize themselves with a variety of advanced building designs and convince them that the new features will not be overly complex or require specialized skills. The initiative for new designs should be firmly in the hands of private builders and designers; government help should be offered in a way that they find helpful. Tests should be conducted in a variety of regions. They should also be designed to allow prospective customers to visit model buildings and allow local utilities to observe the performance of a well-documented building operating in their service areas.

No building in the research program should be subsidized by the federal government unless it is constructed under the auspices of this program.

The federal involvement should be limited to the following:

o Builders would be asked to propose building designs based on a general set of characteristics (e.g., meeting performance levels at low cost using simple passive solar). The proposals would contain both an outline of the approach proposed by the builder and the name of the architect or engineer who would complete the design.

o The federal government would select the best approaches and the best designers, using a peer review process.

o The federal government would provide funding for the architect to complete the design and provide winning candidates with any additional resources available at national laboratories or other federal facilities (e.g., manuals, access to design programs, limited amounts of consulting with specialists in efficiency or solar technology, etc.).

o A management group would provide design review and guidance during construction.

o The federal government would cover construction loan costs of the completed building for two months and assist in advertising the building and conducting "open house" visits during this period.

o In return, the government would obtain detailed records of construction techniques and construction costs and be allowed to conduct unobtrusive monitoring of the building's performance during the following two years.

The advantage of the approach described is that it requires a minimum of direct federal exposure and places the full risk on the builder; the builder has selected both the architect and the design. The builder is thus completely responsible for ensuring that the building is marketable in his region. Other demonstration schemes have failed miserably on these points.

Improved Analytic Procedures for Determining Building Performance

Techniques available to commercial firms for evaluating the performance of efficient buildings, particularly solar buildings, are surprisingly primitive and seldom well validated. The following steps should be undertaken to improve the tools available:

o Information on the performance and costs of solar devices and advanced efficiency systems should be collected on a regional basis.

o A methodology should be adopted which adequately treats building efficiency and solar equipment and these techniques made available to architects and private design firms.

Simplified Design Tools

The Department of Energy currently has a program for developing "manuals of recommended practice" to help builders understand the merits of a variety of relatively simple, efficient building designs. These manuals should provide simple, "cookbook" evaluation techniques, since the vast majority of builders and architects will not have the expertise or be able to afford the time to design efficient buildings and evaluate them using computer simulations. Builders can be expected to comply with industry recommended standards only if the techniques are easy to implement as a part of routine construction practice and if a proven sales record is developed for the buildings. A national program should:

o Develop design guides that are useful both to professional architects and to builders who do not use architects (in close cooperation with professional trade organizations such as ASHRAE).

o Ensure that the development of design guides is carefully integrated with the experimental work conducted in the programs discussed earlier by including successful experimental buildings in the guides.

o Ensure that the design guides are constantly updated so that the results of demonstrations are included as successes are achieved; but care must be taken to leave standard approaches in the guides if they prove to be satisfactory, since continuity is important.

o Ensure that solar approaches are adequately integrated into the guides.

o Develop a simple "performance" approach using look-up tables and simple graphs to determine design energy consumption.

o Include sections in which very efficient building designs are described. The guides should present a strong case for designs which are cost effective and which should be marketable. A variety of solar options should be included in this section.

o Careful attention should be paid to the experience of foreign builders, particularly those in Canada, Sweden, and Denmark, since valuable theoretical and experimental work has been under way in these nations for a number of years.

Building Energy Labels

A building performance rating should be developed in connection with the applied research program and the design guide projects just described. This rating should be constructed with the assistance of the NAHB and other industry organizations, as well as local utilities. The rating could simply indicate whether the building meets industry standards; or it could be a "point" system such as the one developed by the PG&E company in California. The building energy rating system for new homes should be developed in tandem with the program designed to develop ratings for existing homes, and specific attention should be paid to linking this rating system to incentives.

HUD Multi-Family Projects

The Department of Housing and Urban Development is the major influence on new rental and multi-family housing. Through its Section 8 subsidized contracts coupled with FHA 221 D4 tandem program, as well as other programs, HUD could accelerate energy efficient building in this market. Its subsidized mortgage funding is in great demand by developers. For example, in the Sacramento Region HUD office, there were three times more applications than units allocated to that region last year.

A powerful incentive would be created if HUD would place a high priority on the energy efficiency of the buildings qualifying for these programs. A minimal additional cost to HUD, such a "procedural incentive" could result in significant improvements in the energy efficiency of new rental and multi-family units. HUD should not only give preference in funding those buildings with the best design features for energy efficiency and solar, but also accelerate the processing of their applications. Such "fast-tracking" has been shown to be an effective stimulus to the new building market in California.

Incentives for States to Adopt Performance-Based Building Standards

Actions recommended under this program are:

o establish indirect inducements to states to adopt efficient, performance-based building standards through the secondary mortgage market, FUA, or the Federal Power Marketing Authorities;

o encourage states to experiment with new enforcement techniques for performance standards in new buildings; and

o provide direct grants to states to cover the costs of implementing BEPS before Federal sanctions are approved by Congress.

Since efforts to achieve early national implementation of BEPS have failed, it may be advantageous to persuade a number of states to undertake standards equal to or better than BEPS by providing them with incentives. California, New York, Illinois, and other states are now considering the adoption of performance standards (many now have standards based on ASHRAE 90-75). Federal incentives could ensure greater uniformity of codes, and ensure that the standards adopted were well designed, and reasonable. These states could also be used to test and examine a number of approaches to implementing a performance-based code. Specific programs follow.

(1) Provide incentives for states willing to adopt PERFORMANCE-BASED STANDARDS EQUIVALENT TO (OR BETTER) THAN BEPS before a national standard is adopted. These incentives should consist of direct grants to cover the costs of early implementation and indirect credits given to builders and/or purchasers in the states that comply.

(a) The direct grants should be given to cover the following:

o overall management of the development and implementation of the new standard;

o development and adoption of materials needed for local implementation, e.g., MORPS;

o demonstrations for local builders (in excess of those described in other programs);

o training inspectors (or support for other enforcement mechanisms); and

o data collection and evaluation of successes.

(b) Indirect credits should be provided in the following form:

o Mortgages for new buildings in complying states could be subsidized through the conservation/solar bank with appropriate amendment of E.S.A. which does not presently allow the bank to subsidize conservation improvements in new homes, or through FNMA or GNMA by giving the buyer a 1/2-point reduction in mortgage value when the building is purchased.

o States would be allowed to reallocate gas saved through the BEPS standard to industries and utilities in the state. "Finders keepers" gas would not be limited by the restrictions of the Fuel Use Act.

o Participating states could be given preferential access to power available through the Federal Power Marketing Administrations (FPMAs).

o Federal rules assigning priorities to interstate natural gas could assign a higher level of priority to all users in states participating.

Once a state qualified for the "early adoption credits" the credits would be available through 1985 even if a national standard is adopted prior to 1985. No states could enter the program after a national standard was adopted.

The two sets of grants are designed to overcome major classes of barriers. The direct grants to the states would allay state fears that the standards might impose an expensive management responsibility on them. The indirect grants would provide pressure from building customers, industries and utilities within the states to adopt the standards. This would make adoption more politically palatable. The mortgage incentive could asuage the concern of builders worried that the standards would slow building sales.

(2) Experiment with a number of techniques for ensuring performance-base standard compliance. One of the advantages of encouraging early adoption of performance based standards in some states is that an opportunity will be provided for experimenting with different methods for ensuring compliance with the standards. Available evaluations of using code officials to enforce a performance standard in California show that real compliance is far less than 100%, more

realistically 65%. Given that the standard enforcement technique is fairly ineffective (and still very costly) experimentation with innovative techniques which are designed to achieve equivalent levels of energy savings in new homes coupled with careful evaluation of the effectiveness of both traditional and innovative techniques should inprove the effectiveness of implementing performance standards for new homes. A two-year experimentation period would allow the development of federal regulations for state enforcement techniques based upon the findings from the two-year period. A number of schemes have been proposed and several deserve a test, accompanied with a careful program to evaluate the costs and the results. Examples of schemes that can be examined are:

o   Multiple "walk-through" inspections by local code officials (this places a large burden on the skill and integrety of the official).

o   A building audit conducted after the building is complete (perhaps using auditors trained in the RCS program).

o   Utility rates in which the predicted consumption of the building is available at a modest rate with all additional consumption available only at a much higher rate (this would require some technique which allows a customer redress against the builder if the building fails to perform as promised).

o   Builder warranties of design performance and a state/building industry co-sponsored arbitration system to process claims of warranty violation.

o   Issuing building permits which include a rating of the building design (e.g., AA permit is better than the standard, A at the standard's level, B 10% less and C 20% less than BEPS), with a requirement that the building be posted with the design rating at time of sale;

o   Allowing "market-mix" compliance, e.g., the average usage level within a complete subdivision must meet the performance standard, coupled with a Building Energy Rating scheme; and

o   Allowing trade-offs with other energy saving features provided by the builder, e.g., appliances, bike paths, etc.

The information and incentive measures just described should attract participation by a significant percentage of the States. However, complete nation-wide coverage may well require Federal sanctions. The status of the current legislation which would require such sanctions portends a significant delay in the nation-wide implementation of BEPS. However, there may be advantages to delaying full nationwide implementation of BEPS for two or three years if DOE undertakes an aggressive program to induce states to adopt performance-based standards now. This "delay" would allow time for experimentation with innovative enforcement techniques, which should increase the ultimate energy savings achieved upon nation-wide implementation. This time period would also allow the provisional certification of analytical techniques and computer programs, which could then be tested against real world results over a wide range of conservation and solar designs. Upon enactment of Federal sanctions, the analytical techniques and enforcement techniques will be much more refined; avoiding "locking" the Federal regulations into current practice. In addition, the information and demonstration program should have communicated to the building industry the construction techniques necessary to meet BEPS, and to purchasers and financial intermediaries the benefits of purchasing energy efficient homes. If an aggressive program is launched to include energy costs in loan formulations, the real "first-cost" barrier fo BEPS will be completely eliminated, and in fact energy efficient homes will find a larger potential buyer market than other homes. The losses posed by a two or three year delay in Federal sanctions can be more than made up for if the Department looks at this delay as an opportunity to improve BEPS, and assigns the initiatives discussed previously adequate staffing and resources to achieve such improvements.

Incentives for Better Buildings

Performance standards as they are likely to be published may not achieve a mix of buildings optimally designed to minimize national energy costs. This is due in part to the limitations of the methods now used to compute building design performance and in part to the fact that the standards are based on expected energy costs paid by the customer and not on the full social cost of energy.

The current technique for selecting a performance standard uses a computer program to select the building performance which would minimize life-cycle energy costs. The simple technique employed suffers from three basic defects:

o   The analysis relies on national average energy prices and does not reflect regional differences in these prices;

o   The analysis is able to examine only a limited set of technical opportunities; in particular the standard does not consider the use of any active of passive solar design (even though it is permisable to use such designs to meet the performance standards once established); and

o   Repeated analysis has indicated that life-cycle costs are nearly the same for a broad range of building designs. It is often possible to achieve a significant increase in energy saving with a relatively small investment. The computer, of course, selects the theoretical minimum cost point but in fact the range of uncertainties is so great that the real minimum cost point could well be realized in a building design which saved much more energy than the computer's optimum.

An analysis which established a performance standard approaching minimized net social costs would be based on marginal cost of energy from new supplies or on an analysis of full marginal costs plus a premium for externalities.

o   Marginal Price of Supply. The energy costs faced by homeowners, particularly for gas and electricity are based on historical average cost the price of new supply is far higher.

o   Social Premium. There is a value to the country for reducing energy use beyond the cost of supplying the energy. The environmental effects of energy production and the cost associated with possible supply interruptions caused by our dependance on foreign sources of petroleum are two major examples of these externalities. This report assumes that the cost of externalities is the equivalent of $20 per barrel for oil and gas.

It is probable that there will be cases in which it will be to the utilities and society's benefit to finance the installation of on-site equipment, but they will refuse to do so because of uncertainty about the future or simple reluctance to alter past practices. This places the federal government in a dilemma. It would be tempting to assume that in the cases where utilities refused to finance the installation of on site systems up to the point justified by the marginal cost of energy the conservation and solar bank could provide the justifiable subsidy. However, if the conservation and solar bank were to undertake this policy the utilities which refused to subsidize system installations would receive a windfall benefit; therefore, no utility would undertake financing unless forced to.

However, for utilities which face shortages of capacity and for which the marginal cost of new supply is greater than the current production cost, it may be impossible to establish "marginal prices" within the limitations of traditional ratemaking practice. This problem is explained in detail in the utilities section of this Report). In these cases, where there is a calculable amount which the utility can pay to defer the need to build additional capacity, such innovative measures as utility financing of on-site equipment at below the market rate of interest, penalty new home hook-up charges for houses not meeting minimum life-cycle cost (at the marginal price of new supply) criterion, and establishing more than one class of service, must be considered.

PG&E has successfully used incentives to induce builders to go beyond current state building standards for energy efficiency. PG&E's incentive payments are $2.00 per "point," with one point equivalent to a savings of 3 therms/year or 30 kWh/year. Translating this one-time payment into an equivalent cost of conserved energy, PG&E is paying between $0.33 and $0.66 per MBtu saved (at 30-year time horizons, and 3% to 10% real discount rates). To date, builder response has been quite good, with participating homebuilders averaging about 75 "points," for payments of about $150 per dwelling unit from the utility company.

The PG&E program estimates savings of $25 MBtu/yr for each participant. Not all of these savings are due to reductions in heating requirements.

The following specific actions are recommended.

1. The Solar and Conservation Bank should be provided adequate staff and adequate funding to ensure that credits are allocated in a way that allows investors to reach an optimum building design which reflects full social costs. This will require taking care to use the Bank in areas not well covered by existing incentives such as the income tax credits. It would, therefore, be particularly valuable for low income groups, passive solar equipment, etc. Estimates of the total investments which may be required from the Bank are evaluated in the following section. The bank has been selected as the principle vehicle for delivering incentives because it offers great flexibility. It can offer credits to owners or builders, can offer direct subsidies as well as reduced interest, and programs can be adjusted to meet market requirements. Appendix 1 discusses the merits of a variety of techniques for providing incentives.

2. The statute which establishes the Bank should be carefully examined to ensure that the Bank is allowed to provide credits which encourage optimum building design. At present the separation between solar and conservation programs may make it difficult for the bank to encourage the use of efficiency in new houses, allowing only credits for identifiable solar equipment. In an examination of the legislation reveals that the bank is not able to link credits to performance better than the prevailing building code, the statute should be ammended to permit this use.

3. Fund a major program to advertise the programs offered by the Bank. Experience with utility ZIP programs makes the value of such advertisement programs very apparent. In the TVA, for example, an early ZIP experiment yielded little result primarily because it lacked such a public outreach program.

2. Supply Curves for Saved Energy

The analysis assumes the following energy savings and costs for conservation and solar measures. These costs/savings estimates were developed from extensive building energy analyses conducted at LBL and SERI repeated in an appendix to this report.

3. Energy Saving Measures Incorporated into New Homes

It is assumed that new homes are built to conform to one of the five categories in Table 1.55. The "measure numbers" refer to the energy conservation and solar measures in Table 1.56

4. Programmatic Effects

In order to estimate the effects of the programs described in Appendix I, we have estimated the change in new home construction practice which is likely given the implementation of various program levels. Table 1.55 summarizes the estimates of program effects by indicating the proportion of the houses built during the time period which fall into each of the five categories described in Table 1.55.

The estimates in Table 1.57 are based on the best information on current programs and survey data available to the SERI Conservation and Solar Project. References to the data on which these parameters are based are included as footnotes to Table 1.57.

5. Calculation of Energy Saving Impacts

The information in the previous four sections can then be used to calculate the expected energy savings from the implementation of various program levels.

For example, in order to calculate the annual energy savings and cost of "Performance Level A Solar" houses built during the 1980-1985 time period the following calculation would have to be made:

$$6.6 \times 10^6 \times 0.04 \times (5.0 \times 10^6 + 2.5 \times 10^6 = \text{Fuel Savings (1)}$$

$$6.6 \times 10^6 \times 0.04 \times (230 + 242 + 0) = \text{Cost of Measures installed in fuel heated homes (2)}$$

Equation 1 consists of:

6.6×20^6 = Total fuel heated housing stock constructed 1980-85 (from Table 1).

0.04 = Proportion of total fuel heated stock built to "Performance Level A Solar" standard

Table 1.55 NEW HOME CLASSIFICATION

Conventional Construction	1980-1985 No Energy Saving Measures	1986-1990 No New Energy Saving Measures	1991-2000 No New Energy Saving Measures
Influenced by, but not meeting Performance Level A	1	1	5
Performance Level A achieved with conservation measures only	1, 2, 4[a]	1, 2, 4	1, 2, 4, 5
Performance Level A achieved with solar and conservation measures	1, 2, 3[a]	1, 2, 3	1, 2, 3, 5
Performance Level B	1, 2, 3, 4, 6	1, 2, 3, 4, 6	1, 2, 3, 4, 5, 7

[a]Of the two methods of meeting Performance Level A, the one achieved through both conservation and solar measures is more cost effective. This is because the "conservation only" houses do not receive the large benefit of moving 125 sq. ft. of glazed area to the south wall for passive solar gain. The conservation only option is included because some houses may have inadequate solar access (although in new developments, this problem will be minimal) and because some home buyers may not be willing to risk the construction or purchase of a relatively innovative solar dwelling.

during 1980-85 assuming the implementation
of the Information and Demonstration program
(from Table 1.57).

Table 1.56 COSTS AND ENERGY SAVINGS FOR CONSERVATION AND SOLAR MEASURES
(Per house, Fuel = MBtu/yr, Electricity = Kwh/yr. Cost in 1978 dollars)

	1980-1985 Energy Reduction			1986-1990 Energy Reduction			1991-2000 Energy Reduction		
	Fuel	Electricity	Cost	Fuel	Electricity	Cost	Fuel	Electricity	Cost
1. Increase attic insulation from R19 to R38	5.0	512	230	6.2	635	285	6.4	656	295
2. Increase wall insulation from R11 to R19	4.0	411	242	5.0	510	300	5.2	527	310
3. Move 125 sq. ft. of glazed area from N, E, and W walls to south wall	2.5	270	0	3.1	335	0	3.2	346	0
4. Triple glaze all previously double glazed windows	3.0	306	425	3.7	380	525	3.8	393	543
5. Decrease infiltration from 0.6 to 0.3 A.C.H.	Not Available			Not Available			18	1210	640
6. Package of solar features available in 1980-1990	12.3	1258	565-2420	15.2	1560	700-3000	Not Available		
7. Package of solar features available in 1991-2000.	Not Available			Not Available			5.0	665	300-1300

Table 1.57 MEASURES UNDERTAKEN

	1980-1985					1986-1990					1991-2000				
	Conv. Pract.	Perf. A Influ.	Perf. A Con.	Perf. A Solar	Perf. B	Conv. Pract.	Perf. A Influ.	Perf. A Con.	Perf. A Solar	Perf. B	Conv. Pract.	Perf. A Influ.	Perf. A Con.	Perf. A Solar	Perf. B
Information and Demonstration	.9[a,b]	0	.04	.04[i]	.02[c]	0	.8[g]	.09	.09	.02	0[g]	.8	.075	.075	.05
Incentives for States to mandate Performance Standards	.66	0	.16[d]	.16[d]	.02	0	.6[g]	.19	.19	.02	0	.5	.2	.2	.1
Manditory Perormance A	.66	0	.16	.16	.02	0	.2	.39[f]	.39[f]	.02	0	.1	.4	.4	.1
Manditory Performance A Performance B Incentives	.63	0	.16	.16	.05[e]	0	.2	.375	.375	.05[e]	0	.1	.35	.35	.2[h]

[a]Reed, Fred, RCS survey, Lawrence Berkeley Labs. Indicates that 10% of California homeowners consider energy efficiency to be an important factor in the purchase of a home.

[b]The PG&E Award Home Program (a voluntary standard) succeeded in inducing 8.6% of the new homes build in the service area to be built to the standard before introduction of strict California building standard.

[c]Two percent of the population is generally considered to be the "innovator margin," that is, persons who are willing to make full use of innovative technology (Everett Rogers, Diffusion of Innovation, 1962, and others).

[d]Estimates that states which represent 33% of new home construction are seriously considering the adoption of building standards, it is likely that with federal inducement they would adopt BEPS as a standard.

[e]Approximate equivalence to early results of the new home incentive programs in California; Haley, John "PG&E Experience in the New Home Market."

[f]Ten percent improvement in enforcement over projected levels (DOE) and evaluated California experience ___ (ACEEE summer study, draft).

[g]Increased level of thermal integrity is based on effectiveness of proposed programs, and "price effect" increases in consumer willingness to adopt energy saving measures (SERI Conservation and Solar Report, Building Section, p. 58), based upon market data from Denver and Washington in text.

[h]Assumes a "Learning Curve" for builders and designers which delays the major impacts of the Better than BEPS incentives until after 1990, corresponds with current levels achieved by PG&E award program.

[i]Assumes the revision of the Standard to include passive solar gain (as recommended in the policy section) see Balcolmb, Douglas. "Solar and Conservation Working Together," Los Alamos Scientific Labs.

The following measures satisfy the "Performance A Solar" standard (from Table 1.55):

5.0×10^6 = Btu saving from increasing attic insulation from R19 to R38 (Table 1.56)

4.0×10^6 = Btu saving from increasing wall insulation from R11 to R19 (Table 1.56)

2.5×10^6 = Btu saving from moving 125 squ. ft. of glazed area from the north, east and west walls to the south wall (passive solar gain) (Table 1.56).

Equation 2 consists of:

230 = Dollar cost of increased attic insulation (Table 1.56)

242 = Dollar cost of increased will insulation (Table 1.56)

0 = Dollar cost of moving glazed area (Table 1.56)

In order to determine the total energy saving impacts of the laboratories program the savings must be summed across measures undertaken and years. The same calculation must be performed for electrically heated homes and the effects of air conditioning saving must be added to total saving.

6.   Calculation of Financial Incentives to Build Homes "Performance B"

By definition, a home built to the Performance Level B standard minimizes life-cycle costs at the energy prices faced by the consumer. In other words, given the assumed energy prices in this study the consumer does not have a financial incentive to undertake a marginal energy conservation or solar investment with a higher cost than the marginal measure undertaken to meet the standard. However, we have assumed for the purposes of this study that there is a "social premium" of $20 per Bbl. oil equivalent. This value represents the externalities borne by society because of the use of the energy, such as environmental effects and the risk of economic disruption which would be caused by an interruption in our supply of imported oil.

There are two principal methods of inducing investment up to this "social optimum": a subsidy to the consumer for investments above the market price and a tax to set the price of energy equal to the social cost. The tax necessary to reach this price level is approximately 50¢ per gallon.

The "total incentive" presented on Table 1.58 is the summation of the individual incentives for the proportion of the housing stock built to "Performance B" which presented in Table 1.57. The lower bound of this range is a zero incentive since if the lower cost estimate is realized the Performance B package will be cost-effective at energy prices faced by the consumer. However, there would be a package of measures not considered in this study which would require incentives.

Appliances

Residential appliances account for a major share of the residential energy use. The following discussion includes the efficiency of both heating and air conditioning equipment under the term "appliances," thus the following should be kept in mind: the collective group of "appliances" discussed effect nearly 100% of the energy use in Buildings; and, the the usage of furnaces, air conditioners and lighting is not independent of the shell improvements discussed earlier. Excluding space heaters and air conditioners, appliances accounted for over 40% of all residential and commercial buildings energy use in 1977.

Appliance energy efficiency, therefore, is a critical element in any discussion of conservation and renewable usage in buildings. A fact often overlooked by analysts considering buildings related issues. There is limited evidence to date about market approaches to appliance improvements, however, the evidence which is available indicates that while considerable improvements can be induced by information and incentive programs, a large portion of the appliance market will not be directly impacted by labeling, early retirement or other consumer oriented programs because approximately 65% of all appliances purchased are not purchased by their ultimate user. Owners of rental housing, construction firms, and others are responsible for a large proportion of appliance purchases. Such entities can only be indirectly effected by such programs. In addition, the individual energy usage of any one appliance is typically a small proportion of a households annual energy use. Therefore, economic self-interest tends to be down-played in consumer decisionmaking in favor of amenities, such as luxury features, color, style, etc.

For this reason appliances may be the single most important market sector for minimum efficiency standards. Such standards are currently being promulgated for 1981, with more aggressive standards taking effect in 1986. These federal standards will also reduce the difficulties created by a multiplicity of numerous and different levels of state standards which are presently being legislated. However, provisions should be made to allow for state standards which do not interfere with commerce or place undue burdens upon manufacturers. The analysis conducted to date by the Department of Energy on the impact of the appliance standards does not indicate that such standards will have any major negative impacts upon the industry. However, careful attention should be

Table 1.58*

	Saving in 1990			Saving in 2000			Capital Requirements			
	Fuel Saving	Electric Saving (Bil. KwH)	Total Saving (Bbl/day)	Fuel Saving	Electric Saving (Bil. KwH)	Total Saving (Bbl/day)	Investment in Systems	Consumer Investment ($ billions)	Incentive (5-yrs PV) ($ millions)	Incremental 5-Year Program cost ($ millions)
Information and demonstration	66	4.6	58,000	327	18.3	263,000	16.06-17.55	16.06-16.93	0	102
Incentives for States to adopt Performance Standards	96	6.7	85,000	418	24.9	340,000	22.71-24.71	22.71-23.77	0	79
Mandatory Performance A	118	8.3	105,000	514	31.9	430,000	29.5-31.49	29.5-30.55	0	0
Performance B Incentives	128	8.9	113,000	534	33.4	450,000	30.98-35.41	30.98-33.38	345	41.2

*All figures are cumulative

paid to assisting the industry re-tool to produce more efficient appliances, particularly those firms willing to re-tool early and/or produce appliances significantly more efficient than the proposed standards. In addition, the promulgation of standards should not be seen as a cure-all, and should be complemented by various information and consumer incentive programs as discussed.

The following discussion over-laps commercial and residential buildings. However, the figures stated for energy savings relate exclusively to residential dwellings.

The mix of policies recommended in this section are also premised on a basic characteristic of the appliance sector: its diversity. This diversity is expressed in three important ways:

o in the range of efficiency levels among models of a given appliance type, available on the market at any one time;

o in the variety of features and services often provided, beyond the "basic model" of a given appliance (many of which features have important consequences, positive or adverse, for energy efficiency); and

o in the differences in operating efficiencies among appliances in actual use, due to differing installation, usage, and maintenance practices.

In all three cases, recognition of this diversity will affect the mix of actions chosen to improve efficiency in the appliances sector. In particular, a strategy that depends exclusively on standards specifying minimum allowed efficiencies for new appliances, like the proposed federal Appliance Energy Performance Standards (AEPS) or the similar state-level standards already in effect, will overlook important opportunities to affect the sales mix of all new appliance that exceed such minimum requirements, as well as ways to influence the average efficiencies of all the appliance stock that is currently in use. It is important to have an appropriately chosen and well-enforced program of minimum efficiency standards for new appliances, under AEPS, to eliminate the least efficient models from being sold each year. However, other essential elements of an appliance efficiency strategy include:

o information and incentive programs to help shift the distribution of new appliances that already meet minimum requirements, towards the most efficient available models (and to encourage consumer purchase of "close substitutes" such as chest rather than upright freezers, that meet the most important needs while using less energy);

o increased emphasis, through RCS and other utility, state, or local programs, on improved appliance maintenance, energy-efficient operation, and the early retirement of very inefficient models (or seldom-needed appliances, such as second refrigerators).

o provisions for manufacturers to use a "model-mix" approach (similar to the fleet-averaging for new auto efficiency standards), to meet or exceed efficiency targets while still providing for the diversity of consumer preferences; and

o regularly announced upward revision of these standards, adopted (as "long-range standards") well in advance of their effective date to provide long-term guidance to manufacturers;

o research, demonstrations, and manufacturer capitalization incentives to raise the efficiency level of the "best available" models.

Recommendations in each of these areas are outlined in the following sections.

Expanded Appliance Labeling, Disclosure, and Consumer Information Programs

In cooperation with other federal agencies and industry or trade groups, DOE should:

o extend energy performance labeling to incandescent light bulbs and other lighting equipment,

o participate in publishing consumer-oriented lists of appliance performance and costs,

o support demonstration programs to influence appliance purchase decisions by those who are not individual residential consumers, and

o based on marketing studies of consumer response, modify the efficiency indicators and format of the appliance labels, as needed, for increased clarity and impact.

Incandescent light bulbs and fluorescent tubes represent a major category of energy-using equipment that is commonly marketed and purchased on the basis of energy input (wattage) rather than actual light output (lumens) or efficiency (for example, an index based on lumens/watt). Lighting for residences, commercial buildings, and industry currently represents over 400 gigawatt-hours annually, or roughly one-fifth of all electricity consumption in the U.S. For both incandescent bulbs and fluorescent tubes and ballasts there is already a range of efficiency among products currently on the market. In the future, there is even greater technical potential for improvements in efficiency, perhaps by as much as a factor of three for incandescent screw-in bulb replacements, according to studies underway by the LBL Windows and Lighting group. (Note that this technical potential for improved efficiency is in addition to the opportunities for reducing excessive lighting levels.) Well-designed labeling systems that emphasize light output and efficiency will make it

easier to market these high-efficiency products, and at the same time help to protect consumers from misleading or over-stated claims about "energy-saving" lighting products.

There is another way in which the current FTC labeling program should be extended to increase its impact. All appliances now covered under the program (and probably a number of appliances that currently are not) should also be covered by expanded information programs oriented to the appliance purchaser. Such programs, which could be co-sponsored by the federal government, manufacturers and retailers, utilities, and state or local consumer organizations, should make widely available a series of clear, comprehensive, and easily understood appliance performance lists. Organized by brand and model number, these lists should show both rated energy consumption (or efficiency), and annual or lifetime operating costs--arrayed against typical retail selling prices. Where possible, these lists should also indicate which local distributor can obtain which models (at least, those at he upper end of the efficiency range). Comparable information on the energy efficiency of new cars is now fairly easy to obtain, in the form of EPA booklets distributed through many dealers, banks, and other outlets.

As one participant recently testified* in the AEPS hearings: "It is of little benefit to the consumer to find out that the model he is looking at is in the middle of the efficiency range (on the FTC label) if he cannot identify which models are at the more efficient end of the range." The same testimony pointed out that, at present, there is only one such list made available by an industry source (the American Refrigeration Association), for central air conditioners. Other groups, like the Association of Home Appliance Manufacturers, used to publish similar information (for refrigerators and freezers) but no longer does so--apparently due at least in part to the legal and technical complexities introduced by the federal labeling and testing program itself. (Although, reportedly, AHAM will issue a list of room air conditioners sometime soon.)

Unfortunately, many of these industry listings published in the past (as well as a list of models that comply with current California standards, issued by the California Energy Commission) are difficult to obtain, difficult for the lay consumer to read and understand, and make no direct effort to highlight the most efficient models currently avalable, in each category.

DOE is now proceeding with a project to design such a consumer list. We urge the Department to attach a high priority to this effort, work closely with the other organizations listed earlier, and test-market several different approaches before settling on any single approach to appliance efficiency lists.

A third need for broadening the existing appliance labeling program is to identify those types of transactions which do not involve an individual consumer purchasing an appliance from a retail outlet, for his/her own use -- in other words, all those cases where a label physically attached to the appliance may never be seen or read by the prospective buyer.

---

*Comments by Michael Martin of the California Energy Commission.

Examples would include retail catalog sales, appliances that are builder-selected and installed in new homes, and appliances that are bulk-purchased by bids or contracts (like replacement refrigerators or air conditioners, purchased by a public housing authority or motel chain). The volume of such "non-retail-outlet" sales, their share of the total market, the degree to which they represent less (or more?) efficient models than the average, and the information or incentive strategies that might affect purchase decisions should all be examined through DOE-sponsored research studies and demonstration projects-possibly with joint participation by manufacturers or professional and trade organizations.

Finally, in order to assure that the labeling program as currently conceived is effective in reaching consumers and influencing their purchase decisions, DOE should initiate, along with the FTC, a series of marketing studies to test consumer awareness, understanding, and response to the appliance labels. Such studies might lead to simple but important changes in the current label formats, or perhaps to the use of performance indicators other than (or in addition to) those currently provided (i.e., Energy Factors and annual operating costs for energy). Since appliance purchases represent a multi-billion dollar industry nationally, it is probably worthwhile for the federal government to invest a few tens of thousands of dollars for sophisticated marketing analysis, to find out how effective the efficiency labeling program really is, in altering those billions of dollars of purchases in the direction of lower life-cycle costs.

Appliance Operation, Maintenance, and Selective Early Retirement

To help improve the average efficiency of existing appliance stock, DOE should place greater emphasis, within the RCS program, on the proper adjustment, maintenance, and operation of home appliances, low-cost appliance retrofit measures, and recommending the early retirement or replacement of those appliances that are under-used (like second refrigerators) or have very low efficiencies compared to models currently available. Cooperative programs with utilities could also be used to demonstrate the use of incentives to encourage early retirement or replacement of such appliance models.

The RCS audit guidelines and auditor training programs could place increased emphasis on simple tests of the operating efficiency of refrigerators, freezers, air conditioners, furnaces, and water heaters, in order to identify which appliances should be serviced or adjusted, replaced with a more efficient model as soon as the owner can afford to do so, or retrofitted with low-cost measures that are justified by the remaining life of the appliance. Examples of such measures, in addition to water heater insulation jackets, flue dampers, and intermittent ignition devices (already included as RCS program measures) are: clock-timers (especially on pool and spa filter pumps), refrigerator and oven door seal adjustment or replacement, shading or evaporative pre-coolers added to air conditioners, seasonal heat recovery or venting from dryers and refrigerators, disconnecting refrigerator/freezer "anti-sweat" heaters (in humid

climate), and partial insulation of freezers and second refrigerators (especially those in warm locations in the house). Accompanying these physical measures, there should also be an increased effort to educate residents about the energy-efficient use and maintenance of their appliances.

In many cases, testing, adjustments, or low-cost, simple retrofits might be performed by the RCS auditor during the initial on-site visit (as recommended earlier, in the section on RCS). In other cases, the utility might offer special incentives for early retirement of inefficient appliances (as PG&E currently does, with its $25 "bounty" for older, second refrigerators), or include more expensive appliance retrofit or maintenance measures in its overall financing package under RCS.

Many of these same measures would apply, with only slight variations, to the recent expansions of the RCS program into the multi-family residential sector, and small commercial buildings.

Finally, there has been virtually no serious research into cost-effective retrofit measures for existing appliances, or into simple appliance performance tests that can be used by RCS-trained auditors in the field. In both areas, DOE should take the lead to sponsor the necessary field-oriented research over the next one or two years -- perhaps in conjunction with the other "applied research" and measurement activities recommended earlier.

Tighten 1981 AEPS and Allow Future State Standards that are Better than AEPS

DOE should revise its proposed 1981 AEPS regulations to be no less stringent than those already adopted by California (or other states). The regulations should also make clear DOE's intent to grant exemptions, as authorized by law, for any future state appliance efficiency standards that are more stringent than federal standards. The following specific actions are recommended.

o Provide increased staffing and technical support for development and promulgation of AEPS standards.

o In adopting final rules for the 1981 AEPS, require levels of efficiency that are no lower than the most stringent of any existing state standards.

o In the final rule and accompanying comment, provide a clear statement of policy that DOE intends to grant exemptions for any future state standards that are more stringent than AEPS, wherever this is justified based on the intent of federal law (which calls for setting standards at the maximum efficiency levels that are feasible and economically justified).

o Ensure that solar equipment is properly integrated into AEPS standards.

The degree to which the proposed 1981 AEPS exceed or fall short of efficiency levels currently required by California and other states varies among appliance categories, and in some cases (e.g., frost-free vs. manual refrigerators) even varies by sub-types. However, in cases where state-level efficiency standards already exist, there is no justification for federal standards to take effect in 1981 which would replace and substantially weaken these prevailing standards. In fact, this would represent an unfair burden on those manufacturers who have already made changes in their product lines to comply with state standards. In the case of water heaters, for example, several manufacturers testified at DOE's August 1980 hearings in support of stronger 1981 standards. These manufacturers claimed that under the proposed 1981 standards they would be forced to retool in order to replace their current models with less efficient but lower first-cost models.

Since the proposed AEPS requirements (based on federal test procedures) often use different measurements of efficiency than do existing state standards (or ASHRAE guidelines, for that matter), comparisons can be complex. Consider a few specific examples, though:

o   For a 16 cu. ft. top-freezer automatic refrigerator, AEPS '81 proposed standards would allow 4-11% more electricity consumption than existing California standards. For a 22 cu. ft. side-by-side model, they would allow 11-99% more consumption than state standards. (On the other hand, for a 14 cu. ft. manual-defrost model, AEPS levels are 11% lower than those in the California standards.)

o   The differences are even larger for an upright, automatic-defrost freezer. A 16 cu. ft. model could use 18-23% more energy under AEPS than under the existing California standards.

o   On-peak loads for a typical (30,000 Btu) central air conditioner could be 3-7% higher under AEPS 81 than under the California standards, while for a large (15,000 Btu) room air conditioner the gap is over 9%.

Federal law provides that federal appliance efficiency standards, once adopted, are to pre-empt any state standards -- but DOE is allowed to grant exemptions to this pre-emption provision upon petition by a state. The argument for a single nationwide standard for appliance efficiency is that it will reduce confusion, make it less expensive for manufacturers to comply, and ease enforcement problems (since appliances are often shipped across state lines for final sale). On the other hand, DOE's current methodology which used a single value for the (avoided) cost of conserved energy and a national average for the load of a climate-dependent appliance (air conditioners, furnaces) tends to result in efficiency requirements which are too lenient in some states and too stringent in others. This should justify DOE granting exemptions for states that wish to adopt higher standards of minimum efficiency -- especially where climatic factors (high utilization of air conditioners) or high marginal costs of energy (especially on-peak energy) are involved. In addition,

it may be in the interest of the nation as a whole for some states to serve as a "technology-leading" market area, where new models with higher-than-required efficiencies can be initially introduced, with less risk to the manufacturers of being undersold by low first-cost inefficient models.

The industry's concern over a proliferation of separate state efficiency standards may be simply unwarranted -- once federal AEPS levels are adopted and in effect, few states will have the interest or the resources to pursue separate, more stringent standards. In any case, to guard against such proliferation, DOE should consider, in its exemption procedure, whether another state has already been granted an exemption for similar better-than-AEPS state standards, thereby encouraging the five or six most active states to coordinate their activities. Alternatively, DOE itself could announce, in adopting the basic, nationwide AEPS, or two alternative (more stringent) levels that states could adopt as an option within their jurisdictions (based on higher-than-average energy costs, usage levels, or other circumstances). The important thing is that DOE indicate clearly its intent to allow states to continue to adopt standards that may be more stringent than the AEPS requirements that apply nationwide. Finally, in administering the DOE state grants program, there should be no question that development of tighter state-level appliance standards (or innovative enforcement procedures) are eligible program costs.

Long-range Appliance Standards: Technology-forcing and Market-influencing Approaches

The following specific actions are recommended:

o   In revising its initial proposal for 1986 AEPS, DOE should first redefine its criteria for "maximum technically feasible efficiency" levels, set long-range (1986-1991) standards at or close to such levels, with periodic review and updating, and incorporate in the standards additional appliances as well as electricity time-of-use considerations.

o   DOE should seek Congressional approval to introduce (at least on a pilot basis) the use of "market-mix" efficiency requirements, as an alternative or complement to standards specifying minimum allowed levels of efficiency.

o   DOE should establish a joint program, with NBS, to obtain field test data on appliance performance under actual usage and maintenance conditions, and refine the appliance test procedures, as needed, based on these data.

In the course of DOE's hearings on the proposed AEPS rules, a number of comments were received on the 1986 proposed standards and the maximum technologically feasible energy efficiency levels (MTFEEL). The California Energy Commission, NRDC, Alliance to Save Energy, and others pointed out that the proposed MTFEEL levels reflected only technology that was currently available on the market, or else considered likely to become available in the next few years -- even in the absence of federal standards or other actions. In other words, the proposed MTFEEL levels cannot really be interpreted as the maximum level of efficiency that could be achieved, with the introduction of new technology, by the late 1980s or early 1990s.

In our view, it is important for DOE to establish to set true "MTFEEL" levels soon, both to provide longer-range guidance for designers and manufacturers, and to help identify the priority needs for government-assisted and proprietary appliance R&D. Another equally important role for such MTFEEL levels is to establish a framework for setting long-range appliance efficiency standards, which are announced and adopted several years in advance of their effective date. In the interim, such long-range standards would provide a basis for incentive and information programs, aimed at manufacturers as well as consumers, to encourage the production, marketing, and purchase of appliances that substantially exceed the minimum levels of efficiency specified in the current standards. In other words the MTFEEL levels should be "technology-forcing," while the long-range standards and interim incentives can be "market-forcing."

In proposing the 1986 AEPS, DOE has taken a first step toward such long-range standards, but in many cases has not gone far enough beyond current practice, in either the proposed AEPS '86 standards or the MTFEEL levels. We urge DOE to establish, as a matter of policy, that standards that are at least five years away should be set at levels no less stringent than the highest efficiency models currently on the market. In some cases long-range standards should go even beyond this level, at least part of the way towards a true "MTFEEL" target, with lead times of five years or more.

A few examples will illustrate just how cautious both the proposed AEPS '86 and the MTFEEL levels are, for at least some types of appliances. As several participants in the recent DOE hearings on AEPS pointed out, a prototype 16 cu. ft. automatic defrost refrigerator has already been built (with federal funding assistance!) that uses only about 650 kWh/year -- lower than the MTFEEL level and significantly lower than the 1986 AEPS standard proposed for the same model. Even this prototype by no means exhausts the readily identifiable engineering improvements that could be made in the motor-compressor system, with today's technology.

In the case of the freezers (manual-defrost, 15 cu. ft. chest style), there is already a model on the market that uses 700 kWh/year, a level that is 25% lower than the proposed 1986 AEPS standard for the same model, and approaches the MTFEEL consumption. The best central air conditioner now on the market, with an energy efficiency ratio (EER) of 13.0, has performance equal to the proposed MTFEEL level (for a packaged system). A 30,000 Btu unit with this efficiency rating would have a connected load nearly 20% lower than a similar unit that only conformed to proposed 1986

AEPS for central air conditioners. A similar situation exists for room air conditioners -- the highest efficiency (10,000 Btu, 100 volt) model now available has an EER of 11.6, and performs nearly as well as the level specified as the "maximum feasible" by DOE. That best-available model has an on-peak load of 860 kW, which is again 20% lower (nearly 200 kW) than the proposed AEPS rules would call for beginning in 1986.

It is important for DOE not only to upgrade its currently proposed 1981 and 1986 standards, as well as the MTFEEL levels, but to include additional categories of appliances. Also, for all electrical appliances, the economic analysis should explicitly account for the added value of saving on-peak electrical energy (which tends to be supplied by plants using oil or gas at relatively low efficiencies), and of reducing peak generating capacity requirements.

Early priorities for broadening the AEPS standards include dishwashers and clotheswashers, electric motors, and possible fluorescent ballasts and other lighting system components. In the case of both clotheswashers and dishwashers, efficiency standards should reflect both hot water usage and direct use of electricity by the appliance (as the current test procedures used in the FTC labeling program already do). In our analysis of least-cost potentials for reduced water heating energy, nearly one-half of the total potential savings was due to reduced hot water usage, rather than improvements in the efficiency of the water heater itself. (Of course, a significant share of this reduced hot water consumption was attributed to low-flow shower and fauced fittings rather than appliances -- but these plumbing fittings could also be required under federal appliance and/or building efficiency standards, as they already are in California and other states).

The potential for electricity savings through standards for improved electric motor efficiency has already been examined by DOE, at Congress' direction. Strong consideration should be given to extending efficiency requirements to at least some categories of motors (this is already being considered in California). A recent study for the Pacific Gas and Electric Company* indicated a technical potential for reducing industrial electricity use within the PG&E service territory (representing about 40% of the state, or roughly 4% of the nation) by nearly 9%, merely by upgrading the efficiency of electric motors (mainly those over 5 hp.) up to the best efficiency levels commercially available today.

Another obvious area for upgrading efficiency through standards (and related programs) is fluorescent ballasts. The staff of the California Energy Commission, which is now considering such standards on a statewide basis, has recently estimated the potential savings from efficient ballasts at over 1 billion kWh annually by the year 2000 (or the equivalent of 350 megawatts of savings in on-peak electrical generating capacity). These are savings over and above the 40% penetration that is assumed through market forces alone, and these savings (which could be roughly scaled up by a factor of ten for the U.S. as a whole) could easily be three to four times as high if the potential savings from high-frequency

---

*A. D. Little, Inc. "PG&E Estimates of Energy Conservation Potential, 1980-2000." June 1980.

ballasts and automatic daylighting controls were also considered, rather than simply more efficient ballasts of the type that are already widely available on the market.

Another step that DOE should take is to seek Congressional approval to implement "market-mix" type standards, similar to the federal vehicle efficiency standards, for at least some categories of appliances. Rather than specify a minimum level of efficiency to be reached by all appliances sold, such standards would specify a (higher) sales-weighted average efficiency, but leave it up to the manufacturer how to offset the sales of somewhat less efficient models by producing and marketing extremely high-efficiency models. Such market-mix standards would be nicely complemented by the limited-term incentives proposed in Section V.C. This approach might be particularly useful in cases where appliance efficiency varies considerably even within an appliance class (depending on the features chosen), since it would provide DOE with a politically acceptable alternative to either creating new appliance "categories" for features like through-the-door ice service, or else setting very loose AEPS-type minimum standards based on a lowest-common-denominator approach, to avoid excluding all such models from the market.

A final step that DOE should undertake is to establish a program with the National Bureau of Standards to insure that the testing procedures used to determine appliance efficiencies are a valid representation of actual usage conditions. Many of the current testing procedures may not give an accurage rating of in-use performance; moreover, the tests tend to over-rate certain aspects of the appliance and under-rate others. For example, the current test for refrigerator efficiency employs a "hot-box." The additional heat of the appliances's environment is intended to compensate for various operating factors, such as the door opening and closing. However, this test procedure will tend to overstate the benefits of high refrigerator insulatin levels, and under-rate the benefits of high motor and condenser efficiency. A program through NBS should continually upgrade test procedures and check them against in-use performance, to insure that the appliance efficiency ratings for both standards and labeling do not repeat the mistakes made in automobile efficiency rating, where test ratings versus in-use performance have varied by 20% or more.

Federal Incentives for Better-than-AEPS Appliance Purchases and Early Manufacturer Retooling

To complement the "long-range" appliance standards' approach (see above), DOE should design legislative proposals for: (1) limited-term tax incentives for buyers of appliances that conform to long-range standards (and exceed the current AEPS); and (2) federal loans or tax incentives for appliance manufacturers who retool early to produce better-than-AEPS appliance models. The following specific actions are recommended.

o   Design and implement, in cooperation with willing utility companies, a series of pilot-scale consumer, retailer, and/or manufacturer incentive programs aimed at better-than-AEPS appliances.

o   Based on research and demonstration findings, prepare legislative proposals for nationwide incentive programs to complement the minimum efficiency requirements under the AEPS program, and help shift the distribution of appliance purchases toward the highest efficiency models.

In a previous section, it was argued that the federal government, possibly in conjunction with utility companies, should provide financial incentives for builders of homes and commercial buildings that exceed the minimum efficiency levels required under the proposed federal standards. A parallel case can be made for the role of federal (or combined federal/utility) incentives to encourage the manufacture and purchase of the most efficient appliances -- to complement the federal AEPS standards that are designed to eliminate the least efficient models from the market.

One difference between appliance and building efficiency incentives is that in the case of appliance we would recommend a dual approach: (1) incentives aimed at consumers' purchase decisions, and (2) assistance to manufacturers in obtaining, at attractive terms, the capital necessary to retool for the production of improved models, earlier than required by the standards.

The purchaser incentives could take the form of an easily-obtained rebate in the amount of, say, $20/MBtu of annual savings. This is somewhat lower than the incentive suggested in the case of buildings, because of the shorter lifetime of appliances and the added impact (and costs) of proposed incentives for manufacturer retooling. Thus, a typical better-than-AEPS refrigerator, saving an average of 100 kWh/year (slightly over 1 MBtu of resource energy) would be eligible for a $20 cash rebate. The federal government could pay 1/2 to 2/3 of this, for example, with cooperating utility companies paying the rest (to reflect the benefit to all its ratepayers of saving energy, i.e., the difference between average-cost rates and replacement costs for new supplies).

Such incentives should be in effect for only a limited, pre-announced period, linked with the schedule implementing long-range efficiency standards (see Section V.B, above). In designing such incentives schemes, a number of questions need to be resolved concerning consumer response to varying levels and designs of incentives, compared with the normal market responses to current and future energy prices, changes in consumer tastes, etc. Thus, it clearly makes sense to try out some alternative incentive schemes as small-scale pilot projects, before launching a nationwide program. For example, one variant might be to allocate a share of the buyer incentive to the retail outlets or to their sales force -- an approach similar to that traditionally taken by manufacturers trying to boost sales of a new (or slow-moving) product line.

Assistance to manufacturers to retool their production lines could likewise take several forms.

The federal government could offer loan guarantees, up to some limit, for the costs of retooling a manufacturing plant to produce an appliance that was at least 20% better than the current standards (or one that meets "long-range" standards taking effect at least three years in the future). An alternative formulation might involve a federally- subsidized interest rate for such industrial retooling loans, with the level of subsidy varying according to the efficiency of the new appliance model. Again, the task of detailed design of an effective and equitable incentive package requires additional research and pilot scale testing.

Program Costs and Impacts

Table 1.59 summarizes the estimated costs and energy savings, as of year 2000, which could be achieved through each of the programs recommended above, working in conjunction with existing federal and non-federal programs, and the underlying market forces of higher energy costs, technological improvements, and competition in the appliance market. The costs shown in Table 1.59 include the total additional investments required to raise the efficiency of appliances, but do not include the specific costs of government programs, nor separate estimates of any government or utility incentive programs to pay part of the cost of purchasing these more efficient appliances.

It is important to note that, unlike the estimated program costs and energy savings effects for the preceding four program areas, in the case of appliances we have projected that virtually the entire economically- justified technical potential (discussed on pages , above) would be realized, after twenty years, through diligent pursuit of the current programs and the additional measures recommended here. One major reason for this is that, unlike the residential or commercial buildings sub-sectors, the appliances stock will be almost completely turned over, through normal market mechanisms, by the year 2000. A combination of strong, well-enforced appliance standards, coupled with sound R&D to steadily raise the level of "best practice" and with incentives and labeling/information programs to promote purchases of models significantly exceeding the standards, should be adequate to achieve these changes over the next two decades. We have not yet separately estimated the effects of these programs (and the recommendations for improving the existing appliance stock) in the nearer-term, the late 1980's and early 1990's.

Table 1.59 POTENTIAL SAVINGS FROM APPLIANCE STANDARDS

Appliance Category	Annual Savings in year 2000 (QUADS)	Total Investment in all Appliances existing in the year 2000[c] ($ Billion)	Cost of Saved Energy
Electrical Appliances	2.54	39.4	0.4.3¢/kWh
Heat Pumps (Exist. Bldg)	.17	6.5	3.2-3.8¢/kWh
Heat Pumps (New Bldg)	.26	13.3	3.0-6.4¢/kWh
High Effic. CAC (Exist. Bldg)	.27	5.5	1.9-2.0¢/kWh
High Effic. CAC (New Bldg)	.2	5.8	2.5-2.7¢/kWh
Electric Water Heating	.85	6.9	0.7-1.5¢/kWh
Heat Pump H.W.	.2	12.8	6.0¢/kWh
Solar H.W. (Elec)	.35	15.1	2.4-6.5¢/kWh
SUBTOTAL ELECTRIC	(4.84)	(105.3)	
Fuel Furnaces (Exist. Bldg)	.29	9.7	2.2-2.6$/MBtu
Fuel Furnaces (New Bldg)	.13	7.4	2.6-4.4$/MBtu
Fuel Appliances	.25	4.4	0.7-10.2$/MBtu
Fuel Hot Water	.7	15.9	1.6-4.0$/MBtu
Solar Hot Water (Fuel)	.35	15.1	2.4-4.3$/MBtu
SUBTOTAL FUEL	(1.72)	(52.5)	
TOTAL ALL APPLIANCES	6.56	157.8	

[a]Amortized assuming a 3% discount rate for the lifetime of the appliance

[b]Appliance savings includes some items which are also accounted for in other end-uses, discrepancy is attributable to different definition of appliances used in supply curve analysis and the proposed Appliance Efficiency Standards.

[c]The cost is calculated from the supply curves, where an existing appliance is replaced by a more efficient model when it wears out, only the incremental cost is included, where one is replaced by a different system (e.g., heat pump replacing resistance heating) the total cost is included, and some item costs are assumed to change over time (e.g., high efficiency heat pumps).

( ) = non add

Section Five

Least-Cost Technical Potential for Conservation and Renewable Resources in Commercial Buildings

INTRODUCTION

We are now seeing considerable activity and interest in energy conservation in the commercial sector. Recent energy-related and architecture/engineering journals report many articles on successful cases of energy conserving new buildings and retrofits. For example, the Energy User News has reported over 80 cases of successful commercial building retrofits in the last 5 years. Also, examples exist of a few design firms with excellent energy conservation skills now designing at least some new office buildings with estimated annual site energy consumptions in the 40,000-50,000 Btu/sq. ft./year range rather than the recent average of 70,000-80,000 Btu/sq. ft./year. Such design/trends toward lower illumination levels and more efficient lighting systems, more efficient heating, ventilating and air conditioning systems and controls, double glazing and added insulation. However, there is only very partial information on the magnitude of this movement toward conservation. Also, there is substantial evidence of failure to conserve energy at cost-effective levels, and evidence of substantial economic, institutional and social impediments to energy conservation within the sector. If a lot, but not enough, is happening, how can the movement toward conservation be a facilitated, and encouraged to move faster?

Our objective here of assessing technical potentials for new and existing commercial buildings, and of assessing potential least cost ranges relative to current market positions, is severely hampered by the relative lack of consistent data about the technical potential for commercial buildings, in comparison with residences. In addition, commercial buildings generally more difficult to analyze than houses because of their large size and complexity, but also variations in energy use for different building types, mixes of functions, sizes and ownership types in commercial building makes the analysis difficult.

In the residential sector, analysis of technical potential has focused on single family detached houses. To derive similar estimates of technical potential for commercial buildings, about 15 separate analyses are needed simply to address significantly different building types (hospitals, schools, hotels, stores, etc.). Furthermore, since the costs and benefits of retrofitting existing buildings differ from those for new buildings, 2 analyses are needed for each building type. However, focusing on five major types of commercial buildings would permit assessment of about 70% of the floor space and energy use in the sector.

For such analyses, one should not simply analyze a single prototype within each of ten to thirty building types to obtain sufficient results, for there is too much diversity even within each building type. Sensitivity analyses would also be needed to assess the impacts of changes in the key factors mentioned above. Likewise, surveys taken to assess energy-related characteristics need to be fairly large to address all of the variability involved.

Efforts are now beginning to analyze the least-cost potential for new commercial buildings, using life-cycle costing techniques and sensitivity analysis on key variables for a broad range of prototypes. Also, a survey is now underway to determine the magnitude of energy-conservation movement in the new commercial building market from the mid 70's to 1980. For existing buildings, a number of on-going efforts will improve our knowledge of their energy consumption and characteristics, but results from such efforts are generally not yet available. We are not aware of plans for extensive life-cycle cost analyses of retrofit technical potentials for existing commercial buildings similar to that being initiated now for new commercial buildings.

Thus, given the current status of research and data collection activities for the commerical sector, the estimates of technical potential developed here are based on incomplete data and analyses, however the best available case studies and professional judgment were used.

New Buildings

A recent assessment of least cost energy strategies for both new and existing buildings in the commercial sector has been accomplished by the Energy Productivity Center of Mellon Institute (Sant, 1979). The study used 4 buildings - a hospital, a school, an office, and a retail store - in 4 regions of the country, and derived cost and energy savings estimates for 3 additive levels of energy conservation packages. That report indicates that conservation measures for new and existing commercial buildings are cost effective at levels well above what has been published to date.

While there are a number of discontinuities between the methodologies used in formulating the results of that study and the technical potential estimated here, the general conclusions are similar. In comparing for example conservation potentials for new office buildings both the construction costs and the energy savings are lower in the Mellon study than the results described below for the BEPS life-cycle cost (LCC) study of 3 office buildings. The energy analysis conducted for the BEPS LCC study were derived using a detailed computer program for energy analysis of the conservation options, and those estimates are used here.

A more detailed comparison of the methodologies than is possible here would be needed to describe the basis of these differences in energy saving potential.

The analysis of typical new office buildings indicates a technical potential of some 60% to 65% energy savings from recent design practice using existing technology and with potenital life-cycle costs which are lower than typical recent buildings. Initial analytical results indicate that minimum life-cycle cost ranges would result in energy reductions of over 50% (AIA, December 1979).

The Building Energy Performance Standards (BEPS) research has generated considerable information on the energy behavior of new building designs. However, to date reasonably solid and detailed life-cycle cost/ benefit analyses of technical potentials are available only for typical new office buildings. While similar analyses are now being initiated for other types of new buildings as part of the BEPS program, for purposes of this report, it is assumed that the possible levels of technical potential for other building types will be consistent with the percentage reductions observed in the life-cycle cost results of the office buildings. This assumption is based upon 2 sets of data on energy conservation potentials for new commercial buildings (see AIA, January 1979 and December 1979):

o From the BEPS Phase 2 redesign effort on 168 building designs, the average percent energy reduction obtained for all building types

combined was about 40 percent. This was also about the same percent reduction achieved by office buildings.

o The same factors which permitted achieving a substantial energy reduction through the life-cycle cost analysis for the office buildings, should also operate to permit a similar reduction for all new commercial buildings in the aggregate, even though the potential for particular buildings or building types might vary, even considerably.

This generalization from office buildings to the aggregate of all new commercial building types may not apply to specific building types because of the diversity of energy use relative to building function, ownership, operation, and size.

Retrofit

For existing commercial buildings, extensive analyses of technical potentials comparable to those done for BEPS have not been accomplished. While a number of efforts have been accomplished or are underway to improve our understanding of the energy consumption of the commercial building stock, the best available data on retrofit technical potentials is from case studies of recent retrofits. Preliminary results from two different surveys of data on recent retrofits are reported in this section. These results are limited by incomplete and inconsistent reporting of retrofit costs and energy reductions. The results do indicate that most retrofits reported in the case studies have payback periods of less than 2 years, and even more often less than 1 year. It is expected that the energy conservation achieved by such retrofits falls well short of the least-cost potentials of, say, a 30-year life-cycle cost approach.

A survey of knowledgable professionals produced an estimate of possible energy savings by retrofitting of up to 50% by the year 2000.

Limits to Analysis

There are several limits to the scope of this analysis of the commercial sector.

Individual buildings: The analysis examines technical potentials for single buildings only. Conservation potentials for groups of buildings, campuses or communities are not examined; this would include effects of density or of using waste heat from one building, such as a larger computer center, to heat others. Such potentials could be considerable. CEQ in "The Cost of Sprawl" indicated that compact communities could effect about a 40 percent energy reduction over dispersed communities, including construction, operation, transportation, etc.

Hardware fix oriented. The estimated levels of technical potential are primarily hardware-driven. This is especially true for the analysis of new commercial buildings, where the quantitative emphasis is on assessing the energy conservation possible through changing the physical design and characteristics of buildings. Some operation and maintenance (O&M) savings are included in the retrofit case studies of existing buildings but generally O&M savings have not been isolated in the case studies from the savings derived from hardware modifications. This is discussed further in Section 3, "Some Characteristics of the Commercial Sector."

Economic, institutional, and social constraints. These constraints on achieving the technical potential are not examined in any detail. Such constraints include: strong first cost biases in investment decisions; short investment decision time horizons, typically well under 5 years; lack of information on opportunities, costs and benefits; nonavailability of funding; current tax and corporate structures which separate capital and operating costs; and tenant/owner relationships.

Perhaps one of the most important constraints on policy-formulation for the commercial sector is the lack of information and understanding of decision-making processes relating to energy conservation in the commerical sector. Anecdotes abound indicating that institutional and social factors can play key roles in conservation decisions in the commerical sector. The impacts of such factors counter a common assumption that energy-related decisions in the commercial sector are primarily determined by economic rationalism tempered only by imperfect information.

This situation is exacerbated by the wide variability of building types, ownership types, and investment opportunities. Commercial buildings may be owned and operated by private for-profit corporations, not-for-profit groups, and by public institutions. Building owners, especially of many small commercial buildings, may be small organizations with little experience or expertise in building design construction; they maybe involved with only one or two building design or retrofit projects over their entire corporate lifetimes. At the other end of the scale are large corporations with substantial construction and retrofit porgrams. These organizations will often have sizable inhouse professional staff and can be expected to use a much more sophisticated decision making processes.

Yet even in such large corporations, organizational structure can hinder the application of rational economic decision making. For example, the separation of capital budgets and O&M budgets within the organization can create a situation where managers responsible for building construction will be rated on their ability to control first costs of construction regardless of eventual O&M costs, or of total life-cycle costs.

HISTORICAL VIEW OF STOCK AND PROJECTIONS

Commercial buildings now use about 1/3 of the total energy consumption of the building sector, or about 10 quads. Source electricity accounts for 67% of commercial sector energy use.

Figure 1.48 displays the current fuel consumption by end-use. In terms of use by building type, nearly 70% of this energy share is used by retail/wholesale, educational, office and health type buildings.

Growth of Stock

Between World War II and 1978, commercial floor space practically tripled from 9.9 to 29.6 billion square feet. This growth has not been uniformly distributed across building types. Office, educational, retail/wholesale, warehouses, and miscellaneous categories have grown more rapidly than other commercial building types. (See Figures 1.49, 1.50, and 1.51). In 1980, these building types represent over 70% of the commercial floor space.

According to ORNL projections, by the year 2000 commercial floor space is expected to increase 64% reaching 51 billion square feet. By 2000, 54% of the building stock then standing will have been constructed after 1980. This is a 3.4% annual growth rate for new buildings. The existing stock, on the other hand, is projected to decay at a rate of 1.5% per year. As Figures 1.51 and 1.52 indicate, the current distribution of floor space among building types will change little in the next 20 years.

Increase in Energy Intensity

Prior to 1973

Energy intensity (i.e., energy consumption per square foot of floor space) increased markedly. A typical structure built in the early 70's was designed to use almost twice as much energy per square foot as one built in the late 40's.

A number of factors contributed to the increases, including:

o   Decreasing energy prices (Figure 1.53)

o   Increased comfort conditioning requirements made possible by air conditioning innovations which resulted in wide spread use of mechanical cooling systems;

o   Trends in design standards; for example, standards for illumination levels increased by a factor of four during that period.

COMMERCIAL ENERGY USE
BY FUEL AND END USE, 1980, QBTU

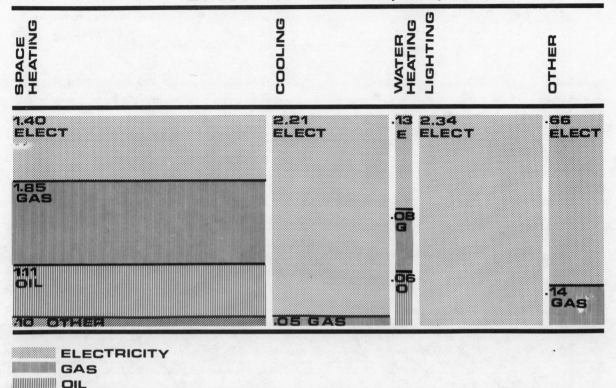

SPACE HEATING | COOLING | WATER HEATING | LIGHTING | OTHER

1.40 ELECT
1.85 GAS
1.11 OIL
.10 OTHER

2.21 ELECT
.05 GAS

.13 E
.08 G
.06 O

2.34 ELECT

.66 ELECT
.14 GAS

ELECTRICITY
GAS
OIL
OTHER

Prior to 1973, energy prices fell except for oil, which increased at a slight rate. It is not surprising that, while floor space was growing at 4.2% annually, energy use increased at a rate 38% faster than floor space growth.

After 1973

This situation was reversed after 1973. Energy use increased at a rate 34% slower than floor space as prices rose between 1973-1978. Table 1.60 indicates the rate of annual growth in prices and consumption for pre-1973 and post-1973.

Projections to 2000

By 2000, the ORNL model projects the commercial sector to use 12.4 quads of resource energy, or a 24% increase. Given the concurrent estimated 64% growth in total floor space, the ORNL projections indicate a 25% reduction in energy use per square foot of space. (See Appendix I for assumed Energy Prices, 1980-2000). These projections assume 1982 as the starting date for the implementation of BEPS. It has been also assumed that the BEPs market penetration for the commercial sector is 80%. Without BEPS, the energy use of commercial buildings would be projected to be significantly higher. The largest reductions are expected in the warehouse, hotel/motel, office, religious and miscellaneous categories (see Figure 1.54, "Energy Intensity Change 1980-2000").

Also, recent increases in market share by electricity are not expected to reverse, since those building types exhibiting the highest floor space growth rates (i.e., retail/wholesale, offices and education) have a percent of electric usage higher than the weighted average for the entire sector (see Figure 1.55). Also, the demand for electricity is relatively inelastic given that few fuel substitutes are available for end uses such as cooling and lighting. These end uses are, however, prime targets for energy conservation.

POTENTIAL FOR NEW BUILDINGS

New buildings are especially important in the commercial sector because of the projected rapid growth rate

Figure 1.49.

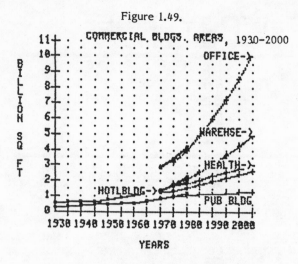

Figure 1.50.

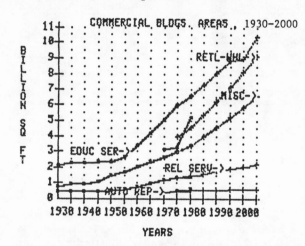

Sources: J.R. Jackson and W.S. Johnson, Commercial Energy Use:
A disaggregation by Fuel, Building Type, and
End Use, ORNL/CON-14 (February 1978)

Figure 1.51.

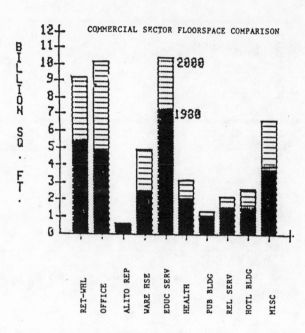

Figure 1.52.

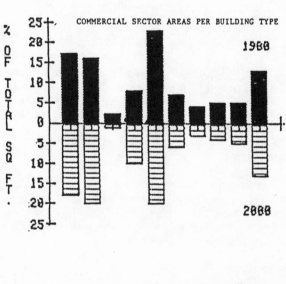

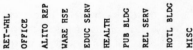

SOURCE: ORNL Commercial Energy Use Simulation, 1970-2000

of new floor space. Recent ORNL projections indicate that, in the year 2000, over 50% of commercial floorspace will have been built after 1980. Even with an expected slowdown of growth in this sector, a reasonable lower bound might be 35% of commercial floorspace in 2000 built after 1980. In any event, there is a major opportunity to maximize the energy efficiency of the 15 to 20 billion square feet of new commercial floorspace to be built in the next 20 years. This is especially important since new commercial buildings have an average life expectancy of 45 to 50 years. The energy efficiency of buildings built after 1980 will have a significant impact on the overall consumption of the commercial sector both in the year 2000 and afterwards.

Results of recent BEPS research indicate that, at least for typical new office buildings, energy

Figure 1.53.

Estimated Commercial Energy Use and Fuel Prices, 1947-1978

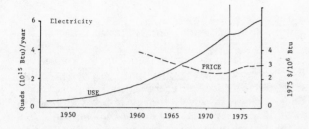

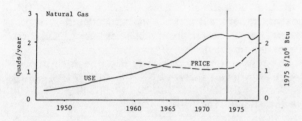

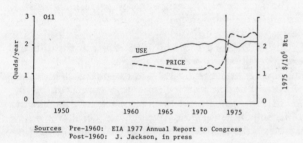

Sources Pre-1960: EIA 1977 Annual Report to Congress
         Post-1960: J. Jackson, in press

Table 1.60.   FUEL PRICE AND CONSUMPTION ANNUAL
              GROWTH RATES

	Prices		Consumption	
	Pre-1973	Post-1973	Pre-1973	Post-1973
Electricity	- 3.6%	4.0%	8.1%	3.7%
Natural Gas	- 1.0%	10.2%	7.5%	-- *
Oil	+ 3.0%	10.5%	.2%	-- *

*Slight decreases

reductions of 60% to 65% from recent practice are possible using existing technology and with potential life-cycle costs that are lower than those for typical recent buildings. Initial estimates indicate minimum LCC ranges would produce over 50% reductions in energy from recent practice. Furthermore, research results indicate that, in some cases, it is feasible to both reduce first costs of construction and obtain energy reductions of 20% to 30% at the same time.

Work is now beginning as part of BEPS research to refine the analysis of office buildings and to extend the analysis of life-cycle cost potentials to other building types.

Available Information on Technical Potentials

For new commercial buildings, recent major efforts as part of the BEPS research program have produced a large body of research results and data. The results provide valuable information about the energy characteristics of recent building designs and indicate significant possible energy conservation potentials through building design.

BEPS is undoubtedly the largest organized research effort and data collection ever undertaken concerning buildings, their energy characteristics and their potential energy uses. It is important to distinguish between the BEPS research activities and results and the BEPS regulatory intent. Because the research has been conducted in the environment of creating a Federal regulation that contains a potentially severe economic sanction, it has been hampered by time and resource constraints from having a positive information impact on the building industry. Although there has been some effort to date to disseminate the considerable information developed by the research, it is minimal compared to the effort to disseminate information about the proposed regulations.

Also, critics of the regulation have buried the value of the research efforts by attacking the proposed regulations and their potential impacts. Incomplete or deficient research results, attributable largely to the time and resource constraints, also have been attacked by critics of the potential regulations. Thus, there has been considerable emphasis on research deficiencies, which has probably been aimed actually at forestalling or eliminating a potential regulation perceived as burdensome or as counter to the interests of a number of groups.

At the same time, resource limits within the BEPS program, coupled with the priority requirement of building a technical base to support a regulation, have essentially precluded DOE from effectively disseminating the considerable results of the BEPS research in formats useful to the building community and to building owners and users.

The BEPS research efforts are indeed incomplete at this point in time, especially for commercial buildings. However, the work already accomplished, even though incomplete, constitutes by far the most substantial base available for assessing the least-cost technical potentials for the commercial sector. Further, a number of studies just beginning will add considerably more to this base.

The remainder of this subsection will briefly discuss some key elements of the BEPS commercial buildings research which has examined in detail estimates of energy performance from design data for buildings in the 1975-76 time frame and means to make such designs considerably more energy efficient. The analysis covers 12-16 building types and a range of climate conditions in the continental United States.

FIG. 2.4 ENERGY INTENSITY CHANGE 1980-2000

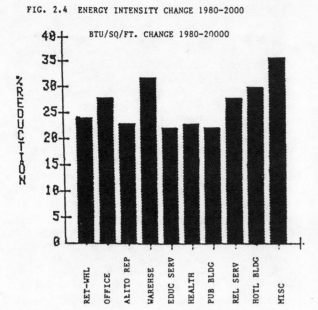

BTU/SQ/FT. CHANGE 1980-20000

SOURCE: ORNL Commercial Energy Use Simulation

Figure 1.55.   Energy Use by Building Type for the Commercial Stock

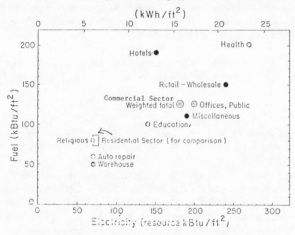

Energy Intensity: The square for the residences, plotted for
comparison, was calculated from the total energy use by residences,
divided by the number of homes and an assumed average floor space
of 1350 feet.

For purposes of assessing technical potentials for
energy conservation in this section, results of special
interest are estimates for four different levels of energy performance:

o Design estimates for recently built buildings;

o Impacts of current component based standards
for commercial buildings (ASHRAE 90-75 type
standards); and,

o Two different estimates of technical potentials: (1) a Phase 2 redesign exercise of 16
building types of 1978, and (2) a life-cycle cost
(LCC) analysis of office buildings in 1979.

These levels of energy performance are all estimates of the potential energy consumption of the
buildings, once constructed. They are not measures of
actual consumption; rather, the estimates are derived
from processing data available at the building design
stage through computer energy analysis models. Additional analysis is needed to provide a better link between design estimates and actual consumption. This
might include numerous comparisons between design
estimates and actual consumption. A good piece of
analysis toward this end is almost completed by Steve
Diamond at LASL for about 6 buildings using DOE-2 as
the estimating tool.

The following sections briefly describe each of the
four energy levels analyzed.

Recent Design Practice

The first level of design energy performance estimated for commercial buildings was for typical recent
design practice.

During Phases 1 and 2 of BEPS research two sets of
energy estimates were made using the same sample of
buildings constructed in 1975-76 to establish a baseline
of current design practice for energy conservation. In
Phase 1 energy calculations were made for a sample of
1661 buildings using relatively little detail for each
building (about 100 data points per building) and a
simplified version of a hour-by-hour energy analysis
computer program (AXCESS). The results obtained
were levels and ranges of building design energy consumption by building type and region (see AIA, Final
Report, January 1978,for tables of these results).
Typical building characteristics were also tabulated
(AIA, Task Report, January 1978).

In Phase 2 a sub-set of 168 of the 1661 Phase 1
buildings was selected and analyzed in much greater
detail. Whereas, the Phase 1 data had been provided
on a voluntary basis by design firms, in Phase 2 the design firms were hired to provide detailed data about
the energy related features of the building designs. A
detailed energy calculation was made using the
AXCESS computer program.

For a summary of energy results of the Phase-2
buildings, See Figure 1.56. The tail of each arrow indicates the average site design energy for the 1975-
1976 sample buildings of each type. The head of each
arrow indicates the average site design energy for the
results of the redesign exercise for each building
types, discussed below.

Much of the remaining discussion and results for new
buildings relate to these 168 Phase-2 buildings.

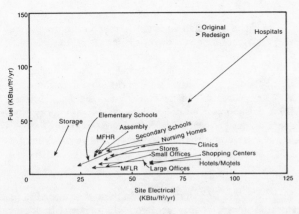

Figure 1.56.

Impacts of ASHRAE 90-75R on Recent Practice

The second level of design energy performance examined was the impact of component performance standard requirements.

One objective in the BEPS Phase-2 analysis was to assess the energy impacts if ASHRAE Standard 90 type component requirements were applied to current design practice. When the Phase-2 buildings were originally designed, few States had energy codes in place. Standard 90-based requirements are now in place in some 45 states.

Approach. ASHRAE 90-75R (Feb. 1978 version) and one hundred twenty-five buildings from the Phase-2 sample were used in the anlaysis. A documented and uniform method applied to each building changed the design characteristics to conform with the mandatory "shall" requirements of 90-75R. Not all mandatory elements could be incorporated due to simulation limitations and/or lack of clear definition of the element.

Results. The analysis produced a measure of a component-based code in terms of its impact on annual energy performance. This is an important result, for component codes do not contain an inherent energy-use estimate.

The application of the component requirements of 90-75R to the 125 sample buildings from 1975-1976 results in an average reduction in site energy of 22%. The impact of 90-75R varied by building type. It ranged from a maximum reduction of 41% for warehouses to an increase of 3% for hospitals. The variability of the impact of these component requirements on individual buildings was even more pronounced.

Observations. The results provide a theoretical measure of the exact compliance to ASHRAE 90-75R minimum requirements. However, the study did not address results that might be obtained in actual implementation due to differences in how different designers would interpret and apply the requirements across

a range of buildings. Designers would choose specific equipment and components with performance requirements slightly better than the exact requirements indicated in the code. Also, a number of component requirements can be interpreted in different ways. Variations in interpretation by each designer could significantly change the energy results for individual buildings. This potential is especially large in the section on lighting requirements.

Information from enforcement of State energy codes would provide further insight into the impact of Standard 90 based requirements on design practice. For example, the Massachusetts Building Code Commission (MBCC) has examined a number of buildings for compliance with the energy requirements of the Massachusetts Code. This code is similar in stringency to Standard 90. While enforcement authority in Massachusetts resides with the local code officials, not the MBCC, the MBCC has technical capability to do detailed compliance checks and provides feedback on code violations to both the local code official and the designer. The following observations have been made from this review experience:

o Envelope Requirements - On average, the commercial buildings examined have envelopes some 20% better than the code requirements. This indicates that current envelope requirements in the Standard 90 type codes are, at least for this region, lenient and have no substantial impact except on the high end of the energy spectrum.

o HVAC systems and equipment requirements - 90% of the plans submitted are reported to have failed to comply with these code requirements. The potential energy impacts of the non-complying elements is not known.

More Stringent Versions of Standard 90 Type Component Requirements. Given 1975-1976 practice, the current energy measures required by Standard 90 are about 1/2 as conservation forcing as the energy levels required by the BEPS Notice of Proposed Rule (NOPR) (November 1979). Also it had been recognized from the BEPS research in 1978 and 1979 that a number of the component requirements of Standard 90 could be made more stringent, so that energy results equivalent to BEPS might be possible. Several sensitivity analyses were conducted on office buildings and warehouses via computer analysis in 1978 to examine possible ways to increase the stringency of certain Standard 90 requirements to produce energy levels. The results indicated that, on average for each of the 2 very different building types examined, that BEPS energy levels could be attained by the measures used. This initial analysis did not include detailed examination of the technical or economic feasibility of these tighter requirements.

Further analysis is now underway as part of BEPS to do the technical and economic (life-cycle cost) feasibility analysis to produce a set of recommendations for Standard 90 type component requirements equivalent on the average, to BEPS in stringency. Results of this work are expected in 1981, however, it might well take from 1 to 2 years for such recommendations to be approved by ASHRAE through its concensus process. In the meantime, some or all of the recommendations may find their way into model codes or into certain State or local codes.

Thus, a more stringent set of compliance performance requirements for commercial buildings is expected to be a major means of achieving the general energy levels indicated by the BEPS Design Energy Budgets (DEB's) for commercial buildings. This, of course, assumes reasonably effective compliance with and enforcement of the requirements at local and State levels. To date, this effective compliance has not been demonstrated on a widespread basis.

Redisigns: A Limited Assessment of Technical Potential for Energy Conservation

The third energy-performance level examined in the research considered how much energy might be saved if designers, using their existing resources, but also given some information, feedback and incentive to conserve energy in the design process would redesign the set of "recent" buildings with the emphasis on energy conservation.

In Phase 2 of BEPS, the original design team for each of the 168 sample buildings was hired by the AIA/Research Corporation, under contract to HUD, to redesign buildings that they had already designed for construction in 1975 and 1976. The Redesign was to emphasize energy conservation. Several important factors were included in the instructions to the designers:

o   They were to maintain the same general cost range for the building as the original building; e.g., a speculative office building should not become an expensive corporate office showpiece, rather it should remain in the general cost level of speculative office buildings.

o   Also, the designers were instructed to use off-the-shelf technology ("if you can buy it, use it").

o   The designers were, however, free to change the location of the building on the original site, its orientation, configuration, number of floors, construction material, lighting, heating, and cooling systems, etc.

o   The redesign had to be responsive to the original program of requirements of the owner.

The designers were provided with some training and assistance: (1) a three day workshop initiating a review of current energy conserving design practices; (2) a workbook containing summaries of energy conserving design practices; (3) two peer-group reviews while developing their Redesigns. The reviews included participation by energy specialists acting in consulting roles.

Results. The Redesigns resulted in an average 40% reduction in site energy use from the original buildings. When the BEPS program was transferred to the Department of Energy (DOE), the results of these Redesigns were used in support of the energy budgets in the Nov. 1979 NOPR for BEPS.

Figure 1.56 indicates the average reduction by building in electricity and fuel use (site energy). The results for most building types tend to cluster, except for hospitals, shopping centers, hotels/motels at the high end and storage (warehouses) at the low end. Furthermore, the relative reductions in fuel and electricity use are consistent for the average results of these building types which cluster in the middle of the chart.

These average results from the sample buildings per type are excellent for showing aggregate trends. However, they could mislead a reader into thinking that all buildings within each type behaved in the same consistent way. In fact, there was considerable variability within the building types. Figures 1.57 through 1.60 show the results for each building in the samples for small offices, large offices, stores, and warehouses.

Note that for small and large offices, the strong differences in conservation trends between fossil fuel and electricity depend upon whether the original 1975-1976 design was heated electrically or by fossil fuel. Some seemingly anomolous trends for individual buildings are caused by HVAC system changes which involved switching heating fuels, which was permitted during the redesign exercise. For example, several small office redesigns switched to efficient hydronic heat pump systems, and as a result, the predominant heating fuel switched to electricity from fuel. Also, note in warehouses, the relatively greater proportion of reduction in fossil fuel use since energy for heating is the predominant use in this building type (see AIA, Task Report, January 1979; and AIA/RC, March 1980 for breakdowns of energy results by end-use within buildings and building types).

Limited Assessment. The redesign exercise provided only a limited assessment of technical potential for

Large Offices

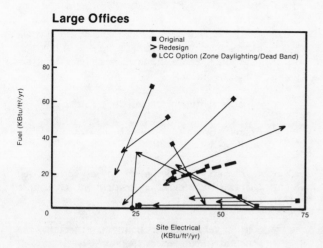

Figure 1.57.

Small Offices

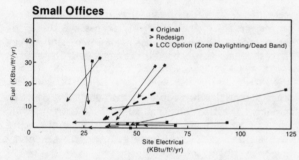

Figure 1.58.

Storage

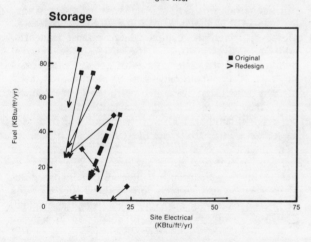

Figure 1.59.

Stores

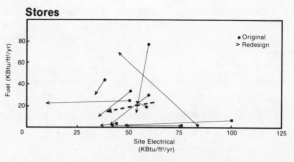

Figure 1.60.

practice. The energy conservation skills and experience of the designers for those buildings ranged from extensive to little or none. In fact, in 7% of the cases the redesigned buildings were calculated to use more energy than the 1976 original buildings. Thus, designer endeavors were not always successful, even though the average energy production of the 168 buildings as a whole was reduced about 40%. It is expected, therefore, that technically achievable energy levels are considerably lower than the averages produced by the redesigns. This expectation is supported by the results of the subsequent life-cycle cost analysis of 3 typical office buildings, described below. Lack of widespread information and expertise within the design professions about conservation opportunities and cost/benefits as well as a lack of reasonably accurate yet cheap and easy to use energy analysis tools are strong impediments toward achieving today the expected least-cost ranges of energy conservation.

Also the computer tools used in Phase 2 could not measure all significant energy conserving features in the Redesigns, especially, (1) Deadband thermostatic controls and (2) Daylighting strategies. Both strategies can result in significant energy savings. However, very few of the original 1975-1976 designs contained these features. The computer program capabilities were later enchanced in 1979 to include both of these features. These enhancements were used to derive the Standard 90 results shown in the previous section and the Life-Cycle Cost results indicated in the next section.

In the following section on Life-Cycle Cost Analyses, some better measures are provided for least-cost technical potentials, at least for typical office buildings.

Life-Cycle Cost (LCC) Analysis

The fourth energy-performance level examined in the BEPS commercial buildings research was a life-cycle cost analysis. To date the analysis has been accomplished only for "typical" office buildings.

Approach. The LCC analysis for three office buildings was conducted to determine the existence of minimum

conservation in new commercial buildings, primarily because of designer knowledge limitations.

The Phase-2 redesign exercise was as much, if not more, an assessment of the current state of design knowledge of energy conservation as it was an assessment of technical potential. The buildings used in the Phase 2 sample were selected using random selection techniques to be representative of the current design life-cycle cost ranges as well as to obtain better cost and financial information. The analysis was a pilot study of one building type, with office buildings selected due to their large expected construction volume in the 1980s. The building design data and energy results from the Phase 2 research provided the starting point with three buildings selected as typical within different climates:

o A 100,000-sq.-ft. six-story speculative office in Denver, CO;

o A 29,000-sq.-ft. four-story owner-occupied office in Minneapolis, MN; and,

o A 92,000-sq.-ft. four-story owner-occupied office in Raleigh, NC.

Additional design strategies were developed by the architect/engineer design teams of record, assisted in depth by the research team and by energy consultants. This LCC exercise can be considered a good assessment of the conservation potentials of current technology because of the expertise applied and the extensive systematic approach of seeking design solutions which would minimize energy use.

Since the three buildings were originally all-electric, most of the design schemes analyzed were all-electric. Comparable solutions for gas heating systems, for energy levels at BEPS or better, were about 2 to 10 MBtu/sq. ft./year (site energy) higher, depending upon building, HVAC system, and climate (see AIA, December 1979).

Detailed first costs were prepared for each scheme as well as replacement cycles and costs for appropriate items. Also, local prices and regional fuel escalation rates were used. Three different economic scenarios were used for the LCC analysis, each with a 40 year study period:

o Real discount rate of 10%; construction costs paid in 1st year of construction; 0% inflation (OMB criteria).

o Real discount rate of 3%; loan term 40 years at 12% interest; 6.5% inflation.

o Real discount rate of 15%; loan term 40 years at 12% interest; 6.5% inflation.

Several important factors not examined in the analysis were:

o Sensitivity analyses for changes in economic parameters, including: study period, construction costs, fuel prices, escalation rates, or complete private sector investment viewpoints (e.g., rate of return after taxes).

o Sensitivity analyses for changes in key technical parameters, including: size; mix of building function. Some climate sensitivity analysis was accomplished.

o Impacts of time of day electric rates.

o Changes in amenity levels were examined on a qualitative basis only.

o Combinations of design strategies were not sequenced by return on investment or simple payback.

o Resale value of the building, especially for speculative builders or the next buyer.

Results. Site energy levels at minimum LCC ranges were from 25-35 MBtu/sq.ft./year. Minimum LCC ranges did not change markedly across the three different economic perspectives. Minimum site energy ranges regardless economic assumptions were from 20-25 MBtu/sq.ft./year.

For these three "typical" office buildings, the technical potential range represents a 60%-65% reduction in energy use from the original 1975-76 designs, while the LCC minimum range represents generally more than a 50% reduction in energy use. These energy levels are also substantially lower than the BEPS NOPR budget levels.

The analysis was not sufficiently complete to derive a single LCC minimum point for any of the 3 buildings. Moreover, given the complex interactions, tradeoffs, and alternatives possible in the design of commercial buildings, and the uncertainties of economic forecasting, an LCC minimum range may be most appropriate.

Even though first costs of construction were in many cases from 7% to 12% higher than the first costs of the original buildings. Life cycle costs in the minimum range were lower than the life-cycle costs of the original buildings.

However, first costs did not necessarily increase as energy levels were reduced. For all 3 buildings, solutions were achieved which both lowered first cost and decreased energy use 20% to 30% from the original. This was primarily because of reduction in lighting levels and associated reduction in required cooling equipment capacity and use.

Table 1.61 indicates the ranges of results for selected building designs analyzed for the 3 typical office buildings as part of the life-cycle cost analysis. Also shown are percent changes in first cost of construction. For the "OMB" economic perspective mentioned above, the life-cycle costs for all alternate solutions shown are lower than the life-cycle costs for the original 1975-1976 design.

Table 1.61. SELECTED ENERGY RESULTS FOR 3 BUILDINGS IN LIFE CYCLE COST ANALYSIS

Different Design Solutions to Same Building Design Problem	Buildings											
	Denver				Minneapolis				Raleigh			
	A*	B*	C*	D*	A	B	C	D	A	B	C	D
Original 1976-1976 Design												
Heating and Cooling	20.2	(41)			40.6	(59)			41.3	(58)		
Lighting	25.9	(52)			18.7	(27)			18.8	(27)		
Total	49.4	(100)			70.1	(100)			70.7	(100)		
ASHRAE 90-75 Exact Applicable Requirements												
Heating and Cooling	12.6	(39)	(38)		31.4	(53)	(22)		26.1	(42)	(37)	
Lighting	16.1	(50)	(38)		20.5	(34)	(0)		13.1	(26)	(30)	
Total	31.9	(100)	(35)	(-1)	59.8	(100)	(15)	(0)	50.2	(100)	(29)	(-1)
ASHRAE 90-75R Requirements Per Component, or 1975-1976 Value, whichever is Better												
Heating and Cooling	8.4	(30)	(58)		28.5	(55)	(30)		9.5	(29)	(77)	
Lighting	16.1	(58)	(38)		15.8	(30)	(16)		13.1	(40)	(30)	
Total	27.7	(100)	(43)	(-1)	51.9	(100)	(26)	(-1)	32.6	(100)	(46)	(-1)
Redesign Exercise Results												
Heating and Cooling	8.0	(30)	(61)		18.1	(45)	(55)		8.5	(28)	(79)	
Lighting	14.8	(56)	(42)		14.1	(35)	(25)		15.4	(51)	(18)	
Total	26.5	(100)	(46)	(11)	40.1	(100)	(43)	(7)	30.0	(100)	(58)	(7)
"Technical Potential"												
Heating and Cooling	6.0	(29)	(70)		5.2	(21)	(87)		9.8	(40)	(76)	
Lighting	11.2	(55)	(57)		12.7	(50)	(32)		9.2	(37)	(51)	
Total	20.4	(100)	(59)	(16)	25.3	(100)	(64)	(11)	24.6	(100)	(65)	(9)

*A = Energy in MBtu/Sq. ft./year.

*B = Percent of total building energy.

*C = Percent reduction from original design.

*D = Percent of first cost increase.

NOTES:
1. Energy results for HVAC fans, domestic hot water, elevators, escalators, and general exhause fans not listed here.

2. "Process" energy is not included in this analysis.

As can be seen from the "technical potential" case, total energy reductions ranged from 59 percent to 65 percent. Combined reductions for heating and cooling systems were from 70 percent to 87 percent. Lighting system reductions were from 32 percent to 57 percent. First costs at this level increased from 6 to 16 percent.

Conservation measures that contribute substantially to the energy reduction included: HVAC systems and control improvements, including thermal storage, "deadband" thermostatic controls, more efficient lighting systems and daylighting (see Appendices A-C, AIA, December 1979 for a description of the costs and energy savings of the conservation strategies used).

Renewable Energy Supply

Daylighting in New Commercial Buildings

In development of the proposed BEPS energy budget levels, there were no direct measurements of passive solar benefits for new commercial building designs. While passive solar design strategies and solutions were encountered in the 1978 redesign exercise, the annual energy calculation program being used at that time-AXCESS-could not model the passive solar features. Such features included daylighting, greenhouse effect, trombe walls, rock bed storage; further, daylighting was a very common strategy and occurred in about 70 percent of the redesigned buildings.

By 1979, an analysis of the impact of daylighting had been conducted and a daylighting algorithm was added to the AXCESS Program. Several daylighting strategies were then examined for each of three typical office buildings designs which were part of the lifecycle cost study just described. Table 1.62 summarizes the results for eight daylighting strategies on the three buildings. For each building, both on-off switching and dimming controls were used; the Denver Buildings used a 3 zone system, while the Minneapolis and Raleigh Buildings used 2 zone systems. For the Raleigh Buildings, two totally new building design concepts were also developed to maximize daylighting and utilizing Task Lighting.

The six daylighting strategies applied as modifications to the redesign buildings were effective means of conserving energy. The average reduction in total annual design energy requirements was 5.5 percent. (Then range was between .5 percent and 14 percent.) In addition, the two totally new designs for the Raleigh Buildings resulted in reductions in total annual design energy rquirements of over 18 percent.

These were reductions from already efficient building and lighting designs. The annual design energy requirements for the three redesign buildings (Base buildings for the daylighting strategiees) were:

- o   Denver   26,505 Btu/GSF/Yr.

- o   Minneapolis   40,076 Btu/GSF/Yr.

- o   Raleigh   30,030 Btu/GSF/Yr.

The impact of the daylighting strategies on the combined energy requirements for heating, cooling, and ventilating was insignificant for the six modification strategies for the two new designs for the Raleigh Building, the increases in HVAC energy requirements (about 1500 Btu/GSF/Yr., or 14 percent) were more than offset by the substantial reductions in daylighting and tasklighting (about 7000 Btu/GSF/Yr. or 46 percent of the lighting energy).

It is difficult to separate the impacts on the heating versus cooling energy requirements from the daylighting, because of the HVAC Systems and computer model used; generally, the heating requirement increased but was offset by a reduced cooling requirement and by reductions in HVAC auxiliary and fan energy.

For the six modification strategies, there is a 1- to 1.5-percent increase in capital cost and less than a 1-percent increase in life-cycle cost from the base buildings. For the two new designs for Raleigh, capital costs increased from 2 to 5 percent, while life-cycle costs both increased and decreased depending upon the economic perspective used.

Daylighting could have a significant impact on commercial energy consumption. It is estimated that the Raleigh Design #100 would use 5565 $Btu/ft^2/yr$ less than the Base Building Design. Using this reduction as a surrogate for commercial buildings generally, daylighting could displace 55 billion kWh or 0.6 quads

if incorporated in the 33.5 billion square feet of commercial space projected to be built between 1980-2000. It seems reasonable to expect that one-fourth to one-half of these buildings would implement this level of daylighting if public policy encourages such action. Thus daylighting is projected to have the realistic potential to displace 0.2-0.3 quads annually in the commercial sector by the year 2000 if aggressively implemented.

Photovoltaics in Commercial Buildings

The potential for use of photovoltaics on commercial buildings is considerable. If very low-cost arrays are developed, it may become feasible to use photovoltaics as vertical wall coverings on commercial structures. However, that has not been assumed. Array costs are expected to be similar to the residential case except it is assumed that the arrays are mounted on racks on a flat roof. The cost of such racks has been estimated to be $0.208/$W_p$ (OTA, 1978, Vol II, p. 689) in 1980 dollars. Thus, the cost of commercial photovoltaic installations is assumed to be:

	High	Low
1986	$2.40/$W_p$	$1.80/$W_p$
1990	$1.80	$1.50
1995	$1.65	$1.40
2000	$1.50	$1.30

New commercial floor space is assumed to follow the estimates (with regional distribution) as predicted by the ORNL Commercial Model. It is assumed that arrays are spaced on the roofs so there is no shading on December 21. Typical latitudes for each region are shown in Table R-25. There seem to be no careful estimates of the number of floors in the average commercial building. It has been assumed that the average commercial building has two floors so the roof area is half of the new floorspace. If this overestimates the roof area, the penetration estimates in this section may be considered to implicitly assume some retrofits as well as new construction.

The fixed charge rates for commercial ownership were calculated using the formalism developed by OTA (OTA, 1978, Vol. II) for commercial buildings. It is assumed that financing is 75% debt and 25% equity, that the owner is in the 50% Federal tax bracket, and that property taxes, insurance, etc. are 2.25%/year. It is also assumed that the owners receive a 10% return on equity (in constant dollars) and that the prsent solar tax credit of 15% for commercial investments applies.

Using these assumptions, the potential range of energy displacement by photovoltaics on commercial

Table 1.62. IMPACT OF DAYLIGHTING (ENERGY)
(Measured in Btu/Gross Sq. Ft./Yr. "DHW" and "Other" Not Shown)

Building Design	HVAC (Incl. Aux. & Fan)	Lighting	Total
Denver Redesign (Base Bldg. Design)	9,361 (100)	14,790(100)	26,505(100)
Denver 3 Zone Daylighting	9,416(101)	12,722(86)	24,510(92)
Denver Dimming Control	9,322(100)	11,185(76)	22,879(86)
Minneapolis Redesign (Base Bldg. Design)	19,461(100)	14,091(100)	40,076(100)
Minneapolis 2 Zone Daylighting	19,429(100)	12,683(90)	38,626(96)
Minneapolis Dimming Control	19,276(99)	12,659(90)	38,462(96)
Raleigh Redesign (Base Building Design)	10,562(100)	15,297(100)	30,030(100)
Raleigh 2 Zone Daylighting	10,576(100)	14,486(94)	29,128(97)
Raleigh Dimming Control	10,644(101)	15,157(98)	29,869(99)
Raleigh New Design #100	12,071(114)	8,349(54)	24,465(81)
Raleigh New Design #101	11,292(107)	9,225(60)	24,564(82)

buildings is 0.157-0.252 quads. For utility ownership, comparable displacement of 0.181-0.246 quads is expected if utilities receive a 40% tax credit. The displacement drops to about half this level or 0.082-0.149 quads if utilities receive only a 10% tax credit.

Further Research on New Buildings

The BEPS research described above represents a substantive set of analyses available for new commercial (and for multifamily residential) buildings. However, it is not complete, nor is it as well developed as comparable cost/benefit analyses of single-family residences.

Several research projects recently initiated within the BEPS program for commercial buildings are significant both for improving our current understanding of new building energy conservation potentials and for providing additional tools and information for achieving further conservation. These activities include:

o A comprehensive LCC analysis of all major commercial and multifamily building types. Results of this analysis are expected to identify cost/beneficial ranges of energy conservation for the diverse conditions encountered in new commercial buildings of different types in different regions. Also, the analysis will address unresolved technical issues relating to process energy, variations in function within building types, building size and climate.

o   Analysis of potential changes to ASHRAE 90 type component requirements discussed earlier, to effect energy conservation levels similar to those now specified by BEPS. This effort also includes a life-cycle cost component. Such requirements, incorporated into standards and energy codes, will provide valuable additional guidelines for increased conservation.

o   A detailed analysis of 1979-80 design practice, aimed at determining how rapidly the commercial sector is responding to rapidly rising fuel prices; in short, to help assess how well the market is working.

o   The identification and description of case studies of successful energy conserving new building designs.

o   The development of simplified energy analysis and evaluation tools, which would be especially useful for small and medium size commercial buildings. Such tools would also be especially helpful to smaller design firms and developers.

Results of these and other research and data-collection activities, when available, will provide a much firmer basis for assessing least-cost conservation potentials for new commercial buildings. Also, if they are disseminated in easily used formats, they will assist in providing needed information to the building community about conservation opportunities and costs.

POTENTIALS FOR RETROFIT OF EXISTING COMMERCIAL BUILDINGS

Extremely little reliable information is on hand to generalize about commercial building retrofit activity even though thousands of buildings have been audited. Several projects now in progress will provide, when completed, valuable information about the energy use of commercial buildings by building category, by size, by age, and by region.

However, in this report we must derive estimates from available data, which is extremely limited. Available data is in the form of case studies, mostly of public and educational buildings. Data content is often descriptive rather than quantitative, and data completeness, detail and presentation format typically varies considerably from case to case. In most

reported cases there is not sufficient data readily available to do even a cursory cost-benefit analysis.

Thus, for the purposes of this analysis, retrofit conservation estimates for the commercial sector are lumped. There is no effort to analyze potentials independently by building type or by region, even though such potentials may vary considerably. Nor is there a detailed effort to relate savings to specific retrofit actions. For the cases where sufficient data is available, retrofit results are simply expressed in terms of estimated overall percent savings as opposed to annualized investment expenditures vs. energy savings per year, (i.e., $/MBtu), since hard data is only sporadically available on pre and post retrofit energy consumption per fuel type, fuel prices at time of investment, dollars invested, life cycles and discount rates.

The available data does support conventional assumptions that commercial retrofit options usually do not focus on shell upgrading and domestic hot water strategies, as in the residential sector. Rather, current emphasis is on operations and maintenance and on mechanical systems, controls and lighting.

Several efforts have been made recently to increase the level and consistency of documentation of current retrofit activity.

As part of preparing this report, two analyses have been conducted. In the first, LBL asked fourteen experienced architects/engineers to provide their subjective estimates of least-cost potentials. In the second analysis, energy related publications were reviewed for the type and completeness of existing retrofit data. Data was obtained for 82 commercial buildings; however, for our purposes the data was very incomplete. Subsequently, owners/operators of buildings identified in the publications were contacted and asked to provide additional information on retrofit measures, costs and related energy savings.

Also, another collection of retrofit case studies is now being assembled by Howard Ross and Susan Weyland of the Buildings Division of DOE. In this unfunded staff effort, they have been contacting state and individual sources to obtain available retrofit data. To date, they have obtained some data on nearly 140 commercial buildings.

Also, as examples, we have included in this section brief descriptions of 4 of the more complete case studies.

The following subsections describe each of the above-mentioned efforts in more detail.

Professional Judgment Survey

In this survey, a set of fourteen experienced architects/engineers were asked to establish least-cost potentials in 1990 and 2000. They were instructed to assume marginal energy prices ($1.50/gal. oil, $1/therm of gas and $0./10/kWh for electricity).

The average of the replies indicated a potential savings of 25% in both fuel and electricity in 1990, and an additional 25% in 2000. Thus, assuming that the 1980 stock of office buildings consumes 28 kWh/sq. ft and 135 KBtu/sq. ft of electricity and fuel (18,135), the 1990 and 2000 levels drops to (14,100) and (10,70) respectively, as plotted in Figure 1.61.

Also, the general magnitude of estimated potential energy savings is similar to BEPS. As observed in Figure 1.61, the electricity-consumption level is down to about the BEPS level. However, fuel use remains much higher. The higher relative fuel use resulting from this analysis is primarily because current estimates of fuel use in the existing commercial stock are higher than design estimates of fuel use in the post 1973 buildings used in the BEPS analysis.

Collection of Existing Retrofit Data from Publications

The objective of this analysis was to obtain estimates of the current levels of retrofit energy savings and investments from data published in energy related publications. The focus was on identifying retrofit measures and on collecting quantitative cost data and pre and post energy use data, by fuel type when possible. This was a small scale effort funded at less than a person-month. The data was collected in two stages by Ray Biley of W. S. Fleming Associates.

First, a literature search was made of energy publications to determine the availability of retrofit data. Of particular value were articles in Energy User News spanning 5 years. Overall, articles reporting 190 retrofits were identified. Of these, 82 cases were commercial buildings; the remainder were industrial facilities.

It was found that the existing information is at best piece meal and do not lend itself to a comparative analysis across buildings or building types since owner/operators reported in different modes and in different levels of detail. These are data on dollars invested and estimated percentages of energy saved but virtually none on building size or pre- and post-energy consumption by fuel type.

Because very few of the published cases included even the minimal data for a consistent cost/benefit assessment, a follow-up collection of additional data was conducted. Owner/operators of the commercial buildings identified from the publications were contacted in State two. To obtain a minimum set of consistent data for analysis purposes, additional data was requested to supplement the information provided in the publications. Such data included: building size, pre- and post- retrofit energy savings per fuel type, pay-back period, etc.

To date, only limited additional data has been obtained. For example, of the 82 retrofits identified, only 40 contain sufficient data to estimate the percent of total energy saved. Only 15 cases provide sufficient

data to estimate percent of energy saved by fuel type. For cost data we are able to estimate the retrofit costs per square foot in only 20 instances. Further, the documentation provided often does not permit isolating savings from operation and maintenance changes relative to savings from hardware. One of the most interesting measures, the cost of the energy saved, can be derived in only 8 cases. Most of the data obtained to date pertains to offices, schools, stores and a few hospitals. Other building types were not extensively reported in the periodicals examined. Thus, the current results presented here are partial and limited to the responses received so far.

All of the above-mentioned case study data provides some measure of current retrofit activity, including current owner willingness to invest in retrofit. However, the data should not be considered an indicator of least-cost potentials for commercial retrofit. It seems rather that owners are correcting gross problems and are concentrating on the big-ticket items with quick paybacks.

Building owners are concentrating mostly on easy to use off-the-shelf measures such as time clocks, demand delimitors, night set backs, increased maintenance, reduced foot-candle levels and in some instances go as far as implementing energy management systems.

The available data does indicate that the investment levels are typically low and very effective in terms of their impact on the energy use. Nearly 90% (17 out of 20 cases) of the investments were less than 50 cents per square foot and produced estimated energy savings in the range of 20% to 35% of the total energy use (See Figs. 1.62 and 1.63). Such strategies result in a low cost of conserved energy (about $2/MBtu assuming a 10% discount rate and 30 year life-cycle (See Figs. 1.64 through 1.65).

A comparison of this figure with average fuel prices ($13.75/MBtu for electricity, $3.17/MBtu for gas and $3.38/MBtu for oil, Ref. C-13) shows that current levels of conservation in the retrofit area are yet to be carried to the point where the marginal costs of added conservation are equal to the weighted price of the energy saved.

The estimated energy savings are substantial for the 40 cases where it is possible to derive such estimates are available from the data:

o   For 22 cases (55%), estimated savings exceeded 25% of total energy use.

o   For 30 cases (75%), estimated savings exceeded 20% of total energy use (See Fig. 1.66).

These levels of savings seem to indicate sizeable potential savings from retrofit and to support the estimates of large potential savings from the professional judgment survey. However, there is no way to know if

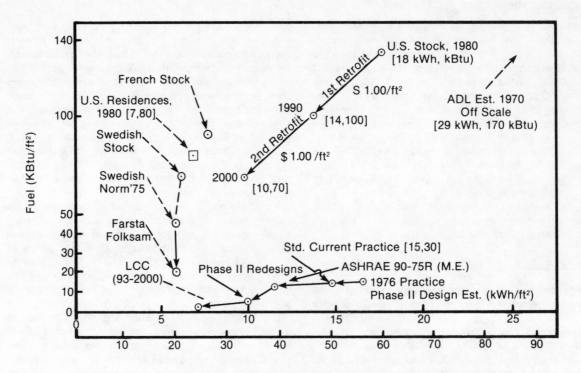

Figure 1.61.   **Energy Use of Existing U.S. Office Buildings and New Fuel-Heated Office Buildings.**

Progress in Swedish building efficiency is shown for comparison. Swedish buildings already use considerable daylighting and electricity is not decreasing, but space heat has dropped from 70 kBtu/ft² for stock to 50 kBtu for Stockholm buildings conforming to the 1975 "Swedish Building Norm" to 20 kBtu/ft² for the Farsta Folksam building which uses thermal storage over nights and weekends. the "U.S. Stock" point comes from the 1980 entries on Table 3 (DOE/EIA 1979 Report to Congress) divided by 32.5 B ft² of commercial space (from the ORNL model) with electricity scaled up by 10% to convert from "commercial sector" to "offices only." It is assumed throughout the text that the whole commercial sector will follow the office buildings trend.

The oblique line represents 100 kBtu/ft² BEPS resource energy, approximately.

the buildings reported in the publications are representative of the population of existing buildings. For example, one might suspect some built-in bias in that the most successful retrofits would tend to get reported, whereas less successful retrofits, or failures, would not be reported. Also, no effort has been made here to verify the accuracy of the information reported.

Case Studies

The following descriptions of several case studies are provided as examples of current retrofit activity.

The relative cost-benefit of retrofits are expressed in $/MBtu of conserved energy. The dollar figure represents the investment converted to annual cost using the uniform capital recovery (UCR) factor at a real discount rate of 10% and a loan period of 30 years (for homeowners a 3% discount rate and 20-year loan were

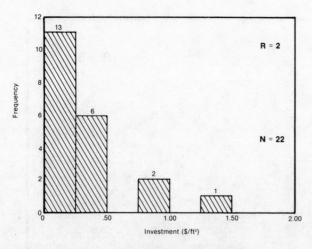

Figure 1.62. **Retrofit Investment ($/ft²)**

chosen). Thus, for commercial buildings these assumptions yield a factor of .106 (UCR for homes is .067).

o Ohio State University (Columbus School)

The Ohio State Columbus campus invested four million dollars in retrofit between 1974-78 reducing fuel and electricity use by 40% and 50% respectively, or the equivalent of twelve million dollars in cumulative bills to date. As can be seen in Figure 1.67, the original buildings were very inefficient compared to the national average, probably because they were operated 24 hours a day for most of the year. The average cost of the retrofit was $0.6/sq. ft., savings in resource energy 300 KBtu/sq. ft./yr and the cost of conserved resource energy $0.20/MBtu, i.e., $0.025/gallon, or $1.16/bbl (these computations have actually doubled Ohio State's published costs, because they were estimated too low).

o State of Minnesota

Hirst, Tyler, Eastes, and Dumagan analyzed the results of detailed engineering audits performed at 41 institutions (including seven office buildings) owned by the State of Minnesota. (Part of the results were published in the June 1980 ASHRAE Journal, page 47). The published results for the seven office buildings were combined.

These seven buildings start well below the U.S. average, yet they can be retrofit for $1/sq. ft. It is projected that their fuel use will be halved and their electric use reduced to 73%, from 12.6 to 9.2 kWh/sq. ft. resulting in a cost of conserved energy of $1.22/MBtu.

o Ebasco

Ebasco is a large engineering company (6700 engineers) and a world leader in power plant development. It has participated in the design of 900 power plants, but has recently decided to diversify into end-use efficiency and has been offering audits and guaranteed savings to commercial buildings, campuses, hospitals, etc. Data on seven of their current proposals are plotted on Fig. 1.68. (The average guaranteed improvements is plotted as a heavy arrow.) Their average recommended retrofit investment is $1.19/ft. for a guaranteed savings of 20% in electricity and 45% in fuel. The average cost of conserved resource energy is $1/MBtu.

o Elementry Schools

For completeness, we quote one discouraging experiment with ten elementary schools. The American Association of School Administrators supported by DOE and assisted technically by LBL, undertook the retrofit of 10 elementary schools around the United States. The experiment started in 1975 when admittedly there was less interest and experience in retrofit. Over the 2-3 years of the experiment, there were indeed about 20% savings, but by both the controls and the retrofitted schools (Ref. C-1). However, there was no single active supervisor (e.g., a utility, a service company, or a qualified engineer emloyed by the school board), the school was left to deal directly with contractors who frequently misunderstood what was needed, thus, installing the wrong hardward or showing poor workmanship.

Summary of Retrofit Potentials

This section has provided a brief overview of the existing retrofit data in an effort to derive some assessment of least-cost potentials for the retrofit of existing commercial buildings. Following is a summary of the trends that could be observed:

o Retrofit data is just beginning to be accumulated in a consistent, publically available manner; the sources are many and so are the reporting formats. Thus, there is not a detailed, articulated and consolidated retrofit data base in the public domain at this point in time.

o State energy offices are a good retrofit data source however, the best quality hard data is in the hands of private consulting firms.

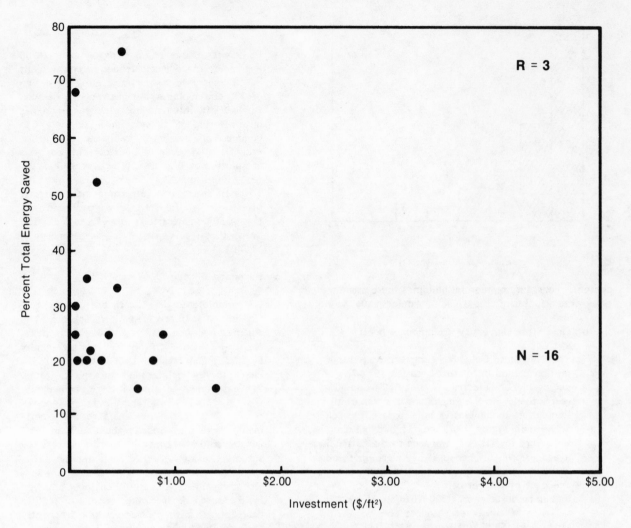

Figure 1.63. **Percent Energy Saved by $/ft² Invested**

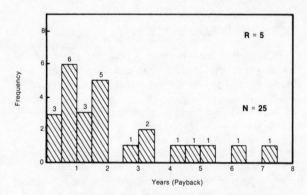

Figure 1.64. **Payback Period of Retrofits**

o Most of the available data covers schools, colleges/universities, offices (both public and private) and hospitals. Good documentation for other building types has not yet been located. For example, it is known that certain restaurant and store chains have active retrofit programs.

o Current retrofit measures cover, for the most part, off-the-shelf solutions yielding very high rates of return and very quick paybacks.

o The cases analyzed to date, suggest the potential, on a cost effective basis for considerably higher levels of retrofit investment than those now occurring. The existing gap between the cost of the conserved energy and the energy prices show that the least-cost potentials are yet to be achieved.

o Because the available case study data is so limited, and also because it does not seem to

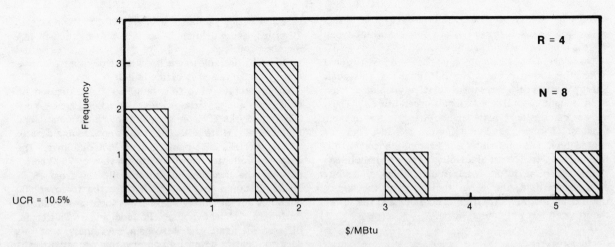

Figure 1.65. **Cost of Saved Energy**

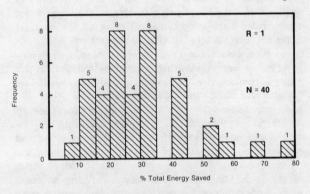

Figure 1.66. **Retrofit Energy Savings (% of Total Energy)**

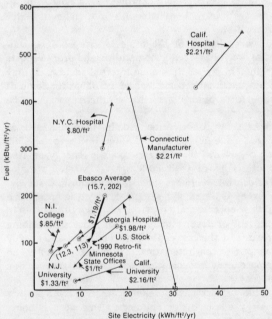

Figure 1.68. **Ebasco Proposals.**
Average Cost of Conserved Energy, Using Capital Recovery Rate of 10.5%/years; Hirst Survey of Minnesota State Offices, $1.22/MBtu, $1.00/MBtu.

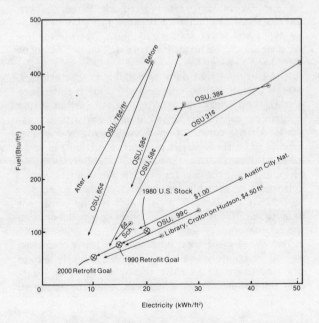

Figure 1.67. **Ohio State University Retrofits (and one Columbus school).**

contradict the estimates derived from the professional judgment survey, the estimates derived from the professional judgment survey have been used for estimates of technical potential for commercial retrofit, in the absense of better data.

An indication that the present levels of conservation are below the actual technical potentials in that nearly 40% of commercial retrofits yield less than a one year pay-back, 70% fall in the two to three year range, meanwhile, 75% of the retrofits analyzed generate savings in excess of 20% of total energy use. This strongly suggests that savings are possible well beyond what are presently being achieved.

There are, however, built-in market and institutional mechanisms that prevent the fulfillment of this goal. Energy costs still represent a relatively small portion of a tenant's total operational expenditure. Thus, owners can easily transfer higher energy costs to their tenants. This process is not limited to non-owner occupied type buildings alone, for example, hospitals bill their operational costs (including energy) to their patients according to pre-set formulas. Also, taxes act as a disincentive, the higher the tax rate, the higher the income tax benefits from deductions and the lower the induced level of conservation.

Further Research

As noted earlier, several projects are now in progress that will provide valuable information about building retrofits.

For virtually all of the projects cited below, substantial work has been accomplished and draft reports are either in preparation or in internal review. It may well be that final reports will be available for some of these efforts before the final of this report will be available.

The Energy Information Administration (EIA) of DOE has recently completed interviewing over 6000 building owners or managers on the energy-related characteristics of their buildings, as part of a major nonresidential buildings energy consumption survey. The survey is based on a sample designed to ensure the statistical representation of the results. EIA is now in the process of developing a series of reports describing the results of this survey, including building and energy consumption characteristics. However, results on building energy consumption are not available at this time.

Another effort is the General Electric (GE) project on energy use in office buildings. This work is being conducted on a cost-sharing basis with DOE, analyses of the 1977 BOMA Experience Exchange Report data base for large office buildings. The analysis includes both commercial and government office buildings and has examined the relationships between energy consumption and such factors as: downtown or surburban location; building height; building age; computer space; air conditioning equipment. The analysis has also examined energy consumption trends from the 1975 BOMA data base to the 1977 data base. While the BOMA data base is considered very comprehensive, it may not be representative because of the types of office buildings that may tend to be entered into the data base and the voluntary means of collecting data.

GE is also conducting another cost-sharing effort with DOE to to define a more representative survey of office buildings. This is an analysis of randomly

selected office buildings over 40,000-sq.-ft. in size in 20 cities, and will include both BOMA and non-BOMA member buildings. Results of this work are expected to add substantially to our knowledge of representative energy conservation in office buildings.

Other studies of office buildings have focused on buildings in specific cities including New York (Tishman/Syska & Hennessy), Baltimore (Hittman), and Philadelphia (Hittman). Hittman is now adding data to a computerized data bank of over 2000 buildings. The data was collected from on-site surveys of 2000 buildings for the commercial sector collected from 1970-1978. Detailed data has been collected for over 900 buildings, of all categories, from locations including Baltimore, Minneapolis, Philadelphia, California, Illinois, Michigan, and Wisconsin. An analysis of this data to identify potential conservation opportunities is just beginning. Furthermore, for some 100 buildings within this data base, retrofit data is available under time-of-day rate conditions. Timing of the retrofit activites is available, but cost data is incomplete.

Another effort for DOE, as part of the BEPS research, has been tabulating current data bases for the energy consumption of existing buildings, including commercial buildings. Work to date has identified numerous potential data sources available from energy audits on many buildings. The overwhelming proportion of the sources identified were state agencies and the buildings were audited as part of the schools and hospitals program. In some cases data for other types of public buildings is also available. A few sources, each among Federal agencies, utilities, energy consultants, private companies, and trade organizations for specific building types were identified. These sources tended to include a broader spectrum of building types and fewer public buildings. This work is being done by the National Institute of Building Sciences (NIBS) and Xenergy Inc.

It is expected that data on building characterictics, energy consumption, operations and maintenance (O&M), and retrofit activity exists within many companies, especially large corporations with professional design, construction management, and operations staffs. However, it may prove difficult in some cases to obtain such information, in publicly available form, especially details for specific buildings.

It would be valuable for the existing data sources to be publicly available in a consolidated and consistent data base. Today most data bases are resident with the originators.

The above-mentioned data, with a few exceptions, even if completely available, does not directly help in the assessment of retrofit technical limits or least-cost potentials. Further analysis would be required; and some alternatives include:

o for buildings with consumption data, examine for reductions over time. If they exist, seek O&M changes (or retrofits) as explanations.

o identify O&M and hardware retrofit opportunities in each type of building. Conduct life-cycle cost analyses and sensitivity analysis to assess technical limits and least-cost potentials.

Some initial efforts toward these ends have been initiated.

The Buildings Division of the Department of Energy is currently compiling and analyzing retrofit data from a wide variety of sources in an effort to consolidate the data into one set of results. So far, this collection effort covers approximately 140 buildings. Most of the buildings are publically owned; only 30 are privately owned. The breakdown of the data base by building type is:

Offices:	38
Schools:	22
Colleges/Universities:	65
Hospitals:	15
Correctional Units:	9
Hotels:	4

The objective of the study is to categorize the different types of retrofit measures in terms of their impact on energy use and cost of energy saved. This information is expected to provide a measure of the energy and dollar returns which might be produced from any given retrofit option.

Following are brief descriptions of the major sources of the data gathered so far:

o Buildings in Philadelphia and elsewhere within Pennsylvania were tested to evaluate the effectiveness of the NECA/NEMA energy conservation manual. These buildings were divided into three sample groups: a) No assistance or consultation was given, b) The conservation manual was handed to the building managers, c) In addition to the manual, assistance from a professional consulting engineer was provided. The energy use of these buildings was monitored for 12 months, also, data from previous years is available, however, retrofit costs are non-existent.

o The Bureau of Energy Conservation of New York State: 20 State facilities including 2 university units, 9 correctional units and 9 psychiatric facilities were audited. Operation and maintenance measures alone were applied. Cost data are not available.

o State of New Jersey Department of Energy: 13 public office buildings were audited, retrofit measures were limited to operation and maintenance measures. Cost data are not available.

o American Council on Education/Association of Physical Plant Administrators of Universities and Colleges/National Association of College and University Business Offices: Data compiled on more than 60 colleges and universities in the United States and Canada. A wide variety of retrofit measures are covered: envelope, electrical, HVAC, O&M, utilities production etc. . .

o Private Sources: Approximately 20 to 30 buildings (offices, hospitals and hotels for the most part) are analyzed. Some of the most complete data sets are found in this category.

At the present time, Ross and Weyland are checking and tabulating the data received. Results are not yet available, but it is expected that this information will be available by April, 1981.

COMPARISON WITH AUTOS

A massive program to improve the energy efficiency of autos is presently being carried out as manufacturers are being persuaded by legislation, consumer preference, and foreign competition. Since cars are replaced every 10 years, their fleet fuel efficiency will double in ten years. The capital cost of this strategy is estimated at $10 billion, to achieve a reduction in fuel use from ten quads annually to five. Buildings are more longlived than cars, but luckily are also easier to upgrade. The main energy saver in the buildings sector is the retrofitting of old buildings, a strategy which we estimate to cost $30-60 billion, and result in our reduction in energy consumption from 10 quads to 7.5 quads by 1990 and 6 quads by 2000 (Figure 1.69).

The comparison between energy savings with autos and commercial buildings may be summarized as follows.

	Automobiles	Commercial Buildings
Investment to 1990 ($ Billions)	100	30+
Saved (Resource Quads)	5	2.5
$ billions/annual quads saved = $/annual MBtu saved	20	12.5
Cost of conserved energy at 10.5% annual CRF		
(in $/MBtu)	2	1.25
(in $/gallon)	0.25	0.15

+ Phase I Retrofit

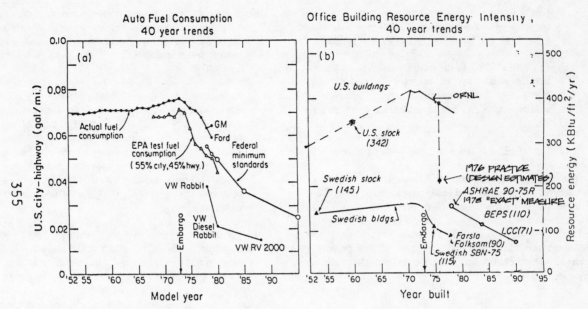

Figure 1.69.   Horsepower Race, Autos vs. Office Buildings.

A 40-year perspective, measurements vs. calculations.

Sources: Society of Automotive Engineering Transactions, 750957; J. Pierce, 1975, Scientific American 232.1 (Jan. 1975); F. von Hippel, 1980, "U.S. Transportation Energy Demand - Draft Report," (July 1980).   Office Buildings: Energy Efficient Buildings - Draft Report, LBL 11300, EEB 80-6.   (See also figures C-2 and C-5.)

This simple calculation shows that energy conservation in buildings has a similar potential and a lower cost when compared with energy conservation by automobiles.   In fact, the balance is even further in favor of buildings than the above table shows, since an energy conserving building has a much longer lifetime, i.e., it saves energy for longer, once a dollar has been invested, than does an automobile.   If we write off the auto in ten years, the CRF rises to 16%, and the cost of conserved gasoline rises to $0.40.

SOME CHARACTERISTICS OF THE COMMERCIAL SECTOR

Diversity of Building Use

A striking characteristic of commercial sector buildings is the wide variety of human activities for which they provide space and services, including energy services.   This diversity of functional use causes considerable diversity of energy use and constitutes a major distinction between the commercial and residential sectors.

Residences have a common set of functions: providing habitation units for one or more persons, including kitchen facilities within each unit.   Even this may take many forms: detached single story, split level, or two story; townhouses; multi-family high or low rise residences for different population segments; mobile

homes; vacation homes; etc.   However, the basic use is similar.

The commercial sector, on the other hand, encompasses many disparate activities.   This is typified by the building classifications by which the commercial sector is subdivided.   Table 1.63 lists a few classifications.   Many other variations exist and are referenced in model codes as well as state and local building codes.

Different uses of commercial buildings can require radically different energy support requirements.   For example, the energy per square foot needed to support a fast food kitchen, a hospital operating room, or a large computer installation is radically different from the energy per square foot required to support routine office tasks, a hospital patient room, a high school gymnasium, or a warehouse.

These differences in energy requirements result from many factors related to the specific use of the space, the hours (duration) and the intensity (occupancy density, etc.) of use.   Different activites require different levels of energy-related support to ensure human comfort, health, and productivity, including temperature, humidity, ventilation, illumination, and energy for typewriters, xerox machines, microphones, kitchens, computers, sales displays, frozen food freezers, etc.

One tends to talk of buildings and Btu's rather than of the human activities and the associated services-- illumination,   temperature,   humidity,   "process"

Table 1.63. BUILDING CLASSIFICATIONS

ORNL	EIA*	BEPS
Retail-Wholesale	Retail & Personal Soc. Food Sales	Shopping Centers Restaurant Fast Food
Office	Office	Small Office Large Office
Automotive Sales and Service	Service	Automotive Sales and Service
Warehouse		Warehouse/Storage
Education Services	Education	Elementary Schools Secondary Schools
Health	Health Care	Hospital Clinics
Public Buildings	Public Buildings	Public Buildings
Hotel/Motel	Lodging	Hotels/Motels Nursing Homes
Miscellaneous	Assembly Other	Assembly Gymnasium Theatre/Auditorium Community Center Multifamily High Rise Multifamily Low Rise

*Nonresidential buildings energy consumption survey

requirements--for which the buildings are designed, constructed, and operated. It has been said that "people use energy, not buildings." This is somewhat of an overstatement, for an energy-efficient building can provide the same amenities and services for much less energy use than an "energy hog," but it is still an important point. It is the human activities that dictate the energy-related service requirements. The many categories and subcategories of commercial sector buildings result from the very wide range of human activities for which commercial buildings are used.

The building classification schemes included in Table 1.63 are not sufficient to isolate the key energy-related factors related to the diversity of uses of commercial buildings and their spaces. The BEPS scheme, while it attempts to isolate buildings into categories with similar energy service requirements, does not include certain types of buildings; e.g., prisons, churches, police and fire stations, etc. The other schemes group building functions under common headings which may have extremely different patterns of use and energy service requirements.

For example, "Health Care" buildings can include functions with extremely different hours of use and energy service requirements: hospitals; mental facilities; rehabilitation facilities for drug addiction or alcoholism; other extended care facilities; veterinary facilities; and clinics (general, dental, mental, veterinary, etc.). A general clinic with doctor's offices is likely to have use and energy patterns more similar to an office building with six day and evening use than it is to a hospital. Even within the hospital designation there can be significant variations. A hospital "building" can be predominantly a patient care facility, an administrative.facility, a surgical facility, a laboratory facility or varying combinations of these. Each of

these facilities will differ in hours and intensity of use, heating, cooling, ventilation and lighting level requirements and in required support services and equipment.

Likewise, "Food sales and service" facilities can include several types of full service restaurants, fast food and carry out restaurants, cafeterias, and retail food establishments like supermarkets, bakeries, small food stores, and fish, meat or farmers markets. While the markets require very little energy for processing food, supermarkets can have substantial refrigeration requirements. Fast-food restaurants, on the other hand can use as much as 600,000 Btu/sq.ft./year or more of site energy, including cooking energy. This is more than 4 times the average actual consumption from recent BOMA data for large office buildings of mixed ages. In fast food restaurants, kitchen "process" energy box cooking can be 60 to 80% of the total energy required for the buildings, can vary considerably with the type of food being cooked (hamburgers, fish, fried chicken, etc.), and can make the energy required for human comfort (heating, cooling, lights) seem relatively insignificant.

The variation within the "Assembly" building category is even more pronounced. The category includes buildings with such activities as town halls, auditoriums, convention halls, gymnasium, skating rinks, indoor pools, libraries, museums, night clubs, and passenger terminals (airport, bus, train, etc.).

Multi-family residences tend to be treated inconsistently by different classification schemes. These buildings have attributes of both commercial and residential sectors. The BEPS research, after Phase 1, has included them in the commercial sector, even though in Phase 1 high rise were in the commercial sector and low rise in the residential. Because of this definitional ambiguity, the issues specific to multifamily residences tend to receive short shrift from analyses in both commercial and residential sectors. A paper on multifamily residential technical, financial, and institutional issues was prepared by D. Bleviss from the Santa Cruz conference held in preparation for this report. The paper, entitled "Retrofitting Mulifamily Housing," is included as Appendix C-4 to this chapter.

Multi-function buildings, which have changing uses over time, and establishment-building differences tend to increase the diversity given through the above examples:

o A small office building that is 50% office, 20% card store and 30% pizza parlor has different energy requirements from a small office with 100% office related space.

o A space in a strip shopping center might house, over time, a dentists office, a pizza parlor, a computer store, a liquor store, etc.

----

o   A 100,000-sq.-ft. surburban office building might contain 30 businesses or establishments. A college campus may have 30 diverse buildings under a single establishment, where those buildings may, or may not, have a central heating or cooling plant. Corporations such as Penneys, Sears, Montgomery Ward, IBM, Safeway, McDonalds, 7-11, etc., may control many buildings in different locations through direct ownership or franchise.

Even within a single building category, variations in the mix of activities from building to building can change energy service requirements. For example, a warehouse may typically contain 10 percent of its space as offices and 90 percent as storage. The office space generally requires greater energy service: more illumination, more stringent temperature, and humidity conditions (including cooling). If a particular warehouse contains 20 percent or 40 percent office space, its energy service requirements will be higher.

Impacts of Diversity

The diversity outlined above complicates energy-related analyses of and policies for the commercial sector, relative to the residential sector. Commercial buildings tend to be larger than residences; that internal loads (people, lights, processes) predominate in the larger, or more complex commercial buildings, but not necessarily in the smaller or simpler buildings; and heating, ventilating and air conditioning equipment (HVAC) or lighting systems often are more critical determinants of energy efficiency and cost than is the building shell. For example, doubling insulation in an already effective office building in Denver resulted in less than a 1% total energy savings, where as changes in HVAC and lighting systems produced much higher savings (C-3).

The diversity of uses of buildings within the commercial sector spawn or relate to several important factors affecting least-cost technical potentials in the commercial sector. These include:

o   intensity of energy use,

o   conservation potentials,

o   ownership types, and associated investment strategies and opportunities,

o   physical energy-related characteristics of the buildings

o   the processes used to design, construct, and operate them.

The following subsections briefly highlight some of these factors. However, an overall structure of the commercial sector relative to these factors is not attempted here.

Intensity of Energy Use

The amount of energy used by a commercial building is highly dependent upon what it is used for, as well as upon the efficiency of the building design and operation.

Figure 1.56 shows how the average site energy varies by building type for new commercial buildings circa 1975-1976. These are BEPS Phase 2 estimates of building potential energy use that are derived from design data using computer analysis (see New Buildings Potential Section). Hours of operation within each building type have been normalized. Within a building type, the estimates were derived with each space use having the same operating schedule from building to building. Figures 1.57 through 1.60 show individual building design energy estimates within 4 selected building types.

Building Characteristics

This subsection highlights some descriptions of commercial buildings as they relate to different building uses. The subsection is in 2 parts: 1) Use Requirements, and 2) Building Physical Characteristics.

Use Requirements

The following are brief descriptions of how some of these factors can vary by building type and within building type. For the most part, data for these descriptions is derived from the BEPS design data for buildings built in the 1975-76 period. In many cases, boundary conditions for these requirements are set by standards or are codified into law.

Occupancy density. The occupancy density estimates derived from the BEPS Phase 2 Research are shown in Figure 1.70 for a few selected building types. As can be observed, building average occupancy densities not only vary significantly across building types but also within any given type.

Lighting. Illumination levels for different spaces are related to the visual difficulty of the tasks involved. For example, typical recent Illuminating Engineers Society standards (1972) for lighting levels at the task surface include 30FC (footcandles) for office conference areas, 70 to 100FC for general, 30FC for gymnasium general exercise areas, 50FC for vocational cooking areas, office tasks, etc.

Such variations in required levels of illumination translate into variations in installed lighting levels in

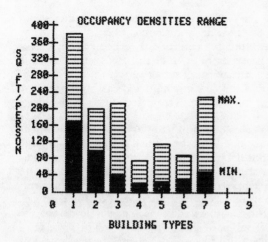

1. HOTEL/MOTEL
2. LARGE BUILDINGS
3. SMALL OFFICES
4. ELEMENTARY SCHOOLS
5. SECONDARY SCHOOLS
6. SHOPPING CENTERS
7. STORES

Figure 1.70.

SOURCE: U.S. DEPARTMENT OF ENERGY, "STANDARD BUILDING
OPERATING CONDITIONS, TECHNICAL SUPPORT DOCUMENT FOR
NOPR ON ENERGY PERFORMANCE STANDARDS FOR NEW BUILDINGS,"
NOVEMBER, 1979

different buildings and building types. The average watts/sq. ft of installed lighting for several building types from the BEPS Phase 1 sample included:

Offices	2.8 watts/sq. ft
Education	2.6 watts/sq. ft
Assembly	2.2 watts/sq. ft
Warehouse	1.8 watts/sq. ft
Multi-family high rise	1.6 watts/sq. ft

The variation within a building type can also be significant. For office buildings installed lighting capacity varied considerably. Of the total square footage of office space in the sample the following distribution occurred:

Watts/sq. ft	Percent of space
1 to 2	13%
2 to 3	35%
3 to 4	35%
4 to 5	17%

In the late 60's and early 70's, installed capacities for office buildings of 4 watts/sq. ft or more were not uncommon. By the mid-70's the average, as listed above, was 2.8 watts/sq. ft. In 1980, while average practice is higher, "good" office lighting design, with emphasis on more efficient lighting systems and reduced illumination levels, is resulting in installed capacities of 2 watts/sq. ft or less. By 1990 or so, installed capacities of 1 watt/sq. ft are estimated to be achievable for office buildings, using advanced lighting system technologies. Such improvements in lighting system efficiencies and in lighting control technologies can cut the energy service requirements for lighting in the commercial section by 1/2 by 1990 (S. Berman, LBL). Further, from an investment viewpoint, government funding to encourage the market penetration of such new techniques has extremely good cost-benefit ratios.

Installed capacities are only part of the variability. The hours per day that various parts of a building's lighting system are used also contributes to differences in energy use. As Figure 1.71 shows, design energy-use levels for lighting in BEPS Phase 2 buildings varied considerably, with warehouses at the low end of the spectrum. Also, observe the typically low energy levels for residential type uses (multifamily high rise-12, 164 Btu/sq. ft/yr average, and multi-family low rise-11,255 Btu/sq. ft/hr average).

Ventilation. Minimum ventilation requirements can vary from 7 CFM (cubic feet per minute) in the case of warehouses to 30 CFM in the case of kitchens and dining rooms. The Phase II data presented in Figure 1.72 shows the composite variation for fourteen building types. As expected, warehouses and multifamily low rise have the lowest averages at 5.9% and 4.4% of daytime outside air respectively, and hospitals and nursing homes have the highest average of 80.3% and 35.8% of daytime outside air.

Process loads. Although BEPS data on energy levels for process loads is very limited, since process loads were not included in the energy analyses, energy loads (peak demand) range from 0.5 watts/ sq. ft. for offices to 56.5 watts/sq. ft. for fast food restaurants. Data at the building level is presented in Figure 1.73, "Process Load Histograms." There is a wide range of variation in the watts/sq. ft. values, which occurs not only among the average values for these three building types, but also among individual buildings within each type. For example, while all small office buildings but one have process loads of less than 1.0 watts/sq. ft., the office building with over 5 watts/sq. ft. of process load has bout 30 percent of its floor area as a restaurant.

Building Physical Characteristics

This section describes variations in several physical characteristics of commercial buildings. In addition to this text, additional information on physical characteristics of the BEPS sample of 1975-1976 new commercial buildings is presented in AIA, Task Report, January 1978 and AIA/RC, March 1980.

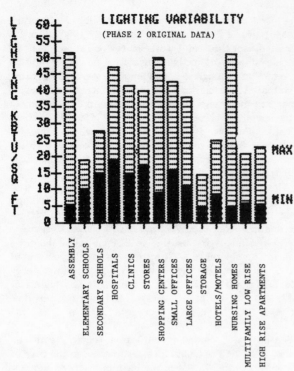

Figure 1.71.

SOURCE: AIA/RC, "ANALYSIS OF DATA ANOMALIES, FURTHER
ANALYSIS OF PHASE II BLDGS", MARCH, 1980

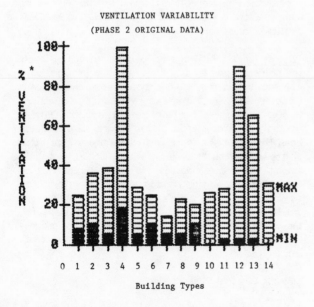

Figure 1.72.

SOURCE: AIA/RC, "ANALYSIS OF DATA ANOMALIES, FURTHER
ANALYSIS OF PHASE II BLDGS", MARCH, 1980

* Data expressed as a percent of daytime outside air

Building size. Floor area is probably the factor showing the most variance in the commercial sector. For
instance within the category of "office buildings," sizes can vary from a 1000 sq. ft. branch bank, a
15,000 sq. ft. one story suburban office, a 100,000 sq.
ft. 7-story suburban multi-tenant building to a
40 story, 1 million sq. ft. office tower. Buildings
within the hotels/motels category vary from the simple strip roadside motel with 20 rooms an office, recreation room, and maybe a small restaurant to a large
conference hotel with guest and meeting rooms, auditoriums, shops, restaurants, etc. Size variation for 16
building types from the BEPS Phase 2 sample is present in Figure 1.74. For this sample of 168 buildings,
large office buildings and shopping centers showed the
widest variation in size. Fast food and other restaurants proved to be most consistent in size. Table 1.64
indicates the size distributions encountered in the
larger Phase-1 sample of 1661 buildings which was aggregated into 12 building types.

Neither Figure 1.74 or Table 1.64 indicates the preponderance of very small buildings in the commercial
sector. Other studies note that over 50% of commercial buildings are less than 5,000 sq. ft. in size. However, in terms of square footage of space, and
implicitly, of relative total energy use within the
office building category, a single 1 million sq. ft.
office building is equivalent to 200 offices of 5,000 sq.
ft. While the above BEPS figures pertain to samples of
buildings circa 1975-1976, the Energy Information Administration (EIA) of DOE has recently interviewed
owners/operators of over 6,000 nonresidential buildings
representative of the building stock.   That effort
should provide a more definitive description of the distribution of building sizes, when that information becomes available.

Building Life. Building life expectancy can vary from
building types to building type:

o   An owner occupied or spec office building can
    reasonably be expected to have a life of 40-50
    years, unless it is in the path of a major devel-
    opment. This expectation is considered in the
    life-cycle cost analysis for office buildings dis-
    cussed in Section 3 of this chapter.

o   Many publicly owned buildings have life expec-
    tancies considerably longer than 50 years.

o   Fast food restaurants, on the other hand, may
    have an estimated lifetime as a fast food
    facility of only 15 years.

Such differences in the expected life of the total facility, coupled with shorter replacement cycles for
building systems and components, may impact investment decisions and opportunities from building type to

HISTOGRAMS: Process Energy Levels

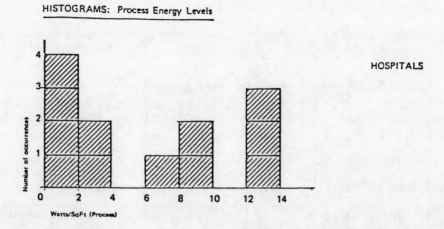

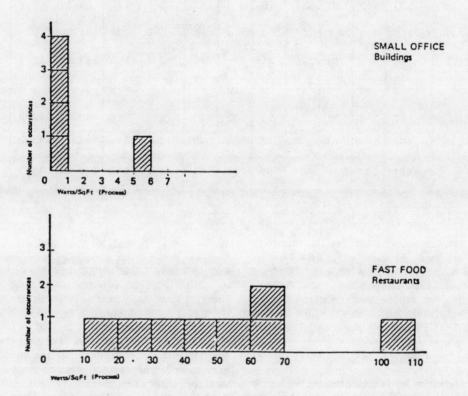

Figure 1.73.

SOURCE: AIA/RC "ANALYSIS OF DATA ANOMALIES, FURTHER

ANALYSIS OF PHASE 2 BUILDINGS", MARCH 1980

building type. The study period for life-cycle analyses of fast food restaurants will undoubtedly be shorter than for office buildings, for example. This is similar to the differences between the life expectancy of a house compared with the shorter life expectancy of a mobile home.

Table 1.64. GROSS FLOOR AREA   (NUMBER OF BLDGS)   SAMPLE TOTALS

BUILDING TYPE	TOTAL NO. OF BLDGS	NO. OF BLDGS. IN EA. CATEGORY (CATEGORIES ARE IN 1000'S OF SQ. FT.)							SUMMARY STATISTICS			
									RANGE		AVERAGE	ST. DEV.
		<5	5-<10	10-<25	25-<50	50-<100	100-<200	200+	FROM	TO		
OFFICE	237	48	63	59	24	19	13	11	1.4	856.1	41.6	96.4
EDUCATION-ELEMENTARY	157	4	8	40	64	40	1	0	1.1	102.6	36.3	20.2
EDUCATION-SECONDARY	172	7	15	30	27	34	46	13	2.4	340.8	77.6	70.9
EDUCATION-COLLEGE/UNIV.	57	1	6	10	13	15	9	3	3.1	304.9	68.3	63.8
HOSPITAL	40	0	0	5	3	7	10	15	10.7	548.7	168.9	135.4
CLINIC	113	12	36	27	19	11	4	4	3.0	235.5	31.0	45.3
ASSEMBLY	167	23	36	87	15	6	0	0	1.7	86.2	16.4	14.3
RESTAURANT	195	119	59	17	0	0	0	0	1.2	21.0	5.1	3.2
MERCANTILE	176	16	30	49	26	25	22	8	1.8	687.0	52.7	76.3
WAREHOUSE	81	3	9	41	16	10	1	1	4.0	230.0	29.6	31.3
RESIDENTIAL NON-HOUSEKEEPING	162	10	8	37	59	30	13	5	2.8	932.0	53.9	85.4
HIGH RISE APARTMENT	104	0	0	2	13	29	47	13	15.1	853.4	141.2	137.4
TOTALS	1661	243	270	404	279	226	166	73	1.1	932.0	49.3	81.0

PRELIMINARY

BUILDING ENERGY PERFORMANCE
AIA RESEARCH CORPORATION
S & H INFORMATION SYSTEMS

Glass area. The amount of glass area also relates to how buildings are used. As shown in Figure 1.75, storage and stores have the least glass (5% and 7.5% of the gross wall area, respectively, while those building types with high vision window area requirements--i.e., offices and multifamily high rise--have the largest percent of glazing (36% and 25% of gross wall area).

Operations and Maintenance

The concentration on hardware in this section so far is in some ways unfortunate, for O&M cost changes are considered to be a source of significant energy savings. Yet, such savings are difficult to implement on a wide spread basis except for the obvious and easy changes that many building owners have already made. This may be primarily because decisions to save energy through the operation and maintenance of buildings involves many small decisions by individuals, the impacts of which are not always obvious to the individual.

At least in one's residence, if energy-conscious actions are initiated, the feedback can be direct--less money out of pocket for fuel and electricity bills. However, an individual using a commercial building may be one of 10, 20 or 1,000 people using the building. If an individual switches from energy wasteful to energy conserving habits in the use of the building, there may be no discernable feedback. The fuel bills may never be seen, or might be paid by someone else, even by another organization. Further, the individual may feel little or no control over the building's energy use. Indeed, this may actually be the case, depending upon the building design.

At the level of individuals using a commercial building, direct incentives to conserve energy through individual action can be lacking. An example of this relates to use of lighting systems. In offices and schools, daylighting has the potential to save significant energy. The cost-effectiveness of daylighting schemes is further enhanced if automatic sensors and controls are not required to monitor light levels and adjust lighting system output. Yet, examples exist of schools, with manual switching for lighting which have signs and

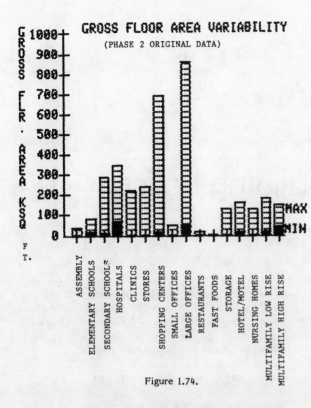

Figure 1.74.

SOURCE: AIA/RC "ANALYSIS OF DATA ANOMALIES, FURTHER

ANALYSIS OF PHASE II BLDGS, MARCH, 1980

the lights were left on all day regardless of available daylight.

Likewise, building operators may not perceive conservation to be positive from a self-interest viewpoint. It is often stated that a main objective of building operators is to above all else avoid complaints. Officials in the building energy office of a major city encountered consistent resistance from the operators of municipal buildings even in the attempts to conduct energy audits. The operators evidently perceived that the pending audit would point out that they were doing something wrong. The city cannot legally offer financial incentives to the operators who conserve energy, which means that other incentives need to be found. Such incentives might include ways to enhance the self-esteem or image of those building operators who conserve energy versus those who do not.

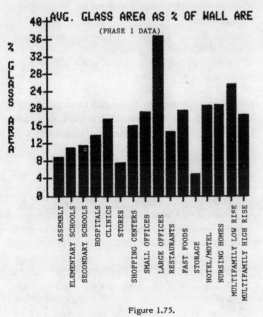

Figure 1.75.

SOURCE: AIA/RC "ANALYSIS OF DATA ANOMALIES, FURTHER

ANALYSIS OF PHASE II BLDGS, MARCH, 1980

instructions installed near the switches to encourage proper use, but the instructions were not followed and

Section Six
Commercial Building Policy

NEW COMMERCIAL BUILDINGS

There is a clear potential for significant cost-effective energy savings and renewable energy applications in commercial buildings. With careful design, it is possible to reduce the fuel consumption of new commercial buildings to between 30 and 40 percent of current practice. Future improvements in our knowledge of commercial buildings may well make further reductions possible.

Our understanding of the the decision process for new commercial buildings energy design is considerably more limited than that for residential dwellings. While at times the evidence available to make policy judgements may seem slim for certain residential sectors (for example rental dwellings) commercial buildings are even less understood, and the complexity of building types, ownership, and operation within the commercial sector compounds the problem.

It is therefore difficult to develop a plan of action for the commercial sector. We need to to learn more about new buildings design and use, disseminate that knowledge to a variety of individuals, and provide incentives for businesses to bypass institutional biases. Beyond basic information and incentive programs, policy recommendations are on very soft ground. However, demonstrations of policy approaches and continued analysis of the commercial buildings market-place could lay the ground-work for policy making in the future. Such experimentation and analysis should attempt to identify where the market-place does work well, under what conditions and for what reasons it does not, and provide examples of policy approaches to those problem areas, learning from both successes and failures.

The major thrusts of the recommended actions are: (1) continue to learn more about commercial building energy design and use; (2) educate architects, engineers, owners, and leasors of commercial buildings about the methods and benefits of energy conserving design; (3) provide reinforcement for the information programs through incentives; and (4) experiment with policy approaches to the commercial sector with careful evaluation of the programs implemented.

Information and Demonstration Programs

Actions examined under this program include:

o increased Federal support for analytical and applied research into commercial building design and use;

o financial and other support for architectural and engineering professional organizations for training and education;

o programs allowing professional business organizations to improve the information available to building owners and managers about energy efficiency;

o a building rating system for commercial buildings, and

o improved BEPS analysis.

Analysis and Applied Research

The program outlined in the Residential Section for Applied Research should incorporate commercial buildings as well.

The design of new commercial structures will require careful attention to integration of shell features and energy using systems, such as heating, cooling, lighting, and computers. In addition there are potential energy saving opportunities in combining process loads. The demonstrations called for should include

170

demonstrations of approaches to these systems such as integration of two or more energy using systems. The research program should also expand efforts to monitor the buildings in the demonstration program.

A specific program should be established at one or more research centers to address the needs of the built-to-lease market. Often, these buildings are designed to be constructed quickly with a minimum of attention to energy.

This program should be complemented by an ongoing effort to analyze the commercial building market by both ownership and building type. This analysis should seek to identify where the market-place is working and why, as well as where it does not.

Anecdotal evidence refutes the assertion that building owners' cost sensitivity will adequately facilitate energy efficient design. Various businesses which build and occupy their own buildings demonstrate institutional biases, such as separation of design and construction responsibility from operation and maintenance responsibility. A careful, documented analysis by ownership and building type coupled with the above program is fundamental to effective policy formulation in the future.

Training and Education

The Department of Energy currently has underway a program to educate faculty members at universities in energy efficiency and to improve the architecture and engineering curriculum at their schools. This program should receive expanded support, allowing it to go well beyond its current two seminars per year. In addition, the professional architectural and engineering associations should be supported in the development and implementation of training courses for the practicing professionals and others. One possible approach would be to establish analytical centers (perhaps in assocation with the applied research centers) where the latest design techniques and analytical tools would be available, and allow these universities to extend this knowledge to local professionals and provide training in the latest design tools.

Careful attention should be paid in this effort to providing analytical tools to both architects and engineers for the design of commercial buildings. Particular attention should be paid to simple accessible tools especially for small and medium builders, while continuing work on complex tools such as computer programs as well as education on their use.

Continued emphasis should be placed on expansion of "cook book" component performance approaches which parallel ASHRAE 90. These should incorporate more information on renewables and attention should be paid to making such approaches more stringent. As these educational and training programs progress, the professional organizations and educational institutions should be encouraged to revise accreditation require-ments to include energy skills for new architects and engineers.

Building Rating System

While some commercial buildings are built to the specifications of an owner-occupant, many are built for sale or lease. The application of a building rating system, which could be used by prospective buyers or leasees and by financial institutions should be undertaken for commercial structures. While the commercial sector will be more complex to approach than the residential in developing such a rating system, a limited effort to develop and test the application of such a system for commercial structures is warranted.

BEPS Analysis

The life-cycle costing analysis necessary for Building Energy Performance Standards is still being undertaken, currently under contract with Oak Ridge National Laboratory. This analysis should receive increased emphasis by the Department of Energy.

Attention should be paid in this analysis to the use of daylighting in commercial structures, since lighting is about 50% of the energy usage in office buildings. In addition, the potential for integration of building shell and internal systems should be investigated. Analysis should also be continued on embedded energy in commercial structures, particularly those meeting the stricter BEPS levels. At this level, nearly one-third of the energy in a structure can be embedded energy. Trade-offs between embedded energy and site energy use should be determined and incorporated into future BEPS and ASHRAE regulations.

Moreover, the information which has been gained to date through the BEPS support program and the knowledge gained through this life-cycle cost analysis need to be widely disseminated.

Commercial Buildings Incentives

Without a careful analysis of the commercial buildings market by ownership and building type, and experience with policy initiatives towards this sector it is difficult to develop incentive programs to accelerate energy efficiency as definitively as in the residential sector. However, the existing incentives should be expanded, and a number of new incentives implemented.

Actions recommended are:

o   Revise Commercial Building Tax credit to include lighting controls, movable insulation and other measures;

o   Provide tax incentives for performance route building design, until the cost of the performance route is reduced to the near conventional prescriptive path;

Commercial Building Tax Credit

The existing Commercial Buildling Tax credit does not include many important energy saving items, such as lighting controls and movable insulation. The Department of Energy should review the existing tax credit list in cooperation with professional architectural, building and engineering associations and produce recommendations for the Department of Treasury to up-date and expand this list to include a more representative list of the energy saving options which should be encouraged in commercial buildings.

Tax Incentive for Performance Design

The design of most commercial buildings using performance tools, such as complex computer program; is presently more expensive than the normal prescriptive route. The Federal government could encourage the use of the performance route in new commercial structures, and eliminate the existing barrier of this additional cost by providing a tax credit for the additional costs incurred. Such a tax credit should decrease over the next two to four years, as the tools developed for the performance route make it comparable in cost to a prescriptive design route.

Standards

The existing ASHRAE 90 Standards are currently being revised, and should produce a more energy efficient base-line for commercial building design. In addition, the Building Energy Performance Standards being promulgated by the Department of Energy are intended to apply to commercial buildings as well as residential structures. However, without the life-cycle cost analysis now being conducted by Oak Ridge National Laboratory, implementation of BEPS could be haphazard.

The incentives proposed in the residential sector for states to adopt BEPS should also require that they include the design energy budgets proposed for BEPS commercial buildings. The continuing programs of DOE, ORNL, and ASHRAE should be coordinated to produce coherent and consistent documents on life-cycle cost minimum designs for commercial buildings by both prescriptive and performance methods. This process will most probably take two years, and in the interim the effectiveness of information and demonstration programs and the experiences of states adopting the proposed energy budgets should be evaluated. Careful attention should be paid to the actual compliance with voluntary standards, such as ASHRAE, and the actual energy use of buildings designed to that standard should be compared with their design re-

quirements. A program within the Department of Energy should be established to undertake this analysis and evaluation, and provide a recommendation to the Secretary within two years on the role of a federally mandated BEPS in the commercial buildings sector.

EXISTING COMMERCIAL BUILDINGS

Existing Commercial buildings currently consume 40% of the total energy used by buildings. It is possible to reduce the average energy consumption of the commercial building stock by 50% in the next 20 years. Through aggressive programs of information, outreach, and financial incentives most of this potential can be realized.

Information and Demonstration Programs

Actions examined under this program include:

o   establishing an applied-research program for commercial building retrofits;

o   establishing a cooperative program with business and professional organizations for dissemination of information; data base development on retrofit costs and savings; and the education of building operators;

o   extending DOE support for feedback billing and metering to commercial sector demonstrations; and,

o   establishing a program to work towards the elimination of restrictive lease provisions.

Applied Research

The applied research program should follow the same general format as the applied research program directed at residential buildings. The program should be conducted in connection with private firms interested in working in the area of providing commercial building audits and provision should be made to allow some of these participants to maintain proprietary interests in specialized retrofit techniques. The major thrust of the program, however would be to:

o   better understand energy flows in commercial buildings by monitoring existing structures, and designing computer simulations which more accurately describe actual energy flows:

o   develop improved diagnostic equipment for measuring heat flows and appliance efficiency in commercial buildings;

o   design improved lighting systems; and

o train auditors in techniques for improving building performance.

The program should be conducted on a regional basis and adequate management provided to ensure that buildings are properly selected, and that information gathered is transmitted to building owners.

Cooperative Programs with Business and Professional Organizations

Cooperation with business and professional organizations as outlined in the new commercial section should also apply to existing commercial buildings. The Department of Energy should develop cooperative efforts to disseminate the results of its applied research and other analytical programs to building owners, operators, leasors, architects and engineers concerned with the existing stock of commercial buildings. In addition, DOE support for these organizations to collect data on the costs and savings of retrofits in a consistent format would provide a very cost-effective method of developing a commercial building retrofit data base.

In addition to these information programs, specific programs need to be initiated to approach building operators. Programs are necessary to provide them with information on improvements in the efficiency of building operation, and in addition programs should be instituted to over-come and reverse the institutional bias which some building operators may have against building audits. Anecdotal experience has shown that in some cases building operators have viewed energy audits as a threat. Not only did audit programs sometimes add additional work requirements on the operators, such as detailed recording of data, but audits were viewed as calling into question their integrity as building managers. The DOE Commercial Buildings program should expand its work with building operators and owners to overcome this institutional bais, including using incentives and other techniques to make efficiency in the building operators direct interest.

Feedback Programs

The feedback metering and billing program discussed in the residential sector should be extended to commercial buildings. As mentioned earlier, PG&E is undertaking a feedback metering program in commercial buildings without federal support. Little federal support for commercial building programs of this nature could be identified. Clearly, feedback metering in buildings with large process loads, or heating and coolings loads should provide a base for significant potential savings. Particularly if such metering is tied to time-of-use utility rates, of increasing block-rate structures.

A building energy Rating System program should be established for commercial buildings, similar to the residential program. Leased commercial buildings and building re-sale should both be points where such a rating system can be used by leasors, prospective purchasers and lenders to evalute the buildings structural and operational energy efficiency.

Restrictive Lease Provisions

A major impediment to energy efficiency improvements in the commercial buildings market (and the residential rental market) is restrictive lease provisions. The Building Owners and Managers Association estimates that at least 20% of the building owners who are willing to udnertake energy efficiency improvements will not do so because of restrictive lease provisions. The Department of Energy should undertake a program to:

o develop model lease forms that allow equitable pass through of energy improvement costs;

o explore the possibilities for constructing test-cases in state and federal courts to remove restrictive provisions of existing leases;

o support legislation to over-ride restrictive leases both on the state/local and federal levels.

Improve Building Audit Program

Actions recommended under this program are:

o increased allocation of DOE staff and resources to the expansion of RCS to commercial buildings;

o initiate an aggressive audit development and auditor training/certification program in conjunction with the applied research program;

o provide a tax credit for the cost of audits performed on buildings not qualifying for RCS, and cost-sharing grants for audits of non-profit buildings not covered by other programs.

RCS Expansion to Commercial Buildings

The Congress, through the Energy Security Act, has extended the Residential Conservation Service to small commercial buildings. While a major expansion to the commercial sector at this time would be premature (without additional understanding of commercial buildings, energy retrofits, and audit procedures) the limited nature of this program may provide the basis for improved commercial building audits and retrofits. However, the DOE support program for this expansion of RCS needs to be expanded in order to carefully monitor and support audit training, audit techniques, resulting retrofit cost and savings, and the institutional problems which are encountered.

The Department of Energy should allot additional staff and resources to realize the full potential of this limited expansion of the RCS program. This program should be carefully linked to the commercial buildings applied research program, audit development program, and the DOE cooperative programs with professional and business organizations. In addition, this program should explore the application of renewable measures to the heating, cooling, and process energy requirements of existing commercial buildings.

Development of Commercial Audits

DOE should initiate an aggressive program of commercial building audits coupled with carefully monitored retrofits, in conjunction with the applied research program. This program should attempt to enhance the data base on retrofits in various commercial building categories, and develop audit techniques and training programs. The skill level and training requirements for auditors should be clearly defined, and certification procedures developed for auditors for each major class of commercial buildings.

Incentives for Non-qualifying Buildings

In order to encourage the retrofit of all commercial buildings, the government should provide a tax credit for non-RCS qualifying commercial buildings for 50% of the cost of the audit.

Many commercial buildings are non-profit, or state and local government owned and operated. The federal government should provide a direct payment program equivalent to the tax credit suggested for buildings not qualifying for the RCS Program, excluding, however, buildings which are funded through the Schools and Hospitals Program, or other Federally subsidized conservation/solar programs.

Finance and Regulation

Actions recommended under this program are:

o expand existing tax credits;

o initiate a detailed feasibility study of limiting the deduction of fuel costs to specified levels in commercial buildings to be forwarded to the Treasury for inter-agency review.

o provide loans and loan guarantees for retrofits through the Small Business Administration.

Expand Existing Tax Credits

The existing commercial building tax credits should be expanded to include items such as lighting controls and movable insulation. A process to develop a more appropriate list of measures is described in the new commercial buildings section.

Fixed Level of Business Tax Deductions

The current tax laws allow the deduction of all energy expenses as legitimate business expenses. There is reason to argue that excessive energy costs should not be accorded this treatment, in the same manner that other excessive business expenses may not be allowed by the IRS. Both owner occupied and leased buildings deduct energy costs, the imposition of a limitation on this deduction would influence these building owners to undertake conservation/renewable measures. However, a system specifying building type, level of energy use, hours of operation, and climate zone would need to be developed for rental residential and commercial buildings. In addition, energy use levels would need to be developed for fuel used for heating versus other energy requirements in order to address those buildings not directly paying all energy costs.

It is suggested that the Commercial Buildings program at DOE, working in conjunction with the General Counsel's office, prepare a report examining in detail the feasibility of constructing such a system for limiting the fuel deduction. This report should be forwarded to the IRS for official comments. Either a regulatory re-interpretation could be undertaken by IRS, or new legislation could be considered if Treasury is reluctant to introduce such a measure without such statutory changes.

Loan for Retrofits

With the exceptions of public buildings, senior citizen centers, and schools and hospitals, the financial incentives for commercials buildings are minimal. The Small Business Administration (indirectly) and the Economic Development Administration provide some financial assistance. EDA provides funding by specific formula for commercial buildings only in "distressed areas." The SBA does include building funding in its loans but when contacted was unable to quantify the impact or exact amount. The current commercial tax credit provides only minimal incentive, given the quick rate of return required by most commercial building owners, and the fact that a 15% investment tax credit improves the rate of return on an investment by only a few percentage points in a discounted cash-flow analysis.

If the capitalization of the Solar and Conservation Banks is significantly increased, as recommended earlier, this should help make attractive financing available for solar and conservation applications in commercial buildings. However, some commercial establishments will be unable to obtain conventional loans.

The Small Business Administration has extensive experience dealing; with businesses that cannot obtain conventional financing. However, competition for the limited funds of the SBA is stiff. In addition, SBA does not have a specific program for energy improvements or solar improvements in existing small businesses. Clearly, some--if not many--of these firms will need assistance given the expected increases in energy costs.

An expanded program of loan guarantees and direct loans for energy related improvements, starting at $100 million for FY 1982, would begin to address this problem. A higher priority could be given for such loans under the existing SBA program, but additional appropriations for the SBA would still be necessary if available funding for other purposes, which is already limited, is not to be severely constrained. Such funding should not be made part of the loan program for energy-related businesses, but should either be a part of the general SBA loan program or a separate program of its own.

CHAPTER 2
INDUSTRY
Section One
Overview

It is easy to sense the importance of energy in industry. Showers of sparks from steel foundries, the noise of mechanical equipment in assembly lines, and columns of steam and smoke from industrial boilers bear a vivid witness to the fact that more than a third of the nation's energy is used directly by factories, mines, farms, and the construction industry.

This report will examine the changes which can be expected in the way American industry uses energy during the next two decades, paying close attention both to changes in markets and changes in technology. It will also examine the role which biomass and direct solar heat can play in meeting industrial needs for energy. Specific programs are proposed for encouraging more efficient use of the financial resources available for industrial energy investment and for attracting more capital to such investments.

With the proper set of programs to encourage efficiency and use of renewable energy, it should be possible for U.S. industry to meet its increased production goals during the next two decades with virtually no increase in industrial energy demand by the year 2000. Meeting plausable goals for use of biomass and other solar resources can mean that industrial use of oil, gas, coal and other conventional energy resources may actually decrease. Industry is already reacting to sharply increased energy costs; industrial energy use per unit of output in the U.S. actually declined 8% between 1973 and 1978, and an additional 13% decline is expected by the year 2000 without any change in present policy.

POTENTIALS

The goals for industrial energy use during the next two decades depend critically on three considerations:

(1) Industrial activity is likely to grow more slowly over the next two decades than the economy as a whole, following a trend which has held for over a decade. (See Figure 2.1) It appears that, as the national income increases, people will spend a larger part of their income for information, communications, recreation, health care and other services requiring relatively little industrial activity.

(2) The number of pounds of product manufactured each year is growing even more slowly than total industrial output. It is estimated in this study that aggregate materials production will decline some 25% with respect to gross national product (GNP) during the period 1978 to 2000. The basic materials industries (metals, paper, chemicals, petroleum refining, ceramics and food), which are now responsible for 70% of industrial energy consumption, are not growing as rapidly as electronics and other industries where the products weigh less on the average and where energy costs are a relatively small fraction of overall production costs. (See Figure 2.2.)

(3) A number of commercial technologies are available for improving the efficiency with which industry consumes energy. These technologies can be and will be used to great economic advantage in new factories, and many of the technologies can be retrofitted into existing facilities. Significantly greater savings can be expected from innovations likely to occur during the next few years. This analysis is conservative in that it projects no savings from such innovations--simply because it is not possible to construct a plausible quantitative estimate of their impact.

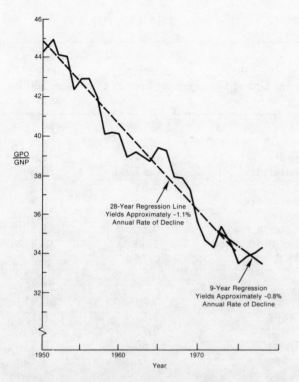

Figure 2.1. **Industrial Value Added (or GPO) to GNP Ratio, 1950 - 1978 (In Current Dollars)**

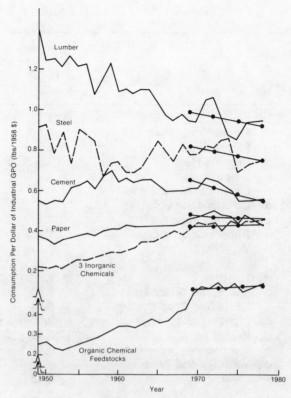

SOURCE: M. Ross and R. Williams. **Our Energy/Regaining Control**, to be published by McGraw-Hill. 1980.

Figure 2.2. **Consumption of Basic Materials by Industry, 1950 - 1978**

Goals for industrial energy use were developed by examining cost-effective efficiency improvements for each major category of energy service. Demand for each category of energy service was projected to the year 2000 using an appropriate growth indicator, and an efficiency improvement target was developed. The results of this projections summarized in Table 2.1. Goals for use of renewable resources in industry are shown in Table 2.2. Specific policy changes to achieve or approach these conditions which encourage energy productivity and alternative energy use are recommended in the policy section. The energy use targets set here apply if those policies are put in place.

The impact of the policies suggested here is summarized in Table 2.3. The bottom line of Table 2.3, shows that if the proposed policies are adopted to accelerate the trend, industrial energy intensity (the average amount of energy consumed per unit of energy service) could drop 26% between 1978 and the year 2000. At the same time, GNP in the year 2000 would be rising 73%. The combined effect would be that the increased industrial output could be achieved with only a very slight increase in total industrial fuel use.

Industrial Efficiency Targets

Energy efficiency in industry can be increased in any of the following ways:

(a) Product-related changes, especially changes in product design which affect the use of basic materials.

(b) Fundamental process changes, changes which require very large investments and impact a variety of factors of production.

(c) Changes in energy conversion equipment or energy conservation technologies which involve significant investment and are primarily connected to energy use and costs.

(d) Changes in management of operations and maintenance, and retrofits with low cost equipment, which affect energy use.

The efficiency targets summarized in Table 2.1 were developed by examining how these kinds of savings applied to each major industrial sector.

A brief examination of the most important energy service categories, process heating mechanical drive, and electrolysis (accounting for 75% of industrial energy use) will demonstrate the plausibility of the targets developed.

Table 2.1 GOALS FOR ENERGY USE IN INDUSTRY AND AGRICULTURE

	Energy Use in 1978[c] (Quads/year)	Demand Growth Indicator[a]	Efficiency Improvement Goal[b]	Goals for Energy Use in 2000[c] (Quads/year)
1. Industry	(29.1)	--	--	(29.4)
o Mechanical Drive	7.8	IVA	0.77	8.8
o Electrolysis	1.5	GNP	0.74	1.9
o Process Heat	12.5	Materials	0.65	10.6
o Space Heat	0.4	IVA	0.61	0.3
o Chemical Feedstocks	2.6	Materials	0.85	2.9
o Construction Asphalt	0.9	GNP	0.94	1.5
o Metallurgical Coal	2.1	Materials	0.85	2.3
o Repressurizing Gas	1.1	Gas Production	1.00	0.8
o Vehicles	0.2	GNP	0.77	0.3
2. Agriculture	(1.63)	--	--	(1.66)
o Space and Water Heat	0.05	Livestock	0.75	0.05
o Grain and Crop Drying	0.11	Crops	0.75	0.12
o Lighting and Refrigeration	0.05	Livestock	0.77	0.05
o Irrigation	0.40	Crops	0.60	0.35
o Mechanical Drive (Statnery)	0.19	Crops	0.77	0.20
o Vehicles and Farm Machinery	0.83	Crops	0.75	0.89
3. TOTAL	30.7	--	--	31.1

( ) = non add subtotals

[a]Demand in each energy service category is assumed to grow at the same rate as a "demand growth indicator." GNP is assumed to grow at an average rate of 2.55% per year over year from 1978 to 2000. Industrial Value Added (IVA) in industry is assumed to increase 0.85% per year more slowly than GNP, and pounds of materials produced per year ("Materials" in the Table) is assumed to grow 0.5% per year more slowly than IVA (see figures 2.1 and 2.2). Livestock production is assumed to grow as fast as U.S. population (0.8%/year from 1978 to 1990 and 0.6%/year 1990-2000) and crop production 1% faster than population.

[b]Ratio of energy used per unit of energy service in 2000 to energy use per energy service in 1978.

[c]Includes losses in generation and transmission of electricity.

Process Heat

The target is to reduce the energy intensity of process heat by 2.0% per year in the period 1978-2000. This target can be put in perspecitve by examining the trends in the chemical industry. In the period 1972-78 the process heating fuel required to produce a pound of product was reduced at an annual rate of over 4% per year. Moreover, projections into the mid 80's by individual firms indicate a continuing rapid rate of decline in energy intensity. Union Carbide projects that its total energy use excluding feedstocks per pound of product will decline 2.7% per year from 1978-85, while DuPont projects that its energy use per pound of product will decrease 2.3% per year from 1979-86. Fuel requirements for process heat alone would undoubtably be reduced more rapidly than these over-all energy requirements. In terms of these current trends and the industry's projections, the energy intensity reduction target associated with process heat can be viewed as modest.

Mechanical Drive and Electrolysis

In making these projections to the year 2000, it is assumed that thee will be a 5% savings due to improved motor efficiency, which should be easily realized due to the relatively short lifetimes of most

Table 2.2 THE POTENTIAL FOR DISPLACEMENT OF CONVENTIONAL FUELS BY BIOMASS AND DIRECT SOLAR HEAT IN THE U.S. INDUSTRIAL SECTOR IN 2000*

(in quads)

Energy Form[a]	1978 Contribution	Present Policy Continued	Policy of Accelerated Implementation
Biomass[b]	.84 – .97	2.4 – 5.1	4.8 – 10.5
Direct Solar Heat	Negligible	0.1 – 1.0	0.5 – 2.0

*Assumes total demand for industrial energy (excluding agriculture) of 30 quads in 2000. Electricity from renewables is considered in the Utilities section. Biomass estimates are given in terms of oil displaced, rather than primary biomass supply.

[a]The economic assumptions are defined in the technical sections on biomass and direct solar heat. Moreover, whereas the efficiency improvement targets consider the delay in implementation associated with new technology, the solar potentials represent the applications advantageous in simple economic terms.

[b]These totals include biomass uses in the Buildings and Transportation section.

Table 2.3   COMPARISON OF U.S. INDUSTRIAL ENERGY REQUIREMENTS IN
THE YEAR 2000 WITH AND WITHOUT POLICIES TO ACCELERATE
ENERGY CONSERVATION*

Relative Energy Intensity[a]	Fuel Use (quads)	Index (1978 = 100)	
1978 Fuel Use	1.00	29.1	100
Constant Energy Intensity[b]	1.00	39.7	137
Normal Energy Intensity Reduction[c]	0.87	34.6	119
Accelerated Energy Intensity Reduction[d]	0.74	29.4	101

*Agriculture not included. Assumes that total U.S. industrial value added in 2000 is 46% higher than in 1978.

[a]Average amount of energy required per unit of energy service.

[b]Fuel requirements in 2000 assuming energy intensities remain at 1978 values.

[c]This projection applies with present policies and presently expected energy price increases.

[d]These results apply if the policies proposed in this report are in place. The energy intensity reduction factor 0.74 is what would be obtained as a pure price response, if the average price of primary energy (weighted by fuel type for the 1978 mix of fuels) rose as projected in this study from the 1978 level of $2.00 per million Btu to $5.30 per million Btu by the year 2000, and if the price elasticity were -0.3. Such a price elasticity is often used in government policy analyses. However, it is at the low end of a range of elasticities that have been estimated recently for the industrial sector.

industrial motors. An additional 5% is achieved through better management, and a 15% savings is associated with systems improvements. Both of these latter two projections are conservative, given the technical opportunities in these areas. Thus, a 23% savings is projected relative to 1978, corresponding to a 1.2% per year reduction in the energy intensity of mechanical drive in the period 1978-2000.

Among electrolytic processes, primary aluminum production is the major electricity consumer. A major opportunity exists to reduce electricity use in aluminum production from the 1976 level of 7.5 kWh per pound to 5 kWh per pound, by shifting from the predominant Hall-Heroult process (introduced in the late 1880's) to the new Alcoa smelting process. The prospect of drastically higher electricity prices for the primary aluminum industry and for other electrolytic industries as well, will be a powerful incentive to shift to more efficient technologies. In this analysis, we assume an overall reduction of 26% in energy use per unit of output in the electrolytic industries by the year 2000, corresponding to an annual reduction in energy intensity of 1.3% per year, 1978-2000. Given the assumption of doubling in electricity prices by the year 2000, this projection may be conservative.

Figure 2.3 shows the projection of electricity use per unit of industrial GPO. This figure also indicates the percentage of industrial GPO spent on electricity which this would imply. It is interesting to note that even with the targeted efficiency improvements, industrial expenditures on electricity per unit of output increase dramatically.

Industrial Cogeneration

Projected savings resulting from increased use of industrial cogenerations are not included in the of energy efficiency improvements shown in Table 3.1 because of uncertainties about future national demands for electricity (see the chapter on Utilities). Fuel savings of 15-35% relative to the fuel requirements for the production of heat and electricity in separate facilities are possible with cogeneration.

The economics of cogeneration appear particularly attractive in the basic-materials industries. These industries tend to be capital intensive, with year round, round-the-clock operations. Cogeneration in these industries would thus involve primarily the generation of baseload electricity. Under a wide range of conditions the fuel savings associated with cogeneration in these circumstances would result in a busbar cost lower than the cost of electricity delivered to an industrial customer from a new central station baseload plant.

If cogeneration is to become a major fuel saving option, then cogeneration technologies characterized by a high electricity to steam output ration (E/S ratio) must be encouraged. For a given process steam load, such technologies can generate 4-8 times as much electricity and 3-5 times as much primary fuel savings as the more familiar low E/S ratio cogeneration technology, the steam turbine unit. Table 2.4 indicates the potential for cogeneration in six major steam using industres, if the high E/S ratio cogeneration technologies were deployed for the steam loads that are technically suitable for association with cogeneration, at the 1976 levels of steam demand. The cogeneration potential in these six industries would result in annual fuel savings of about 2 Quads through the displacement of almost 100 GW (e) of baseload electricity from conventional central station power plants--corresponding to approximately one quarter of total U.S. electricity production in 1979.

PUBLIC POLICY

A major objective of this analysis will be to find some way to attract the resources needed to rebuild these critical and energy inefficient industries in the U.S.

The policies examined will be designed to meet two fundamental principles.

o First, they should improve the overall productivity of U.S. industry instead of focussing exclusively on energy savings.

o Second, they should make maximum possible use of private decisions instead of public decisions; the market should be used wherever possible.

To meet these criteria, specific policies are designed to: (1) encourage capital formation in industry, particularly in the basic materials industries, (2) make the price of energy reflect its true economic cost, and (3) improve the climate for industrial research.

There is a fundamental tension in the design of programs designed to encourage industrial investment through market mechanisms. If market forces are to work effectively, energy prices must be allowed to increase to reflect the cost of producing new, expensive supplies. In the long run, such price increases, plus industrial tax reform, can send proper signals to the market. However, such measures should be phased in slowly to avoid major dislocations on the part of firms which cannot easily adjust.

Finding Capital for Corporate Investments

Analysis conducted by American industry frequently concludes that the major barrier to improvements in industrial productivity and industrial solar energy equipment lies in its inability to find capital even when the economics of improved efficiency are clearly

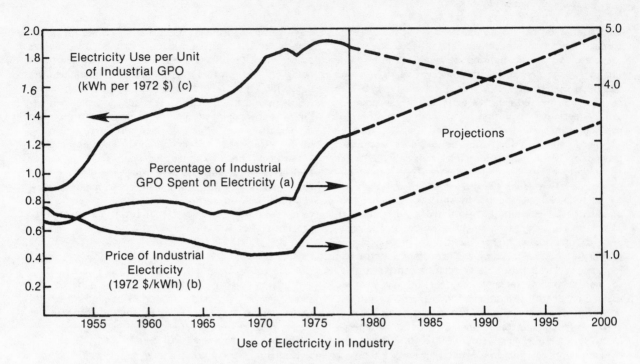

Use of Electricity in Industry

NOTES: (a) Industrial GPO is the gross product originating in the agriculture, mining, construction, and manufacturing sectors. Industrial GPO data were obtained from the July issues of **Survey of Current Business** for
the period 1975-1978 and from Bureau of Economic Analysis, **The National Income and Product Accounts
of the United States, 1929-1974, Statistical Tables,** January 1976 for the period 1950-1974. The number
given for 2000 is based on an average growth rate of 1.7%/year, 1978-2000.

(b) The price of the electricity given here is the average price for large light and power customers. For
the period 1950-1970 electricity prices were obtained from Bureau of the Census, **Historical Statistics of
the United States, Colonial Times to 1970,** September 1975; prices for 1971-1978 were obtained from
"1980 Annual Statistical Report, **Electrical World,** March 15, 1980. The price given for 2000 is the industrial
electricity price projected for 2000 in the Sawhill project. It is the marginal cost of electricity delivered from
a baseload coal power plant (@ a capital cost of \$750/kW and for coal costing \$1.90/10^6 Btu in 1978 \$) in 2000.
This cost, 5.1¢/kWh in 1978 \$, becomes 3.4¢/kWh in 1972 \$.

(c) Electricity use includes both purchased electricity and electricity generated on site. Historical data
(1950-1978) were obtained from Energy Information Administration, **Annual Report to Congress 1979,**
Volume Two. The value given for electricity use per dollar of industrial value added in 2000 is obtained
by projecting energy use in mechanical drive and electrolysis per dollar of industrial GPO. The result is
that by 2000 electricity use per dollar of GPO is about 20% less than in 1978.

(d) Industrial GPO and the price of electricity are expressed in constant 1972 dollars using the GNP deflator.

Figure 2.3.

favorable. This conclusion is strongly supported by the
findings of this study.   Artifically low energy prices
and heavy taxes on industrial corporate income have
led to an historic underinvestment in industrial plant
and equipment designed to improve energy productivity. This has been accompanied with regulatory programs which may provide too much protection for
investments in utility plant and equipment which has
led to over investment in these sectors.

The basic materials processing industries face particularly   serious   problems   because   energy   price
increases are especially important for these industries,
because with the prospect of slow growth in these

industries the capital needed for improvements will be
difficult to find, they are responsible for the bulk of
the energy consumed in American industry. Without
significant changes in existing trends, nearly half of
the equipment now used by the US materials industry
will still be in use in the year 2000.   Among other
things, this will mean that US industrial facilities will
continue to be older and less efficient than most
European and Japanese plants for the rest of the century.

Over the past few decades, capital formation has
been a relatively constant fraction of GNP, but the
share of capital available for industrial productivity

Table 2.4 PROCESS STEAM USE, POWER CONSUMPTION AND THE COGENERATION POTENTIAL IN SIX INDUSTRIES

Industry	Process Steam Use, in 10^{12} Btu/year[a]	Process Steam Use Technically Suitable for Cogeneration, in 10^{12} Btu year[b]	Power Consumption in 10^9 kWh/yr.[a]				Target Cogeneration via high E/S Ratio Cogeneration Technologies, in 10^9 kWh/yr.[d]
			Purch. El[c]	Onsite Prod.[a] Mech.	Elect.	Total	
Food	307	198	39.1	4.1	2.6	45.8	40
Textiles	125	75	34.8	0.6	1.5	36.9	15
Pulp and Paper	1210	859	43.5	1.8	23.4	68.7	172
Chemical	1516	1059	145.4	14.4	18.8	178.6	212
Pet. Ref.	584	364	26.3	4.1	1.5	31.9	73
Steel	366	300	54.6	10.5	10.0	75.1	60
Total	4108	2855	343.7	35.5	57.8	437.0	572

[a]For 1976, as given in P. Bos, "The Potential for Cogeneration Development in Six Major Industries by 1985," Report to the Department of Energy by Resource Planning Associates, Inc., 1977.

[b]See Table 4-6.

[c]For 1976, as given in the Annual Survey of Manufacturers 1976: Fuels and Electric Energy Consumed.

[d]Assuming the process steam use technically suitable for cogeneration is associated with cogeneration with an average E/S ratio of 200 kWh/10^6 Btu.

investment has been gradually eroded. One reason for this trend is that pollution control has been absorbing a substantial share of available capital. But far more important is fact that a growing fraction of capital is going to the energy supply sector. Although retail energy sales in 1977 accounted for only 12% of GNP, the energy industry accounted for over 40% of all new plant and equipment expenditures (see Figure 2.4). The energy-producing sector has generated growing demands for capital, placing increasing pressures on the capital markets at a time when the energy-using sector also has critical capital needs, including those for investments in energy efficiency and production of renewable energy.

From the point of view of individual businesses, the result has been a sharp increase in interest rates, making borrowing difficult and expensive for even those companies which are able to borrow. Many companies, unfortunately, already have such high debt-equity ratios that conventional borrowing is no longer feasible. For these companies, new equity issues or internal profits are the only possible sources of new capital. Often both sources are not practically available for use by the very companies most in need of modernization and energy conservation. The basic materials processing industries have experienced shrinking profits, thus diminishing their ability to finance projects with internal funds. Lowered profits have also led to lowered price-earnings ratio for corporate stocks, making it undesirable for a company to issue new stock. For such companies all the traditional means for raising capital are either precluded or very expensive.

One result of the lack of capital is that investments in energy productivity are expected to achieve a rapid payback, while the payback period for energy supply can be relatively long, often because of regulatory protection. It is not suggested that the payback period should be the same for a heat-recovery device and for a new electric generating plant. However, the criteria for determining the particular financing conditions should be the same.

The total demand for industrial capital over the next two decades will be enormous. Calculations performed for this study indicate that the nation's industries may need to invest an extra $170 billion (1978$) over the next 20 years to improve the productivity of the U.S. industrial base. This amount is in addition to the historic rate of capital investment in industrial plant and equipment. Because many industries will find it difficult to accumulate the capital needed to make improvements, however attractive, an aggressive capital formation policy is needed.

Three categories of programs are examined here. (a) Programs designed to reduce or eliminate the burden of federal taxation on corporate income. Several alternatives are discussed, including reduction in the corporate income tax rate, liberalized deprecia

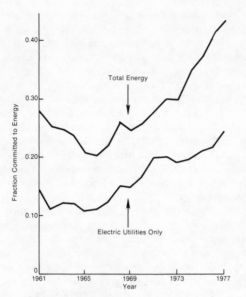

SOURCE: M. Ross and R. Williams. **Our Energy/Regaining Control**, to be published by McGraw-Hill, 1980.

Figure 2.4. **Fraction of New Plant and Equipment Expenditures Committed to Energy Supply**

tion, and additional tax credits. (b) Programs designed to remove the regulatory barriers which make industrial investments by utilities or third parties difficult. (c) Programs designed to ensure that receipts from the windfall profits tax (and possibly a corporate fuels tax) are equitably available to investments in energy supply and energy efficiency. The cornerstone of this last program is a project designed to provide funds for industries to scrap existing obsolete facilities and rebuild.

Reducing Corporate Income Taxes

The 46% tax on corporate income and unfavorable depreciation policies have the effect of discouraging industrial investment in energy efficiency and renewable energy technologies. A fundamental review of the need for a federal tax on industrial production and of the logic which motivates such a tax is long overdue. Techniques for reducing corporate taxes include reducing the corporate tax rate, liberalized depreciation and investment tax credits.

There have been a number of analyses of the industrial capital formation that would flow from such kinds of tax reform. A study by Brimmer and Sinai considered increased general investment tax credits, a reduced corporate income tax rate and indexed depreciation (Brimmer 1976). The ratios of dollars of investment to dollars of tax subsidies were calculated to be 0.7, 0.3, and 1.7 for the three respective measures. In other words two of the three measures would not even produce as much investment as the loss in Federal revenues. Several other studies also suggest

that such measures may have a low capital-to-subsidy ratio. On the basis of these studies, it cannot be concluded that liberalized depreciation would be more effective than the other options. However the studies do indicate that indexed depreciation would be relatively effective in stimulating the longer-term investments required by the energy-intensive industries. The studies did not include the effect of accompanying corporate tax relief with a fuel tax. Such an analysis is urgently needed.

Industrial Investments by Utilities or Third Parties

Some of the institutional problems which have retarded industrial investments in energy efficiency and solar energy could be alleviated if companies that sell, rent or operate energy-related equipment could be encouraged to provide energy services at an industrial site.

Energy-Service Firms. In some cases firms specializing in energy services may own, operate and maintain facilities at industrial plants. Other firms may specialize in financing or renting components-- including components completely integrated into plant equipment, such as major automatic control systems and furnace retrofits. A variety of relationships can develop, and many targets of opportunity have not been examined. Some legal questions may need resolution. Exploratory funds should be offered for studies of legal and financial issues and, more importantly, to encourage actual institutional experiments.

Utilities and Utility Subsidiaries. An alternative to the creation of a new energy-service industry involves utility investments in industrial equipment. Utility investments in industrial equipment could be made either directly or through an unregulated subsidiary. The advantages of utility investments (in industrial renewable energy production, cogeneration equipment located at an industrial site, and perhaps energy efficiency devices at an industrial site), are discussed in the Utility section of this study.

Current federal law does not prohibit utility investments in industrial energy equipment, but such investments are often inhibited by a number of regulatory difficulties. Present tax laws should be changed so as not to discriminate against utility investments in energy efficiency and renewable resources. There should also be reexamination of the need for price regulation of utilities in areas where competitive conditions exist.

There is uncertainty about the most appropriate institutional relationship between an industrial customer and the utility as a supplier of energy services. There is also uncertainty about the tax and regulatory limits on utility investments in industry. The DOE can begin to resolve these uncertainties by cooperating with state public utility commissions and relevant federal regulatory and taxing organizations, in the investigation of existing examples at the state level, and by developing other applications as are necessary to fully determine the effectiveness of such utility-industry relationships.

Equitable Capital Subsidies for Renewable Energy and Energy Conservation In Industry

The Energy Security Act of 1980 created an independent Synthetic Fuels Corporation, wholly owned by the federal government, to subsidize industrial investments. The existence of such a body will distort a fair allocation of resources among competing industrial demands for capital. In particular, since the credits available from the corporation cannot be used for industrial investments in energy efficiency or solar energy, such investments are made relatively unattractive.

One option to redress this imbalance would be to broaden the scope of the Synthetic Fuels Corporation (Energy Security Act of 1980 Title I) so it could also consider investments in energy efficiency improvement and solar energy in industry. The corporation should be given the freedom to choose between investments in supply and investments in demand purely on the basis of economic merit.

Scrap and Build Incentive. A specific role for a revised Synthetic Fuels Corporation, would be to provide funds for a Scrap and Build Program for energy-intensive industries. Industrial productivity and fuel saving opportunities exist both in retrofitting of present equipment and in building new facilities. Since the gains from creating new facilities are typically much greater than from retrofit, the "scrap and build" program is designed to ensure that adequate funding is available for buildings new facilities. The potential impact of the programs designed to accelerate the replacement of older plants and to encourage use of more efficient equipment in new and retrofit plants is summarized in Table 2.5 for four basic materials processing industries.

If existing facilities are replaced at 5% per year, double the current rate, half of the existing plants would last beyond 1990. Even with a very rapid replacement of capital, many existing facilities will still be used for many years and should therefore be retrofitted.

DOE could support Scrap-and-Build subsidy legislation to reward the scrapping of major obsolete facilities and replacing them with new facilities. Reconstruction at the same site should be encouraged, so as to preserve the value of the existing infrastructure and minimize worker dislocation. U.S. industries using a 10% per year real discount rate should find a scrap and build investment profitable in terms of energy savings alone with an investment incentive equivalent to about 13% of the value of a new plant. This level of subsidy, would make scrap and build projects profitable but would not necessarily enable firms to proceed with projects because they might not have adequate access to capital. For a variety of reasons already discussed it is clear that a number of firms most in need of the scrap and build program will find it difficult to raise capital and therefore may have high discount rates that they will not be able to justify scrap and build projects without an even higher subsidy. The required subsidy is very sensitive to the discount rate. It is proposed that, at least initially, the subsidy be some 20%, or roughly $1.7 billion per year on the average.

It is interesting to observe that 20% subsidy, when expressed in terms of dollars per barrel per day of oil equivalent savings is close to the credit available for synthetic fuels plants in the Energy Security Act, approximately $20,000 per barrel per day of oil equivalent energy. The subsidy level is also comparable to the 25% capital cost reduction which Charles Berg has argued would lead industry to scrap antiquated facilities.

One option for obtaining the funds for the Scrap and Build subsidy would be to use the revenues that could be made available by removing the present subsidies for the routine provision of energy (some $6.7 billion per year). The level of these energy supply subsidies is much larger than the requirements of the Scrap and Build program. Windfall profits taxation on oil (or natural gas) would also be a legitimate source. Still another source would involve the levy of a tax on industrial energy. Raising the average price of industrial energy 2% would be adequate to provide the level of subsidy needed.

An interesting feature of the Scrap and Build Program is that the total required investment and energy implications are relatively well defined. The major uncertainties are whether the estimated near term subsidy level would be adequate to generate the desired level of capital investment, and, for the longer term, what is the most effective combination of measures for a more market oriented strategy involving energy price decontrol, the energy tax and corporate tax reduction. Further analyses to clarify these issues should be given high priority. However, the uncertainties about costs and benefits of this program are no greater--and may well be less--than those which characterize the synfuels and other supply programs.

Small Business Assistance. At present the Small Business Administration (SBA) has a small program to provide capital assistance to small companies. This program should be expanded in scope and given more adequate funding.

Table 2.5 THE EFFECT OF PROGRAMS DESIGNED TO ACCELERATE THE REPLACEMENT OF OBSOLESCENT PLANTS

	Growth Rate in National Capacity (% per year)		Fraction of Capacity in Year 2000 in New Plants		Average Energy Intensity of Plants Operating in 2000 (1978 = 1.0)		
	1975-1979 Statistics	1978-2000 Forecast	Business As Usual	Accelerated Turnover of Older Plants	Business As usual	Accelerated Turnover of Older Plants	Incremental Investment (billions of dollars)
Paper	2.0	2.0	0.35	0.64	0.8	0.65	18
Chemicals	4.4	2.0	0.35	0.64	0.8	0.65	50
Cement	--	-0.5	0	0.50	0.86	0.66	3
Steel	0	0	0	0.45	0.86	0.67	44

Notes: See Chapter on Industry for details.

Venture Capital. The incentives described above are designed primarily to encourage industrial investments in technologies whose commercial worth has already been demonstrated. There is also need for a venture capital program designed to encourage more innovative ventures and projects that involve market tests of systems whose acceptance can not be easily evaluated.

The Venture Capital Program should be built around the following elements:

(1) Funds for solar and energy efficiency investments should be drawn from the same resource and held to the same programmatic tests.

(2) All proposals should be unsolicited and, the availability of funding and generic priorities should be widely advertised, but specific equipment designs should not be requested.

(3) Proposals should be selected using general criteria combined with the use of review panels chosen in a way that would not compromise the proprietary nature of the proposals.

(4) The proprietary rights of all proposals should be protected.

(5) The federal government should be reimbursed by the participating company, with interest, if the product supported becomes a successful commercial venture.

(6) The program should encourage joint ventures.

(7) While no precise formula should be required, indication of industrial interest in the project, as evidenced by substantial cost sharing, should be a major criterion in selection.

(8) One objective of the program should be to channel funds to small innovative firms as well as to large firms.

Energy Prices

Energy is still priced well below its marginal cost. Three techniques are examined for ensuring that industrial fuel prices send a proper signal: (1) further reductions in the regulations of energy prices, (2) elimination of energy supply subsidies, and (3) a fuels tax.

Reducing Price Regulation

While dramatic steps have been taken toward the full deregulation of energy prices in recent years, prices are still distorted by a variety of state and federal controls. The section of this report dealing with the utility industry examines the impact of these programs in some detail and suggests techniques which can be used to remove the remaining distortions in utility energy prices.

The important point for the purposes of the present argument is that a history of underpriced energy has left industry with investments which make no economic sense in today's price environment. These facilities are vulnerable to sudden changes in prices which may result from a disruption of foreign supplies or a sudden decision to remove regulatory price controls.

Eliminating Energy Supply Subsidies

The discrepancy between market prices and full marginal costs results in part from price controls and in part from federal subsidies to energy suppliers. It has been estimated subsidies to conventional energy sources are at least $6.5 billion per year. These subsidies are historically out-dated and should be reduced or eliminated. If all these subsidies were eliminated and revenues were allocated to the economic sectors on which the burden of higher energy prices fell, the revenues allocated for the industrial sector would be more than adequate to pay for all the programs proposed here for that sector.

A Tax on Industrial Energy

The price of energy delivered to industry does not reflect the national security costs of continuing heavy dependence on foreign oil, the environmental costs of energy production and use, or macroeconomic effects resulting from massive energy imports. These effects make the total social cost of energy even higher than the cost which would result if regulatory and subsidy distortions were removed completely. Regulations, targeted incentives, or energy taxes could supplement deregulation and removal of subsidies to effectively bring prices in line with full social costs. Of these alternatives, energy taxes have the advantage that they impact the general economic climate relating to energy decisions, without involving the government in detailed planning. They may be a more attractive way to tax industry than conventional taxes on corporate income.

One major difficulty with the energy tax concept is that many energy-intensive industries are already facing severe financial difficulties. Any program designed to improve energy efficiency which includes an energy tax must be accompanied by measures such as the scrap and build program outlined above, which can provide these heavily impacted industries with the capital they will need to make new investments in response to higher energy prices. The fuel taxes must also be imposed slowly so that industry will have time to react without unproductive haste.

A well designed program of energy taxes could, however, lead to a net increase in the capital available for energy intensive industries. For example, a tax package could include both a reduction in corporate income taxes and direct rebates. The rebates could be given directly to industry on the basis of a formula which rewarded investments in new plant and equipment or corporate research and development or the rebates could be administered through a special fund which would support such investments.

The issues associated with energy taxes are complex. As a first stage, a careful review of these issues should be carried out in the context of a general analysis of energy price policy. The need for and the phasing in of energy taxes should be assessed in combination with the measures which warrant the highest priority in the rationalization of energy prices: the elimination of subsidies and energy price decontrol.

Improving the Climate for Industrial Research Investments

Three federal actions can encourage industrial support of research:

o increased tax credits for industrial research investments;

o federally-sponsored research facilities; and

o a liberalized patent policy.

In addition, direct federal support of relevant fundamental research is recommended.

Increased Tax Credits

It is proposed that there be enacted a tax credit for corporate research investment which is 125% of actual expenses, in lieu of the present tax deduction.

Two special issues should be considered in reviewing the merits of this proposal. First, it is critically important to construct a much more careful definition of "research and development" than is now used in the tax law. It appears that under the current definitions, many routine functions can be technically qualified as R&D. Secondly, since the incentive will not be attractive for firms that have no major tax liability, it may be desirable to allow the credits provided under this program to be refundable.

Federally-Sponsored Research Facilities

DOE should begin to provide federally-sponsored research facilities jointly sponsored with industry. The arrangement could ensure a close working relationship between teams in a government research facility and their counterparts in industry. Such research facilities could be managed as an extension of a state university or a national laboratory. Federal sponsorship of such a facility could be limited to providing land and a building, or a more active role could be taken. Any analysis of such a program should begin with a careful examination of European and Israeli experience with such facilities, as well as U.S. experience in defense-related industries.

Patent Policies

The following seven steps are suggested to remove some of the problems confronted when DOE attempts to transfer title to a patent:

o   Expedite the process for acting on requests for waivers of patent rights to 30 days or less.

o   Provide potential contractors with better notice of the standard to be employed by DOE in exercising its patent rights by:

 - providing more definite standards regarding patent rights in the DOE procurement regulations;

 - clarifying the standards with liberal use of examples, such as those used by the Internal Revenue Service in its publications;

 - publishing DOE decisions regarding the exercise of march-in right; decisions to grant exclusive license and waivers,

o   Encourage innovative contracting and ensure that small, innovative companies are given preference wherever possible.

o   Remove bureaucratic barriers to contract approval by decentralizing the authority to enter into subcontracts and cooperative agreements and to make grants for solar energy projects.

o   Take immediate steps to reduce the time required to process a contract for development of solar energy or energy conservation equipment.   A strict requirement to let all contracts within three months should be enforced;

o   Provide more funds to support a formal program for funding unsolicited proposals.

o   Provide special programs for socially and economically disadvantaged contractors.

Direct Federal Support of Fundamental Research

DOE should significantly increase the now minimal federally-supported research in fundamental areas which fall between basic science and product-specific development programs.   Examples of such research include fundamental studies (appropriate to heavy industry) of subjects like corrosion, lubrication, combustion, physical separation of materials, heat transfer, phase transitions (boiling, solidification), multiphase flow and fluid transport. In addition, studies reexamining fundamental approaches to energy-intensive industrial processes, such as the manufacture of steel, paper, glass, and cement are needed. Research in all these areas was once a major element of university engineering programs, but has been badly neglected in recent decades.

Programs with Specialized Objectives

Removing Regulatory Constraints to Cogeneration

The Fuel Use Act (FUA) of 1978. This Act has the effect of discouraging some of the most promising approaches to cogeneration. Since most of the circumstances which prompted the passage of the act have vanished, it is now in order to have a fundamental reexamination of the need for most of the controls it imposes. While a complete revision of the act would be preferable, a number of relatively minor modifications in the Act and its related regulations could remove most of the major barriers to adoption of cogeneration.

The Public Utility Regulatory Policy Act of 1978. This Act (PURPA) is a major milestone in removing institutional barriers to cogeneration for qualifying cogeneration facilities. The major shortcoming of the Act, insofar as it deals with cogeneration, is that it is overly restrictive as to the requirements for qualifying status, set forth in Section 201.

Technical and Financial Assistance for Industrial Audits

The development of industrial energy auditing firms needs to be enhanced. The Department of Energy should initiate a program to refine the methodology involved in an industrial audit and to develop and implement training programs for auditors. The results of this program should be made widely available to industrial firms, utilities and audit firms. In addition, DOE should provide technical and financial assistance to private firms that provides industrial audits . DOE should also consider a program of loan guarantees and limited equity financing for corporate interests entering this field.

Solar and Conservation Offsets Under the Clean Air Act

The Clean Air Act, as amended by the Clean Air Act Amendments of 1977, requires that a state program which issues permits for new or expanded staionary sources of air pollution must stipulate certain requirements. One of these is that, in an area that has not met ambient air quality standards (a nonattainment area) a permit to construct or expand a major air polluting plant may be granted if total emissions of air pollutants by the new or expanded plus existing plants in the area will be less than previous total emissions (42 U.S.C. 7503(1)(A).   This is the so-called offset policy.

Because they do not pollute the air, solar and conservation devices can be useful in implementing an offset strategy. Industry may comply with the Act by installing solar technology, thus achieving a net reduction in air pollution for the area. However, the inclusion of solar and conservation offsets is left up to the states, which must draw up the final implementation plan. The Department of Energy, in conjunction with the Environmental Protection Agency, can encourage states to include offset policies in their planning for growth in non-attainment areas through education and direct technical assistance.

SOLAR TECHNOLOGIES FOR INDUSTRY

The potential for displacement of conventional fuels by biomass and direct solar industrial process heat was summarized in Table 2.2. The following sections present an abridged discussion of each of these technologies.

Use of Biomass

Biomass is the largest potential source of renewable energy. The best available analyses indicate that the total sustainable biomass supply in the U.S. may be between 7 and 17 quads in the year 2000 (see Table 2.6). It is possible that biomass supplies could exceed this level if advanced forest, agricultural or aquaculture management techniques are developed. The supplies of biomass which will be economically available depends critically on how the resource is managed. Analysis presented in this study leads to an estimate that between 7 and 14 quads of biomass will be economically available in 2000 for conversion to energy, under ideal conditions.

As Table 2.6 indicates, wood is by far the most important biomass resource. Access constraints, environmental restrictions, land ownership objectives and competition with alternative forest uses (such as recreation) are likely to limit the quantity of wood available for conversion to energy to between 5 and 9 quads. This estimate has been made assuming that the forest product industry is supplied with all wood likely to be required during the next 20 years. The amount of wood economically available for conversion to energy will depend on the demand for forest products, the price of competing fuels, and the successful integration of wood fuel and conventional wood products harvesting. In order to safely reach the 9 quad per year goal, forest management programs must be undertaken which ensure that enviromental damage is avoided and that wood fuel supplies are harvested as a part of sound forest management and forest harvesting programs.

The nation's farms can produce between 0.5 and 5 quads per year of biomass in the form of starch or sugar crops (such as grains) or grasses. The supplies are limited primarily by the fact that at some level of production increased use of energy crops will increase food and feed prices. Uncertainties about the availability and productivity of cropland, domestic and international food and feed demands, and weather fluctuations all make it difficult to construct sound energy policy based on this supply source. Crop residues can be used for energy without inflating food prices, but supplies of these residues are limited because these residues must be used to prevent soil erosion.

Table 2.7 indicates potential biomass useage under the assumption that the policies explained later in this section are adopted. The allocation of biomass among different competing markets is not an exact science and markets may develop very differently from those shown in the table. It is clear, however, that major markets for wood will continue to be found in the forest product industries and industries in close proximity to wood resources. Common industrial applications will be direct combustion or cogeneration systems and on-site gasification. It is assumed that homeowners will continue to use wood for heating but will begin investing in advanced highly-efficient wood stoves.

A variety of techniques will soon be available for converting biomass to methanol for use in transportation. Methanol will probably be the most attractive synthetic fuel since it can be produced cheapest and from a variety of feedstocks (including coal).

The goals in Table 2.7 assume that almost 28 billion gallons of ethanol (2.4 quads) is produced from grasses and legume herbage estimated to be available (see Table 3.5). This is optimistic given the uncertainties about biomass supply and the problems of bringing lignocellulose-to-ethanol technologies into widespread commercial use by 2000. These same biomass supplies could be used to produce methanol or amonia.

Agricultural feedstocks will be used in other applications. A small amount of grains will undoubtedly be converted to alcohol on-farm and used to run tractors and other farm equipment. Animal manure (which now poses a serious waste disposal problem in some areas) will be converted to methane on-farm with anaerobic digestion technologies. Low-btu gasification systems may also be used on-farm for grain or crop drying.

The chemical sector is also likely to provide major markets for biomass. Uncertainities associated with future bioconversion technologies (those producing ethylene, for example) and limited ethanol and methanol markets (methanol is being produced relatively inexpensively from natural gas) make quantitative estimates difficult. In Table 2.7 the conservative assumption has been made that little biomass will be used to manufacture chemicals. The successful development of advanced pyrolysis or gasification systems, however, could change this projection dramatically.

Table 2.6 BIOMASS SUPPLIES AVAILABLE IN 2000 (Quads of primary fuel per year)

Feedstock Source	Total Potential (irrespective of price)
Forestry Resources[+]	
Annual Growth Increment Available (net annual growth less projected harvest)	3.5[a]-9.3[b]
Enhanced Growth Increment from Intensive Management	2.9[c]-11.9[d]
Mortality	1.8[e]-4.1[f]
Rough, Rotten and Salvable Dead Trees*	1.2[g]-1.7[h]
Logging Residues	1.9[i]
Land Clearing*	.2[j]-.3[k]
Mill Residues**	1.3[l]
Non-commercial Forest Land	.8[m]-.9[p,n]
Subtotal less nonsustainable categories	12.2-29.4
Subtotal (likely) Available[o]	5-9
Agricultural	
Intensive Cropland Management[p]	
- Grain & Sugar Crop Production to produce ethanol (not completely additive with category below)	0-1
- Grass and legume herbage production (not additive with above)	0-5
Intensive Management of Pasture and Hayland[q] to product grass and legume	0-5
Crop Residues[r]	.8-1.2
Animal Manure[s]	.1-.3
Agricultural Processing Wastes[s]	.1
Subtotal Agriculture[t]	1.0-6.6
Municipal solid Waste[u]	1.23
Sewage Sludge[v]	.1
Aquatics	***
Total Supplies Available	7.4-17

See notes to Table 2.6 at the end of this section

Encouraging the use of the vast supplies of biomass is not without risk. For example, using wood for energy offers the opportunity to either increase forest productivity and the availability of forest products. Improved forest management, however, could devistate thousands of acres of forest land. Poorly managed use of crop residues for energy can increase soil erosion in certain parts of the country. The success of a national biomass program requires that flexible policies be instituted which can be readily adapted to changing economic or environmental circumstances. Better management and coordination between the agencies participating in a Federal biomass program is an absolute necessity.

Specific policies to encourage the use of biomass are explored in detail in the Industry chapter. Important policies are summarized below in two basic areas: wood energy and agricultural energy.

Table 2.7a. ENHANCED CASE FOR BIOMASS ENERGY SUPPLY IN 2000[a] (Quads)

	Primary Supply				
Use Sectors	Ethanol[m]	Methanol[l]	Methane[q]	Direct Combustion /Gasification[k]	Total[j]
Transportation	.25–2.4[f]	0–4[h]	0	0	.3–4.0
Food Processing	0	0	0–.05	0–0.1	0–.15
Forest Products	0	0	0	3–4	3–4.0
Chemicals[n]	0[n]	0	0	0	0[n]
Residential	0	0	0	1–2	1–2
Agriculture	0–.2[f]	0–.1	.1–.3	0–.1[e]	.1–.7
Other	0[o]	0–.5[i,j]	0	1–2[g]	1–2.5
Total	.25–2.4	0–4	.1–.35	5–9.2	7.4–14.4

Cost per
Million Btu $10.70 – $17.80[r] $5.00 – $15.00
                    $8.00 – $20.90[s] $3.00 – $6.00[u]
(1979$)

Table 2.7b ENHANCED CASE FOR ENERGY DISPLACED BY BIOMASS IN 2000[a] (Quads)

	End Use Energy Displaced (Oil Equivalent)				
Use Sectors	Ethanol[m]	Methanol[l]	Methane[q]	Direct Combustion /Gasification[k,p]	Total[j]
Transportation	.38–1.8	0–3.72	0	0	.38–3.0
Food Processing	0	0	0–.05	0–.01	.0–.1
Forest Products	0	0	0	2.7–3.6	2.7–3.6
Chemicals[n]	0[n]	0	0	0	0
Residential	0	0	0	.8–1.6	.8–1.6
Agriculture	0–.2	0–.1	.1–.3	0–.1	
Other	0[o]	0–.2[i,j]	0	.8–1.6	.8–1.8
Total	.38–1.8	0–3.72	.1–.35	4.3–7.0	4.78–10.52

Cost per
Million Btu $10.70 – $17.80[r] $5.00 – $15.00
                    $8.00 – $20.90[s] $3.00 – $6.00[u]
(1979$)

See notes to Table 2.7a and Table 2.7b at end of this section

Wood Energy

It should be possible for the U.S. to produce 9 quads of wood per year by the year 2000 without interupting the production of conventional forest products. (Canada plans to use 7 quads of wood for fuel by 1985).

Wood fuel use can leverage investments in forest management by providing a relatively short-term return on forest management practices (most forest management investments not linked to producing wood fuel do not yield returns until timer is harvested--sometimes 5 to 30 years later), thereby increasing the productivity and availability of all forest products. The goals for use of wood as an energy source are fully consistent with projections for use of wood for other purposes.

The most important elements of wood energy policies are forest management, research and capital assistance programs. These programs should be built around the following objectives: developing and communicating techniques for optimizing the production of wood for energy in each region in a way that will lead to profitable businesses in the near-term as well as the long run; promoting the development of new technologies in wood harvesting, transportation and conversion; ensuring that the many small industries likely to grow up around wood energy are not inadvertently disadvantaged by regulatory programs or existing tax policy.

Forest Management.   Policies to encourage proper forest management of forests should be the centerpiece of national, state and local efforts to develop wood energy.   Forest management programs can increase the availability of all forest resources (including fuel wood), link forest products production with sound environmental practices, and encourage the development of fuel wood supply infrastructures (an important requisite for wood energy development). Forest policies should be designed to allow the market freedom to determine the mix of energy and conventional forest products derived from the forests.

Forest management policies can be divided into two kinds: those which affect private forest lands, and those which affect National forest lands.

Sixty percent of all commercial forests belong to private, non-industrial owners. In the East, nearly three-fourths of the commercial forests are privately owned.   These forests can be expected to play an increasingly important role in wood energy development and forest product production as old growth timber inventories in the West continue to decline.

Unfortunately, neither market forces no past federal policies have simulated many of the four million individuals who own these woodlands to invest in proper forest management. A recent industry report revealed that, while five million acres of forest lands are being harvested annually, only 2.2 million acres are being regenerated through planting or purposeful site preparation.   An even lower ratio--one out of nine harvested acres--is being purposefully regenerated on private, non-industrial forest lands (NFPA 1980). Industry estimates that, in the 25 major timer-growing states, almost 80 million acres of private, non-industrial forest lands need silvicultural treatments to boost productivity. The industry estimates that a $6 billion investment over the next 50 years in increased forest productivity would increase timber growth by six billion cubic board feet per year (NFPA 1980).

Most small, non-industrial forest landowners lack the training and capital resources necessary to maximize the wood fuel and forest products potential of these lands. The average forest landowner does not posses the marketing and forest management skills, or the resources necessary to develop a wood fuel supply infrastructure.

The New England Fuelwood Pilot project (recently initiated by the USDA's Agricultural Stabilization and Conversion Service) is an excellent model program for overcoming these barriers.   This program provides small, non-industrial forest landowners with technical assistance and cost-sharing opportunities to improve the management-and profitability-of their forest lands.   Landowners receive both technical and financial assistance in the marking of trees for cutting, evaluating forest stands, the construction of access roads, etc.   Cooperative forest management is also encouraged, where appropriate, to realize the economies-of-scale, maximize labor and gain access to capital resources. The program has, in fact, leveraged supplies of fuelwood at a cost of only $3.50 per barrel of oil displaced. Programs of this kind need to be strengthened considerably. In addition, tax and regulatory policies can be designed to aid the private forest landowner.

The federal government is directly responsible for the management of the National Forest System, which comprises 18 percent of total U.S. commercial forest land.   The preponderance of national forest lands are located in the Northwest. In 1976, 16 percent of total roundwood supplies and 20 percent of total sawtimber supplies came from the national forests (RPA 1980). In 1978, nearly $900 million in timber was removed from the National Forests (Stat Abst 1979).

Because of the scale of resources harvested from the National Forests and the federal government's ability to ensure their proper management, these forest lands can play an important role in wood energy development. Important initiatives designed to stimulate wood energy development and forest management on National Forest lands include:

o The Forest Service, which is responsible for managing the disposal of logging residues on Natinoal Forest lands, should move aggressively to establish programs for making productive use of residues from foresting operations. Concentration yards should be used to gather wood at a common site for marketing. These residues, which now pose a serious forest fire hazard and impede regeneration in many cases, are a significant potential energy feedstock. About 2 quads of logging residues are presently burned or returned to the soil every year.

o U.S. Forest service planning activities should incorporate wood energy programs as an integral part of forest management programs.

o The Forest service should improve its biomass survey methods to facilitate adequate national planning for use of this resource.

Capital Assistance and R&D. The programs, described earlier in this section for encouraging research and innovative energy investments in industry can also be used to help industries find the capital needed for new ventures involving the production and use of biomass. The most important research priorities include the development of low-cost harvesting equipment, air-blown gasifiers, and small-scale biomass-to-methanol systems.

Agricultural Energy

Alcohol Fuel Programs. Existing policies designed to encourage the production of alcohol do not adequately reflect the risks inherent in using large amounts of cropland to produce fuel. The most important alcohol fuel subsidy--exemption of the 4¢/gallon federal gasoline tax on gasohol--could lead to higher than desired farm prices, uncomfortably low food/feed reserves, or unjustified windfall subsidies. As a consequence, the alcohol subsidies should be restructured to allow sound market principles to determine the best type and extent of alcohol fuel production. Moreover, alcohol policies should give equal support to methanol as well as ethanol production.

Forgiving the 4¢/gallon federal gasoline tax on gasohol mixtures is roughly equivalent to a $20 per barrel tax on crude oil. It would be far preferable to tax fossil fuel directly; a $20/barrel tax on oil or on gasoline would have as great an effect on the markets for gasohol as the current program, but would have the great advantage of allowing the market to choose between investments in transportation efficiency and investments in various synthetic fuels, including the choice between methanol and ethanol.

Because alcohol fuels policy is intimately related to both agriculture and energy policy, future alcohol policy should be coordinated more closely with existing agricultural policies in both USDA and DOE. For example, farm program management decisions in USDA should incorporate projected demands for crops used for alcohol. The crop reserve programs will need to be reconsidered and perhaps expanded to include an energy reserve as well as the food/feed reserve. Also, alcohol production plans will need to consider projected feed/food demands from USDA.

If the current structure of the alcohol incentives is to continue, it would be desirable to develop greater flexibility in administering them. An interagency committee could be given responsibility for monitoring the effects of alcohol fuels production on food and feed prices and exports, and providing an 'early warning' of any potential supply or price problems. Authority could be provided to the task force to react appropriately to any market imbalance by: increasing food/feed reserves, altering regulation of cropland, restricting permits for alcohol production, and even altering or removing subsidies given to the production of alcohols.

If the policy objective is net premium fuels production, oil and natural gas should not be used in producing alcohol. The Energy Security Act of 1980 however, allows certain sugar refineries producing alcohol, and some existing distilleries producing fuel grade alcohol to use natural gas on a priority basis (gas not subject to curtailment).

Capital and Technical Assistance Programs. In order for the agricultural sector attain some degree of energy self-sufficiency and perhaps produce some of the nation's transportation fuels, programs to provide technical and capital assistance to the farmer need to be strengthened. The Department of Agriculture's Farmers Home Administration and Rural Electrification Administration can play key roles in providing capital assistance.

The Department of Agriculture is uniquely equipped to implement a technical assistance program directed toward on-farm energy production technologies. The USDA has a long record in the area of agriculture technology transfer emanating from its strong research support system and the extension service arm, the Cooperative Extension Service (CES). A most pressing need is to coordinate its research efforts with DOE and the private sector.

The technical assistance available from the Agricultural Stabilization and Conservation Service should also be utilized to provide direct technical assistance by trained personnel for the installation and operation of on-farm technologies. With a minimum of effort these technicians could be trained in the nuances of construction and operation of on-farm energy producing technologies, such as anaerobic digestion and on-farm distillation. (Research priorities are discussed in the Industry Chapter.)

Solar Industrial Process Heat

About 40% of industrial energy is used to provide heat for industrial processes. Solar energy could make a significant contribution to meeting this need. However, the attractiveness of solar collectors for providing process heat depends heavily on the temperatures required, and the schedule of the demand, the available solar radiation at or near the industrial site, and most importantly the system costs.

The fraction of IPH energy that is technically compatible with direct solar technologies is estimated in Table 2.8. However, the attractiveness of these devices depends critically on whether costs can be reduced. The technologies are expected to have a significant impact only when installed system prices can be reduced to less than half that of the current generation of experimental units the cheapest of which cost approximately $40-$50 per square foot. Contributions above about one Quad of energy annually are possible

Table 2.8 TEMPERATURE RANGES FOR MAJOR CLASSES OF COLLECTORS

Process Temperature (°F)	Total Demand (Quads)			Fraction Compatible With Direct Solar[a]	Potential Direct Solar Market in 2000 (Quads)	
	1977 (Quads)	2000 existing programs	2000 new programs		Existing Programs	New Program
0-212	0.96	1.0	0.89	0.75	0.76	0.63
212-350	3.55	3.7	3.0	0.64	2.4	1.9
350-550	2.75	3.0	2.4	0.81	2.4	1.9
550-1100	2.50	2.6	2.1	0.30	0.81	0.66
1100-2000	0.97	1.0	0.85	0.25	0.26	0.21
2000+	1.48	1.6	1.3	0.15	0.24	0.20
Total	12.20	12.9	10.5	---	6.8	5.6

Demand data from Section 3. Industry compatible fraction based on a concensus of analysts at SERI. See Table 4.3 for relative energy intensities.

[a]Compatible fraction for the 0-212 and 212-350 ranges are reduced to account for the nearly complete use of biomass by the paper industry, which accounts for .3 Quad of the demand in these ranges.

if costs can be reduced to a third of current system costs. There are sound reasons for believing that such cost reductions are possible, but a continued major development program is plainly necessary to achieve them.

The economic contribution of solar IPH systems under a variety of assumptions about system costs, fuel cost, financing and land availability are displayed in Table 2.9. The expected potential is calculated in

Table 2.10. It can be seen that with high fuel prices and low-cost financing as much as 1.4 quads could be supplied with solar IPH, even with the success of aggressive programs to increase industrial energy efficiency.

This large potential clearly justifies a serious research effort designed to greatly reduce the costs of solar collectors, storage equipment, and systems to transport thermal energy.

Table 2.9 POTENTIAL CONTRIBUTION OF SOLAR PROCESS HEATING SYSTEMS IN 2000 (QUADS)

System Cost	Fuel Cost	Finance Option	Economical % Displacement	Current Policies		New Policies	
				50% land	75% land	50% land	75% land
Optimistic	HIGH	Utility	.97	1.6	2.5	1.3	2.0
		Industry	.94	1.6	2.4	1.3	1.9
	LOW	Utility	.93	1.6	2.4	1.3	1.9
		Industry	.36	0.61	0.91	0.50	0.74
Expected	HIGH	Utility	.92	1.6	2.3	1.3	1.9
		Industry	.42	0.70	1.0	0.57	0.86
	LOW	Utility	.40	0.69	1.0	0.56	0.84
		Industry	.03	0.065	0.078	0.053	0.06
Pessimistic	HIGH	Utility	.46	0.78	1.2	0.64	0.95
		Industry	.19	0.33	0.48	0.26	0.39
	LOW	Utility	.14	0.25	0.36	0.20	0.30
		Industry	.02	0.039	0.052	0.032	0.042

Notes: High fuel costs imply oil reaches 60$/BBL   2000
Low fuel costs imply oil reaches 40$/BBL   2000
75% land implies that 25% of industrial sites cannot use direct solar heat because of land use problems.
See Table 4.8 for collector costs and characteristics.

Table 2.10.   EXPECTED POTENTIAL OF DIRECT
              SOLAR HEAT (Quads)

Finance Options	Fuel Costs	New Programs	Existing Programs
Industry	HIGH	0.73	0.90
	LOW	0.14	0.17
Utility	HIGH	1.41	1.73
	LOW	0.76	0.93

Assumptions about probabilities
(1) Solar equipment costs   probability
     --optimistic   0.15
     --expected   0.60
     --pessimistic   0.25
(2) Solar fraction
     57%   0.20
     42%   0.80
(3) Land availability
     50%   0.60
     75%   0.40

Notes to Table 2.6

+Conversion factors for forestry estimates:

    85 cubic feet per cord
    1.25 ODT (oven dry tons) per cord
    17 million Btu per ODT

   *Nonsustainable category

 **This category will be used for energy and other purposes within the near term.

***Should on-going research prove successful, the production of biomass from this category could be significant.

aDerived from USFS estimates of the net annual growth (total growth less mortality) and harvest levels in the year 2000. Assumes increased efficiency of product yield/unit input and increased use of tops and branches (biomass multiplier revised from 1.6 to 1.4). (USFS 1979a, 1979b)

bDerived from OTA estimates of net annual growth and USFS projections of harvest levels in the year 2000. OTA estimates are higher than USFS because USFS projections consistently underestimates growth. This growth increment represents an amount above that needed by the forest product industry.

cAssumes modest growth increases resulting from intensive management as predicted by the Forest Service under free-market conditions. (USFS 1979b). The utilization efficiency of this growth increase is assumed to be that projected by OTA (OTA 1980).

dAssumes achievement of growth increases estimated by OTA. (OTA 1980, USFS 1979a) OTA enhanced growth estimates show that potential growth is two to four times that typically predicted by the Forest Service. Given this, and the fact that current Forest Service growth estimates are about half the growth that could be achieved under fully stocked conditions, the total potential growth could be between 18 and 36 quads/year. However, how much enhanced growth can be achieved economically is not known. Therefore, the enhanced annual growth increment is arbitrarily reduced by 50%.

eProjected from current levels (USFS 1979a) due to increased amounts of overstocked stands. Mortality increases should not be assumed to occur under "intensive management."

fThis represents the upper range of the OTA mortality estimate which results from an underestimation of gross growth in USFS timber stand yield tables. (OTA 1980)

gAssumes that the stock of rough, rotten and salvable dead material (2.06×10^9 ODT) is drawn down over a period of 30 years; following the method of SRI (1979).

hSimilar assumptions to "g". Estimates of rough rotten and saluable dead material taken from OTA (1980). This estimate includes low quality hardwoods on former conifer sites.

iComparable figures to USFS-derives stimates of logging residues were not available from OTA (1980) due to the inclusion of rough, rotten, low-grade and savable dead material in the OTA logging residue estimates. This figure includes inventoried commercial material wasted in harvesting assuming that, as a percentage of total removals, this waste is proportionally 50% less than the same figure in 1977. (USFS 1978, 1979b) Assumes that logging residues composed of tops and branches represent an amount equal to 40% of the harvested and removed material. The 40% figure assumes more complete utilization of tops and branches than is currently prevalent.

jOTA 1980

kEstimated on the basis of an annual decline of .05% decline in commercial forest land area and an average of 38.6 ODT/acre cleared. (USFS 1978, 1979a)

lEstimated on the basis of utilized and unutilized mill residues together as a percentage of total removals in 1970 and projected removals in 2000 (USFS 1973, 1979b)

mEstimated from OTA (1980) assumption of 205 million acres of noncommercial forest producing an average of 10 ft^3/yr.

nEstimated from USFS (1978) inventory of 228 million acres of non-commercial (including 107 million acres in Alaska). Productive-deferred or reserved forest land is not included.

oConstraints related to access environmental restrictions, land ownership objectives and competition with alternative forest uses (e.g., recreation) are likely to limit this quantity to perhaps 40% of the total supply. This assumption consistent with reliable regional and national estimates.

pOTA 1980. There is considerable uncertainty about the amount of land which can be brought into production, crop switching possibilities, etc. The two options presented for intensive cropland management are not additive.

qOTA 1980. This category is not completely additive with above agricultural categories.

rOTA 1980. This supply is likely to increase in proportion to crop production.

sOTA 1980.

tIt has been assumed that the upper limit uses crop residues, animal manure, agricultural wastes, pasture and hayland, and approximately 1.5 quads from intensive cropland management.

uBased on per capita generation rates of 2.7 lb/day and value of 5,,000 Btu/lb. (wet) for a population of 250 million persons in 2000.

vAssumptions outlined in Sec. II. Septic systems not counted but may be potentially available because pumping (and dumping) is required anyway. About .02 q is used to incinerate sludge and may be a source of conserving natural gas.

Notes to Table 2.7a and 2.7b

aThis scenario assumes an aggressive national program for biomass utilization. Intensive management of croplands and wood-producing forests is assumed to be national policy.

bHigh range (OTA 1980) assumes feedstock costs as high as $1.26 (gallon ethanol. Costs could be higher if feedstock production costs (on potential cropland, for example) are higher.

cAssumed price for gasoline in 2000--equivalent to $1.98 per gallon including delivery charge.

dHigher range assumes wood feedstock costs as high as $80.00/dry ton.

eGrain and crop drying.

fEthanol categories not necessarily additive. Optimistically assumes 5 quads lignocellulose produced yielding 85 gal/DT. Table indicates quads of ethanol, not quads of feedstock input.

gAssumes the use of wood-fired boilers and gasifiers in the non-forest product industry and utility sectors.

hMethanol production is dependent upon wood availability and competition from coal. High range assumes that only 5 quads of wood will be used in the direct combustion/gasification category. Methanol could come from gasification of agricultural feedstocks.

[i]This methanol could be used in turbine, fuel cell, or chemical applications. This methanol is not necessarily additive with other methanol categories because of restrictions on the wood resource base. Conversion efficiencies from Pefley, 1979.

[j]Subtotals are not additive but are reduced by those resource categories (e.g., wood) which are not additive. It is assumed here that at least 7.0 quads of wood are used.

[k]In this category about .89 quads of oil equivalent is displaced per quad of feedstock input, assuming an average direct combustion/gasification conversion efficiency of 76% replacing oil burned at 85% efficiency. The product energies shown are .76 times the primary energy inputs.

[l]Using advanced technology, it is assumed that for every Btu of wood (primary supply) used, approximately .93 Btu of oil equivalent is displaced. This conversion efficiency will be achieved if future methanol plants achieve 70% conversion efficiencies and if future automobiles are engineered to exploit methanol's unique characteristics (by utilizing automotive engines which disassociate methanol into hydrogen and carbon monoxide, and by increasing compression ratios) thereby increasing end-use efficiency by at least 20%. 1.1 Btu of crude oil is required to produce 1 Btu of gasoline.

[m]For this category, assuming that most of the ethanol is produced from lignocellulose, between 98.8 and 240 million barrels of oil are displaced for every quad of product. Lignocellulose could also be used to produce methanol. This range is a function of whether or not the ethanol is used solely as an octane booster (a limited market) or as a blend or stand alone fuel. (OTA 1980.) Future octane booster markets are expected to be very limitted.

[n]The chemical sector is a key potential market for biomass. However, because of the uncertainties associated with future bioconversion technologies (those producing ethylene, for example) and limited ethanol and methanol markets (methanol is being produced relatively inexpensively from natural gas) biomass is not projected to penetrate this sector measureably. Technological advancements could change this appreciably.

[o]Some ethanol could be used in peaking turbines. However, it has been assumed arbitrarily that methanol will be used in these applications.

[p]For the residential sector it is assumed that .8 quads of oil are displaced for every quad of wood input.

[q]Quads of methane are interms of product. It is assumed that using biogas in on-farm applications has the same process efficiency of conventional fuels.

[r]See Appendix for ethanol cost estimates. This cost range assumes that economically attractive lignocellulose conversion technologies will become commercial by the year 2000.

[s]See Appendix for methanol cost estimates

[u]See Appendix for description of cost estimates (OTA 1980).

Section Two
Public Policy Recommendations

There has been an historic under-investment in industrial energy productivity because of public policies that have led industrial investors to place a lower value on energy productivity than they would have in a market environment free of distortion. Investment decisions have not reflected the true costs of energy because energy prices have been kept artificially low. Moreover, in the competition between purchasing energy and making investments in efficiency, the structure of the present system of corporate income taxes has the effect of discouraging industrial investments in productivity while encouraging capital formation in regulated utilities to a point where they may be over-investing in energy supplies.

Any change in public policy designed to improve the efficiency with which industry uses energy and the rate at which renewable resource technologies are adopted must be carefully integrated with a comprehensive national program to increase industrial productivity. Narrow concentration on energy can be counterproductive, distorting investment decisions and, on occasion, even discouraging imaginative improvements in industrial processes. There can be little doubt that some form of comprehensive program to encourage industrial capital investment and to increase research and the rate at which process changes are introduced into U.S. industry is long overdue; the U.S. is lagging behind its western trading partners in the technical sophistication of its industrial base. This has happened in spite of the fact that the U.S. maintains world leadership in most areas of basic science and in some areas of industrial technology. The problem is two-fold: a lack of modern facilities as a result of inadequate investment in basic materials production capacity, and a lack of technical leadership in many areas related to industrial productivity. It is unlikely that the rate of adoption of solar and efficiency technologies can be significantly accelerated without tackling these fundamental problems.

Conversely, a program capable of accelerating industrial investment in productivity would significantly reduce industrial consumption of conventional energy resources. National policy should be built around a clear set of goals for improving industrial productivity.

An effective industrial energy policy must meet the following criteria:

o It should create an investment climate that facilitates an even-handed choice among investments in improved efficiency, renewable energy, and other future energy supplies.

o It should create incentives designed to encourage industrial investment in technologies that increase productivity and reduce production costs generally, and should not concentrate exclusively on energy-related investments. Many of the technical approaches likely to improve energy productivity will be attractive for a number of reasons having little to do with energy, e.g., reducing demands for capital, reducing demands for valuable materials, and improving labor productivity. A properly balanced program is more likely to lead to large energy savings than a narrowly focused program, which may lead only to retrofitting obsolete equipment with energy-saving devices.

o It must allow the focus of decision-making to be the industrial energy user. Industrial decision-makers can take into account relationships among inputs, products, and production capabilities that may vary strongly from plant to plant, government agencies and the energy supply industry cannot.

196

o It should encourage technical innovation and foster the possible leapfrogging of the basic-materials production technology of Western Europe and Japan. One benefit from fundamental innovation could be reduced scale of (economically competitive) production facilities. This would enable incremental creation of new facilities in relatively small modules which, in turn, could be introduced rapidly, with relatively little risk in each investment.

o It should provide direct incentives for industry to replace outmoded facilities. Reconstruction should be encouraged at existing sites whenever this is practical since major plant relocations can inflict great hardship on communities.

The strategy that follows is designed to eliminate the distortions that have been introduced into energy markets by years of federal intervention. The discussion is organized around the following major themes:

(1) Moving toward rational energy pricing, so that investment programs can be designed with recognition of the long-term costs of energy.

(2) Making adequate capital available for investments in industrial facilities.

(3) Encouraging industrial investments in research and development (particularly applied research) and supplementing this private investment with direct government support of fundamental research relevant to industrial productivity.

(4) Improving the transfer of information on energy efficiency and renewable energy technologies from research organizations to industry, and improving the educational programs that train scientists, engineers, industrial designers and operators in these areas.

(5) Improving the regulatory environment that controls activities related to energy use in industry. This will include both (a) removing burdensome regulatory barriers that have no clear function under present circumstances and (b) using existing environmental and other regulations more effectively to encourage and support investments in energy efficiency and renewables.

Missing from this list of themes is a program of commercial demonstrations, a major element of con-

ventional energy planning. With the exception of engineering development tests designed to demonstrate technical performance and feasibility, federally sponsored commercial "demonstrations" are recommended only if proposed as a program that can be supported by the venture capital initiative, as described below. This deemphasis on federal involvement in commercial demonstrations is an important characteristic of the market-oriented strategy presented here. This strategy emphasizes the creation of an economic climate that is conducive to private-sector initiatives. It focuses direct federal involvement on desirable activities the private sector is unlikely to pursue, even in an "improved" economic climate.

Targets for industrial energy consumption to the year 2000 are developed in Section 3 of this chapter and are summarized in Table 2.3. These targets are not projections; they are goals the authors believe are consonant with the future energy prices that are estimated in this study and with the policies described below.

With the quality of information currently available, quantitative estimates of the costs and benefits of individual programs are often speculative. However, estimates have been made in important cases where there is a basis for credible analysis. The combined effect of all the policies and programs described in this chapter would total 5 Quads per year by the year 2000.

The evidence presented in the following discussion argues strongly that the programs recommended would, if taken together, capture the efficiency potentials summarized in Table 3.3. It would be a mistake to believe, however, that these estimates can be read as confident forecasts. What is important is that a better case can be made for the effectiveness of this approach than can be made for alternative programs to meet U.S. energy needs.

In any program of this sort it is critical that policies be continually adjusted in response to experience and changing market conditions. Such adjustment requires a well-designed program for collecting information and evaluating programs. It is important that legislation that supports initiatives of the kind described here should provide for subsequent evaluation and also allow flexibility so that program administrators can adjust policies as evaluations come in. Although data collection, evaluation, and feedback are necessary for any public policy, the prospects for adjustment are far better for energy demand than for other types of energy policies, since most individual projects dealing with industrial efficiency or use of solar energy will involve much smaller-scale efforts. Mistakes are inevitable, but a multi-million dollar mistake is much less painful to admit than a multi-billion dollar mistake.

FINDING CAPITAL FOR CORPORATE INVESTMENTS

The present analysis agrees with industry's frequent assessment that capital scarcity is the major obstacle to rational investments in energy productivity and in overall industrial productivity.

Over the past few decades, capital formation has been a relatively constant fraction of the GNP, but the share of capital available for industrial productivity investment has been gradually eroded. One reason for this trend is that pollution control has been absorbing a substantial share of available capital. But far more important is the fact that a growing fraction of capital is going to the energy-supply sector. Although retail energy sales accounted for only 12% of the GNP in 1977, the energy supply industry in that year absorbed over 40% of all new plant and equipment expenditures (see Figure 2.4). The energy-producing sector has generated growing demands for capital, placing increasing pressures on the capital markets at a time when the energy-using sector also has critical capital needs, including those for investments in energy conservation and production of renewable energy. Needs for new products, for maintaining or increasing the production capacity for present products, and for meeting new regulatory requirements all compete with investments in cost-cutting. Cost-cutting therefore often has low priority, and the result is an extreme reluctance to abandon old or obsolescent plants and replace them with newer facilities.

From the point of view of individual businesses, the result has been an sharp increase in interest rates, making borrowing difficult and expensive even for the companies that are able to borrow. Many companies, unfortunately, already have such high debt-equity ratios that conventional borrowing is no longer feasible. For these companies, new equity issues or internal profits are the only possible sources of new capital. Often both sources are not practically available for use by the very companies most in need of modernization and energy conservation. Companies with large demands for energy have often experienced shrinking profits, thus diminishing their ability to finance projects with internal funds. In addition, a low price-earnings ratio for corporate stocks has frequently made it undesirable for a company to issue new stock. For such companies all the traditional means for raising capital are either precluded or very expensive. Capital shortages may particularly affect low-growth industries and industries facing severe foreign competition. Many of these industries are also energy-intensive. It is clear that mechanisms are needed to make investment capital available from outside sources (such as government, private financial institutions, and the energy industry) for improvement in those industries.

One result of the lack of capital is that investments in energy productivity are expected to achieve a rapid pay-back, while the pay-back period for energy supply can be relatively long, often because of regulatory protection. It is not suggested that the pay-back period should be the same for a heat-recovery device and a new electric-generating plant. However, the criteria for determining the particular financing conditions should be the same.

The total demand for industrial capital over the next two decades will be enormous. Calculations presented later in this discussion indicate that the nation's industries may need to invest an extra $170 billion over the next 20 years to improve the productivity of the U.S. industrial base. This amount is in addition to the historic rate of capital investment in industrial plants and equipment. Because many industries will find it difficult to accumulate the capital needed to make improvements, however attractive, an aggressive capital formation policy is needed.

Four categories of programs are examined here: (1) Programs designed to ensure that industrial energy prices are set at levels that encourage adequate investment in energy efficiency and renewable resources; (2) Programs designed to reduce or eliminate the burden of federal taxation on corporate income; (Several alternatives are discussed, including reduction in the corporate income tax rate, liberalized depreciation, and additional tax credits); (3) Programs designed to remove the regulatory barriers that make industrial investments by utilities or third parties difficult; (4) Programs designed to ensure that receipts from the windfall profits tax (and possibly a fuels tax) are equally available for investments in energy supply and energy efficiency. The cornerstone of this last program is a project designed to provide funds for industries to scrap existing obsolete facilities and rebuild.

Energy Prices

Energy is still priced well below its marginal cost. For example, the average price paid by industry for natural gas in 1980 was $2.80 per million Btu, but gas costing $3.50 per million Btu is being imported from Canada. In 1978 the average price paid for electricity was 2.5¢/kwh, but the cost of electricity from new plants was about 5¢/kwh.

Three techniques are examined for ensuring that industrial fuel prices send a proper signal: (1) further reductions in the regulations of energy prices, (2) elimination of energy supply subsidies; (3) a fuels tax.

Reducing Price Regulation

While dramatic steps have been taken toward the full deregulation of energy prices in recent years, prices are still distorted by a variety of state and federal controls. Oil prices have been entirely deregulated, The Natural Gas Policy Act of 1978 (P.L. 95-621) will deregulate the price of gas by the mid-1980's, and the Public Utility Regulatory Policy Act of 1978 (P.L. 95-617) encourages the use of marginal-cost pricing by electric utilities. The section of this report dealing with the utility industry examines the impact of these programs in some detail and suggests techniques that can be used to remove the remaining distortions in utility energy prices.

The important point for the purposes of the present argument is that a history of under-priced energy has left industry with investments that made economic sense in an era of artificially low energy prices. These facilities are vulnerable to sudden changes in prices, which may result from a disruption of foreign supplies or a sudden decision to remove regulatory price controls.

Eliminating Energy-Supply Subsidies

The discrepancy between market prices and full marginal costs results in part from price controls and in part from federal subsidies to energy suppliers. It has been estimated that special tax advantages for electric utilities are $3 billion per year (Cone, et al., 1980). (These are accelerated depreciation for investor-owned utilities and corporate and investor tax advantages for publically owned utilities.) Substantial federal enterprises in support of the routine provision of electricity are over $1/2-billion-per-year activity (see Table 2.11) (Cone, et al., 1980; General Accounting Office, 1979). Subsidies for routine development and provision of oil and gas were also near $3 billion per year in the late 1970's (Cone, et al., 1980). These subsidies are historically out-dated and should be reduced or eliminated. If all these subsidies were eliminated and revenues were allocated to the economic sectors on which the burden fell, the revenues allocated for the industrial sector would be more than adequate to pay for all the programs proposed here for that sector.

A Tax on Industrial Energy

The price of energy delivered to industry does not reflect the national security costs of continuing heavy dependence on foreign oil, the environmental costs of energy production and use, or macroeconomic effects resulting from massive energy imports. These effects make the total social cost of energy even higher than the cost that would result if regulatory and subsidy distortions were removed completely. Regulations, targeted incentives, or energy taxes could supplement deregulation and removal of subsidies to effectively bring prices in line with full social costs. Of these alternatives, energy taxes have the advantage that they impact the general economic climate relating to energy decisions, without involving the government in detailed planning. All other alternatives require some federal agency to decide which technologies are the most appropriate for investors. Energy taxes have received increasing attention in recent years (Page, 1977; Bullard, 1974; Forrester, 1979; Hannon 1979).

One major difficulty with the energy tax concept is that many energy-intensive industries are already facing severe financial difficulties. Any program designed to improve energy efficiency that includes an energy tax must be accompanied by measures that can provide these heavily impacted industries with the capital they will need to make new investments in response to higher energy prices. The fuel taxes must also be imposed slowly so that industry will have time to react without unproductive haste. Unless such a program of capital assistance is provided, an energy tax could result in a net decrease in the funds available for improving the energy-intensive industries.

A program of energy taxes could, however, lead to a net increase in the capital available for energy-intensive industries if coupled with a program that provided relief from other kinds of taxation on corporate income. Attention must be paid to the details of such a program to insure that energy-intensive industries are not subsidizing other industries. For example, a tax package could be designed around a program of rebates that returned the energy tax revenues to industry. The rebates could be returned directly to firms on the basis of a formula that rewarded investments in new plants and equipment or corporate research and development, or the rebates could be administered through a special fund that would support investments leading to increased productivity.

A number of special aspects of energy taxes need to be considered:

(1) Energy taxes could disadvantage U.S. importers and exporters of energy-intensive products unless an adjustment is made at points of entry. It may be necessary, for example, to forgive the energy tax for products manufactured for export (e.g., manufacturers would not be taxed for the energy used to manufacture export products so that the U.S. products would not be overpriced on world markets.) Similarly, a tax reflecting the fuel value embodied in energy-intensive products imported into the U.S. would need to be levied.

(2) The tax could be procedurally complex and expensive to administer unless care is taken in its design. One simple technique for administering the tax would be to use the existing corporate income tax forms; energy use is already reported on these forms since a

Table 2.11 ESTIMATED FEDERAL GOVERNMENT SUBSIDIES IN SUPPORT OF THE
ROUTINE PROVISION OF ENERGY SUPPLIES IN 1977

Federal Activities	
	(million 1977 dollars)
Low interest appropriations plus tax exemptions for hydroelectric and transmission facilities[a]	290
Enrichment services[b,c]	180
NRC regulatory costs[b]	146
Privately Owned Utilities	
Liberalized Depreciation[a,b]	2000 (approx)
Publically-Owned Utilities	
Exemption from federal income taxes[a]	
Tennessee Valley Authority	130
State power authorities and municipal utilities	80
Rural Electrification Administration (REA)	294
Tax exempt bond subsidies[a]	260
Loans and loan guarantees provided by REA	260
Oil and Gas Industries	
Percentage depletion allowance[a]	
Oil	550
Gas	1150
Expensing of intangible exploration and development costs	
Oil	740
Gas	460
TOTAL	6540

Notes:

(a) B.W. Cone et al., (An Analysis of Federal Incentives Used to Stimulate Energy Production), second revised report prepared by Battelle Pacific Northwest laboratory, PNL-2410 Rev. II, February 1980.

(b) General Accounting Office, "Nuclear Power Costs and Subsidies," June 13, 1979.

(c) The subsidy for enrichment services is the estimated difference between what the private sector would have charged and what the government actually charged.

(d) The liberalized depreciation allowance for utilities translates into an outright tax savings (instead of just a deferral) for a utility growing sufficiently rapidly or for a utility that is not growing, as long as there is significant general inflation.

Source: Robert H. Williams, Princeton University
Testimony before the Energy Conservation and Power Subcommittee, U.S. House of Representatives, February 24, 1981.

deduction can be claimed for expenditures on energy purchase.

(3) Differential taxation of different energy forms is a very sensitive issue. Analyses and assumptions that seemed valid in the 1970's are no longer adequate guides for policy. The magnitude of the tax on different energy forms should conform to generally perceived external social costs associated with these forms. The high national security costs associated with our high level of oil use suggest that a tax on oil should be relatively high. On the other hand, while reduction of natural-gas use by industry has been a focus of past legislation, new information suggests a policy reappraisal for this fuel. The new information is: major gas savings are potentially available, especially in the building sector; and gas supplies at the now-developing prices may not be short, although it is currently uncertain. The levy of energy taxes would have to be based on new evaluations of these situations.

(4) The rate of phasing in an energy tax should be carefully considered. If a significant tax is deemed socially desirable, the tax should be raised to the desired level slowly over a period of several years to minimize dislocations. Even if a tax is introduced slowly, it could have immediate effect because anticipation of a substantial tax some years in the future would induce firms to make appropriate adjustments in their investment strategies.

The issues associated with energy-price regulation and energy taxes are complex. As a first stage, a high-level review of these issues should be carried out in the context of a general analysis of energy-price developments. The need for and the phasing-in of energy taxes should be assessed in combination with the measures that warrant the highest priority in the rationalization of energy prices: elimination of subsidies and price decontrol.

Reducing Corporate Income Taxes

Working in concert with artifically low energy prices in discouraging industrial investment in energy efficiency and renewable energy technologies is the 46% tax on corporate income and the deduction of energy costs from taxable income. A fundamental review of the need for a federal tax on industrial production and of the logic that motivates such a tax is long overdue. The current tax structure is a monstrously complex accumulative collection of programs with overlapping and often conflicting objectives. It seems plausible to argue that a program that had the effect of shifting some of the industrial tax burden from corporate income to corporate fuels would encourage industrial investments in energy productivity. The following analysis is premised on this rather modest assumption.

Reducing the Corporate Tax Rate

The most straightforward approach to reducing corporate taxes involves reducing the corporate tax rate. This would encourage capital-formation in industry, while avoiding most of the problems of definition and administration that seem to plague other forms of corporate tax relief.

One possibility is to abolish the corporate income tax altogether. It has been argued (Thurow, 1980) that this change would not only be an incentive to invest, but would also eliminate the present biases in the structure of capital toward debt and away from equity, and it would be more socially equitable. According to Thurow:

The corporate income tax should be abolished, regardless of whether you are a conservative or a liberal. Based on our principles of taxation, the corporate income tax is both unfair and inefficient. In a country with a progressive personal income tax, every taxpayer with the same income should pay the same tax (horizontal equity), and the effective tax rate should rise in accordance with whatever degree of progressivity has been established by the political process (vertical equity). The corporate income tax violates both of these canons of equity. Consider the earnings that are retained in the corporation on behalf of the individual stockholder. Low-income shareholders with personal-tax rates below the corporate rate of 46 percent are being taxed too much on their share of corporate income. To the low-income shareholder the corporate income tax is unjustly high. Conversely, high-income shareholders with personal-tax rates above 46 percent are being taxed too little on their share of corporate income. To the high-income shareholder, the corporate income tax is a tax shelter or tax loophole. As a consequence, vertical equity is being violated. Horizontal equity is also being violated, since two individuals with exactly the same income will pay different taxes, depending upon the extent to which their income comes from corporate sources.

A possible weakness of reducing the corporate tax rate is that it may not be the most effective way to promote investment, since it applies uniformly to all income, whether derived from increased productivity or from some other source (Eisner, 1975).

Accelerated or Indexed Depreciation

A more directed program for encouraging capital investments would provide liberalized depreciation for broad categories of industrial equipment. The proposal that has received the most attention in the recent Congressional session has been the proposed Capital Recovery Act (Jones, Conable) which would allow 10-year, 5-year or 3-year depreciation for industrial investment. An alternative that should be considered seriously is indexing the depreciation, i.e., evaluating depreciation in terms of the inflated prices that apply at the time the depreciation occurs. For long-lived capital this could be a particularly effective reform. Since liberalizing depreciation would significantly encourage the use of efficiency technology and solar energy in industry, it deserves careful evaluation and support.

Investment Tax Credits

An even more tightly targeted program would provide tax credits for a selected list of energy investments. The National Energy Conservation Policy Act of 1978 and the Windfall Profits Tax Act of 1980 grant a number of credits for industrial investments in energy conservation and solar-energy devices.

The net effect of the credits allowed under existing programs is very unbalanced, however. At a minimum, credits should not discriminate in the choice between efficiency and solar investments and should not be limited to a brief restrictive list. Another major deficiency is that the conservation credits apply only to retrofits and do not cover new industrial construction. There is an obvious risk in designing a program that leaves in the hands of federal policymakers decisions more appropriately made in the market place. A further difficulty with tax credits is that many firms for which incentives are appropriate have little tax liability; for this reason refundable tax credits should be considered. A thorough review is needed of the net impact of the existing tax credits, and the steps needed to improve them.

Evaluation

There have been a number of analyses of the industrial capital-formation that would flow from different kinds of tax reform. A study by Brimmer and Sinai considered increased general investment tax credits, a reduced corporate income tax rate, and indexed depreciation (Brimmer, 1976). The ratios of $ investment to $ tax subsidies were calculated to be 0.7, 0.3, and 1.7 for the three respective measures. In other words two of the three measures would not even produce as much investment as the loss in Federal revenues. Several other studies also suggest that measures of this kind

may have a low capital-to-subsidy ratio. The specific predictions vary greatly, however, because there are many theoretical uncertainties and a great variety in the assumptions, including the source of revenues to make up the subsidy (From, 1971; Eisner, 1975; Senate Budget Committee, 1978). On the basis of these studies, it cannot be concluded that liberalized depreciation would be more effective than the other options. However, the studies do suggest that indexed depreciation would be relatively effective in stimulating longer-term investments required by the energy-intensive industries. The studies did not include the effect of accompanying corporate tax relief with a fuel tax.

Industrial Investments by Utilities or Third Parties

Some of the institutional problems that have retarded industrial investments in energy efficiency and solar energy could be alleviated if companies that sell, rent, or operate energy-related equipment could be encouraged to provide energy services at an industrial site. Examples of such relationships include:

o An electric utility or its subsidiary that provides process steam, (by means of fuel-based cogeneration or solar collectors), or mechanical drive services involving introduction of more efficient motors.

o An independent firm that provides compressed-air services.

o An oil-industry subsidiary that provides process-heat services by renting equipment to a chemical industry.

o A turbine or engine manufacturer that finds it desirable to finance or lease its equipment at an industrial site.

The kinds of relationships illustrated by these examples are important since, in many cases, the company that needs an energy service is reluctant to invest in solar or conservation or solar technology for several reasons: (1) an acute shortage of capital, (2) a concern about entering an unfamiliar area of investment, (3) demands placed on its capital for investments more closely related to improving the company's primary product, and (4) a prevailing attitude that investments that reduce the cost of producing a product are not as attractive as investments in other areas. Such problems can be overcome if a company offers to provide both expertise and capital for a solar or energy-efficiency investment at the industrial site, perhaps offering a contract to provide certain energy services at an agreed price.

Energy-Service Firms

Some energy-service firms are envisioned as owning, operating, and maintaining facilities at industrial plants. A more limited kind of activity that should also be available is the financing and renting-out of components, including perhaps components that are completely integrated into plant equipment, such as major automatic control systems and furnace retrofits.

A variety of relationships can develop, and the targets of opportunity are not clear. Some legal questions may need resolution. Management systems and financial arrangements have been explored thoroughly in only a small number of special cases (for example, when steam from electric-power stations is sold for process purposes to industries in the close vicinity). Given the primitive state of the industrial-energy services industry, funds could be provided by government to explore the possiblities.

Exploratory funds should be offered for studies of legal and financial issues and, more importantly, to encourage actual institutional experiments. The present DOE initiative in the buildings area, which experiments with a variety of institutional approaches, provides a model for the activity. Note that while exploratory assistance should help speed the process, it is not critical; government subsidy is not a necessary feature of the energy-service industry concept. Federal action to help remove regulatory barriers may, however, be needed. (See the discussion entitled "Leasing (for the Residential and Commerical Sectors)" in the Attachment to the Domestic Policy Review of Solar Energy.)

Utilities and Utility Subsidiaries

An alternative to the creation of a new energy-service industry involves utility investments in industrial equipment. Utility investments in industrial equipment could be made either directly or through an unregulated subsidiary. The advantages of utility investments (in industrial renewable-energy production, cogeneration equipment located at an industrial site, and perhaps energy-efficiency devices at an industrial site), are discussed in the Utility section of this study. For utilities stockholders, investments in industrial energy equipment may in many cases be preferable to investments in conventional generating equipment. If a utility is compelled to make a relatively uneconomic investment in new conventional generating capacity instead of in energy efficiency, utility customers and utility stockholders are not well-served. While utility capital is in relatively short supply in many cases, there may be circumstances in which a specific utility would find an investment at an industrial site attractive, although the industry itself would not be attracted to the same investment. This

situation could be due in part to the regulatory protection given utility investments, which allows utilities to invest in equipment with relatively long pay-back periods. The difference in outlook, however, may be primarily psychological, since utilities consider their primary business to be energy, while industrial management must consider energy a peripheral issue.

Current federal law does not prohibit utility investments in industrial-energy equipment, but such investments are often inhibited by a number of regulatory difficulties. For example, energy equipment purchased by a utility that becomes a non-mobile part of an industrial facility may not qualify for either the standard 10% investment tax credit or for any additional tax credits given for energy efficiency or renewable resource equipment. Such a utility investment also may not qualify for the accelerated depreciation allowance provided for certain classes of industrial equipment. Present tax laws should be changed so as not to discriminate against utility investments in energy efficiency and renewable resources. There should also be reexamination of the need for price regulation of utilities in areas where competitive conditions exist.

There is uncertainty about the most appropriate institutional relationship between an industrial customer and utility as a supplier of energy services. There is also uncertainty about the tax and regulatory limits on utility investments in industry. The DOE can begin to resolve these uncertainties by cooperating with State Public Utility Commissions and relevant federal regulatory and taxing organizations in the investigation of existing examples at the state level and by developing other applications as necessary to fully determine the effectiveness of such utility-industry relationships.

Equitable Capital Subsidies for Renewable Energy and Energy Conservation In Industry

The Energy Security Act of 1980 created an independent Synthetic Fuels Corporation, wholly owned by the federal government, to subsidize industrial investments in the conversion of coal, shale, peat, tar sands, and heavy oil into useful fuels and chemical feedstocks. Investments in magnetohydrodynamic (MHD) topping cycles for commercial generation of electricity would also be supported. Subsidies available under the corporation include: direct loans and loan guarantees, joint venture agreements, and agreements to purchase products manufactured by subsidized firms at an agreed price.

The existence of such a body will distort a fair allocation of resources among competing industrial demands for capital. In particular, since the credits available from the corporation cannot be used for industrial investments in energy conservation or solar energy, such investments are made relatively unattractive.

One option to redress this imbalance would be to broaden the scope of the Synthetic Fuels Corporation (Energy Security Act of 1980 Title I) so it could also consider investments in efficiency improvement, solar energy, or even a much broader range of industrial investments. The corporation should be given the freedom to choose between investments in supply and investments in demand primarily on the basis of economic merit.

A major issue for the design of capital subsidies is deciding which types of projects should be encouraged. Distinctions are often made between projects targeted at energy projects and more general projects, and between retrofit of existing equipment and the creation of new facilities. In the following discussion of the "scrap-and-build" incentive there is a discussion of whether non-targeted financial support can yield adequate energy benefits.

Scrap-and-Build Incentive

A specific role for the revised Synthetic Fuels Corp., would be to provide a Scrap-and-Build Program for energy-intensive industries.

Industrial productivity and fuel-savings opportunities exist both in the retrofitting of present equipment and in building new facilities. Since the gains from creating new facilities are typically much greater than from retrofit, the "scrap-and-build" policy is designed to ensure that new facilities receive the emphasis they merit. (However, it is not possible to create new facilities extremely rapidly; even with a very rapid replacement of capital, such as 5% retirement per annum, or twice the expected rate, many existing facilities will still be used for 10 to 20 years and should therefore be retrofitted).

The lack of growth and the capital shortages that afflict most firms in mature industries cause them to be slow in creating new facilities. Because of the high cost of capital, the marginal operation of antiquated facilities still competes favorably with building of much more productive new facilities. In other words, although present antiquated facilities have higher labor and energy costs per ton of product, the residual capital cost is almost zero; meanwhile new plants have low labor and energy costs but high capital costs, especially at the capital charges that mature industries have to pay.

If construction of new facilities in basic materials-processing industries were to be emphasized, it might become possible to reestablish the technological leadership of the U.S. by leap-frogging the technology of foreign competition. At present the opportunity is largely foreclosed by the extremely slow pace with which new capacity is being created.

DOE should support scrap-and-build subsidy legislation to reward the scrapping of major obsolete facili-

ties and replacing them with new facilities. Reconstruction at the same site should be encouraged, so as to preserve the value of the existing infrastructure and minimize worker dislocation.

The form of subsidy (grants, loan guarantees, interest subsidies, tax benefits, joint ventures, etc.) is a complex question that must be explored in terms of the financial conditions of firms and the needs of the private financial market. The subsidy should typically be provided in the initial stages of the project and it should preferably be handled by a financial intermediary with some flexibility. Projects undertaken should be profitable in terms of provision of capital at favorable rates, taking into account the opportunity costs associated with scrapping the old facilities. Thus some of the initially injected funds could be recoverable for later projects.

Possible examples of scrap-and build-subsidies would be:

o   purchase of the facility to be scrapped

o   off-balance-sheet financing of the new facility, such as a joint venture with the participating firm, with the expectation that the participating firm will later be able to purchase total ownership at a fair price.

A preliminary evaluation of the "Scrap-and-Build" proposal, aimed at completely renovating the basic materials-processing industries in the U.S. by the year 2000, is presented below.

Small Business Assistance

At present the Small Business Administration (SBA) has a small program to provide capital assistance to small energy companies. This program should be expanded in scope and given more adequate funding.

The current program allows the SBA to make both direct and guaranteed loans to qualifying small businesses. While firms receiving funding under the program can use some of the funds for research and working capital, the combined amount used for these purposes must be less than 30% of the total value of the loan.

It is expected that this program will be expanded in FY 1980. An expansion of the program is clearly desirable; the expected addition of $10 million in direct loans and $32.5 million in loan-guarantee authority may not be adequate, given the burdens being placed on small business today. The program should have adequate funds to provide active support to small businesses that manufacture energy-efficiency and renewable-resource equipment, and to small companies that wish to provide auditing services and perform retrofit installations.

The SBA should also encourage the investment companies it supports to give priority to small businesses entering or expanding in the energy-efficiency and solar fields.

In addition, the SBA provides some assistance to small firms that are suffering economic injury as the result of energy shortages. This program is clearly inadequate, given the sometimes fatal combination of increasing energy prices and an impending recession. All SBA loan programs should give priority to firms that undertake retrofits for on-going enterprises and other energy-related productivity investments.

Venture Capital

The incentives described above are designed primarily to encourage industrial investments in technologies whose commercial worth has already been demonstrated. There is also need for a program designed to encourage more innovative ventures and projects that involve market tests of systems whose acceptance cannot be evaluated easily.

The rate of venture-capital formation was very low in the 1970's, as indicated in Table 2.12, which shows the recent history of stock offerings by new firms. S. J. Friedman of the Securities Exchange Commission has stated.

> ". . . We are acutely aware of . . . the slowing of capital flows to small and growing firms in the 1970's. The trend has resulted in impediments to innovation. . ." (Friedman, 1980).

Some growth in venture capital investment seems to have occured in 1978 and 1979 in firms specializing in venture capital although their total contribution to national venture capital assests is still relatively small. Three categories of venture capital firms invested about $600 million in 1979 (E.F. Heizer, Jr., 1980). The businesses supported were highly diverse, with electronics and real estate ventures prominent.

A shortage of venture capital has been cited by a number of sources, particularly by the recent U.S. Domestic Policy Review on Industrial Innovation, as a major impediment to entry of new firms. This DPR recommended a major program of federal funding for venture capital, administered through the states. The proposal would also improve industrial energy productivity and use of renewable resources. It should be promptly implemented. However, the currently proposed funding level of $4 million will not make a significant impact on the market. Additional funds could also be made available through channels other than the states, as recommended by the Review.

Table 2.12   COMMON STOCK OFFERINGS FOR WHICH THERE WAS NO PRIOR MARKET

(New Issues)

Year	Share Value ($Millions)	No. of Issues
1968	1,742	649
1969	3,545	1,298
1970	1,451	566
1971	1,917	446
1972	3,301	646
1973	1,872	177
1974	117	55
1975	236	25
1976	271	45
1977	276	49
1st half 1978	54	18
2nd half 1978	160	40
1st half 1979	256	59

Source: Investment Dealers' Digest.

The Department of Energy's Office of Industrial Programs (OIP) has already instituted a series of projects that have similar objectives, and these projects should be encouraged and expanded as described below. The state of California has also begun to experiment with programs of this nature.

The Venture Capital Program should be built around the following elements:

(1) Solar and energy-efficiency capital should be drawn from the same capital resource and held to the same programmatic tests. Programs now supported in the DOE Office of Industrial Programs have many of the characteristics of the approach recommended here. These characteristics should be integrated into the solar-industrial program.

(2) All proposals should be unsolicited, and the availability of funding and generic priorities should be widely advertised, but specific equipment designs should not be requested.

(3) Proposals should be selected using general criteria (see below), combined with the use of review panels chosen in a way that would not compromise the proprietary nature of hte proposals. Reviewers should include academics and staff from National Laboratories. Definitive steps should be taken to make procedures simple and the overall process, from proposal to denial or approval, short. It is critical not to inadvertently inhibit small firms from participating in the program.

(4) The proprietary rights of all proposals should be protected.

(5) The federal government should be reimbursed by the participating company, with interest, if the product supported becomes a successful commercial product. Products should be supported only if it appears unlikely that they would be able to attract venture capital from private sources. The program must assume support of high-risk investments, and caution should be excercised to guard against overly conservative criteria in the selection process. If program funding is adequate, the selection should become venturesome. It might be desirable to increase funding and adopt criteria to avoid super-safe projects.

(6) The program should encourage joint ventures, such as proposals involving both the suppliers of equipment and potential customers for the equipment.

(7) While no precise formula should be required, indication of industrial interest in the project, as evidenced by substantial cost sharing, should be a major criterion in selection.

(8) One objective of the program should be channeling funds to small innovative firms as well as to large firms.

A venture capital program rising to $500 million a year, roughly ten times the present program in the Office of Industrial Programs, would be appropriate. A program of this scale would reflect the specific opportunities now evident to OIP staff (Harvey, 1980). The scale is also reasonable in terms of the general national level of venture capital activity (well over $1 billion per year) and the fraction of investment now going into energy supply (over 40% of all plant and equipment expenditures). This scale is also not unreasonable when compared with the total level of capital investment by energy-intensive industries (about $20 billion per year at present without petroleum refining, and $35 billion when petroleum refining is included).

The energy benefits of such a program are very difficult to define. Analyses by the Office of Industrial Programs suggest that fuel savings resulting from their venture-capital program are presently one barrel of oil equivalent saved per 30¢ of program expenditures, a savings/cost ratio that might decline to one barrel saved per dollar spent in the expanded program proposed here (Harvey, 1980).

Such savings achievements would probably not continue for the long-term. This suggests that adding $50 million per year to a venture-capital program that continues into the 1990's would yield savings of something less than 1.2 million barrels of oil per day or $50 million a year.

IMPROVING THE CLIMATE FOR INDUSTRIAL RESEARCH INVESTMENTS

Three federal actions can encourage industrial support of research:

o   increased tax credits for industrial research investments;

o   federally sponsored research facilities; and

o   a liberalized patent policy.

In addition, direct federal support of relevant fundamental research is recommended.

Increased Tax Credits

It is proposed that there be enacted a tax credit equaling 125% of actual expenses for corporate research investment, in lieu of the present tax deduction. (Under present tax law, industrial investments in research are deductible from gross income for tax purposes.) Under this proposal the subsidy for research would be increased. The premise of such a tax credit subsidy is that companies themselves are in the best position to determine what kinds of research are best suited to their needs. (If they have funds for long-term investment, they will usually support the research whose benefits will accrue to them rather than to the industry in general.)

Two special issues should be considered in reviewing the merits of this proposal. First, it is critically important to construct a much more careful definition of "research and development" than is now used in the tax law. It appears that under the current definitions, many routine functions can be technically qualified as research. Secondly, the incentive will not be attractive for firms that have no major tax liability, it may be desirable to allow the credits provided under this program to be refundable.

Federally Sponsored Research Facilities

DOE should begin to provide federally-sponsored research facilities jointly sponsored with industry. The arrangement could ensure a close working relationship between teams in a government research facility and their counterparts in industry. Such research facilities could be managed as an extension of a state university or a national laboratory. Federal sponsorship of such a facility could be limited to providing land and a building, or a more active role could be taken. Any analysis of such a program should begin with a careful examination of European and Israeli experience with such facilities, as well as U.S. experience in defense-related industries.

Patent Policies

The Federal Nonnuclear Energy Research and Development Act of 1974 requires that title to all inventions conceived or reduced to practice during the course of any DOE contract by any person who "was employed or assigned to perform research, development or demonstration work and the invention is related to the work he was employed or assigned to perform . . ." (42USC 5908 (a) 91) must vest in the government. In addition, DOE acquires certain rights to "background" patents developed prior to or during the performance of the contractor's work for DOE. The statute gives DOE wide discretion to waive these patent rights under standards set forth in the statute and supplementary regulations; however, even when the waiver is granted, the government retains significant rights, including worldwide license, "march-in" rights, etc. (42 USC 5980 (h)).

A contractor with significant expertise in an energy-related field must weigh the experience and financial rewards to be gained from a DOE contract against the possible effect that acceptance of such a contract will have on its patent portfolio and its competitive position.

In addition to the potential risk, there is considerable uncertainty about the nature and continuity of DOE's patent policy. For example, the regulations that define the government's right to acquire licensing of background patent rights, to "march-in," or to grant or revoke an exclusive license are so broad as to render these government rights unpredictable. Further, the only way to gain knowledge of prior unlitigated, administrative decisions by DOE on patent rights issues is by word of mouth. Numerous contractors have criticized DOE's patent policy for failing to provide prospective contractors with sufficient notice defining the actual bounds of the government's rights.

Besides the extensiveness and uncertainty of the process, DOE's exercise of its discretionary waiver authority is so slow and unpredictable as to negate the benefits of the policy. The power to grant waivers is concentrated only in Washington and appears to be highly discretionary; furthermore, the process frequently takes four to six months or longer.

The Department of Energy has recently issued a small number of contracts that allow the recipient to maintain proprietary rights to the developed product in return for agreeing to repay the government grant with interest if the product is marketed as a commercial product. While this type of contract is in itself a valuable step forward, the processing time can still be long enough to threaten the participation of small companies. The value of this kind of contract is also limited by the relatively low fiscal resources allocated to genuinely unsolicited proposals.

The following six steps are suggested to alleviate this critical problem at DOE:

o   Expedite the process for acting on requests for waivers of patent rights to 30 days or less.

o   Provide potential contractors with better notice of the standard to be employed by DOE in exercising its patent rights by:

-   providing more definite standards regarding patent rights in the DOE procurement regulations;

-   clarifying the standards with liberal use of examples, such as those used by the Internal Revenue Service in its publications;

-   publishing DOE decisions regarding the exercise of march-in rights and decisions to grant exclusive license and waivers,

o   Encourage innovative contracting and ensure that small, innovative companies are given preference wherever possible.

o   Remove bureaucratic barriers to contract approval by decentralizing the authority to enter into subcontracts and cooperative agreements and to make grants for solar energy projects.

o   Take immediate steps to reduce the time required to process a contract for solar development of energy or energy conservation equipment. A strict requirement to let all contracts within three months should be enforced;

o   Provide more funds to support a formal program for funding unsolicited proposals.

o   Provide special programs for socially and economically disadvantaged contractors.

Direct Federal Support of Fundamental Research

DOE should significantly increase the now minimal federally supported research in fundamental areas that fall between basic science and product-specific development programs. Examples of such research include fundamental studies (appropriate to the basic industries) of subjects like corrosion; lubrication; combustion; physical separation of materials; convective heat transfer; phase transitions (boiling, solidification);

multiphase flow; and fluid transport. In addition, studies reexamining fundamental approaches to energy-intensive industrial processes, such as the manufacture of steel, paper, glass, and cement are needed. Research in all these areas was once a major element of university engineering programs, but has been badly neglected in recent decades.

With a clear mandate from Congress, DOE should establish a program to support such research. (It could be an expanded role for the present Office of Energy Research in the Department of Energy.) Selection, management and review of grants should be performed by outside organizations, so that procedures appropriate to this type of research (in contrast with existing programs, which support basic science on the one hand and demonstrations on the other) can be employed. One possible management arrangement would be joint industry-engineering society groups (Gray, 1977).

The success of the program will depend on how it is administered. Care must be taken to ensure that a diverse set of institutions are supported and that particular attention is given to grants for small research groups and to individuals. The tempation to over-manage and over-concentrate research (e.g., in only in a few centers or National Laboratories) should be resisted. To the maximum extent possible, programs should be responsive to proposals received from outside the government. Funding should not be tied to government requests for proposals in narrow areas of research that are interesting to Federal program managers. In working papers for the U.S. Domestic Policy Review on Industrial Innovation (Industrial Advisory Subcommittee, 1979), industrial consultants singled out the Office of Naval Research Program of the 1940's and '50's as a model for research administration:

> "The main reason for ONR's success as a funding agency was its relative freedom from elaborate procedures for submitting proposals and its rapid decision-making process."

It is recommended that a new program or programs to support research that is relevant to energy and industrial productivity productivity (i.e. emphasizing basic-materials processes and generic technology) within the Department of Energy be funded at a level that will rise to $200 million within a few years. This level of funding is appropriate for several reasons: comparability with other major research programs; the opportunities in terms of research topics; comparability with foreign research; manpower implications; and the relationship with capital and venture capital investment levels recommended in this study. The program should be closely coordinated with related programs at the National Science Foundation (NSF), the Department of Commerce, and the National Aero-

nautics and Space Administration (NASA). In some cases, it would be appropriate to add support for existing programs managed by those agencies. In other cases those agencies could provide guidance on the design of a new DOE program.

It has become painfully clear that industries overseas have developed better energy-efficiency technologies (and in some cases renewable-energy technologies) than have U.S. firms. It would be rewarding to monitor foreign research and design efforts involving energy. A major program should be undertaken to ensure an adequate flow of information about these technologies across national borders. One technique for encouraging a transfer of information would be to establish a set of joint research programs in energy efficiency and solar energy. Programs are now underway involving Spain, Italy, and Israel, and the first steps have been taken in programs with Brazil, Mexico, South Korea, and Australia. Most of these projects are limited to solar energy and are very small. They can, however, be used as models for a considerably expanded program.

COSTS AND BENEFITS

The programs just described can be expected to have two major impacts on energy use in industry: they will encourage the development and adoption of new technologies, and they will increase the degree to which existing industrial plants are refurbished or scrapped and replaced with new facilities.

The proposed programs involve (1) rationalizing the price of energy through price deregulation, removal of subsidies to energy supplies, and possibly through a direct tax on energy; (2) increasing the capital available for industrial investment through reductions in the federal corporate income tax burden, fostering utility and third party investments in industrial equipment, support for small business, and creating a "scrap-and-build" program that would provide a special fund for older energy-intensive industries; and (3) spurring technological innovation through support of research and a venture-capital program.

The "baseline" estimates of future national energy prices assume that oil prices will reach $40 per barrel in the year 2000. The "baseline" estimates, taken in combination, result in a forecast that the average price of industrial energy will increase by a factor of 2.7 between 1978-2000. The "average price" is the weighted average of industrial oil, gas, coal, and electric prices. This is a significant increase but is perhaps less surprising given the fact that the average price of industrial energy increased 50% between 1978 and 1980.

It must be remembered that the goals established for industrial efficiency were not tied precisely to this "baseline" energy forecast. They were, instead, based on a rather conservative estimate of what could be done with technology, most of which has already been proven to be cost-effective with current prices. Higher energy prices and advancements in industrial technology may mean that these estimates are much too cautious.

A number of recent studies have examined the elasticity of industrial energy demand to increases in industrial energy prices. Many of these indicate that if energy prices increased by a factor of 2.7, the reduction in industrial use would be much greater than the 26% assumed in the development of the goals. In light of this, it can be seen that the goals may be quite conservative.

Despite this confidence about the aggregate results set forth in Part III, there is a high level of uncertainty about the impacts of individual programs. Most of the programs work to support each other in complicated ways. The analysis is made more difficult by the fact that many of the most critical variables, such as future fuel prices, cannot be predicted with any certainty.

The major unresolved problem is one of determining how aggressive a federal role is needed to achieve the goals. It is important to recognize, however, that any federal program designed to encourage investments in the kinds of technical improvements described here involve very few risks. The technology required is typically well-established and documented. The major risk is that the federal government might subsidize an investment that would have occurred without any federal intervention. If these funds were derived from windfall profits taxes or fuel taxes, however, the overall effect would still be to reduce consumption and encourage capital formation.

A few specific examples will be examined to illustrate the kinds of costs and benefits that can be calculated for different combinations of federal industrial programs. The analysis will focus attention on the basic materials processing industries (chemicals, petroleum refining, primary metals, paper, ceramics, and food). These industries are important for energy analysis because they account for over 70% of all industrial energy consumption and because they represent a particularly difficult problem for industrial energy policy. Many of the plants in these industries are relatively old and inefficient but are unlikely to be replaced unless a national program is developed to motivate greater investment.

A comprehensive analysis of the feasibility of reaching the goals established in Part III of this report requires a careful evaluation of each major class of industry. Such analysis is urgently needed, but it was not attempted in the current study.

In the analysis that follows, it is assumed that in the absence of any new programs, approximately 2.5% of the 1978 plant capacity in the basic-materials industries would be refurbished each year. One important goal of the programs proposed here is to encourage

industry to retire all plants in use in 1978 that are not refurbished and to replace all the retired facilities with new facilities by the year 2000. If this is achieved, and if the programs designed to improve corporate research, development, and venture funding succeed in ensuring that the average new plant built between 1978 and 2000 is as efficient as the best available new plant, this reconstruction program should lead to energy savings in the basic materials industries of 3.5 quads per year in 2000 or 70% of the savings associated with the programs recommended here. If the accelerated turnover is achieved without using the best available technologies, the savings would be on the order of 2.4 quads per year (see Part III).

The total investment required to encourage the accelerated turnover of plants in the materials-processing industries just described is expected to be $173 billion ($8-$9 billion a year for 20 years). This was calculated by estimating the gross incremental investment required to scrap and rebuild plants in the paper, chemicals, cement, and steel industries ($115 billion) and scaling the result by 1.5--the ratio of the total energy required in these four industries to the total energy consumed by all basic-materials industries (see Table 2.21 below).

In evaluating the effect of policies designed to accelerate the turnover of older plants, the key question is the circumstances under which investors will find it profitable to scrap an older plant and build a new one. As the previous discussion has indicated, one of the reasons that fully depreciated older plants are not replaced is that investors must compare the cost of building and operating a new facility with the cost of operating an older one. The capital cost of the older plant is effectively zero, and this holds the proposed new plant to a severe test: can the new plant pay for itself with savings in operating costs? If the industry is expanding, investors may expect the savings from a new technology to cover only the difference in capital costs between an efficient and an inefficient plant. The outcome of the analysis therefore depends critically on future increases expected in energy prices, expected savings in areas not related to energy (most modern plants will lead to savings in labor and materials as well as energy), the financing available for the new plant, and tax schedules.

In the following analysis it will be assumed that the investor will expect a 10% return on equity (in constant dollars), provide 70% equity financing for the new facility, and finance the remainder with debt costing 3% (in constant dollars). Future costs are discounted at 10%. When energy savings are calculated, it will be assumed that the savings will be valued at the average delivered cost of all kinds of industrial energy and that energy prices will increase according to the Baseline assumptions. (This may somewhat understate the value of the savings since it can be expected that industries will make a major effort to ensure that most energy savings are oil savings. These oil savings are more valuable than average energy savings.)

It will be assumed that an industry will find an investment in a new plant attractive if the capital costs associated with a new plant are less than the levelized value of the fuel and operating savings. Using the baseline costs of energy projected in this study, and the assumed costs of capital described above, it can be shown that the basic-materials industry should find an investment in a new plant profitable if the non-energy benefits have only 14% of the value of the energy-savings benefits.

One specific technique proposed in Section 2 above for encouraging industries to retire older plants and invest in new ones is the "Scrap-and-Build" program. Under this proposal, the federal government would make a direct subsidy to targeted industries using the same mechanisms allowed in Title I of the Energy Security Act. The methods just described can be used to show that industry should find a scrap-and-build investment profitable in terms of energy savings alone if it receives the equivalent of a grant covering about 12% of the value of a new plant. This level of subsidy, although it would make scrap-and-build projects profitable, would not necessarily enable firms to proceed with projects, because they might not have adequate access to captial. A shortage of capital means that firms will have anomalously high discount rates and thus would require higher subsidies to justify scrap-and-build projects. (The required subsidy is very sensitive to the discount rate.) It is proposed that, at least initially, the subsidy be about 20%.

It is interesting to observe that this credit is close to the credit available for synthetic fuel plants in the Energy Security Act, which is approximately $20,000 per barrel per day of oil-equivalent savings. That is, a subsidy of $20,000 per barrel per day (oil-equivalent) for saved energy would come to about 20% of the capital cost of scrap-and-build projects (as evaluated in terms of cost and energy savings shown in Section 3, below). The subsidy level is also comparable to the 25% capital-cost reduction that Berg has argued would lead industry to scrap antiquated facilities (Berg, 1978). The analysis presented here also shows that if society discounts future costs and benefits at a rate of 8% or less, the energy savings alone would justify the investments. This provides a clear rationale for some kind of federal support for industrial reconstruction even if society determines that (decontrolled) market prices of fossil fuels adequately reflect their full cost to the nation.

One option for providing the Scrap-and-Build subsidy (or $1.7 billion/year) would be to use the revenues that could be made available by removing the present subsidies for the routine provision of energy. The level of these energy-supply subsidies is potentially larger than the subsidy required for the Scrap-and-Build program (see the section on Subsidy and Price Regulation above). Windfall-profits taxation on oil (or gas) would also be a legitimate source. Still another source would involve the levy of a tax on industrial energy. Raising the average price of industrial energy 2% would be adequate to provide the level of subsidy needed.

Over the longer-term (1985 and beyond, say), it would be desirable to make the Scrap-and-Build program more market-oriented. Industry would have the incentive for the reconstruction effort without the Scrap-and-Build subsidy if energy prices rose to the marginal costs estimated in this study and if an excise tax were levied on petroleum in the amount of about $7 per barrel of oil (assuming a 10% discount rate and no credits for non-energy benefits). These energy price measures (decontrol and fuel taxation) could be offset by a corporate income-tax reduction. It should be noted that a corporate income-tax reduction also makes investment more attractive because it would reduce the advantage the purchase of energy has with respect to investment in efficiency improvement. This advantage results from the fact that operating expenses are deducted from taxable income, while investments are deducted from income only via later depreciation).

Windfall profits tax revenues associated with price decontrol and the revenues from the petroleum tax would be adequate to offset about 1/3 of the corporate income tax. (If the initial subsidy were continued, it would represent only a tiny fraction of the total revenue flow.) It would be undesirable to decontrol energy prices and to introduce the petroleum tax too quickly. These measures should be phased-in slowly to avoid major dislocations on the part of those who can't adjust and to avoid premature commitments to sub-optimal alternative technology on the part of those who can. But an announced long-term policy to move gradually in this direction, coupled with an initial effort involving a direct subsidy, would give industry the right price signals and would channel capital to the Scrap-and-Build effort.

An interesting feature of the Scrap-and-Build Program is that the total required investment and energy implications are relatively well-defined. The major uncertainties are whether the estimated near-term subsidy level would be adequate to generate the desired level of capital investment, and, for the longer-term, what are the most effective approaches to price decontrol, the energy tax, and corporate tax reduction. Further analyses to clarify these issues should be given high priority. However, the uncertainties about costs and benefits of this program are no greater--and may well be less--than those that characterize the synfuels and other supply programs.

PROGRAMS WITH SPECIALIZED OBJECTIVES

Removing Regulatory Constraints to Cogeneration

The Fuel Use Act (FUA) of 1978

This Act has the effect of discouraging some of the most promising approaches to cogeneration. Since most of the circumstances that prompted the passage of the act have vanished, it is now in order to have a fundamental reexamination of the need for most of the controls it imposes. While a complete revision of the act would be preferable, a number of relatively minor modifications in the Act and its related regulations could remove most of the major barriers to adoption of cogeneration.

Specifically, the Fuel Use Act puts restrictions on the use of oil and gas for cogeneration thus inhibiting its adoption. The cogeneration systems that offer the greatest fuel-savings potential--and also the greatest potential for cogenerated electricity production--are those in which large quantities of electricity are produced per unit of process steam i.e., those with what is called a high E/S ratio: diesel engines, dual fuel engines, spark-ignited gas engines, gas turbines, and gas/steam turbine combined-cycle units. While high-E/S-ratio cogeneration systems could be fired with abundant, low-quality fuels (like coal and biomass) using advanced technologies,* they must be fired today with oil or natural gas. Because industry has had much less experience with high-E/S-ratio cogeneration technologies than with the more familiar low-E/S-ratio steam Rankine cogeneration cycle, and because largely unfamiliar institutional arrangements appear to be most conducive to the deployment of high-E/S-ratio cogeneration technologies,** it is important that we now gain technological and institutional experience with high-E/S-ratio cogeneration technologies based on natural gas and oil until advanced technologies are ready for widespread commercial use.

The Fuel Use Act was passed at the time when natural gas prices were maintained at artificially low levels, and there was a widespread belief that the nation was rapidly running out of new gas supplies. There was concern that as a result residential customers would be deprived of the gas supplies needed to provide heat, hot water, and cooking in many areas. All of this has changed. The price of natural gas will gradually be deregulated during the next few years, and prices have already begun to rise toward "market-clearing" prices. Moreover, it now appears that while the nation is certainly running short of supplies of low-cost gas, there may well be ample supplies of gas from higher-cost sources (including gas from biomass) to sustain production at or near the present level. Moreover, the analysis presented in the Building Chapter of this study indicates that a significant amount of gas can be saved by improving the way energy is used in buildings; this saved gas would represent a substantial increase in the industrial gas supply, even if the overall supply remained constant. This saved gas should be made available to the industries that are able to use it efficiently, presumably the most efficient users would be in a position to make the most attractive bids if the market were freed of artificial constraints. Industries deploying high-E/S-ratio cogeneration technologies should therefore be in a good position to bid successfully for the gas saved by improving energy efficiency in our nation's buildings.

The restrictions the FUA places on the use of residual and distillate fuel oil should also be removed. Cogeneration offers a much more efficient use of both fuels than either central-station electric generation or the direct production of industrial-process steam. While there is clearly a need to conserve distillate, the price of this fuel should serve as an adequate deterrent to large-scale use. Distillate is likely to be used only as a pilot or as backup fuel for certain cogeneration applications. For example, natural gas or synthetic low- or medium-Btu gas can be utilized reliably at high thermodynamic efficiency in dual-fuel, compression-ignition, reciprocating, internal-combustion engines, where pilot oil (amounting to 5-10% of total fuel use) is used for ignition. (Gas by itself will not ignite in a compression-ignition engine.) Also the availability of distillate for backup fuel during periods of gas supply cut-off makes the dual-fuel engine a flexible, highly reliable cogeneration option. Oil consumption associated with both pilot oil use and backup would be small and would be aimed at facilitating a high-efficiency use of natural gas.

Resid-based cogeneration could play an important role in the process of upgrading present refineries to get more higher-quality products out of the "bottom of

---

*The low-quality fuel might be used directly (e.g., in a pressurized fluidized bed combustion/gas turbine system), or the low-quality fuel might be refined before use in cogeneration (e.g., a medium-Btu gas might be produced at a centralized gasification facility and piped to industrial sites).

**As shown in Section 3 of this chapter, high-E/S-ratio cogeneration technologies often involve production of electricity in amounts far in excess of on-site needs. Because industrial firms are usually more interested in product-line investments than in investments aimed at the sale of electricity, utility, third-party, or joint-ownership arrangements appear to be more promising than ownership by the steam-using industry, for situations where on-site electricity production would greatly exceed on-site needs.

the barrel." Because of the trend toward heavier crudes and the higher value of gasoline and distillate relative to residual fuel, there are compelling reasons to upgrade domestic refineries to "crack" the bottom of the barrel.* One of the products of upgrading these refineries would be low-Btu gas (or alternatively medium-Btu gas). Cogeneration facilities burning residual fuel oil or gasified resid could be switched to such gas when the upgrading is completed. The cogeneration potential from use of this gas could amount to about 50 billion kwh per year, or more than 10% of the present electricity and mechanical power needs of the 6 major steam-using industries (see Section 3).**

The FUA does not give blanket prohibition of oil and gas use for cogeneration. The lawmakers recognized the importance of oil- and gas-fired cogeneration units and constructed exemptions in the Act that would allow the use of oil and gas under appropriate circumstances. Unfortunately the many complex provisions of the Act have combined to limit this flexibility. Exemptions, even where they are permitted, can be expensive and time-consuming to obtain, and the need to apply for exemptions has had the effect of discouraging the kinds of cogeneration units the authors of the Act appear to have wanted to encourage. One important exemption is that major fuel-burning installations (MFBI) classified as nonboilers are exempt*** (although the Economic Regulatory Administration may in the future prohibit, by rule or order, use of oil or gas in certain nonboiler MFBI applications). This exemption is favorable to high-E/S-ratio cogeneration because these technologies are classified as nonboilers. This exemption is not as generous as it might appear, however, because a cogeneration unit for which more than 50 percent of the electricity produced is sold or exchanged is not classified as an MFBI, but as an electric power plant, for which the nonboiler exemption does not apply. This is a serious restriction, in part because in some important applications involving high-E/S-ratio cogeneration technologies much more electricity is produced than is needed on-site. Moreover, this exemption would not be available for utility

or third-party-owned cogeneration facilities, where most or all of the produced electricity would usually be sold or exchanged.

The importance of eliminating the FUA restrictions on cogenerators was recognized by the Senate Energy Committee, when it passed on June 16, 1980, an amendment to S.2470 (The Powerplant Fuels Conservation Act of 1980). This amendment exempts all cogenerators from FUA by declaring that cogenerators that are qualifying facilities under the definitions of PURPA section 210 are neither MFBIs nor "Powerplants." This amendment was subsequently approved by the full Senate on June 24; House action is still pending. Whether this change in FUA becomes law depends not only on favorable house action on this amendment but also on passage of the Powerplant Fuels Conservation Act. The FUA amendment should be adopted even if this particular parent bill is not enacted.

The Public Utility Regulatory Policy Act of 1978

This Act (PURPA) is a major milestone in removing institutional barriers to cogeneration for qualifying cogeneration facilities. The major shortcoming of the Act, insofar as it deals with cogeneration, is that it discriminates aginst utility ownership.

First, Section 201 specifies that if more than 50% of the entity that owns a cogeneration facility is a utility, the facility cannot be a qualifying cogeneration facility. Such a facility would thus be denied certain benefits. One of these lost benefits relates to FUA. The FUA cogeneration amendment to S.2470 passed by the Senate removes the restriction on oil and gas use only for systems that qualify as cogeneration units under Section 201 of the Public Utility Regulatory Policy Act of 1978 (PURPA). Another lost benefit relates to the denial of the energy tax credit for cogeneration, granted under the Crude Oil Windfall Profit Tax of 1980 but denied for electric utility property.

A statutory change should be sought that would eliminate this constraint on utility ownership, because

---

*It has been estimated that for an investment of $18 billion U.S. refineries could be upgraded over a 4-10 year period to reduce residual fuel production by 1.6 milion barrels per day and to increase gasoline and diesel fuel production by 0.5 and 0.6 million barrels per day respectively. A byproduct of this upgrading would be the annual production of 0.5 Quads of low-Btu gas. See An Analysis of Potential for Upgrading Domestic Refining Capacity, report prepared by Pervin and Getz, Inc., for AGA, 1980.

**If 0.5 Quads of low-Btu gas were used in combined cycle cogeneration plants, some 50×10^9 kwh per year would be produced and some 90,000 barrels of oil-equivalent fuel that would otherwise be required to produce process steam would be saved. In 1976 the electricity and mechanical power used in 6 major U.S. Industries was 437×10^9 kwh.

***Also units too small to be classified as MFBI's are exempt, and cogeneration installations can be granted exemptions where net oil or gas savings can be demonstrated or where the exemption is needed to employ a technical innovation or where the facility would result in retaining industry in urban areas.

utility ownership (like third-party ownership) appears to be a particularly promising arrangement for fostering the implementation of high-E/S-ratio cogeneration technology.

PURPA also precludes a significant portion of the cogeneration potential through its technical definition of a qualifying cogeneration facility. Section 201 does not include facilities that produce both mechanical power and steam. In 1976, 38% the power cogenerated in six major process-steam-using industries was mechanical power (see Section 3). The definition in Section 201 should be amended to eliminate this artificial restriction.

The interim exclusion of diesels from qualifying cogeneration status under Section 201 of PURPA should also be eliminated. Diesel engines are especially efficient high-E/S-ratio cogeneration technologies that can maintain high efficiencies at part-load and in large and small sizes. The interim exclusion appears to have been the result of concern about air quality. Instead of singling out the diesel engine, all cogeneration facilities should be required to conform to appropriate air-quality regulations. Unfortunately very little research has been done to determine the appropriate standards for diesel emissions. Research on this issue should be given a high priority by Environmental Protection Agency, and standards for both particulate and Nitrogen oxides should be developed as soon as possible. In addition, air-pollution regulations bearing on cogeneration should be carefully based on energy output (e.g., the steam, electricity, and/or mechanical power produced) instead of on the amount of fuel consumed.

State Regulations

Certain state regulations may also limit development of cogeneration. Twenty-four states regulate steam sales, a limitation that may inhibit third-party and utility-ownership arrangements. In order to facilitate these arrangements at industrial sites, state regulations of steam sales to industrial customers should be examined to determine whether deregulation of such steam sales is called for. It is the opinion of FERC that "The commission does not have the authority to exempt cogenerators from state regulations as a steam utility." (Federal Register, July 3, 1979, p. 44, FR 38865, footnote No. 5.)

Information

Current Federal programs for communicating with the industrial/agricultural sector consist, to a large extent, of a potpourri of relatively small and diverse activities. These activities have not been viewed as part of a comprehensive and integrated program to satisfy industry/agriculture needs for technical training and information about renewable energy and conservation. A formal evaluation mechanism should be established to determine whether the most effective and efficient mix of programs is in place, whether there are information needs not being addressed, and whether there is an unnecessary duplication of outreach services. There is also a related need to establish an explicit mechanism by which information outreach program managers from involved Federal agencies can coordiante their efforts with offices implementing them. Programs are needed to gather and disseminate the information and data that are useful to indusry. A mechanism for coordination should be created by the Energy Coordinating Committee.

Energy Information Gathering and Reporting

Public policy relating to industrial energy use must be based on adequate information. While the government's present involvement in corporate energy decision-making (through its price control and fuel allocation programs) should be greatly reduced, government will continue to formulate policies that affect the general economic climate in which industrial decisions are made. Public policy should be based on data and analytical capabilities that enable reliable estimation of the aggregate impacts of alternative policies.

Section 301 of the Energy Security Act requires the annual preparation of energy consumption targets for 1985, 1990, 1995, and 2000. This mandate underscores the need for continued systematic development of data bases and methods of analysis that will provide sound projections of changes in energy efficiency and hence in the demand for energy services.

As discussed below, Federal efforts in data and analysis, should include:

o   accurate reports of demand for energy in industry, such demand broken down according to the quality of the energy required (e.g., mechanical drive, process heat at different temperatures, etc.)

o   case studies of plants that have adopted energy-efficiency or solar technologies

o   continued development of economic models that can anticipate the impact of the energy-efficiency and solar energy technologies that are likely to be introduced in the industrial sector during the next two decades.

Data. Only grossly uncertain information is now available on energy-service demands. Data on the specific functions for which energy is used should be the basis for all projections of energy, energy equipment, and energy-related activities. The more aggregate energy-consumption time series are incomplete, and, in the case of the census information, are also infrequent and slow in becoming available. The DOE is moving ahead in this area, with its tightened reporting of industrial energy performance and its support for a broad industrial energy-use survey (Westat project) and exploration of possible private collection of data based on the French (CEREN) system.

A part of the problem with energy-service data is a lack of metering within many plants. The present considerable uncertainty about allocation of fuels to functions will not be cleared up without a detailed technical effort, including metering.

Other kinds of needed energy-related data are microeconomic information about technical change and information about industrial decision-making. Many studies of technical opportunities have been carried out. Nevertheless, the nation simply does not know with any precision what costs would be incurred and what benefits would flow from various general kinds of technical improvements in industrial energy use, or how rapidly a program of improvement would be implemented under given conditions.

Industry leaders should be asked for assistance in obtaining cooperation for improved data collection efforts. Many industrial users are reluctant to provide such information because they are concerned that proprietary information might be jeopardized and that the availability of improved data could provide a basis for regulation (viz., equipment performance standards, targets for use of secondary materials and possible mandatory aggregate energy-efficiency standards for each industry).

This problem differs radically from industry to industry. Initial discussions to achieve some accomodation might be conducted separately for different industrial groups.

One technique for improving information is an annual survey of industrial energy-use. The survey should have the following characteristics:

o   It should integrate all federal-reporting requirements now imposed on industry and should be included as an element of a single annual census document. At present, the number of reports required of industries has reached a level where it is a serious burden for all companies and may be a prohibitive burden on some small ones. The government should, through an interagency agreement, establish a goal of a single report per year. The addition of an energy census should not be a substantial additional burden since much of the relevant data is already part of current reporting requirements levied for other reasons.

o   The information should be gathered in categories that represent the final use of energy as well as categories representing

primary energy demands. For example, the report should include demands for mechanical drive, steam at the pressure actually required for the process, direct heat, and electricity for electrolysis.

o   Information about the use of biomass and other renewable energy forms should be explicitly included in the survey. The use of renewable energy is now not reported in any systematic way.

Such a revised survey should be the goal of the DOE data-collection initiatives mentioned above.

Analysis. Two types of analysis should be conducted routinely: (1) development and application of models designed to evaluate and project energy use in major industrial sectors and in U.S. industry as a whole, and (2) case-study analysis. While economic models must be treated with caution, they can provide useful insights into the potential impacts of policy options and can provide some information about the interactions among industrial sectors. Their utility is greatest if their limitations are clearly understood.

Systematic case-study analysis based on experience or on design in the field is essentially missing from the U.S. program of energy analysis. Given the extremely poor data base and the difficulties with assumptions, modeling currently can have only very limited validity. Policy-making, both public and private, is, as a result, partially crippled. Systematically gathered case studies of technology applications would help in this situation. Many topics suggest themselves: automatic control systems; sophisticated optimization systems; furnace retrofitting; motor controls for throttleless moving of fluids; and implications of optimization to increase capacity, thereby enabling retirement of less-productive facilities. Institutional analysis by means of case studies is also of interest, e.g., the response of various types of firms to different kinds of economic conditions.

Energy Journals

The Department of Energy could finance the publication of a series of monthly or bimonthly energy journals dealing with of industrial energy efficiency and the use of renewable resource technologies in industry. The publications should contain a "News and Review" section giving brief descriptions of successful installations in industry, research breakthroughs, and new laws and regulations. There should also be a series of technical articles describing novel energy technologies in some detail. The journals could also provide timely accounts of the data and analysis generated in the previously described program of information aquisition. Journals could be published on the following topics:

o   motors and controls

o   process heat (including solar-process heat)

o   industrial applications for biomass

o   on-site electric generation (including cogeneration, industrial wind, and photovoltaics)

o   agricultural energy consumption

o   other major categories of industrial-energy demand.

Technical and Financial Assistance for Industrial Audits

The development of industrial-energy auditing firms needs to be enhanced. The Department of Energy should initiate a program to refine the methodology involved in an industrial audit and develop and implement training programs for auditors. The results of this program should be widely available to industrial firms, utilities, and audit firms. In addition, DOE should provide technical and financial assistance to private firms engaged in industrial audits . DOE should also consider a program of loan guarantees and limited equity financing for corporate interests entering this field.

The Office of Industrial Programs in the Department of Energy has initiated three pilot "Energy Analysis and Diagnostic Centers" (EADS), which have begun to experiment with workshops, symposia, and the preparation of manuals for industrial firms interested in industrial energy efficiency. The concept of regional diagnostic centers should be encouraged and expanded to include information about renewable energy investments. In particular, the regional diagnostic centers should coordinate their work with the Regional Solar Energy Centers and should have full access to information that is available in the Solar Energy Information Data Bank.

An information bank of data relevant to appropriate industrial energy users, constructed in a format consistent with the SEIDB, should be developed and made available through the Centers. The information in the data bank should be summarized in a series of annual publications that list manufacturers, products and product characteristics, research activities, and other information judged useful to industry. Some of this kind of information is presently available through the publications of the SEIDB.

One type of industrial-energy audit program that has particular promise is a training program in which engineers working in a given industrial sector are invited to participate in an on-site workshop where a person trained in industrial energy design will perform a careful audit of a particular factory, recommend a series of efficiency and solar retrofits, and tell how to oversee their purchase and installation. Designers from other plants could profit from the experience of watching an audit designed specifically for their industry and return then to perform similar work for their employers.

Education and Training

The turmoil in energy technology that will almost certainly characterize the next two decades will require a continuing supply of individuals capable of designing, evaluating, installing, and operating novel equipment. There is already an acute need for scientists, engineers, designers, systems operators, maintenance personnel, and other professionals who can design and operate increasingly complex systems. In addition, some skills that were once rather commonly available, operating an industrial power plant for example, have become increasingly scarce.

A National Energy Education Act

The shortage of skilled technical personnel justifies adoption of a National Energy Education Act designed to support education and training at many different levels:

o   scholarships for graduate studies in areas related to applied research, mechanical design, chemical engineering, materials, and other areas directly related to renewable resources and energy efficiency.

o   postdoctoral fellowships (for highly promising individuals) that are not overly restrictive and do not require an onerous burden of reporting

o   mid-career training for industrial engineers and designers to make them aware of advances in energy technology

o   training for system operators and maintenance personnel conducted through local trade schools and industrial colleges

o   a program modeled after the Fulbright Fellowships designed to encourage international exchanges between engineers and scholars working in solar energy and energy conservation. A special program for scholars from less industrialized nations would be desirable

o   specialized training sessions for system engineers working in the same industry.

Other aspects of the education program would apply to people working in the building sector; such programs are discussed in the chapter on building energy.

Summer Study Seminars

The Department of Energy could sponsor a series of summer study programs in each of the major areas covered by the journals. These seminars could bring together university, national laboratory, and industrial personnel to share ideas about trends and developments in the industrial sector. A compendium of work developed in the seminars could be published annually in the relevant DOE journal.

Solar and Conservation Offset Under the Clean Air Act

The Clean Air Act, as amended by the Clean Air Act Amendments of 1977, requires that a state program that issues permits for new or expanded stationary sources of air pollution must stipulate certain requirements. One of these is that, in an area that has not met ambient air-quality standards (a nonattainment area) a permit to construct or expand a major air-polluting plant may be granted if total emissions of air pollutants by the new or expanded plus existing plants in the area will be less than previous total emissions (42 U.S.C. 7503(1)(A). This is the so-called offset policy.

Because they do not pollute the air, solar and conservation devices can be useful in implementing an offset strategy. (See Regulagory Programs for the Industrial Sector, Policy Option I, Attachment to the Domestic Policy Review of Solar Energy.) Industry may comply with the Act by installing solar technology, thus achieving a net reduction in air pollution for the area. However, the inclusion of solar and conservation offsets is left up to the states, which must draw up the final implementation plan.

The Department of Energy, in conjunction with the Environmental Protection Agency, can encourage states to include offset policies in their planning for growth in non-attainment areas in the following ways:

Education

The use of conservation/solar offsets is not well-comprehended by state planning agencies that are responsible for developing plans under the Act. DOE could help by providing information and technical assistance on all economically and environmentally viable renewable energy alternatives. DOE could also work with business and industry associations by providing information and technical assistance in utilizing conservation/solar technologies in their building plans.

Direct Technical Assistance

DOE could provide technical teams to help states or industries in nonattainment areas analyze the potential for utilizing efficiency or solar technologies when large plants are being built or expanded.

The most promising area for solar is that it can help avoid the need to invoke the offset policy when a new plant is being built or an existing one expanded. The added pollution may often exceed the ceiling limit by only a small amount. In such a case, the industry may do one of three things:

o   reduce the capacity of the new plant or planned addition so that it will emit only permissible amounts of pollutants;

o   go through the long and often expensive route of negotiation with other local plants that are producing similar pollutants in order to strike an offset deal; or

o   substitute solar energy for a source burning conventional fuels.

This last alternative is particularly useful where the pool of possible offset candidates has been nearly exhausted or where simple and inexpensive emmission offsets are not obtainable.

Evaluation of Conservation and Solar Alternatives

Permits, Environmental Assessments, and Environmental Impact Statements: Former Section 1500.8(a)(4) of Title 40, Code of Federal Regulations, contained a requirement of the Council on Environmental Quality (CEQ) that conservation be considered in the alternatives section of an Environmental Impact Statements covering new electrical-generating capacity. That section expired on 29 July 1979, when the new CEQ guidelines went into effect. The new guidelines do not contain any such specific requirements; instead they require that agencies "rigorously explore and objectively evaluate all reasonable alternatives," including those not within the jurisdiction of the lead agency. Under this authority, DOE could initiate proceedings with EPA to require that energy efficiency and solar be given full consideration before the issuance of an environmental permit; DOE could also instigate requirements with other agencies to require consideration of solar energy and conservation when drafting their regulations implementing the CEQ guidelines for the EA/EIS processes. At a minimum, DOE should revise its own interim NEPA guidelines (44 F.R. 42136, July 18, 1979) to require consideration of solar and conservation alternatives. This approach could be particularly effective in situations where industrial processes will employ fossil fuels in nonattainment and PSD (prevention of significant deterioration) areas. In such areas, new plants and major modifications of older plants almost always require an EA/EIS or environmental permit. DOE would be particularly suited to provide the analytical tools needed to perform the analysis and to monitor the completeness and accuracy of any applications submitted. (See Regulatory Programs for the Industrial Sector, Attachment to the Domestic Policy Review of Solar Energy.)

Section Three
Industrial Energy Consumption Targets

Setting targets or goals for energy efficiency in the industrial sector is an uncertain process. Projections of U.S. industrial consumption in the year 2000 have changed dramatically during the past decade.

The analysis presented here employs simple, transparent assumptions and methods to arrive at a set of goals. Given the imprecision with which many variables critical to the analysis are known, it is not clear that a more sophisticated approach can be justified. The techniques used are resilient and can be easily used to show the impact of new insights about trends in industrial growth, fuel prices, and improvements in industrial technology.

The analysis begins with a careful examination of U.S. energy consumption in 1978. Consumption is divided into a number of "energy services" which are defined by the way the energy was actually used. Energy services include mechanical drive, electricity for electrolysis, direct heat for industrial processes, space heat for buildings, and chemical and metallurgical feedstocks. This division allows consumption goals to be established on the basis of past trends in the growth of each category of consumption and expected improvements in technologies specifically designed for each service category. These categories reflect the kinds of energy industry actually requires: an industry may require heat to dry paint but does not necessarily require natural gas to operate a drying oven.

The goals are established in four steps: first, a trend for growth in industrial value added is established by examining the history of the past decade (it is determined that industrial value added is growing 0.8% more slowly than the GNP); second, the growth rate for each industrial service category is estimated by examining historic correlations between growth rates in the sectors and growth rates in industrial value added (it is determined, for example, that growth in materials has risen 0.5% more slowly than industrial value added during the past decade); third, techniques for improving efficiency in each energy service category are reviewed and a goal established for improvements possible by the year 2000; and, finally, growth rates for demand in each service category are combined with goals for efficiency improvements to determine the amount of energy required to meet industrial demand in the year 2000.

BASE YEAR (1978) INDUSTRIAL ENERGY CONSUMPTION

The U.S. Department of Energy defines the industrial sector as establishments engaged in agriculture (Standard Industrial Code numbers 1-9), mining and petroleum extraction (SIC numbers 10-14), contract construction (SIC numbers 15-17), and manufacturing (SIC numbers 20-39). In 1978 these industries required about 30 quads or 37% of the total energy consumption of the United States (see Figure 2.5). It is important to notice that six basic materials industries (food, paper, chemicals, petroleum refining, stone-clay-glass, primary metals) account for about 85% of the energy demand in manufacturing. This simplifies the process of establishing overall goals since a careful analysis of these six industries covers a significant fraction of all industrial energy use.

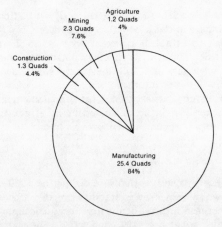

Source: Appendix A

NOTE: Distribution of energy consumption is based on fuel and power requirements of the various categories as a fraction of total industrial demand in 1974 (exclusive of motor gasoline). The apportionment shown i based upon a total demand of 30.2 quads, which includes 1.0 quad of biomass allocated to manufacturing.

Figure 2.5. Allocation of Industrial Energy Consumption among Major Economic Categories (1978)

More detail on the fuel use in these industries is shown in Table 2.13.* The basic materials industries dominate consumption in all areas except space heat. They consume 91% of industrial fuels and 63% of industrial electricity.

Table 2.14 divides industrial energy demand into "service categories" (agriculture is treated separately). It shows that roughly 40% of industrial energy is used for process heat, 20% for feedstocks (chemical feedstock, construction asphalt, and metallurgical coal), and that over 30% for services provided primarily by electric power-notably machine drive. Table 2.15 gives a similar breakdown for agricultural energy use in 1978. The data used to prepare these tables is presented in Appendix A.

RECENT TRENDS

Industry is ultimately associated with the harvest or extraction, processing, treatment, and assembly of materials; transactions in material goods form the basis of an industrial economy. In recent years, this country has experienced a gradual but decisive growth in an economy of exchange and demand for services as contrasted with goods. This reorientation of the thrust of our growth is almost certain to continue. This has significant ramifications for the industrial sector, particularly in a time of rising energy and material prices, and occasional disruptions in the supply of both.

While the long-term historical record suggests that material growth in the future might continue apace as in the past, recent trends strongly indicate that indus

Table 2.13   MANUFACTURING ENERGY USE, BY MAJOR MANUFACTURING ELEMENT AND SERVICE CATEGORY (1978).*

Total Use[a] (quads)

				Element				All Mfg.
SIC	20	26	28	29	32	33	-	20-39
Service Category	Food	Paper	Chemicals	Petroleum	S, C, & G	Metals	Other	Total
Mechanical Drive[b]	0.51	0.68	1.59	0.41	0.38	0.88	2.63	7.08
Electrolysis			0.44			1.08		1.52
Process Heat	0.75	1.85	2.02	3.30	1.12	1.47	1.59	12.10
Space Heat	0.03	0.03	0.01		0.02	0.04	0.24	0.37
Chemical Feedstock			2.52			0.07		2.59
Metallurgical Coal						2.10		2.10
TOTAL	1.29	2.56	6.58	3.71	1.52	5.64	4.46	25.76

[a]Energy consumption figures include energy losses in the generation and transmission of purchased electric power. Approximately 1.5 quad of wood residues are allocated to the total consumption of the paper and other forest products industries. Figures shown are exclusive of motor gasoline consumption in manufacturing.

[b]"Mechanical Drive" includes lighting and other overhead electrical uses.

Source: See Appendix A

---

*The existence of detailed numbers does not mean that the numbers are at all accurate. The information on steam illustrates the problem. For 1974 or 1976, total fuel use for steam has been variously given in "official" studies as 4, 6, 10, and 13 quads. (Boercker, 1979; Energy and Environmental Analysis, 1978; and Hamel & Brown, 1980; Farmer, et al., 1976; and Resource Planning Associates, 1977, respectively.) These numbers are not quite as disparate as they appear because of accounting differences and because the two high numbers include only casual estimates, presumably much too high, for fuel use in smaller boilers in the less energy-intensive industries.

Table 2.14   INDUSTRIAL (Mining, Construction, Manufacture)
                ENERGY USE, BY SERVICE CATEGORY (1978)

Service Category	Total Use[a] $(10^{15}$ Btu)	Percent of Total Use
Machine Drive	7.8	27
Electrolysis	1.5	5
Process Heat	12.5	43
Space Heat	0.4	1
Chemical Feedstock	2.6	9
Construction Asphalt	0.9	3
Metallurgical Coal	2.1	7
Repressurizing Gas	1.1	4
Vehicles	0.2	1
TOTAL	29.1	100.0

[a]Motor gasoline consumption is not included in this table. The amount of gasoline consumed in mining, construction and manufacture as reported by EIA in their Energy Consumption Data Base amounts to about 0.39 quadrillion Btu, according to our rough estimates.

[b]"Mechanical Drive" includes electric power consumption for lighting and miscellaneous electric overhead. All consumption amounts for purchased electric power include energy lost in conversion and transmission.

Table 2.15 AGRICULTURAL ENERGY USE, BY SERVICE CATEGORY
            (1978 estimates)

Service Category	Total Use, Exclusive of Motor Gasoline (quads)	Total Use (quads)
Space and Water Heat	0.05	0.05
Grain and Crop Drying	0.11	0.11
Lighting and Refrigeration	0.05	0.05
Irrigation	0.38	0.40
Mechanical Drive (stationary)	0.16	0.19
Vehicles and Field Machinery	0.38	0.83
TOTAL	1.13	1.63

trial society has reached a turning point in material consumption. There are four major factors working together that point to a fundamental change: the diminished capacity of technology to keep down the prices of raw materials (including energy), saturation of material inputs in the production of economic products, the need to establish stringent controls to limit the adverse side effects of technology on society, and demographic changes that will inevitably slow down economic growth in the future.

The major effects these trends have on industrial energy consumption will be examined in an effort to determine if they will continue to have an impact.

Energy Use

Industrial energy consumption rose during the 1960's, but has shown a remarkable stability in the 1970's despite an increase in total domestic energy consumption of nearly 5% between 1973 and 1979. The share of total U.S. energy use accounted for by industry has as a result declined since the 1960's, from about 39% to the present 37% (see Table 2.16). This decline can be expected to continue if an increasing fraction of national economic growth continues to occur in the service sector and in secondary processing industries, which are less energy-intensive per unit of output.

The history of industrial energy consumption is most effectively illustrated by analyzing the energy consumption relative to the "gross product" originating in industry. This is commonly called an "energy-value added ratio."* Figure 2.6 illustrates this ratio for the years 1950 to 1978. The energy-value added ratio varied erratically through the 1950's and most of the 1960's. The ratio rose dramatically between 1967 and 1970. This trend was reversed in the early 1970's, and the ratio has been declining ever since. In 1978 the ratio reached its lowest value in modern history falling to 88% of the value of the ratio in 1970. Since the ratio removes the effects of inflation (being measured in "constant dollars"), it must be concluded that the energy productivity of American industry increased rapidly during the 1970's.

The ten most energy-intensive industries also achieved remarkable improvements in efficiency.

---

*A considerable quantitative difference exists between specifying a time series for industrial value added using Federal Reserve Board indices and Department of Commerce Gross Product Originating. See Jack J. Gottsegen and Richard C. Ziemer, "Comparison of Federal Reserve and OBE Measure of Real Manufacturing Output, 1947-64," in K.W. Kendrick, Ed., The Industrial Composition of Income and Product, National Bureau of Economic Research, Conference on Research in Income and Wealth, Vol. 32, 1968. Historically, the FRB index for industry overall has been growing about 1% p.a. faster that the DOC index. It should be understood that neither index measures physical output-we examine physical (materials) output separately. The difference between FRB indices and physical production is discussed further in the Conclusions, Section 5.) This report uses the DOC evaluation because it is constrained by the methodology to be a part of GNP. The GNP deflator is used to obtain industrial value added in real terms from the DOC current dollar series, rather than the value quoted in constant dollars by DOC. The index of industrial output chosen is the slowest growing of all the indices mentioned and, thus, the energy-output ratio used in this analysis declines less rapidly than the ratios published by others. Another reason for the smaller apparent decline in industrial energy use per unit of output here (compared to other recent reports) is that industrial energy consumption data for the 1970's was significantly revised in 1979 and early 1980 by EIA. Data used in this report is taken from the Monthly Energy Review of February 1980.

Table 2.16 ENERGY CONSUMPTION COMPARISON
INDUSTRIAL SECTOR VS. UNITED STATES

| | Quads Btu Consumed | | Industrial Percentage |
	Industry	U.S. Total	
1973	29.14	74.60	39.1
1974	28.43	72.76	39.1
1975	26.21	70.71	37.1
1976	27.92	74.51	37.5
1977	28.92	76.54	37.8
1978	29.25	78.44	37.3
1979	29.15	78.19	37.3

Source: <u>Monthly Energy Review.</u> Energy Information Administration. U.S. Department of Energy. Feb. 1980.

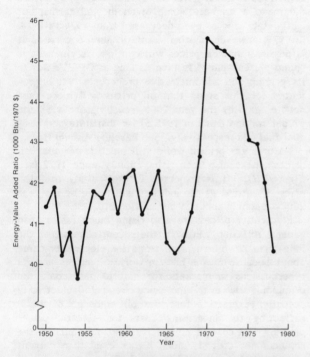

Figure 2.6. **Energy-Value Added Ratio for Industry, 1950 - 1978.**

Energy per unit of production (corrected for changes such as quality of inputs and outputs, capacity utilization, and pollution controls) for these industries fell 14% from 1972-1978. (Office of Industrial Programs, DOE, 1979).

Basic Materials Production

The U.S. economy is becoming less and less materials-intensive, both because growth in services has been more important than growth in industry in recent years and because industry has used less material per unit of value added. The first trend can be seen in Figure 2.1 which plots the ratio of value added in industry (or "Gross Product originating" in industry-- GPO) to Gross National Product (GNP). The trend is obvious. Analysis of the statics shows that the ratio has declined by an average of 1.1% for the past 28 years and declined an average of 0.8% per year for the past 9 years.

In the past few years American industrial growth has been increasingly achieved by adding more value to each pound of primary material consumed rather than simply processing more material.

Between 1925 and 1950 consumption of important basic materials generally grew about as fast as total industrial output. Figure 2.2 shows, however, that there has been a marked shift away from major basic materials-intensive activities during the past decade. The trend is particularly pronounced in the steel and cement industries. Even for the chemical industry, often viewed as a major growth industry, the consumption of basic materials has grown only slightly faster over the last 10 years than total industrial value added. Figure 2.2 indicates that national demand for organic feedstocks and three important inorganic products have been almost level since the late 1960's. Regression fits for 1969 - 1978 are shown in Table 2.17. (In most cases, the data shown in Fig. 2.2 and Table 2.17 are for consumption, i.e., they are corrected for imports of steel, paper, and lumber). It is noteworthy that this relative decline in consumption of basic materials began to develop in a time when the price of basic materials was falling. The saturation of demand for materials in the economy is therefore independent of energy shortages although increasing energy prices are likely to slow growth in the demand for materials even further.

This saturation should hardly be surprising in a society where each person consumes daily his weight in "stuff." Although the U.S. is still a consumer society, less materials-intensive services are accounting for a growing share of total consumption and the goods consumers buy are tending to contain more "value added" and less raw materials. Such shifts away from material-intensive products throughout the economy are reflected in the industrial sector by a trend away from basic materials processing toward a greater emphasis on finishing and fabrication.* In the steel industry, for example, where heavy steel rails represented a major component of production in the past, steel output today is made up more of thin-rolled steel of special qualities. Thus the decline in steel production relative to total industrial output does not mean that the U.S. is doing without steel. Far from it. Rather it means that in today's products steel is used more judiciously and more delicately.

---

*This result is supported statistically by comparing value added series for downstream materials finishing sectors with that for upstream basic materials processing sectors.

Table 2.17 WEIGHTED AVERAGE MATERIALS CONSUMPTION
1969-1978

Product	1974 Energy Use (Quads)	Product Consumption Trend[b] (% p.a.)
Steel	5.0	-1.3%
Cement[a]	1.6	-2.1%
Paper	2.7	-0.6%
Chemicals	6.4	+0.4% organics +0.4% inorganics
	Weighted Average	-0.6%

[a]In the stone, clay and glass products group, SIC 32.

[b]Determined through a linear regression analysis of the data shown in Figure 2.3 for the period 1969 to 1978. The growth rates are for industrial output (measured in pounds) per dollar of total industrial value added (GPO). Industrial value added is expressed in constant dollars using the GNP deflator. The weights for averaging are energy use.

Does this observation of materials saturation smack of a feeble imagination? What about new consumer products? We find there are no product innovations on the horizon that might create major requirements for materials and fuels so as to sustain energy and materials consumption relative to overall economic activity. Fresh applications of petrochemicals (aside from substitutions for other highly processed materials), major household appliances such as swimming pools and hot tubs, private boats and airplanes, huge enclosed malls, and space travel are not being implemented widely enough, and many are not, in fact, very intensive in their consumption of materials and energy relative to their cost. Most of the innovative products that are taking off, such as light equipment for home and recreational use, information technologies, biomedical technologies, and other services have low material and fuel requirements relative to the cost.

Could rapidly increasing investments in energy supply increase the role of basic materials in the economy as some here suggested (Bossong, 1978)? Since it is doubtful that total plant and equipment investment would increase as a result of increasing energy-supply investments, the question should be restated: Would energy-supply investments be more materials-intensive than the plant and equipment investments they replace? Probably not, because energy-supply projects have high labor and design costs just as do most other plant and equipment projects. In the case of power-plant construction materials, for example, use per dollar is comparable to most plant and equipment investments (RTI, 1976).

Industrial Energy Prices

The real increases in the prices of industrial fossil fuels since 1960 are shown in Figure 2.7. A graph of actual fuel prices paid (including all taxes and delivery charges) would have shown a steady increase in fuel prices from 1960 to 1971, followed by a dramatic and unabated climb in prices following the first OPEC oil embargo shock in 1973. By instead plotting industrial fuel prices, which are normalized by dividing by the producer price index for industrial commodities, we have shown that fuel prices were remarkably stable from 1960 to 1972. Between 1972 and 1974, oil prices did make a dramatic jump, even in real terms, but again stabilized or even decreased from 1974 to 1978. In 1979, however, another dramatic jump occurred. It is possible that oil prices will continue to rise in this "jump-plateau-jump" fashion as long as OPEC holds to its pricing and administrative procedures. It is projected for this study that oil prices will more than double again by the year 2000--reaching about $8 and $7 per million Btu (in 1978 $) for distillate and residual fuel oils respectively. See Baseline Assumptions.

Natural gas prices, while still much lower than oil prices*, have also been rising rapidly since 1972 (see Figure 2.7). It is projected for this study that the price of natural gas to industry will triple by the year 2000--reaching about $6 per million Btu.

Electricity prices are also rising much faster than general inflation. In part, the recent reversal of the long-term downward price trend for electricity came about because of the hike in world oil prices and its repercussions in markets for natural gas, coal, and uranium. However, the price trend for electricity actually reversed before the oil embargo of 1973, reflecting the increasing costs for electric power plants. The cost of electricity from new plants, the replacement cost, is much higher even than current prices. It is estimated (Table 2.18) that the cost of providing electricity to industry from a new central station baseload plant is approximately twice the present price.** Since the replacement cost is so much higher than the present price, we can expect a continuing steady rise in the price of industrial electricity.

The overall real price increase for energy purchased by industry is uncertain because the mix of energy carriers has changed a great deal in the 1970's. If we adopt the mix that characterized 1978, and weight electricity according to its primary fuel equivalent,

---

*In 1979 the price of natural gas to industry averaged $1.78 per million Btu (1978 $), compared to $2.77 for residual fuel oil and $3.74 for distillate.

**The replacement cost is much greater than the cost of electricity from existing plants for several reasons. New requirements for air pollution abatement technologies are driving up the price of electricity based on coal. Quality control problems and toughening safety regulations are driving up the cost of nuclear electricity.

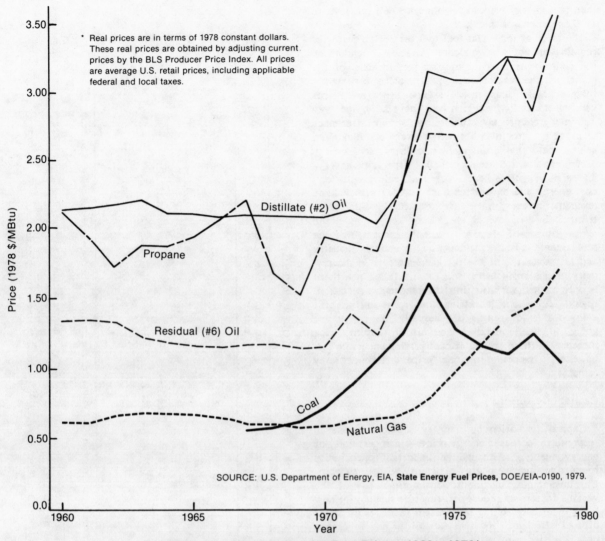

* Real prices are in terms of 1978 constant dollars. These real prices are obtained by adjusting current prices by the BLS Producer Price Index. All prices are average U.S. retail prices, including applicable federal and local taxes.

SOURCE: U.S. Department of Energy, EIA, **State Energy Fuel Prices,** DOE/EIA-0190, 1979.

Figure 2.7. **Real Industrial Fuel Prices, 1960 - 1979***

Table 2.18 INDUSTRIAL ELECTRICITY PRICES
1978¢/kwh[a](Current ¢/kwh)

SIC CODE	Major Industrial Group	1971	1975	1976	1977	Ratio of 1977 Price to the Replacement Cost[d]
28	Chemicals and Allied Products[b]	1.23(0.78)	1.72(1.44)	1.79(1.57)	1.94(1.81)	0.40
2812	Alkalies and Chlorine[b]	0.93(0.59)	1.49(1.25)	1.57(1.38)	1.71(1.59)	0.36
2819	Industrial Inorganic Chemicals,[b] n.e.c.	1.14(0.72)	1.42(1.19)	1.47(1.29)	1.56(1.29)	0.33
33	Primary Metals Industry[b]	1.25(0.79)	1.64(1.37)	1.76(1.55)	1.88(1.75)	0.39
3334	Primary Aluminum[b]	0.68(0.43)	0.83(0.69)	0.96(0.84)	1.05(0.84)	0.22
	All Manufacturing[b]	1.55(0.98)	2.08(1.74)	2.16(1.90)	2.32(2.16)	0.48
	Large Light and Power[c]	1.63(1.03)	2.30(1.92)	2.35(2.07)	2.50(2.33)	0.52

Notes

(a) Prices are converted to 1978 dollars using the GNP deflator
(b) Bureau of the Census, "Fuels and Electric Energy Consumed," Annual Survey of Manufacturers, various years.
(c) "1980 Annual Statistical Report," Electrical World, March 15, 1980.
(d) The replacement cost is the cost of electricity delivered to industry from facilities ordered in 1980, estimated to be 4.8¢/kwh (1978$) for this study. See Baseline Assumptions. This is the cost of delivered electricity from a coal-fired baseload plant (for a capital cost of $750/kw(e) and for coal costing $1.45 per million Btu).

then real energy prices rose almost 90% from 1972 to 1979 as shown in Table 2.19.

The price of industrial fuel to a particular customer remains subject to significant variation depending on the contract term, date of issuance, location of the plant, and the type of industry. The prices reported in the preceding discussion are national averages. Prices vary substantially by region (see Table 2.20), and even within regions; recent studies have shown differences of up to a factor of 3 between average and maximum prices paid by industrial customers for energy (Grumman Data Systems, 1980). Especially energy-intensive industries have located in regions with low-cost energy and have often been favored with volume discounts in rates, in the cases of natural gas and electricity. In the case of electricity, for example, the average price of electricity is only about 1/3 of the replacement cost in the inorganic chemicals industry and less than 1/4 of the replacement cost in the primary aluminum industry (see Table 2.18). Since there is only modest variation in the replacement cost from region to region, industries that have in the past enjoyed the advantages of unusually low prices are going to be especially hard hit by future price increases. There are significant pricing uncertainties, however. Important regional price differences may remain.

Capital Investment

Capital formation by industry should increase in response to increased energy prices and uncertainity of energy supply. Although an important opportunity remains to improve energy efficiency without substantial investment, moving toward optimal use of the various factors of production requires substantial new investment both to improve present facilities (in terms of fuel efficiency and fuel flexibility) and to create new facilities using state-of-the-art processes.

Despite the pressing need for capital investment in the energy-intensive basic materials processing industries, recent investment trends are unfavorable to the task of meeting these needs. As pointed out in 1979 by George Hatsopoulos, president of Thermo Electron Corporation:

Table 2.19 THE REAL PRICE INCREASE FOR ENERGY PURCHASED BY THE INDUSTRIAL SECTOR, 1972-1979

	Portion of Btu's 1978	Real Price Increase factor (1972-1979)
Electricity	0.34[a]	1.32
Natural Gas	0.29	2.52
Coal	0.12	1.53
Petroleum products	0.25	2.00
Weighted real price increase		1.87

[a] Primary fuel equivalent

Table 2.20 REGIONAL INDUSTRIAL FUEL AND POWER PRICES (1977)[a]

Fuel	New England	Middle Atlantic	South Atlantic	East North Central	West North Central	East South Central	West South Central	Mountain	Pacific	U.S. Average
Distillate (#2 oil) ($/MBtu)	3.53	3.50	3.42	3.37	3.32	3.32	3.10	3.24	3.38	3.34
Residual (#6 oil) ($/MBtu)	2.33 (2.23-2.46)	2.28	2.15	2.51	2.27	1.81	1.92	2.10	2.37	2.22
Natural Gas[b] ($/MBtu)	3.04	2.18 (2.74 NJ)	1.28-2.68	2.11 (1.64 IN)	1.11 (0.82-1.64)	1.56	1.05-1.37	1.17 (2.14 ID)	2.11	1.53
Purchased electricity[b] (mills/kWh)	24-14	32-42	24-30	22-28	21-32	19-29	16-32	8-31	5-40	25
Coal ($/MBtu)										

[a] Although the relative magnitude of the prices shown reflects conditions in 1977, all prices are shown at 1978 price levels in order to maintain the consistent use of the 1978 base year currancy in this report.

[b] Prices of natural gas and electricity, in particular, show a wide degree of difference between states within a given region. Where there is a large difference, a range is shown or a particular deviation is shown in parentheses.

SOURCE: U.S. Dept. of Energy, EIA, State Energy Fuel Prices DOE/EIA - 0190, 1979. Regional Averages are approximate.

"In 1978 all industries spent $153.8 billion in new plant and equipment, or 7.2% of GNP. This fraction is about the same as in the 1960's. But if we subtract the capital used in the energy sector, which has been steadily increasing in the 1970's, and the capital used for pollution abatement, which was almost zero in the 1960's, we find that the capital investment in all industries outside of energy supply and pollution abatement was $98.76 billion in 1978, or 4.6% of GNP. This fraction is substantially lower than the average of 5.6 percent for the 1960's" (Hatsopoulos, 1979).

The shift in capital investment toward the energy supply industry is shown in Figure 2.4. The share of total U.S. new plant and equipment expenditures committed to energy supply rose from about 25% in the 1960's to over 40% in 1977. This capital-hogging by the energy supply sector is a major factor behind the "capital scarcity" problem faced by all non-energy supply industries. Of course, these arguments about aggregate capital availability may be less germaine to the issue of capital formation in basic materials-processing industries than the present weak growth and profitability of many firms in these industries.

In order to analyze some capital issues later in this report, for the basic materials processing industries some recent data on industry performance and capital formation are shown in Table 2.21.

The retirement/replacement rates shown in Table 2.21 are calculated, not directly observed. The calculations are based on an analysis of equipment retirement for the period 1935-1970 prepared for the Office of Emergency Preparedness (Jack Faucett, 1979) which led to an estimated equipment life of 18 to 22 years in the energy-intensive industries. The "retirement" under discussion is economic in nature. In most cases the major equipment items are not scrapped and replaced but are rebuilt or refurbished, a distinction which is important for energy analysis.

A final comment on Table 2.21 concerns the chemical industry. Aggregate physical data such as production or average unit capital cost is a problematic concept in this case because of the great variety of intermediate and final products. Chemical industry physical growth was obtained above by examining organic feedstocks and three major inorganic products. The chemical industry unit capital cost was calculated as the ratio of recent plant and equipment expenditures to capital turnover as explained in the notes to Table 2.21. (This method yields results in rough agreement with the actual unit capital cost for steel and a factor of 2 too high for paper.)

EFFICIENCY: CONCEPTS AND TECHNOLOGIES

Basic Concepts

Three kinds of evidence are available to make a persuasive case that energy productivity in industry can be increased dramatically. The first is the historical record. There has been a long-term decline in industrial energy consumption per unit of GNP, (albeit not per unit of industrial value). This trend has accelerated since 1973. Much of this long-term improvement is associated with change in the mix of products, especially, the increasing importance of services relative to goods. The more recent decline in the energy consumption per unit of industrial value added is also associated with improved energy efficiency in the making of industrial products. (Office of Industrial Programs, DOE, 1979). The rapid rate of decline in energy use per unit of industrial value added, approaching 2% per year since 1973, would, if sustained, solve industry's energy problems because it would enable production to increase at levels projected for a healthy economy while absolute energy consumption by industry would decline.

The accelerated decline in energy use per unit of aggregate production since the oil embargo is not really surprising in view of the prices industry has been paying for energy. Industry has experienced the greatest energy-price increases of any major group.

The second category of evidence is a group of studies (Gyftopoulos, 1974; Berg, 1974; CONAES, 1979; Sant, 1979; Reid and Chiogioji, 1979) that suggest that substantial technical and economic opportunities remain to cut energy-related costs and enable industry to regain control over its energy services.

Finally, a comparison of the U.S economy with that of Europe and Japan iindicates that the U.S. uses 1.5 to 1.8 times more energy per unit of value added in industry. Clearly there is room for improving U.S. efficiency.

In order to put these technical opportunities in proper perspective, it is important to present certain basic concepts about industrial energy demand, which, unless re-emphasized, are often lost in discussions of our energy crisis.

Energy Service

EEA, 1978; Sant, 1979, 1980) is critical. Energy service is an important concept because:

Table 2.21 CAPITAL FORMATION IN CERTAIN MAJOR BASIC MATERIALS INDUSTRIES

Industry	Capacity[1] 1978 (million tons/yr)	Retirement/ Replacement Rate[3] 1975-1980 (% p.a.)	Growth Rate in Physical Capacity 1975-1979 (% p.a.)	Annual Average Plant & Equipment Investment[5] 1975-1979 (billions 1978$/yr)	Unit Capital Cost (1978 $/ton/yr)
Paper	67	2.1	2.0[2]	3.3	600[6]
Chemicals	100*	2.3	4.4	7.5	1100[7]
Cement	56	2.6	--	----	115[8]
Steel	115	2.5[9]	0.0	2.7[10]	850[9]

* Nominal capacity of 100 units.

1. Department of Commerce, Survey of Current Business, Current Business Statistics, July 1980.

2. American Paper Institute, "Paper/Paperboard and Woodpulp Capacity," 1979.

3. Energy and Environmental Analysis 1978; ISTUM Vol. I, p. IV-114. These estimates are for energy equipment and are based on analysis of equipment life as mentioned in text, and not on actual retirement during this period.

4. The product consumption trend relative to GNP, plus the growth rate of real GNP of 3.3% p.a. + decrease in capacity utilization of 1.5% p.a. for the period 1974-78.

5. Survey of Current Business. Corrected to 1978$ using the deflator for producer's durable equipment. For steel and paper 17% (note 10) and 12% reductions for pollution abatement have been made.

6. Based on $150,000/ton/day cost of typical new integrated plant in 1974. A.D. Little, "Analysis of Demand and Supply for Secondary Fiber on U.S. Paper and Paperboard Industry", Oct. 1975.

7. Based on estimated total capacity created and total industry investment using data in this table for the late 1970's. The unit cost is I/A(G+R) where I is the total investment rate, A is the capacity, R is the annual retirement/replacement fraction and G is the annual growth in capacity. The unit capital cost for chemicals is millions of 1978$ for a 1% increase in 1978 capacity.

8. Private communication, George McCord, Portland Cement Association.

9. American Iron and Steel Institute, "Steel at the Crossroads: The American Steel Industry in the 1980's", Jan 1980. This is the average for expansion at an existing site and replacement at existing site. For a greenfield plant the unit cost is $1000.

10. American Iron and Steel Institute "Statistical Highlights 1970-1979."

1) It is these services that are needed, rather than energy as such; as a result, projections of energy needs should rest on analyses of the needs for energy services.

2) The usefulness of any energy carrier* depends sensitively upon the energy services needed.

3) The opportunities for cost reduction through efficiency improvement and the appropriateness of policies to support them depend sensitively on energy service.

The purpose of the energy service concept is to focus attention on the combined system of energy carriers and the means of their use, in contrast to considering the energy itself. The manufacturer has a variety of opportunities to change the way in which energy is used. He can change the design of the product without changing its essential character; he can fundamentally change the manufacturing process; he can also modify the energy conversion equipment and improve energy management practices.

Consider the final three options in terms of the energy service of drying or curing paint on new automobiles. This service has been provided by burning gas

---

*The form in which energy is delivered to the point of use, usually refined fuels or electricity. In some analyses forms such as steam and compressed air would be considered as carriers.

in conventional ovens. Many changes could be made in the provision of this service, for example:

o The existing ovens could be operated with the same fuel, but managed more carefully, perhaps with sophisticated add-on equipment enabling recovery of the fuel value of the emitted hydrocarbons, leading to substantial reductions in gas requirements.

o The ovens could, in time, be switched to one of a variety of coal or biomass-based schemes. The high capital cost of such options typically implies substantially higher costs for the service then in the past, but the price of oil and gas is also rising. Under favorable conditions technologies such as cogeneration might limit the cost increase.

o A process change might be adopted if the product remained satisfactory. This could range from a simple change, like higher-consistency paint with reduced drying requirements, to a fundamentally new process like ultraviolet curing resins (now under development) with sharply reduced energy and capital requirements. Substantial variations in cost and completely different energy consumption regimes characterize these different means of providing the service.

This single case only begins to illustrate the complex possibilities for optimization by the firm in the face of new economic conditions.

If an energy service is broadly defined in terms of function, it is seen that a whole complex of changes may be made to provide the energy service at less cost and to change fuel requirements. Analysis of energy services is, however, difficult when fundamental process changes are considered: because of lack of information, because all factors of production tend to be intimately involved, and because in such cases decision making by firms is usually not driven by energy considerations. For practical purposes we will adopt narrow definitions of energy service will be adopted in this report. This approach allows explicitly for only certain kinds of technical change. The approach will be corrected only to account for process change relating to process heat. In other cases it has not been possible to explicitly consider many interesting opportunities for technical change. The energy services considered are listed in Tables 2.14 and 2.15. The most important services are:

o Process Heat. Process heating services are the conversion of input product streams into output product streams in certain desired states as achieved by heating. Industrial process heat at low and moderate temperatures, whether by steam or direct firing, is at present predominantly produced by burning chemical fuels, expecially oil and gas with some coal and biomass (the latter mainly wood wastes and by-products used by the paper industry). Industrial heating at high temperature levels tends to be constrained by the process, with fluid fuels, coal and electricity being dominant in various cases. While electricity is not economical for most process heating applications, it is important in particular applications. In the steel industry, for example, use of the electric furnace has led to major fuel savings, because the electric furnace allows use of much more scrap in steel making than the basic oxygen furnace. This represents only a minor use of electricity, however. In 1976 19% of steel was produced in electric furnaces, but this use accounted for only 1.6% of industrial electricity use (Myers and Nakamura 1978).

o Mechanical Drive and Electrolysis. Machine drive is primarily the service of moving materials, or compressing, grinding, or shaping them. Electrolysis involves the use of electricity for the chemical separation of materials, the most important application being the primary production of aluminum. Mechanical drive is dominated by electric motors, although direct engine-driven shaft power is also significant. Although there are somen seasonal and diurnal variations in these services, they are largely provided by baseload electricity.

o Feedstocks. Use of metallurgical coal in steel-making, organic feedstocks in the chemical industry, asphalt in the construction industry, and miscellaneous industrial feedstocks make up this category. The service is defined as transformations of inputs into certain classes of final products where the inputs are fuels used as materials (rather than as sources of heating). (Many processes involve both feedstock service and heating service.) The consumption of feedstock services should be distinguished from the consumption of feedstocks. For example efficiency in the use of virgin feedstocks to meet the service demand can be increased by, e.g., increased use of secondary materials and by reducing the loss of materials from the product stream during the course of manufacture.

Energy Intensity and Energy Efficiency

The consumption of energy carriers for any partic-ular energy service depends on the level of service and the energy intensity of the service. Thus, if the level of materials production service is specified in terms of tons of production, the intensity would be described in terms of fuel energy per ton. There are two ways to achieve fuel conservation: (1) to reduce the service demand, e.g., to reduce production; and (2) to reduce the energy intensity.*

The analysis of energy intensity is illuminated by introducing the concept of energy efficiency. The energy efficiency of a system is the ratio of the energy intensity of a postulated "ideal" system to that of the actual system (which performs the same task).** The energy efficiency is seen to be a number between zero and one which indicates the limits of performance of the system (under stated assumptions) in carrying out the given task.

The fuel consumption in fabricating a particular product can be found as the number of units produced, times the fuel per unit required by a postulated ideal production system, divided by the efficiency of the actual system.

There are two characteristics of energy efficiencies which are important to this discussion: (1) The energy efficiency of a system is sensitive to the definition of the service it performs, i.e., to the definition of an ideal production system and to the boundary selected for the system. (2) The efficiencies of the complex systems that provide final products in our society tend to be very small.

Consider the making of final products based on steel as an example. Steel-making is, in itself, relatively efficient, but the overall efficiency with which steel is provided in most final products is low, because: (1) In the finishing, shaping, and assembly processes consi-derable energy is used which, from a thermodynamic standpoint, is not essential; and (2) In the manufacture from raw steel to final product a great deal of scrap is generated. (Typically only about 1/3 of raw steel production goes directly into final products, while most of the remaining 2/3 is recycled to the steel-making step.)

This is a general pattern, as pointed out by Ayres & Narkus-Kramer (Ayres, 1976): When efficiency is defined in terms of the fundamental task of a compre-hensive system, it will tend to be very low; concomi-tantly, the possibilities for technical change are multiple, as discussed above in connection with the definition of energy service.*

One correctly concludes that in the long-run exten-sive technical improvements can be made. Long-term reductions in overall energy intensities of the produc-tion of final products are not physically constrained. It is still too early in the history of technology for such constraints to be critical.

On the other hand, the efficiency concept must not be oversold. While it correctly gives a notion of even-tual opportunities, it does not necessarily provide the best criterion for selecting the best near-term targets for improvement. (B. Hamel et al 1979.) Thus it may well be that near-term improvements to the already highly efficient blast furnace or to the relatively effi-cient steel reheat furnace are more cost effective and more significant than improvement of some operations which have very low efficiency. Thus, while a general awareness of low overall efficiencies is important, evaluation of energy performance can be properly made in terms of energy intensity, e.g., in million Btu per ton, without explicit determination of the effi-ciency.

Conservation Technologies

The kinds of technical change that may strongly affect the cost and efficiency of energy services can be roughly categorized according to the kind of deci-sion involved in their adoption. We choose here the categories:

---

*Considerable analysis has been devoted to the indirect contributions to energy intensity, i.e., accounting for the energy embodied in the non-energy products used in a process, such as the energy involved in the manufacture and installation of capital equipment. This analysis reveals that even where capital costs are high, embodied energy is usually low: for energy intensive processes indirect energy use is only a correction, often a small correction, to direct energy use. In this report indirect energy use will not be included in the accounts for particular industrial subsectors; conventional accounting of energy use at point of use is adopted. Another important ambiguity in accounting is associated with the direct consumption of primary fuels to produce carrier energy. This important distinction will be taken partially into account: for electricity the energy values quoted will be units of fuel consumed at the generator, unless electrical energy is specified. On the other hand the energy to refine petroleum is allocated to the petroleum refining subsector and that to transport petroleum to the transportation sector rather than to the final user.

**With this definition, the efficiency is often what is known as a "second-law efficiency." (American Physcial Society 1975).

(a) Product-related changes, especially changes in product design which affect the use of basic materials.

(b) Fundamental process changes, changes which require very large investments and impact a variety of factors of production.

(c) Changes in energy conversion equipment or energy conservation technologies which involve significant investment and are primarily connected to energy use and costs.

(d) Changes in management of operations and maintenance, and retrofits with low cost equipment, which affect energy use.

Product Changes

From an energy standpoint the predominant issue in product design is the requirement for basic materials, since basic materials industries consume 5/6 of all energy consumed in manufacturing. Product design is guided largely by marketing considerations; the cost of energy in manufacturing is usually a minor issue. Nevertheless there is leeway for cutting energy and materials costs by reducing materials requirements. In addition, changes are occurring in materials use because of external factors such as environmental, safety, and energy performance standards.

Design determines the weight of materials and substitution among them in the final product. (The fabrication and assembly process also impacts the weight of materials in the final product and the loss of materials from the product stream.) Product design also influences the potential for materials recycling of the used product and the suitability for incorporation of materials with a high secondary (i.e. recycled) materials content. Product design also influences the desirability of a product for prolonged use, and the potential for repair, remanufacture and reuse. (Hill, 1979; Lund, 1977) New technologies can enable products to wear better, e.g. through improved corrosion resistance (Council on Materials Science, 1979).

The remarkable changes taking place in the design of automobiles illustrate several of these effects. The changes in auto weight, primarily to improve fuel economy, are taking the average car of 4000 lbs. of the mid '70's to about 3000 lbs. in 1985 and, perhaps, to under 2000 lbs. by the millenium. At present about 1/3 the weight reduction is being accomplished by materials reductions as such, and 2/3 by reduction of the vehicles' external dimensions. Use of stronger but thinner sheet steel is an important part of materials reduction. Increased use of plastics including strong reinforced plastics (composites), and aluminum is also important.

Use of secondary materials can be of major importance. (Bever 1976) Recycled steel requires on the average only 35% as much energy (to create finished steel) as use of virgin material. (Chapman, 1974) Recycled aluminum requires less than 10% as much as virgin material. The use of waste paper in making newspaper products is increasing. However, recycling of waste paper of low quality may have to compete with its use as a fuel.

Use of secondary materials can again be illustrated by the automobile. Perhaps 90% of discarded automobiles are now being recycled, a much higher fraction than for most steel products. (Bever, 1978) Auto design is important to recyclability both to the extent special, e.g., low impurity, steel is required for manufacture of new cars, and to the extent that impurities will tend to become incorporated in the scrapped vehicle. For example, copper and nickel make steel brittle, so avoidance of the use of these materials in cars, design to make their removal from the obsolete vehicle easy, and the use of secondary processing which effectively extracts these materials, are strategies to improve the quality of scrap steel.

Design for durability, ease of repair, and remanufacture is an issue attracting increasing attention. Improvement of these qualities is an important option for buyers of products, especially if the underlying technology is stable. The remarkable achievements which are possible are illustrated by products which are designed for lease: certain computers, copiers, and telephone equipment. Returning to our example of the automobile, during this time when auto technology is changing very rapidly, the durability option should, and probably will, have relatively low priority. In the future this option will offer an intriguing challenge to auto makers and marketers.

Process Change

Process change is an industry-specific issue. We briefly describe change in terms of examples in several of the basic materials industries.

Steel. Process and equipment change options have been considered in several studies (Woolf, 1974; Gordian, 1975; Battelle, 1975 and 1976; A.D. Little, 1978b; American Iron & Steel Institute, 1976; OTA, 1980). In iron and steel, process change can be discussed in terms of the four stages of the overall process: the making of iron from largely virgin inputs, steelmaking, semi-finishing and finishing. In the reduction of iron ores the blast furnace remains supreme even while coke is a problem because of air pollution and difficulties in the coal industry. One major change is larger higher-pressure blast furnaces. These are being built, almost exclusively overseas, and offer substantial cost and energy savings. In the U.S. a remarkable improvement in the capacity of existing blast furnaces is being achieved through a variety of improvements. This development means retirement of inefficient furnaces as well as savings at the furnaces in question. Among the improvements are changes in blast furnace feed, including a move not only to ore concentrates, but toward pellets combining concentrated ore, flux, and carbonaceous material. Direct reduction of iron oxides is a different approach which, although it is not well established in the U.S., has high potential (OTA, 1980).

Steel-making is evolving, with basic oxygen plants and electric arc furnaces well on their way to replacing open hearth furnaces, with significant benefits. More radical changes in iron- and steel-making are in the R&D phase. (Szekely, 1979; OTA, 1980)

The semi-finishing step is changing with the introduction of continuous casting and, especially overseas, of direct rolling. These operations replace ingot casting, reheat and rolling, and a second reheat and rolling, with substantial energy and materials savings. The immediate creation of semi-finished products creates serious inspection and scheduling problems, but offers the potential in time for more effective quality control.

The finishing step, consisting, e.g., for sheet steel, of surface preparation and of cold roll and annealing to produce desired hardness, is labor-intensive and not very energy-intensive. Automatic control and continuous product flow are important developments. A radical type of change that could eventually take a major market share with substantial cost savings is direct forming of products as, for example, with powder metallurgy.

Energy use by the steel industry per ton of shipped steel was approximately 36 million Btu (including utility losses) in 1976. (It might have been somewhat lower if the industry had operated near capacity.) The Japanese industry used an average of 30% less energy per ton (OTA, 1980). Consonant with that fact, a new integrated steel mill incorporating dirert rolling and other improvements would require some 35% less fuel than the present average (Battelle, 1972). A reduction in overall fuel use of 25% has been projected for the industry (Chiogioji, 1979).

Paper. In pulp and paper we focus on substantial changes occurring in pulp making, several of them associated with initiatives being taken in Scandinavia. More complete use is being made of the tree, including transport of forest and sawmill wastes to the mill and use of branches and possibly of the tree root. Much of the added material is being used as fuel in boilers, with substantial reduction in use of fossil fuel. There is a move toward continuous pulping with sophisticated process controls (Thermo Electro 1979). Some change in the mix of chemical and mechanical pulping processes will occur, with uncertain energy impact.

Dramatic changes in the paper industry are occurring in the use of water (TAPPI 1979, International Pulp Bleaching Conference, 1979). Both closed and almost-closed systems, and processes with reduced presence of water are actively being implemented. As an example of the latter, bleaching can be accomplished by successive stages of diffusion of chemicals through the pulp rather than successive stages of soaking and washing. Reduced intake and discharge of water will have substantial impacts on both energy requirements and water pollution.

The 1978 average energy use in the paper industry was 39 million Btu per ton of paper produced (including utility losses), of which 42% was internally generated energy (American Paper Institute, 1979). New integrated bleached kraft pulp and paper mills incorporating process improvements such as continuous digestors, vapor recompression concentration of digester by-product liquor, new bleaching techniques and high pressure paper drying presses, use 40 to 50% less energy than this average, and further improvements through process optimization and cycle closure are possible (Hamel, 1980). Of the energy required by such a mill, 85% or more will be internally generated. For the industry as a whole 20-40% reductions in total energy use per ton of production are anticipated (Chiogioji 1979) with the higher figure appropriate to extensive implementation of integrated kraft mills as now anticipated.

Chemicals. The chemical industry is distinguished from the other major energy-intensive industries by its recent history of frequent fundamental process changes. This change has been fueled by rapid growth, product innovation, and process research. Although growth and introduction of new products are expected to slow, the pace of process innovation is expected to continue with changes in feedstocks and continued innovation, e.g., in catalysts and in shortening the chains of intermediate products. A promising area for radical change is the development of biological production processes.

Process change and other efficiency improvements are so rapid in organic chemicals manufacture that 40% reductions in fuel requirements per ton of production (fuel burned plus purchased electricity including utility losses) are typical for plants now being built compared to plants built even as recently as ten years ago (Wishart 1980). This extraordinary achievement is reflected in corporate-wide reductions in fuel required per unit of chemicals production. See Figure 2.8. A corporate-wide 17% reduction, 1978-1985, is targeted by Union Carbide (Wishart, 1980) and a comparable goal has been established by DuPont (Energy Daily, May 12, 1980).

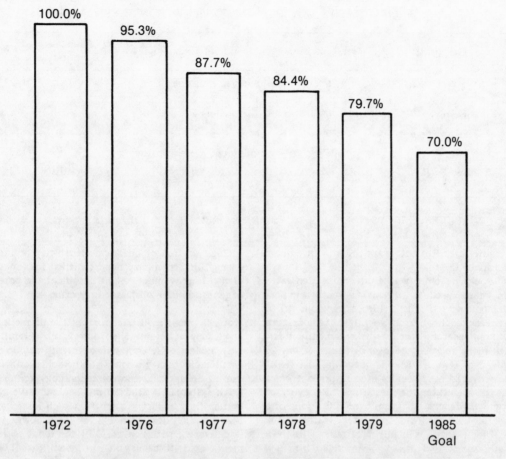

* This is fuel burned plus 10,000 Btu/kWh for purchased electricity.

Figure 2.8. **Energy Used Per Pound of Material Produced, Union Carbide Corp. 1972 = 100***

Other industries. A few examples from other industries illustrate that remarkable reductions in energy intensity are being achieved elsewhere. See Table 2.22 Most of the fuel reductions noted are plant wide and are per unit of production. In all these cases news articles suggest that firms achieving major savings have considerable technical expertize and provide strong management support for energy productivity improvements. It is seen in these examples that process change is often the key to major savings but energy technology improvements and energy housekeeping measures, discussed immediately below, are also very important.

Energy-Efficient Technology for Process Heaters and Electric Motors

Process Heaters. In very general terms process heat equipment can be improved by controls, by insulation and by heat recovery; and further improvement may be feasible by more "specific" means of heating or by cogeneration of heat and work.

Heater controls offer relatively low-cost opportunities for substantial savings in money and fuel, primarily by: (1) Optimizing the timing of start-up and shut down or of idling, (2) controlling the air-fuel mixture to reduce excess air, and (3) optimizing combustion during part-load operation. On point (1), improved refractories, or furnace linings, enhance the practicality of cooling down high-temperature furnaces when not in use. On point (3), like many other energy-conversion devices, even fuel-intensive heaters are often designed and continuously operated for full-load conditions; substantial savings are usually possible with automated adaption to less than full loads.

Improved insulation of furnace walls is becoming practical both because of higher energy prices and

Table 2.22 ILLUSTRATIVE EXAMPLES OF CONSERVATION ACHIEVEMENTS

Industry	Corporation	Fuel Reduction	Principal Technologies	Payback Period	Reference
paper	Western Kraft	36%	higher pressure drying press	2 - 3 yrs	1/28/80
tires	Firestone	20%	heat recovery		2/4/80
cement	Kaiser Cement	30%	dry process		11/26/79
food	Vlasic Foods	26%	equipment modernization		11/26/79
furniture	Inter-Royal	50%	cold water solvents reduced steam pressure	months	3/24/80

*All from Energy Users News

improved materials. Reflectors for radiation are being more extensively used.

More thorough heat recovery from combustion gases (and from heated-product streams) is another important development. (Dryden, 1975; Chiogioji, 1979) While recovery of heat at a very high temperature from high temperature furnaces is technically difficult, much more thorough recovery of heat at low to moderately high temperatures is becoming justified-- with the recovered heat usually used to preheat combustion air or feedwater for the device. Recovery of most of the heat from stack gases over 500°F (and of a substantial part of the heat from stack gases in the 300° to 500°F range) is highly profitable with new heaters at present fuel prices. (American Iron and Steel Institute, 1976; Energy and Environmental Analysis, 1979). A degree of heat recovery used to be standard on most higher temperature furnaces, but during the '60's and early '70's many were installed without heat recovery, an interesting commentary on the cost of fuel and the general engineering outlook of those times. Unfortunately, extensive retrofit with heat recovery equipment of existing heaters is often impractical because installation is too difficult.

Perhaps the prime area of concentration by the U.S. government has been fuel-switching, e.g., the introduction of coal-fired boilers. At many plants this is an awkward step in terms of available skills and space and costly even in terms of present fuel prices. Fluidized-bed combustion of coal may bring space and cost improvements that will make such a change more attractive. Coal-oil slurries may also prove a convenient form in which to burn coal.

Specific heating by electromagnetic means, e.g., induction heating of metals and microwave heating of particular materials, is an attractive alternative in special cases to general heating in furnaces. (American Physical Society, 1975). A recent example is ultraviolet curing of specially formulated inks and coatings. (BDM Corp., 1980)

Cogeneration of heat and electricity or mechanical energy offers major opportunities for energy efficiency improvement. Some details of the potential for cogeneration are discussed in Section 4.

Motors. Most industrial motors are of moderate size and fairly efficient (See Table 2.27). Significant but only modest efficiency improvements can be made. A greater electricity savings can be achieved with better use of linkage and control technology for mechanical drive systems. A study of mechanical drive systems in British light industry quantified such losses and found that typically less than 1/2 the input power to a plant is delivered, with about 1/3 of the input power lost in mechanical transmission. (Ladomatos, Lucas, and Murgatroyd, 1978) Systems changes can be effective in reducing such losses. One possibility, involving hydraulic power transmission as an alternative to an electromechanical transmission, would reportedly lead to a 20% energy savings. (Ladomatos, Lucas, and Murgatroyd, 1979). Another opportunity is offered by variable-load situations. For example, gas-moving systems are usually over-powered and motion is controlled with baffles; similarly, liquid motion is over-powered and throttling valves are used for control. A widely applicable improvement is to employ semiconductor motor controls, such as the alternating-current synthesizer being developed by Exxon, for which it has been estimated that as much as half of a-c motor usage could be economically affected by 1990, with an average energy savings of 30% for the motors affected (BenDaniel and David, 1979).

Energy Management Practices

Most of the 8 to 14% reduction in industrial energy use per unit of aggregate production (1972-78) has been due to shifts in the mix of products (already discussed) and to energy management. Energy management as defined here is a heterogeneous category of low-cost and high pay-off measures. Presumably because of very low energy prices in the past, there are many opportunities to save energy and control energy costs through zero-investment managerial actions or through low-investment high-payoff actions.

Good energy management means planning and goal-setting at high corporate levels and at the plant, organizing, including the appointment of well-qualified staff to energy conservation, effective auditing and reporting procedures, and top management leadership and support. Among specific practices and technical changes are:

(1) Management practices such as:

o inspections to encourage conservation activity;

o training programs for operation of energy-intensive equipment;

o improved maintenance programs;

o accounting procedures to charge energy cost to production departments, not to general overhead, and;

(2) low-level investment programs which may include:

o direct metering of major energy-using facilities;

o sophisticated inspection and maintenance equipment such as infrared scanners;

o implementing sophisticated automatic controls to optimize complex utility systems;

o sophisticated evaluation of downstream operations to optimize use of energy-intensive equipment;

Very interesting technical surveys of low-cost opportunities for efficiency improvement in ten energy-intensive industries were made for FEA (Federal Energy Administration) in 1974-5 leading to the voluntary energy efficiency targets for 1980. As good as much of that work was, the condition of the nation's industrial plants and, therefore, energy management opportunities are highly varied so that without more extensive surveys it must be concluded that it is not known how much energy might be saved by energy-management improvements, nor how much of the opportunity has already been exploited. Well-managed firms with technical expertise have achieved savings on the scale of 20% and more. Further

improvements are requiring more effort and ingenuity, but it is clear that many important oportunities remain in both energy-intensive and light industries. Many of the practices and programs listed above have not been thoroughly implemented even in the best-managed facilities.

INDUSTRIAL ENERGY CONSUMPTION TARGETS

The Approach

Because good data on the details of industrial energy consumption at the point of end use are not available and because industrial decision-making in the specific area of energy productivity investments is not well-understood, it would be inappropriate for this study to determine targets by projecting industrial energy needs over 20 years and more by means of detailed or microeconomic analysis.

Targets developed here are likewise not based on an aggregate macroeconomic analysis, where energy use is projected into the future on the basis of past patterns (which determine a elasticity for energy) and on estimates of future energy prices (Hitch 1977). The elasticities for such a projection are not well enough known to support such an analysis. This study will examine technical opportunities in much more detail than is now possible with macroeconomic analysis.

The efficiency improvement targets set below are estimates based on the data for cost-effective efficiency improvements presented in Section 3 and on judgments of the changes that could reasonably occur within the stated time. The assumptions and calculations are simple and are fully exhibited. The estimates are based on the nation moving effectively toward conditions enabling rational cost-minimization in the industrial sector. Briefly this means that:

o The energy price signals seen by decision-makers continue to move decisively toward the full social cost of energy.

o The financing criteria for energy efficiency and productivity improvement investment be much more even handed vis-a-vis those for energy supply, and capital is available.

o The R&D and marketing of productivity and efficiency improvements proceed vigorously.

o New Federal regulations do not place heavy burdens on efficiency improvement.

The impacts the first three developments would have on the fuel savings associated with cost-minimization are shown qualitatively in Figure 2.9. The figure is designed to show that the cost minimum is broad, i.e. that the optimal level of fuel savings is sen

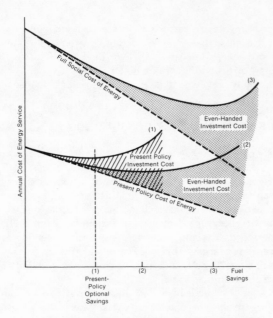

Figure 2.9. Cost of Energy Service with Respect to Increasing Investments to Improve Efficiency (Sketch of Average General Behavior)

Cost of energy service with respect to increasing investments to improve efficiency (sketch of average general behavior). The cost is made up of two parts as shown: cost of energy and cost of investment (shaded). The three cases are: (1) present policy, (2) reduced investment cost associated with even-handed financing criteria and new technology, but continued present energy pricing, and (3) reduced investment cost as in (2) but full social cost pricing of energy.

sitive to decision-making conditions, and thus establishment of the conditions listed could lead to substantial fuel savings such as those targeted in this report. It is not possible to provide quantitative detail for such a figure given present information about current and potential future industrial energy use. Goals shown in this paper are based on an appeal to empirical evidence; they are not based on forecasts of new technology needed to develop figures such as Figure 2.9.

Specific policy changes to achieve or approach conditions that encourage conservation and alternative energy are recommended in the policy section. <u>The energy use targets set here apply if those policies are put in place.</u> Of special importance are: (1) the Scrap and Build Program for energy-intensive industries, which brings the subsidies the nation is offering to energy supply industries to bear on saving energy; (2) a greatly expanded venture capital program to speed the construction and evaluation of prototypes of new equipment and processes; (3) further moves toward rationalization of energy prices; (4) a vigorous research program aimed at basic-materials processes and generic phenomena; and (5) programs to assist less energy-intensive businesses financially and technically. Explicit calculations will be made in this section relating to the creation of new industrial facilities through the Scrap and Build Program.

In setting the efficiency;mprovement targets, we assume a lag in innovation and capital turnover. Roughly one half to two thirds of industrial facilities (in the basic-materials industries) are assumed to be new (post-1978) in the year 2000. From the discussion

of capital investment in Section 2 it is clear that this level of investment is not occurring under present circumstances, but that it is reasonable to argue that such investments are cost effective.

The procedure we adopt for the industrial sector is to consider energy use as the sum of terms of the form:

$$\begin{matrix} \text{energy} \\ \text{consumption} \end{matrix} = \begin{matrix} \text{energy} \\ \text{service} \\ \text{demand} \end{matrix} \times \begin{matrix} \text{energy} \\ \text{intensity} \end{matrix}$$

We then separately project service demands and target energy intensity improvements.

Projection of Service Demands

Goals are established for each service category discussed in Section 1 by associating each with a single basic indicator. In the case of agriculture, the basic indicator selected is population growth, since it appears that agricultural production is more closely related to total population than to GNP. Industrial production is linked indirectly to GNP growth, on the other hand. Service demands are assumed to be related to GNP by extending certain past trends.

In Section 2 the past trend in industrial value added with respect to GNP was examined (Figure 2.3) and regression fits determined. On this basis, it is assumed that the ratio of industrial value added (Gross Product Originating) to GNP will continue to decline by 0.8% per year following the trend which has held since 1968. It is reasonable to extend this trend to the year 2000 because the gradual shift in the economy from goods to services, which explains most of this past trend, shows no sign of diminishing. Trends in consumption of basic materials in physical units were also examined in Section 2. In establishing goals it is assumed that the ratio of basic materials production to industrial value added will continue to decline at 0.5% per year following the 1969-1978 trend. (This becomes a 1.3% p.a. decline with respect to GNP.) It is reasonable to extend this trend to the year 2000 because the shift from materials-intensive industrial production to lighter more sophisticated products appears likely to continue. The decision to use only recent trends in materials use was made deliberately. Higher growth, projection would result if goals were based on 20-30 year trends in national use of materials. Such a projection would be unrealistic however, since the real price increases for basic materials and materials saturation effects which characterized the 1970's more than the 1950's and '60's appear likely to characterize the 1980's and 1990's.

Thus we take

| Growth rate for industrial value added | = | growth rate for GNP | -0.8% p.a. |

and

| growth rate for physical units of basic materials | = | growth rate for GNP | -1.3% p.a. |

Likewise , we take

| growth rate for crop production | = | growth rate for population | +1.0% p.a. |
| growth rate for livestock production | = | growth rate for population | |

A uniform set of assumptions has been adopted for GNP and population growth in this Study. Briefly, real GNP is assumed to grow at an annual rate of 2.7% between 1978 and 1990 and at an annual rate of 2.3% between 1990 and 2000. The average annual rate of growth over the entire period 1978-2000 is approximately 2.5% p.a. Total U.S. population growth is assumed to 0.8% p.a. between 1978 and 1990, and 0.6% p.a. from 1990 to 2000.

Narrow choices for energy service such as those adopted here (like process heat and mechanical drive) could lead to overlooking important avenues of technical change. Thus the projection of service demands on the basis of historical trends overlooks some of the responses to higher energy prices and energy shortages in the areas of product mix and design and fundamental process change. Although no attempt is made to deal with product change (except by extrapolating the past declining trend in the materials to production ratio), an assessment of the impact of process change in the case of the biggest energy service, process heat, is attempted. This is done by considering, below, the energy and capital implications of creating new facilities in contrast to modifying existing ones. (The information is not available at this time to disaggregate these energy implications in terms of a reduction in process heat service requirements and a reduction in the energy intensity (of providing process heat, so the overall performance improvement is accounted for below as an energy intensity reduction.)

One problem with narrowly defined energy services is that a reduction in demand for one kind of service may increase demands for another category of energy service. For example, chemical reduction of aluminum may substitute for electrolytic reduction or mechanical drive operations may substitute for process heat. No attempt has been made to consider such trade-offs explicitly in this analysis.

Targets for Reduction of Energy Intensity

The targets adopted for average annual rate of reduction of energy intensity for energy service and the corresponding 22-year reduction in energy intensity per unit of product are shown in Table 3.10 for industry (excluding agriculture) and in Table 3.11 for agriculture. The characteristics of the target-setting are discussed in terms of the targets for three important categories involving three fourths of industrial energy use in the following subsections.

Process Heat

The target is to reduce the energy intensity of process heat by 2.0% per year in the period 1978-2000. Because of the significance of industrial process heat, process-heat energy intensity will be considered separately for: (1) new facilities using state-of-the-art processes and equipment, (2) old facilities with rebuilt equipment, and (3) old equipment with add-on energy efficiency improvements. Strong evidence indicates the importance to efficiency as well as total productivity of the new process and of thoroughgoing implementation of modern control systems which will characterize new facilities. Moreover, evidence is strong that only limited exploitation of such tools as increased heat recovery can occur in retrofitting existing energy-conversion equipment because of typically high costs of installation. Thus, the scope for energy efficiency retrofits, once the cream has been skimmed, i.e. simple energy management improvements have been made, is limited compared to the improvements which can be made in a new facility or even with completely rebuilt equipment.

In the case of process heat, the reductions in energy intensity that would occur in the absence of new initiatives to accelerate them, are estimated as well as the reductions in energy intensity that would occur

Table 2.23. INDUSTRIAL ENERGY CONSUMPTION GROWTH INDICES, BY SERVICE CATEGORIES, 1978-2000

Service Category	Growth Indicator	Growth in Energy Service		Efficiency Improvement Target (% p.a.)	Energy Intensity in 2000 Relative to 1978	Composite Growth Index	
		1978 - 1990 (% p.a.)	1990 - 2000 (% p.a.)			1990 - 2000 (% p.a.)	1990 - 2000 (% p.a.)
Mechanical Drive	IVA[a]	1.9	1.5	1.2	0.77	0.7	0.3
Electrolysis	GNP[b]	2.7	2.3	1.4	0.74	1.3	0.9
Process Heat	Materials[c]	1.4	1.0	2.0	0.65	-0.6	-1.0
Space Heat	IVA	1.9	1.5	2.3	0.61	-0.4	-0.8
Chemical Feedstock	Materials	1.4	1.0	0.75	0.85	0.65	0.25
Construction Asphalt	GNP	2.7	2.3	0.3	0.94	2.4	2.0
Metallurgical Coal	Materials	1.4	1.0	0.75	0.85	0.65	0.25
Repressurizing Gas	Gas Production	-1.0	-1.5	-	1.00	-1.0	-1.5
Vehicles	GNP	2.7	2.3	1.2	0.77	1.5	1.1

[a] Indicates growth of industrial value added, U.S. average, taken to be GNP growth rate - 1% per annum.

[b] GNP is projected to grow at 2.7% per year from 1978-1990 and at 2.3% per year from 1990-2000, for average of 2.5% per year for the period 1978-1990.

[c] Growth in physical units of basic materials; taken to be GNP growth rate - 1.5% per annum.

[d] These improvements do not take into account changes associated with shifts in fuel types. For example, if lower quality fuels (such as organic wastes or coal) are substituted for a high quality fuel (such as natural gas), additional fuel may be required to produce process heat. This effect is not taken into account here. This will be a minor effect in the scenarios evaluated in this study.

Table 2.24 AGRICULTURAL ENERGY CONSUMPTION GROWTH INDICES, BY SERVICE CATEGORIES, 1978 - 2000

Service Category	Growth Indicator[a]	Growth in Energy Services		Efficiency Improvement[c] Target (% p.a.)	Composite Growth Index		Energy Intensity in 2000 Relative to 1978
		1978 - 1990 (% p.a.)	1990 - 2000 (% p.a.)		1978 - 1990 (% p.a.)	1990 - 2000 (% p.a.)	
Space and Water Heat	Livestock Production	0.8	0.6	1.3	-0.5	-0.7	0.75
Grain and Crop Drying	Crop Production	1.8	1.6	1.3	0.5	0.3	0.75
Lighting and Refrigeration	Livestock Production	0.8	0.6	1.2	-0.4	-0.6	0.77
Irrigation	Crop Production	1.8	1.6	2.3	-0.5	-0.7	0.60
Mechanical Drive (stationary)	Crop Production	1.8	1.6	1.2	0.6	0.4	0.77
Vehicles and Field Machinery	Crop Production	1.8	1.6	1.3	0.5	0.3	0.75

[a] Indicates growth in physical units of N.S. agricultural Production.

[b] Efficiency Improvement Tartets are annual average compound rates of decrease in energy used per unit of production based upon general estimates given in Friedrich, R.A. Energy Conservation for Americal Agriculture. 1978. Cambridge, MA: Ballinger.

under the policies recommended in this Study. Table 2.25 estimates in terms of four energy-intensive industries the energy-intensity reduction factors (1978-2000) for process heating in two cases: a "normal" or present-policy course and "accelerated" conservation associated with the policies spelled out in the Policy Section of this report. In both these cases a pessimistic "retirement" rate of 2.5% per annum is assumed. (Retirement corresponds to what are called here refurbished facilities. See discussion of Capital in Section 2.) Major pre-1978 vintage equipment is assumed to be rebuilt at this linear rate (2.5% of 1978 capacity per year for 22 years). In the normal course, new capacity is created at the capacity growth rate, i.e., it is assumed that there is no scrapping of facilities. In the accelerated case, additional new facilities are built at the linear rate of 2.25% of 1978 capacity for 20 years in connection with the Scrap and Build Program described in the Policy Section. With this program, all pre-1978 vintage capacity is either rebuilt or scrapped and replaced with new plants by the year 2000. (See Note 8, Table 2.25 for a worked example.)

The average energy-intensity reduction factors assumed for new facilities for the period 1978-2000 are 0.60 and 0.70 compared to 1978 average performance, in the accelerated and normal cases, respectively. The average 40% reduction assumed for new state-of-the-art facilities in the fuel required for process heating per unit of production is justified by performance achievements with modern mills as presented in the process change subsection in Section 3. These references typically apply to fuel use for a number of energy services including process heat; the relative reductions for process heating tend to be significantly larger than the overall reductions (see below), so the estimated 40% reduction in process heat intensity may even be modest. The 30% reduction in the normal course in this and the other cases below are rough judgmental estimates of the long-term differences between the two cases, due to the differences in the levels of capital formation, venture capital activity, energy prices and research.

The energy-intensity reduction factors for facilities with rebuilt equipment (0.73 and 0.83 for accelerated and normal cases, respectively) are estimated by considering a variety of improvements including: (1) management (e.g., controlled turning on and off of equipment and improved maintenance); (2) furnace improvement (e.g., improved insulation and air/fuel controls); (3) waste heat recovery; and (4) more sophisticated controls (e.g., improved flame control, opti-

mized part-load operation, and optimized operation with respect to downstream needs). In a typical situation these types of improvements might lead to the following sequential savings for the accelerated case: management 5%, furnace improvement 5%, waste-heat recovery 10% and sophisticated controls 10% to obtain an overall savings of 27% by the year 2000. (This estimated intensity reduction factor is an average; it will vary from case to case.) The postulated 10% savings resulting from sophisticated automatic controls (discussed in Section 3) is probably much smaller than the potential. However the full potential will not be reached in refurbished plants. One check can be made on the waste-heat recovery component of this target. Of the almost 12 quads of fuel used for process heat in 1976, 3.6 quads or 30% are estimated to have been lost in the stack gas at temperatures over $300^{\circ}F$ (1.8 quads at 300° to $500^{\circ}F$, 0.9 quads at $500^{\circ}F$ to $1000^{\circ}F$, and 0.9 quads above $1000^{\circ}F$). (General Energy Associates 1979) Savings of 10% would correspond to one third the total, or about 1.2 quads.* Recovery of an additional 1/3 of waste heat in stack gas is a reasonable target. See Section 3.

The capital cost of the Scrap and Build Program in the four sample energy-intensive industries is presented in Table 2.26.

The combined impact of improvements arising from new and refurbished facilities in the accelerated conservation case for the 4 selected industries is an energy-intensity reduction of 35% for process heat (weighted average of final column of Table 2.25) or a reduction of 2.0% per annum. We adopt this result for industry as a whole.

The 22-year goal of a 35% reduction in the fuel needed to accomplish an average unit of process heating does not take into account changes associated with fuel type. If lower quality fuels such as organic wastes or coal are substituted for natural gas, increased Btu content will be required. This effect is not included. Similarly if synfuels should be employed, the conversion losses, e.g., converting coal to medium-Btu gas, are not included.** The aggregate result presented here would apply if the dominant fuel for process heating remains natural gas. This assumption is not unreasonable for this overall study since, if the targets set in this study are approached, industrial fuel consumption is likely to be dominated by natural gas and direct use of coal, and the introduction of synthetics is likely to be very gradual.

The energy-intensity reduction target we have set forth for process heating (2.0%/year, 1978-2000) can

---

*In addition, waste heat may be recoverable from product and by-product streams.

**Upon making particular assumptions for the mix of energy carriers one could translate the goal given here into a projection for primary fuel by assuming efficiency changes associated with moving from the present carrier mix to the projected mix.

Table 2.25 ENERGY INTENSITY REDUCTION FACTORS FOR PROCESS HEAT, 1978-2000, IN CERTAIN BASIC MATERIALS INDUSTRIES

Industry	Projected Growth Rate for Capacity[1] (%p.a.)	"Normal Projection"			Accelerated Conservation Including Scrap & Build	
		Refurbished Capacity[5] fraction in 2000	New Capacity[6] fraction in 2000	Energy Intensity reduction factor[7]	New Capacity[6] fraction in 2000	Energy Intensity Reduction Factor[7]
Paper	2.0[2]	.36	.35	.80	.64[8]	.65[8]
Chemicals	2.0[3]	.36	.35	.80	.64	.65
Cement	-0.5[3]	.50	0	.86	.50	.66
Steel	0.4[4]	.55	0	.86	.45	.67

1. The GNP growth rate is projected to be 2.5% p.a. and the recent growth rate for domestic consumption in physical terms is shown in Table 2.21 for a period when the GNP growth rate was 2.9% p.a. The projected "basic" capacity growth rates are obtained by subtracting the difference between these GNP growth rates, 0.4% p.a., from the consumption growth rates shown in Table 2.21. These are then modified on an industry by industry basis as discussed in notes 2 through 5.

2. The basic domestic consumption growth rate in physical terms for paper is 1.0% p.a. Because recent capacity utilization is high, raw materials costs will probably be relatively low and export markets are promising, we make a 2.0% p.a. projection for capacity growth.

3. The basic domestic consumption growth rate in physical terms for chemicals is 2.0% p.a. Potential import competition and increasing raw materials costs suggest a figure near this level. We make a 2.0% p.a. projection for capacity growth. Note that growth in economic terms may be much faster because the product mix in chemicals is shifting "downstream" toward more processing; energy use is dominated by upstream processing, however. The basic domestic consumption growth rate in physical terms for cement (-0.5%) is also adopted for capacity growth.

4. The basic domestic consumption growth rate in physical terms for steel is 0.3% p.a. Strong import competition suggests a capacity growth rate projection of 0% p.a. For comparison the American Iron and Steel Institute calls for a 0.6% p.a. growth in raw steel capacity 1978-1988 on its "Aggressive Revitalization" scenario in "Steel at the Crossroads: The American Steel Industry in the 1980's". Note that GNP is projected to grow slightly more slowly in the entire 1978-2000 period than in the first decade (0.2% p.a. more slowly) because of demographic factors, so the AISI projection is only slightly higher than the projection used here.

5. This is the pre-1978 capacity which is refurbished (i.e. all major equipment is rebuilt), based on an assumed rate of 2.5% p.a. It is reduced in the case of cement to account for the decrease in total capacity.

6. The fraction of new capacity in 2000 which is post 1978 vintage is $((1+G)^{22}-S+20)/(1+G)^{22}$ where G is the average annual growth rate for 22 years, is the annual scrap and build rate (if applicable) assumed to apply for 20 years. See Note 8 for sample calculation.

7. The energy intensity reduction factors are 0.6 and 0.7 for new capacity, and 0.73 and 0.83 for capacity with rebuilt equipment, in the accelerated and normal programs, respectively. In the normal projection the residual pre-1978 capacity is retrofitted to 0.90 the energy intensity which applied in 1978.

8. As an example of the calculation we take the paper industry in the accelerated case. The results are calculated as follows: "Retirement" (equipment rebuilding without change of capacity) is assumed to go on at 2.5% p.a. of the 1978 capacity, it encompasses $0.25 \times 22 = 55\%$ of the 1978 capacity by 2000. The fraction of year 2000 capacity which is new is the sum of capacity additions plus scrap and build capacity all divided by total capacity, i.e. $(.546+.45)/1.546 = 0.64$. The energy intensity reduction factor is 0.6 for the new capacity fraction and 0.73 for the refurbished, or $0.6 \times .64 + 0.73 \times .36 = 0.65$.

Table 2.26 GROSS COST OF SCRAP & BUILD PROGRAM FOR CERTAIN BASIC MATERIALS INDUSTRIES

Industry	1978 Capacity (million tons/yr)	Unit Cost (1978$/ton/yr)	Total Scrap & Build Investment[1] (billion 1978$)
Paper	67	600	18
Chemicals	100*	1100*	50
Cement	56	115	3
Steel	115	850	44
TOTAL			115

*Nominal physical capacity of 100 units in 1978 and 1.1 billion dollars capital cost per new unit.

1. The investment by year 2000 is for 45% of 1978 capacity at the unit cost quoted in the second column.

be put into perspective by looking at trends in the chemical industry. In the period 1972-1978, the process heating fuel required to produce a pound of product was reduced at an annual rate of over 4% per year.** Moreover, projections into the mid 1980's by individual firms indicate a continuing rapid rate of decline in energy intensity. Union Carbide (Figure 2.8) projects that its total energy use excluding feedstocks per pound of product will decline 2.7% per year, 1978-1985, while Dupont projects that its energy use per pound will decrease 2.3% per year, 1979-1986 (Energy Daily, May 12, 1980). Fuel requirements for process heat alone would undoubtedly be reduced more rapidly than these energy requirements.** In terms of these current trends the energy intensity reduction target associated with process heat is seen to be modest.

Mechanical Drive and Electrolysis

Most industrial electricity consumption is for mechanical drive and for electrolytic processes. While 70% of mechanical-drive services are concentrated in the six basic-materials industries, continuing electrification leads us to project the demand for mechanical-drive services to grow, as industrial value added (GPO), 0.5% per year faster than basic materials production. Also the demand for electricity services for electrolytic processes is projected to grow as GNP, to reflect the growing importance of such processes in industry. (Between 1969 and 1978, U.S. aluminum production increased at 2.7%/year, only slightly slower than the GNP growth rate, while industrial GPO grew at only 1.8%/year).

*This estimate is obtained as follows. According to data provided by the chemical Manufacturers Association (CMA) non-electrical energy consumption (other than feedstock energy) in the chemical industry fell from 2.74 Quads in 1972 to 2.46 Quads in 1978. Some of this (0.27 Quads in 1976, according to Resource Planning Associates 1978) was used for onsite generation of electricity and shaft power. However, according to the Annual Survey of Manufacturers, onsite generation of electricity declined at an annual rate of 6.2%/year, 1971-1977. Assuming this rate applies for the period 1972-1978, one obtains fuel requirements for process heat declining from 2.39 Quads in 1972 to 2.22 Quads in 1978. CMA data also indicates a 19% increase in output, 1972-1978. Thus fuel requirements per lb. of product for process heat declined at an annual average rate of 4.2%/year.

**For the chemical industry as a whole energy requirements were reduced 3.5% per year per pound while process heat requirements fell 4.2% per pound in the period 1972 to 1978, according to CMA data.

Only relatively modest motor efficiency improvements are possible. (See Table 2.27). But an average level of savings on the order of 5% could easily be realized by the year 2000, because of the relatively short lifetimes of most industrial motors.*

Far greater savings can be achieved with systems improvements, as discussed in Section 3. In making a projection to the year 2000, it is assumed that there will be a 5% savings due to improved motor efficiency, a 5% savings arising from better management (e.g., control of off-and-on timing, and improved maintenance), and a 15% savings associated with systems improvements (taken in sequence). The projection therefore shows a 23% savings for the year 2000 relative to 1978, corresponding to a 1.2% per year reduction in the energy intensity of mechanical drive in the period 1978-2000 (see Table 2.23).

This process of estimation should not be interpreted as a literal model of the change that will occur. The structure of change will be complex, as suggested by the discussion of Section 3: Energy management at plants will be improved. Much equipment at existing facilities will be replaced or retrofitted, but with less efficiency improvement than just indicated. Finally, new facilities will be built incorporating state-of-the-art production processes, in which mechanical-drive efficiency improvements, as just indicated, and extensive process improvements will be achieved. (More than half of the basic-materials production facilities in the year 2000 are projected to be post-1978 vintage.)

Among electrolytic processes, primary aluminum production is the major electricity consumer. A major opportunity exists to reduce electricity use in aluminum production from the 1976 level of 7.5 kwh per pound to 5 kwh per pound, by shifting from the predominant Hall-Heroult process (introduced in the late 1880's) to the new Alcoa smelting process.

The prospect of drastically higher electricity prices for the primary aluminum industry, and for other electrolytic industries as well, is a powerfull incentive to shift to more efficient technologies such as the Alcoa process. The average price of electricity to the aluminum industry was 1.05¢/kwh and that to the inorganic chemical industry was 1.56¢/kwh in 1977 (in 1978 $). These prices were far below the average for all of industry, about 2.50¢/kwh, which in turn is only half the average replacement cost. Since replacement costs vary little from region to region, rising electricity prices will be especially burdensome on electrolytic industries. Marked efficiency improvements are essential to the continued well being of these industries. In this analysis we assume an overall reduction of 25% in energy use per unit of output in the electrolytic industries by the year 2000, corresponding to an annual reduction in energy intensity of 1.3% per year, 1978-2000 (see Table 2.23).**

The energy-efficiency targets set forth here for industrial electricity use imply that the amount of electricity used to generate a unit of industrial value added would be reduced 22% between 1978 and 2000 (see Figure 2.3). This means that the long-term trend toward increasing electricity intensity in industry would be reversed, a trend reversal which actually began after 1976, when electricity use per real dollar of output began to decline after reaching a peak which involved twice as much electricity per dollar as in 1950 (see Figure 2.3.). Despite this reversal, relative expenditures on electricity would increase dramatically with this projection, because of rising electricity prices.

From the 1950's through the early 1970's, industrial expenditures on electricity, expressed as a percentage of industrial output remained remarkably constant at about 2%. Then with the sharp 60% real electricity price increase between 1970 and 1978 this percentage shot up to 3%. Because we expect another 100% real increase in industrial electricity prices by 2000, electricity expenditures can be expected to continue to rise to a level equivalent to 5% of industrial output. (See Figure 2.3). Thus rising electricity prices will be a powerful incentive to make electricity efficiency improvements. The efficiency targets set forth here may well be too modest.

Industrial Cogeneration

Not included among the energy-efficiency improvements targets for the industrial sector shown in Table 2.23 are the potential savings associated with cogeneration. Fuel savings of 15-35% are possible with cogen

---

*While very large motors (above 50 hp) can "go on forever," it appears that most small and medium-sized motors are relatively short lived. According to a recent study by A.D. Little, "there appears to be a consensus that in the small and medium-size motors and driven machines, the average useful life is in the range of 7 to 10 years. However, it appears that the process industries generally expect a somewhat lower life expectancy based on the adverse conditions under which the motors must operate and the rapidity of process obsolescence. It is not uncommon for motors and their associated driven equipment to become obsolete in 3 to 5 years in a fast-moving process industry. In some cases the motors are salvaged and put into a pool, but in a number of instances the entire package of machinery is scrapped." (A.D. Little 1978)

**A different but reasonable scenario is that domestic aluminum smelting will grow much more slowly than projected here because of the high replacement cost for electricity.

Table 2.27 CURRENT AND FUTURE MOTOR EFFICIENCIES*

HP Category	Ave. Efficiency Now	Potential Efficiency Next 2-10 Years	Potential Energy Saving
1	65.0	75.0	0.15
1-5	76.5	85.5	0.12
5.1-20	82.5	89.5	0.09
21-50	87.5	92.5	0.06
51-125	91.0	93.0	0.02
125	94.0	94.2	0.0002

*Integral AC Polyphase Motors.

Source: (A.D. Little 1978).

eration, relative to the fuel requirements for the production of heat and electricity in separate facilities.

Cogeneration is an old technology that is about to enter a period of growth. While production of electricity at industrial establishments has been declining, a renewed interest in cogeneration can be expected because the economics have become more attractive.

The drift away from cogeneration in the past was caused in part by the increasing attractiveness of central-station power generation. As shown in Figure 2.3, the price of central-station electricity was reduced nearly in half between 1950 and 1970. But since 1970, the price of electricity has been rising. This rising price trend can be expected to continue, since the replacement cost of electricity to industrial customers is about twice the present average price of industrial electricity and is three to four times the present price for some industries (see Table 2.18).

The economics of cogeneration appear particularly attractive in the basic-materials industries. These industries tend to be capital-intensive, with year-round, round-the-clock operations. Cogeneration in these industries would thus involve primarily the generation of baseload electricity. Under a wide range of conditions the fuel savings associated with cogeneration in these circumstances would result in a busbar cost lower than the cost of electricity delivered to an industrial customer from a new central station baseload plant; in some favorable circumstances the busbar cost of cogenerated electricity would be less than the busbar cost of electricity from a new central-station baseload power plant as well.

The most familiar cogeneration technology involves the steam-turbine. With this system high-pressure steam is generated and used to drive a steam-turbine for generation of shaft power or electricity, and low-pressure steam is exhausted or diverted from the turbine at the temperature and pressure appropriate for industrial process applications. Alternatively, with a system based on a gas turbine, a spark-ignited internal combustion engine, a diesel engine, or a dual fuel engine, combustion gases are used to drive the engine for power generation, and the hot engine-exhaust gases are used to raise steam for process use in a waste-heat boiler. In a combined-cycle unit (a turbine gas/steam turbine combination) the steam generated in the waste-heat boiler is used to drive a steam turbine, from which the exhaust steam is delivered to industrial processes. In all cases the amount of fuel needed to produce the electricity, beyond what would be needed to produce steam alone, is approximately one half of what would be needed at a conventional central-station power plant.

If cogeneration is to become a major fuel-saving option, cogeneration technologies characterized by a high electricity-to-steam output ratio (E/S ratio) must be encouraged.* For a given process steam load, high E/S ratio technologies (gas turbine/waste heat boiler units; gas turbine/steam turbine combined cycles; internal combustion engine/waste-heat boiler units) can generate 4-8 times as much electricity and 3-5 times as much primary fuel savings as the more familiar low E/S ratio cogeneration technology, the steam turbine unit.

While such high E/S ratio technologies can eventually be fired directly by coal or biomass (e.g., using pressurized fluidized-bed combustors) or by low- or medium-Btu gas or other synthetic fuels derived from

---

*More generally, one should speak of a high ratio of electrical or mechanical energy to heat output. Some cogeneration units produce mechanical instead of electrical power, and sometimes hot water or hot gas is delivered to process instead of steam. In what follows here, high E/S ratio systems should be interpreted in this more general sense.

coal or biomass, these technologies today must be fired with oil or gas. An evolutionary strategy to maximize the long-run benefits of cogeneration would involve gaining experience now with high E/S ratio technologies based on oil and gas and shifting to low-quality fuels as advanced technologies become available. (Alley, 1980)

Table 2.4, shown previously, indicates what the potential for cogeneration would be in 6 major steam-using industries if high E/S ratio cogeneration technologies were deployed* for the steam loads shown in Table 2.28 that are technically suitable for association with cogeneration, at the 1976 levels of steam demand. The cogeneration potential in these six industries would result in annual fuel savings of about 2 Quads through the displacement of almost 100 Gw(e) of baseload electricity from conventional central-station power plants--corresponding to approximately one quarter of U.S. electricity production in 1979.**

Over the medium term (roughly a decade) this evolutionary strategy would have major oil- and gas-savings benefits, as high E/S ratio cogeneration technologies utilizing coal- or biomass- or waste-based fuels become available. But even in the near term, this strategy would lead to net oil and gas savings. This prospect arises because oil- and gas-based cogeneration would be economically attractive today in those areas where oil and gas are major fuels for central station power generation.*** If oil- or gas-based industrial boilers in such areas are replaced by oil-or gas-based cogeneration units, substantial oil or gas savings could result through the displacement of conventional central station electricity based on oil or gas. The dozen or so states where half or more of the electricity is oil or gas based offer the best opportunities for high E/S ratio oil or gas based cogeneration in the near term. (See Table 2.29) Very high levels of cogeneration could be developed in the near term with net oil and gas savings in those states that are also

highly dependent on oil and gas for industrial boiler fuel (e.g., Louisiana, Texas, California). In regions served by utilities which burn oil and gas to generate baseload electricity, the implementation of cogeneration by industry with the sale of electricity to the utility at its "avoided cost," under Section 210 of PURPA, will probably be highly advantageous. It would also probably be advantageous for such utilities to participate in ownership and management of cogeneration facilities. It is to be hoped that this opportunity will lead to orderly implementation of cogeneration in terms of electrical needs and will lead to full exploitation of cogeneration resources through timely adoption of high E/S technologies.

Because there would be only very slow, zero, or even negative net growth in electricity demand throughout the economy if the conservation measures targeted in this study are achieved in all sectors, the full 100 Gw(e) cogeneration potential targeted here would probably not be needed by the year 2000. Nevertheless, it is important to adopt a cogeneration policy that would facilitate the evolutionary strategy set forth here, gaining experience now with high E/S-ratio cogeneration technologies based on oil and gas to facilitate a transition to high E/S-ratio technologies based on low-quality fuels. Not only would such a policy help cogeneration evolve into a major fuel-saving option for the long term, but it would also facilitate the creation of a "cogeneration electrical reserve." While the present analysis indicates that the full 100 Gw(e) of potential cogeneration capacity would probably not be needed by 2000, demand could grow faster than projected in this report because of unforseen developments. Cogeneration should be a favored technology for coping with the uncertainties in demand forcasts, because in the event of an unexpected surge in demand, cogeneration capacity could be brought on-line much more quickly (and in much

---

*The potential is estimated for an average E/S ratio of 200 kwh/10^6 Btu--a characteristic value for a gas turbine system. For a steam turbine system the E/S ratio would be much lower - typically 50 kwh/10^2 Btu. But it could be 300 kwh/10^6 Btu for a combined cycle unit of up to 400 kwh/10^6Btu for a diesel unit. In practice of course there would be a mix of technologies, for which the E/S ratio could be on the order of 200 kwh/10^6 Btu if high E/S ratio technologies were encouraged.

**The electricity that potentially can be produced in these industries via cogeneration, some 570 x 10^9 kwh, represents an increase of about 530 x 10^9 kwh over the 1976 level of cogeneration (assuming about half of the mechanical and electrical power produced onsite in 1976 was cogenerated (see Table 3.15).) This is the amount of electricity that could be produced by 93 Gw(e) of central station baseload generating capacity, operating at an average capacity factor of 65%. U.S. electricity consumption in 1979 was 2075 x 10^9 kwh.

***The economics of high E/S ratio cogeneration (which often involves electricity production far in excess of onsite needs - see Table 2.4) are sensitive to the price a cogenerator can obtain for the sale of excess electricity. According to Section 210 of PURPA this price must be equal to the utility's avoided cost. Because of the excess generating capacity problem in most parts of the country today, the utility's avoided costs are high today primarily in areas heavily dependent on oil and gas for central power generation.

Table 2.28 PROCESS STEAM LOAD IN 1976 TECHNICALLY SUITABLE FOR COGENERATION IN SIX MAJOR INDUSTRIES

(10^{12} Btu/year)

	Food	Textiles	Pulp and Paper	Chemical	Petroleum Refining	Steel	6 Industry Totals
A. Process Steam Load[a]	307	125	1210	1516	584	366	4108
Load Fluctuations	(23)	(10)	(94)	(92)	(35)	(12)	(266)
Low Steam Load	(52)	(24)	(85)	(74)	(0)	(0)	(235)
Low Pressure Waste Heat Steam	(8)	(0)	(0)	(143)	(132)	(35)	(318)
Declining Demand	(26)	(16)	(172)	(148)	(53)	(19)	(434)
B. Steam Load Technically Suitable for Cogeneration[b]	198	75	859	1059	364	300	2855
C. Steam Load Presently Associated with Cogeneration[a]							
Electrical	27	15	236	190	15	101	584
Mechanical	41	6	17	143	41	106	354
Total	68	21	253	333	56	207	938
D. Percentage of Suitable Steam Load Presently Associated with Cogeneration (100 x C/B)	34%	28%	29%	31%	15%	69%	33%
E. Distribution of Existing Cogenerated Steam Load (Electrical)[a]							
Steam Turbine							
Coal	10.5	3.0	35.4	38.0	--	10.1	97.0 (17%)
Resid	14.6	10.8	59.0	95.0	9.8	15.2	204.4 (35%)
Waste Fuel	--	--	118.0		3.0	60.5	181.5 (31%)
Heat Recovery	--	--	--	9.5	0.8	5.2	15.5 (3%)
SUBTOTAL	25.1	13.8	212.4	142.5	13.6	91.0	498.4 (85%)
Gas Turbine							
Distillate	--	0.3	--	19.0	0.7	--	20.0 (3%)
Natural Gas	1.9	0.9	23.6	28.5	0.7	10.0	65.6 (11%)
SUBTOTAL	1.9	1.2	23.6	47.5	1.4	10.0	85.6 (15%)
Total	27.0	15.0	236.0	190.0	15.0	101.0	584.0 (100%)

[a]The numbers presented here are obtained from Peter Bos et. al, "The Potential for Cogeneration Development in Six Major Industries by 1985," prepared by Resource Planning Associates, Inc., for the Department of Energy, December 1977. The various technical factors judged by RPA to reduce the size of the process steam load suitable for association with cogeneration are as follows:

*load fluctuations: arises because electric power and steam demands are out of phase. This factor was important in the RPA analysis because in that study the export to the utility grid of power produced in excess of onsite needs was not considered. Using the RPA numbers understates the cogeneration potential in the present analysis, where export is allowed, because steam and electric power demands do not have to be in phase to cogeneration when the cogeneration unit is tied to the utility grid.

*low steam load: for steam loads less than 50,000 lb. of steam per hour.

*low pressure waste heat steam.

*declining demand: based on management expectations for future reduced steam demand because of conservation, process or product changes, plant obsolescence or shutdown.

These factors collectively reduce the steam base technically suitable for cogeneration by 31% or 1253 x 10^{12} Btu. RPA also excluded from consideration for cogeneration another 5% or 212 x 10^{12} Btu of the steam base because cogeneration based on this steam base (even with steam Rankine cycles) would have resulted in more electricity than could be consumed onsite. This restriction is unappropriate for the present analysis, where export of excess electricity to the utility grid is allowed.

[b]This includes the steam load presently associated with cogeneration.

Table 2.29   OIL AND GAS USE FOR ELECTRICITY GENERATION AND INDUSTRIAL COM-
BUSTORS IN MAJOR OIL AND GAS USING STATES

	% of Electricity in State Generated By Oil and Gas (1978 [a]	% of Total U.S. Electricity Generated by Oil and Gas Gas (1978)[a]	Oil and Gas used in Large Industrial Combustors as a percentage of the U.S. Total (1974)[b]
1. Louisiana	100	7.7	10.8
2. Oklahoma	88	5.3	1.4
3. Massachusetts	84	4.6	0.5
4. Texas	82	22.3	24.0
5. Mississippi	81	2.3	2.2
6. Delaware	71	0.8	0.6
7. California	65	13.5	5.8
8. Florida	65	9.0	2.0
9. New Jersey	55	2.5	2.5
10. Kansas	50	1.8	0.9
11. Arkansas	48	1.3	1.2
12. New York	46	7.7	2.1
13. Connecticut	45	1.7	0.4
14. Virginia	37	2.3	1.4
15. Maryland	32	1.7	1.4
16. Michigan	17	1.9	4.2
17. Pensylvania	13	2.4	4.8
18. Illinois	10	1.5	4.3
	Subtotal	90.3%	Subtotal 70.6%

[a]Edison Electric Institute, Statistical Year Book of the Electric Utility Industry for 1978, November 1979. In 1978 670 x 10^9 kWh were generated from oil and gas sources.

[b]Large combustors are those with a capacity of 100 mllion Btu per hour or more. Oil and gas use in such combustors was 4.8 x 10^{15} Btu in 1974. On this 2.7 x 10^{15} Btu was consumed by boilers. See M. H. Farmer et al, "Application of Fluidized Bed Technology to Industrial Boilers," report to the Federal Energy Administration by Government Research Laboratories, Exxon Research and Engineering Company, 1976.

smaller increments) than conventional central-station generating capacity.

CONCLUSIONS

If the conservation targets set forth above for the industrial sector are met, the result would be (Table 2.30) essentially constant energy demand in the period 1978-2000. Agricultural requirements, shown in Table 2.31, increase slightly.

If the targets set in this study are approached, then the more expensive substitutes for the load now carried by oil and gas, e.g., electricity or synfuels, will be slow to assume the load because less costly domestic resources, especially natural gas, will be able to cover demands for a much longer period of time than has been anticipated. It seems likely that if these targets are approached that natural gas could remain a significant industrial fuel.

Table 2.30   INDUSTRIAL ENERGY CONSUMPTION TARGETS BY SERVICE CATEGORY[a] (Quads)

Service Category[b]	1978 Base Consumption	Energy Consumption Targets[c,d] 2000
Mechanical Drive	7.8	8.8
Electrolysis	1.5	1.9
Process Heat	12.5	10.6
Space Heat	0.4	0.3
Chemical Feedstock	2.6	2.9
Construction Asphalt	0.9	1.5
Metallurgical Coal	2.1	2.3
Repressurizing Gas	1.1	0.8
Vehicles	0.2	0.3
	29.1	29.4

[a]Manufacturing mining and construction. Includes conversion and transmission losses associated with purchased electricity.

[b]Gasoline - powered mechanical drive and mobile equipment is excluded. The figure for vehicles includes only off-road vehiclesa nd diesel fuel.

[c]See Table 3.10 for the composite growth index used for calculating targets.

[d]Potential savings associated with cogeneration are not included here.

Table 2.31 AGRICULTURAL ENERGY CONSUMPTION TARGETS[a]

Service Category	1978 Base Consumption	Energy Consumption Targets for 2000
Space and Water Heat	0.05	0.044
Grain and Crop Drying	0.11	0.12
Lighting and Refrigeration	0.05	0.045
Irrigation	0.40	0.35
Mechanical Drive (stationary)	0.19	0.20
Vehicles and Field Machinery	0.83	0.89
	1.63	1.66

[a]Includes motor gasoline consumption

Source: Tables 3.3 and 3.11.

In Table 2.3 the projected energy savings for the year 2000 are shown in relation to energy requirements if no efficiency improvements were to occur from 1978 to 2000, and also in relation to a crude estimate of savings which would "normally" occur, i.e., which would occur with present policies and expected energy price increases. These estimates show the energy benefit flowing from the policy changes recommended in this report to be an intensity reduction of (87-74)/87 or 15%, or about 5 quads or 2.3 billion barrels per day oil equivalent.

How reasonable are these targets? The projected overall energy intensity reduction factor of 74% (Table 2.3) corresponds to a 1.4% reduction per annum 1978-2000. This can be compared with the 1.6% p.a. reduction in the energy value-added ratio 1973-1978, and the 2.3% p.a. reduction in energy (adjusted) per unit of production in ten industries 1972-1978 as discussed in Section 2. These comparisons with experience in the 1970's may not be very useful, however, since, on the one hand, many easy improvements were possible in the '70's and, on the other, very little investment to save energy occurred in the '70's.

In Table 2.32 energy intensities in the form of energy-use per dollar of industrial production are shown for other highly industrialized nations in relation to that for the U.S. The decrease in U.S. energy intensity by 2000 to approximately 2/3 of its value in 1972 as targeted here would still leave the U.S. energy intensity higher than or comparable to that in most of the other nations in 1972. It would be comparable to the 1972 intensity of Japan and about 10% higher than that of Western Europe.

It is important to realize that the projected stable demand for industrial energy is as much a result of underlying shifts in the economy, toward services and toward less materials- and energy-intensive industrial activities in the economy, as it is a result of efficiency improvements. If only 1/2 of the targeted industrial energy-intensity reduction were met by 2000, (i.e., if only the "normal" energy-intensity reduction in Table 2.3 were achieved), industrial energy demand (without credits for cogeneration savings) would grow only about 0.8%/year, 1978-2000.

Qualifications

The results do not include any credit for savings associated with cogeneration in spite of the highly advantageous opportunities that have been created in some regions under Section 210 of PURPA. The reasons are the low projected demand for electricity in this study overall, and the extensive power-plant construction activity still going on. If the full economically justified potential associated with cogeneration (neglecting the electricity-demand constraint) were achieved and credited to the industrial sector, indus

Table 2.32 INDUSTRIAL ENERGY USE--OUTPUT RELATIONS (1972)[a]

Nation	GDP Originating in Industry as a % of GDP	Industrial Energy Use Per $ of Industrial GDP (U.S. = 1.00)
United States	31.1	1.00
Canada	29.0	1.24
Six Western European Countries	41.8	0.57
France	48.1	0.40
West Germany	50.7	0.52
Italy	33.3	0.69
Netherlands	35.6	0.70
United Kingdom	34.8	0.80
Sweden	25.7	0.84
Japan	40.9	0.65

[a]Darmstadter et al. How Industrial Societies Use Energy.

trial energy use in 2000 would be 7% lower than the value shown in Table 2.20.

The result does not include any fuel penalties associated with changing fuel types, i.e., the increased Btu content typically required if biomass, coal, or oil-shale-based synthetics are substituted for oil and gas as primary fuels. This neglect is justified in the accelerated conservation case by the very low requirements for fuel projected in the overall study.

Estimated Uncertainty of the Target-Setting Exercise

How much larger or smaller might the energy requirements of industry, excluding agriculture, be likely to be in 2000? While many informational and theoretical uncertainties beset the target-setting, our judgment of the issue may be of interest. We examine the question assuming the recommended policies are in place.

High-side estimate

An alternative estimate of energy service demands can be based on the 1950-1978 value-added and basic materials trends with respect to GNP for the period 1950-1978 instead of the period 1969-1978. We have argued that such rapid growth is not likely, but we cannot say it is extremely unlikely. As seen in Figure 2.3, the value added/GNP trend for the period 1950-1978 is essentially unchanged from the 1969-78 trend which was used. As seen in Figure 2.4, however, the materials consumption ratio/GNP for the period 1950-1978 grows more rapidly, about 1% faster on the average, than was found for 1969-78. Thus, we can take growth rates for the period 1978-2000 to be faster by 1.0% p.a. for the three materials-related energy services (see Table 2.23). In addition, the growth in machine drive may have been underestimated because new electrical uses may be a characteristic of process change as mentioned above, so we also adopt a 1.0% p.a. faster growth rate for machine drive relative to Table 2.23. The assumed energy-intensity reductions are not so uncertain on the high side, so we do not consider a higher estimate for them.* The aggregate effect of the increased growth rates for energy services mentioned above would be a 6-quad or 20% increase in energy use in 2000 for this alternative "high side" estimate, corresponding to an annual average growth rate, 1978-2000, of 0.9% per year.

Low-side estimate

The estimates made here for both energy service demand and energy efficiency improvement are not lower limits; they are not radical. If there is one clear fact about the uncertainty of energy demand projections, it is that there is an extraordinary potential for technically induced reduction in demand. Less growth in energy services relative to GNP than projected in this report could be consistent with a vigorous economy and high living standards; indeed it could help the achievement of a healthy economy. In particular, the area of product change briefly explored in Section 3 offers options in recycling, remanufacture and reuse, and redesign that could reduce both energy requirements and total costs. Such changes, beyond extrapolating recent trends, were not considered. In addition, extensive energy efficiency improvements beyond these targetted could be cost effective. The issue is primarily one of the timing of new technical developments and of their implementation. More rapid improvement of technology and skills than cautiously assumed for the target setting could substantially further reduce energy intensities by 2000. A 0.5% p.a. additional reduction in energy intensity as well as 0.5% p.a. less rapid growth in energy services seem reasonable estimates for a low-side projection based on the range of elasticities shown in the note at the bottom of Table 2.3. This would imply a further 20% reduction in industrial fuel requirements by 2000, or 6 quads. Such an outcome seems to us as likely an outcome as the high-side estimate just considered.

Comparison with Other Department of Energy Forecasts

A very brief examination of other forecasting efforts suggests that substantial differences exist in forecasts of the levels of production (industrial production or basic materials production). These differences probably outweigh those associated with projections of efficiency improvement. In the following we concentrate on the forecasts of levels of production, not only because of their importance but also because two of the three other DOE forecasts are not available in final versions which would allow meaningful comparison of the forecasts of efficiency improvement (energy-intensity reduction).

The forecasting effort being made by the Office of Policy and Evaluation does not disaggregate the basic materials industries, but a critical input to its industrial analysis is the Federal Reserve Industrial Output Index which is cited as rising 0.7% p.a. faster than GNP. In contrast, this study is based on industrial production value added (GPO) growing 0.8% p.a. slower than GNP (See Figure 2.1), with some important energy services growing even more slowly; i.e., the basic materials index grows 1.3% p.a. slower than GNP. It should be kept in mind that, for example, a 2% p.a. discrepancy would imply a 50% higher level of energy services, i.e., of production, for the year 2000 than in the present report! The use of Federal Reserve Board indices as direct measures of the level of energy services is discussed further below.

The Energy Information Administration's Report to Congress (1979) also bases production levels on Federal Reserve Board Indices, although in this case they are indices for separate 2-digit SIC categories. In terms of a typical example, on the basis of FRB indices and a macro-economic analysis, they project production by the pulp and paper industry growing 0.7% faster than GNP. In contrast, in this study materials production (including paper production) is taken to be 1.3% slower than GNP. This discrepancy is associated with two factors which move in the same direction. It is partly associated with the use of physical indices in the present study versus the use of FRB indices to represent physical indices in the EIA analysis. For pulp and paper, the latter has been rising an average 0.7% p.a. faster than paper production in tons/year, primarily because the FRB index for the 2-digit SIC paper and allied products is weighted towards the finishing steps (e.g. making bags and boxes) which are relative growth areas in industry. (The FRB weights are based on value added.) The second factor is the weight given in the present study to recent (1970's) trends in contrast with older trends. As can be seen in Figure 2.2, basic materials production was much stronger, relative to aggregate industrial proudction, in the 1950's and 1960's than it was in the 1970's. Thus, while paper production (tons/year) kept pace with GNP in the '60's,

---

*It is important to stress that the quoted energy intensities do not include possible increased energy-conversion losses that would follow from, e.g., use of synfuels. They also do not include effects of possible reduced capacity utilization, more stringent environmental requirements than at present, or deterioration in the quality of input materials. Although these effects could be large in an economy with growing consumption of energy, as remarked above, the pattern of industrial energy use targeted in this report would sharply limit these problems.

it fell in the '70's. In the present study, 1970 trends have been the basis for forecasting production levels; the model used by EIA relies heavily on earlier trends.

The differences in forecast production levels are much less between this study and the Office of Conservation and Solar Energy's Conservation and Solar Strategy study. For example, the C/S Strategy study has paper production growing 0.6% p.a. faster, chemicals 1.3% p.a. faster and steel at about the same rate as in the present study (after removing the difference in assumed GNP growth rates).

The conclusions to be drawn from this rather limited comparison is that truly major uncertainties in projections of industrial energy demand are associated with projections of the level of basic materials production.

This subject should be carefully studied. A second conclusion is that simply because there is such wide variation in the underlying projections of industrial energy service, or production, the impact of conservation policies and associated efficiency improvements which would occur in these different situations are rather different. For example, an analysis in which production grows 2% p.a. faster than another will deal with a much higher proportion of new plant, other considerations being the same. Another aspect of this is that energy savings quoted in absolute terms (e.g., quads or barrels per day) will be relatively large with more rapid production growth than in the present study.

Section Four

Industrial Solar Technologies

The equipment and techniques described in the previous section were designed to improve the efficiency with which energy is consumed. A variety of systems designed to convert direct sunlight and biomass resources into useful energy at an industrial facility are analyzed in the present section. (Solar electric systems such as wind machines and photovoltaic arrays could also be owned and operated by industry but they will not be treated here. They are examined in detail in the Chapters dealing with utility issues.) It is important to bear in mind that renewable resource technologies should be evaluated in precisely the same way as efficiency technologies. Both must serve as elements of a repretory of technologies available to be combined in a system engineered to serve a specific industrial energy requirement at the lowest possible overall cost. Designing such an optimum system, however, is extremely difficult given our present state of knowledge. A separate analysis should be carried out for each major industrial process and for each climatic region. Moreover, an optimum design must consider the solar unit and novel industrial processes simultaneously rather than simply adding the solar equipment to an otherwise standard system. It may be desirable, for example, to alter a basic industrial process to make it more compatable with renewable energy. This could be done by reducing process temperatures, or by building storage into the heat delivery and control system. Without such an analysis all statements made about the utility of renewable resources in industry must be considered preliminary; the following discussion must be read in this light.

While a variety of different renewable technologies can play a role in industry, primary attention will be given here to the use of biomass and to the direct use of solar energy to provide industrial process heat.

BIOMASS

Potential Biomass Resources

Biomass--primarily in the form of wood burned in the forest product industry and in residential stoves--now supplies between 1.55 and 1.8 quads of energy in the U.S. yearly, excluding wood burned in fireplaces (Tillman 1978) (WEI 1979) (Glidden 1980). Much greater contributions are possible. The best available analyses project that the total available sustainable biomass supply in the United States can conservatively be between 7 and 17 quads in 2000 (Table 2.6). The amount of biomass available depends critically on how the nation's resources are managed. It is possible that the 17 quad limit could be exceeded given advanced forest and agricultural management techniques. The amount of this potential which can be used to supply energy at competitive prices is extremely difficult to estimate. At least 3 and perhaps as much as 5 quads of wood will probably be used in the year 2000 even if the federal government does nothing to promote wood energy. Biomass can be used as a means of providing synthetic liquid fuels for transportation, as feedstocks for certain chemical processes, and as fuels for process heat and electricity for selected industries that already have access to biomass. The near- and long-term technologies available for making use of biomass are discussed in detail in an appendix to this chapter.

Wood Energy Potential

Potential U.S. wood energy supplies are summarized in Table 2.6. This table indicates that in 20 years 12-29 quads of wood may be available for use as biomass.

Access constraints, environmental restrictions, land ownership objectives and competition with alternative forest uses (e.g., recreation) are likely to limit this quantity to 5-9 quads; less than half of the total potential supply (Glidden 1980). This estimate is supported by regional and national assessments. The amount of wood available for use as an energy product will depend on the demand for lumber and price of fuel success of policies products, elsewhere in the economy. The wood energy resources shown on Table 2.6 has included an allowance for growth in the forest product industries which is consistent with the analysis of industrial energy consumption addressed earlier in this chapter.

It will be difficult to produce even the amounts of wood energy shown in Table 2.6 unless greater attention is paid to forest management and unless industry makes heavy use of integrated tree harvesting or whole-tree harvesting techniques which remove relatively low quality wood along with conventional forest products. The low quality wood can be used as a source of energy and can be relatively inexpensive if it is removed during logging operations. Programs to encourage the use of forest management (including reforestation, road building, and timber stand improvement) and integrated harvesting operations will be needed if national consumption of wood energy is to approach 9 Quads per year by 2000; specific programs will be discussed in a later section.

Agricultural Energy Potential

The nation's farms can also produce significant amounts of biomass but the potential contribution of this energy source is extremely difficult to predict. The most fundamental issue is the extent to which farmland can be used to produce energy crops without increasing food prices.

This issue has been examined by a number of authors and there is a considerable amount of disagreement (see the section on biomass policy). All agree, however, that food prices may be forced up if the nation attempts to extract too much energy from agriculture. The problem is extremely complex because of uncertainties about the availability and quality of potential farmland not presently in use, and uncertainties about domestic and international demands for different kinds of food products. Fluctuations in the weather can, of course, make major and unpredictable changes in the productivity of farms. The issue can not be resolved conclusively. The only practical solution is to develop flexible programs which continuously evaluate the impact of agricultural/fuel programs and take action if problems appear.

Energy sources in the agricultural industry can be divided into five major categories: crops from conventional farm land (producing corn, sugar cane, etc.); pasture and hay lands; crop residues (corn stalks, etc.);

animal manure; and wastes from agricultural processing (tomato pulp, peach pits, etc.).

Cropland available for the production of energy crops could either increase or decrease by the year 2000, depending upon the future demand for food, crop yields, and the price rises needed to induce farmers to bring new land into production. As much as 1 quad equivalent of fuels could be produced from grains or sugar crops, or about 12 billion gallons of ethanol (Tyner 1980). On the other hand, if some of the same cropland is available and is used instead to produce fast growing grasses, legumes, etc., as much as 5 quads (primary fuel value) could be produced, yielding 2.4 quads of ethanol, or a greater amount of energy through gasification (OTA 1980). This 5 quad estimate should be considered an absolute maximum (Tyner 1980.)

Intensive management of existing and potential hayland and pasture land also can produce 0 to 5 quads of biomass by 2000. The availability of these lands is dependent on the same factors outlined above for conventional cropland; some grasslands by 2000 may be converted to grain production. Biomass production from these lands could increase by applying fertilizers (and increasing the number of harvests per year). Yields from crop and pastureland could be increased with improved cultivation techniques and crops which are bred to produce both food and energy. This category is not additive with above.

Residues from rice, sugar cane, and grain crops can provide a significant amount of energy, even though about 80% of the crop residues need to be left in the field to protect the soil from erosion. It appears that about 1 quad per year of crop residues could be used for energy without exceeding soil erosion standards, although there is considerable controversy as to whether current standards are too restrictive or lax. It has been estimated that local variations in crop yields limit the useable supply to 0.7 quads at present crop levels (OTA 1980). The supply of crop residues will change as crop production changes.

Waste products offer another potential energy source. Animal manure and agricultural processing wastes could contribute between 0.2 and 0.4 quads (OTA 1980). Byproducts from the food processing industries are another potential source although most of the waste will be used as animal feed, chemical production, or other uses. Municipal solid waste and sewage sludge could provide a small amount of energy, perhaps as much as 1.3 quads (SRI 1979).

Biomass production from algae ponds, aquatic plants, oil bearing plants and other exotic sources have not been accounted for because their technical and economic uncertainties (e.g. yields, harvesting techniques) are too great. Table 2.6 is intended to provide a conservative view of the theoretical resource potential.

Biomass Conversion Technologies

Biomass can be used directly as a source of heat by direct combustion or converted to a gas or liquid. Two major types of conversion technologies are available for this task: those that utilize biological enzymes--biotechnologies, and those that gasify or pyrolyze biomass--thermochemical technologies. The former is the preferred method (at this time) for making ethanol, while the latter appears the most straightforward method for making methanol. Biotechnology also offers possibilities of revolutionizing the chemical industry by producing basic feedstocks that are now produced through chemical processes (usually starting from petroleum).

It is likely that large amounts of the biomass energy consumed in the nation, particularly in the next decade, will be consumed in simple boilers. The technology is relatively straightforward and will not be reviewed here in any detail; several recent reports provide excellent reviews of the subject (OTA 1980, TVA 1980).

Biotechnology

When first introduced following the work of Louis Pasteur on fermentation, industrial microbiology for producing chemicals developed vigorously. There was a thriving fermentation chemicals industry both in Europe and the U.S.A. Cheap petroleum and a highly efficient cracking technology soon displaced microbiological industry. Bulk chemicals were no longer produced biologically; instead the focus was on compounds not easily formed by chemical synthesis: antibiotics, vitamins, enzymes.

Hydrocarbon costs are mounting, however, as the fraction of U.S. oil supplied by other countries becomes larger. To seek substitute commodities in response to market forces is the lifeblood of the chemicals industry. It is likely, therefore, that the microbiological production of chemicals will again become a commercially profitable technology.

For example, fermentation ethanol can serve as a fuel and as a basic chemical for synthesizing other products. Ethylene, produced at present from petroleum, is a prime chemical feedstock; its major derivatives are polyethylene, ethylene oxide, vinyl chloride, and styrene. The supply is becoming tighter, but the demand is unabated (more than 10 million metric tons per annum in the U.S.A.) and thus the price is rising. It is possible, however, to produce ethylene from ethanol. Ethanol can be a feedstock for producing butanol acetone, butanediol, glycerol, lactic acid, propionic acid, and acetic acid. Thus, a plastics industry based on renewable resources is technically feasible. The feasibility of producing these materials from biological feedstocks is discussed in detail in an appendix to this chapter.

There is a need to promote this development as rapidly and efficiently as possible (Villet 1979). Two general avenues are recommended:

1) The near-term revival and improvement of conventional, although presently dormant, fermentation technology. Such fermentations are based on readily fermentable materials, e.g., sugars and starch.

2) The long-term development of a new biotechnology based on cheaper feedstocks which, however, are not easily converted (woody biomass; forage crops; agricultural residues). Also in this group are algal and plant systems for producing hydrocarbons and other chemicals.

Process Economics. Except for waste materials, the cost of easily fermentable substrate is high: 60-75% of the cost of production. Processing costs could be reduced by developing more efficient strains of microorganisms and by improved energy efficiency of product recovery.

The cost of more refractory feedstock (lignocellulose) is relatively lower. But the processing technology to produce chemicals and fuels is not at a commercial scale as yet.

Ethanol from Lignocellulose. The Gulf Oil/University of Arkansas process has been used for converting wood polysaccharides into ethanol at a rate of 1 m.t. of feedstock per day. Cellulose conversion is accomplished by using a mutant strain of Trichoderma reesei to produce a cellulose multienzyme system. Sugars from the hydrolysis are converted by a mixed culture of yeasts. The simultaneous hydrolysis and fermentation eliminates glucose inhibition. The pretreated slurry contains 7.5% to 15% cellulose. Ethanol is recovered by steam-stripping and rectification. Residue combustion could provide thermal energy and motive power to operate a plant. A commercial facility processing 1000 tons of cellulose feedstock per day is estimated to cost $70 million and yield 250 tons ethanol/day (about 25 million gal./yr). The production cost is estimated to be $1.16/gal. By-product molasses can be used for animal feed. If a plant is designated a waste conversion facility, municipal bond financing is possible. How far the process design and operation is from optimum cannot be judged at present, but it is very promising. It would be of interest to evaluate the economics of the process at a considerably lower throughput of, for example, 100 tons/day.

Ethanol from Industrial Wastes. Waste products such as whey from the food industry and sulfite waste liquor from the paper pulp industry are potentially rich sources of fermentable sugars. For example, 1500 million pounds of lactose are produced from whey per annum in the United States; the potential yield of ethanol is approximately 100 million gallons.

Potential for the Future. A new manufacturing industry, based on resources which are renewable by photosynthesis, could be established. To develop efficient biotechnological processes for the production from biomass of chemicals and fuels would require a combination of chemical engineering with various areas of biological science. It is not too optimistic to forecast that biotechnology is capable of becoming the new "Wunderwirtschaft" with a potential akin to that of the computer and micro-electronics industry several decades ago, but the potential of this new industry is not to be ignored.

Thermochemical Conversion Technologies

Gasification for Heat Production. Air gasification is the simplest means of turning biomass into a gaseous fuel by partial combustion with air to form a gas containing carbon monixide and hydrogen (and sometimes undesirable oil vapors). The gas that is produced has a relatively low energy content--typically 150 Btf/scf-- that is not suitable for pipeline transport. However, it can be burned in existing equipment with minor modifications, and so is most likely to be used in the "close coupled" mode in which biomass is fed into the gasifier and the gas produced is used directly in a boiler or engine.

Air gasification is likely to find application in two areas. The first is the retrofitting of existing oil/gas boilers. The cost of converting these boilers to use intermediate-Btu gas is estimated to be half to 2/3 of the cost to install a replacement wood fired system. In addition, the use of the gasifier permits return to oil/gas when available and is generally cleaner and more efficient that the direct combustion of biomass. Presently a number of manufacturers are developing gasifiers in the 1-100 MBtu/hr range. There is also a need for smaller and larger gasifiers.

A million or so small gasifiers were used in Europe during World War II to retrofit cars, trucks, and buses to wood chips. It requires about 22 pounds of dry wood to replace a gallon of gasoline using a gasifier, but the operation of the gasifier is less convenient and there is a 20% reduction in horsepower. Gasifiers are well suited for operation of stationary engines such as irrigation pumps and particularly small power generation systems.

A higher energy gas (300-500 Btu/scf) can be generated using oxygen or various pyrolytic-recycle methods to gasify the biomass leading to a number of attractive applications. Such (from coal) was widely manufactured and used in the U.S. and Europe until the advent of low cost natural gas. This gas can be burned in existing boilers where natural gas is now used with no de-rating of the boiler. Although not suitable for long distance pipelines, it can be distributed, like natural gas, in industrial areas. A particularly attractive use is for the operation of turbines for peak power generation.

Thermochemical Production of Methanol. Oxygen gasification of biomass produces a mixture of carbon monoxide and hydrogen called synthesis gas (syngas), which is the basis of a number of important fuels, chemicals, and ammonia. This gas is now used to produce about 1 billion gallons per year of methanol from natural gas and about .45 quad equivalent of ammonia per year, also starting with natural gas. Most of the processes that are considered for producing methanol from biomass operate at relatively low pressures and temperatures, using a variety of different copper-based catalysts and a variety of types of chambers for gasifying the biomass. In the process syngas is produced by heating the biomass, the gas is then purified to remove carbon dioxide and contaminants (such as sulfur and chlorine) that would poison the catalyst, then the composition of the $CO-H_2$ mixture is shifted by reaction with water vapor to a ratio (2 to 1 hydrogen to carbon molar ratio) suitable for methanol production, and finally syngas is compressed and fed into the reactor with the catalyst to produce methanol. There are a number of methanol synthesis processes which are either off-the-shelf, or can be with relatively little development.

Reported capital costs of methanol plants utilizing different processes are in general agreement, though the efficiencies assumed in the different processes vary considerably.

If large markets are to be found for methanol derived from biomass, it must compete with methanol manufactured from coal. With presently available technology methanol produced from biomass could be roughly competitive with methanol production from coal, given that the methanol plants have the same level of output and that the feedstock costs (in dollars per million Btu) are comparable (see Appendix C). However, this situation is not likely to occur. Coal-to-methanol plant would (at least initially) tend to be mine-mouth plants, sitting atop a concentrated source of feedstock. Because the feedstock is concentrated, scale economies can be exploited by buiding very large plants. Plants with coal inputs as large as 74,000 tons/day* have been proposed, although initial plant capacities will probably be in the range 10,000 to 20,000

---

*The annual coal thoroughput of such an enormous plant would be about 0.6 Quads; the estimated capital cost would be $3.9 billion.

tons per day. On the other hand, biomass-to-methanol plants must be much smaller, because the resource is dispersed; the costs of transporting biomass to the methanol plant offsets the gains in scale economy. Biomass-to-methanol plants on the order of 1000 tons/day are usually regarded as "large" for biomass, although plant sizes of 5000 tons/day or even larger may be competitive in some circumstances if very large quantities of feedstocks can be obtained relatively cheaply. Because coal plants built today would be so much larger, methanol produced from coal would be much cheaper than methanol produced from biomass. But the long run situation is not so clear, largely because presently available biomass-to-methanol technology is not optimized to exploit the unique features of biomass; it is simply modified coal technology. Two important advantages are that biomass is low in sulfur and inherently easier to gasify. As shown in Appendix C biomass gasifies at a lower temperature and over a narrower temperature range than coal. These properties favor rapid gasification, which could lead to reduced capital costs for a methanol plant.

A recent study by Science Applications, Inc. for technological innovation in biomass-to-methanol conversion technology (see Appendix D) has concluded that by the mid 1990's, capital costs of biomass plants could be reduced by about 50% by exploiting the chemical characteristics from biomass.* With these improvements, methanol from biomass could be competitive with methanol from coal (as projected by Exxon for new coal-based technology in the 1990's), even if the coal plant is 4 times as large as the biomass plant and the feedstock cost is 1.5 times as large (on a $ per milion Btu basis) as the cost of coal. This assumes, however, that the biomass plants are very large. Of course coal-based technology could improve even more; nevertheless, it seems clear that the prospects are good that with advanced technology, large biomass-based methanol could be roughly competitive with coal based methanol.

One way for the economics of biomass/methanol systems to appear more attractive is by improving production efficiencies. Conventional wisdom, based on very limited experience with coal gasification, has projected thermal efficiencies of 40-50% for biomass conversion to methanol. Estimates were typically derived at a time when energy costs were low relative to capital costs, so that no effort was needed to optimize efficiency. In addition, the gasifiers studied had not been developed specifically for methanol production.

However, recent research may prove conventional wisdom wrong. One fixed bed downdraft gasifier now in the testing stage cracks wood tars and hydrocarbons to a very low level (less than .5% hydrocarbon, less than 300 ppm organics in scrub water), thus eliminating external gas and water processing; the gasifier is designed to operate at high pressure to reduce or eliminate compression costs. High pressure steam is generated from the exothermic steps of the process to supply more of the energy for the plant operation. The waste energy thus supplies almost all of the process energy in the plant and the internal energy balance yields an efficiency of methanol production of 83% on a dry wood basis (Reed 1981).

Advanced biomass conversion technologies or mass production strategies could offset the economies of scale normally favoring coal technologies. If small scale wood-to-methanol plants can be mass produced it is possible that the methanol produced might compete with that derived from coal. Small plants offer a number of advantages, particularly during periods when demands for methanol are uncertain and construction costs are increasing rapidly. Small plants, of course, also can be located close to wood resources and reduce the cost of transporting wood to the plant.

Not the least of the advantages of methanol derived from biomass is that is production and use on a sustainable basis would lead to no net increase in atmospheric CO_2 and the associated climatic problems. In this connection, it is indeed fortuitous that if it becomes necessary to shift from fossil to biomass fuels because of the CO_2 problem, the transition would be much easier with methanol than with most other synthetic liquid fuels, because the production of methanol is an especially attractive way of utilizing both fossil fuel and biomass feedstocks.

In light of these prospects, it is clear that biomass-based methanol technologies should receive high priority in the nation's R&D program.

Pyrolysis of Biomass. This approach can be used to produce methanol, a variety of other products, and in some specialized processes a mixture of high molecular weight alcohols (for instance, methanol, ethanol, and propanol) which approach the density of gasoline. (Per Btu, methanol and ethanol are as good automobile fuels as gasoline or better.

When biomass is heated, the rates of several chemical reactions are increased which change the nature of the biomass. Different reactions are favored depending upon the length of time used to heat the biomass and the final temperature and pressure achieved. The different chemical reactions can thus be manipulated to maximize the production of char, wood alcohol and acetic acid, tars, syngas, or ethylene (Shafizadeh 1968).

---

*A 50% increase in yield means that typically a plant could become 4 times as large without increasing the unit transport cost of biomass residues.

The ancients recognized the clean, hot burning nature of charcoal and learned to optimize its production by very slowly heating biomass over a period of several days to a fairly low final temperature of 400°C (750°F). This favors a dehydration reaction which produced water and char.

The byproducts of the charcoal kiln include organic materials which are fairly small fragments of the original molecules. By using a shorter time of a few hours to reach the final temperature of about 400°C, the production of these organic compounds, e.g., methanol and acetic acid, are maximized as well as the production of a large amount of tars. This type of pyrolyzing was commonly called wood distillation. Unfortunately, the yield of methanol by this manner in commercial sized pyrolysis reactors is very low in the 1 to 2% by weight range. From one ton of dry wood, 5 gallons of low boiling organic liquids and 33 gallons of tars were produced in the early 1930s. By the addition of ethanol to the acetic acid formed, nearly 16 gallons of acetate solvents were coproduced per ton of dry wood. The overall product distribution was about 30% water (by difference) (Nelson, 1930). This type of pyrolysis has been proposed to convert relatively bulky biomass into char which could be briquetted to a high density material and to tars which could be shipped in bulk in stainless steel tanks. (The tars are corrosive to mild steel.) Mixing the tar and char together to form a very dense moist material has also been proposed. The tars produced by pyrolysis at these temperatures have a very strong pungent odor so that they would probably need to be shipped in sealed containers to avoid both the loss of the volatile components and adverse environmental impacts. The tars contain polyaromatic compounds many of which are carcinogens so that adequate precautions must be taken to protect the workers involved. The water formed as a byproduct of pyrolysis at these temperatures has a high organic content which must be removed prior to disposal to avoid an adverse environmental impact. To convert biomass to a more easily transportable material for long distances, on-site pelletizing, baling, or chipping may have economic advantages over char and tar forming pyrolysis.

As the heating rate of the biomass increases, the relatively slow, char-forming reaction progressively has less time to act on the feedstock. At temperatures of 500 to 600°C (925 to 1100°F), the predominant cellulosic pyrolysis reaction breaks the long polymeric chains into fairly small tar molecules often referred to as levoglucosan. Yields of 40% by weight tar and 20% char have been reported. This pyrolysis-tar material has flow to 80-100°C (175-230°F) for good atomization and subsequent combustion in a furnace (Mallon, 1972). Ths substitution of this pyrolytic tar for #6 fuel oil has not been enthusiastically accepted because the tar contains organic acids corrosive to steel and the chemical reactivity of the

material causes it to polymerize and progressively thicken during storage.

At high pyrolysis temperatures, the tars are no longer chemically stable, and they become a very minor byproduct even at very short residence times. When the biomass is heated very rapidly to temperatures above 700°C (1300°F) in the presence of diluent gases such as steam, a relatively simple product slate can be obtained which initially contains significant quantities of unsaturated hydrocarbons such as ethylene and virtually no char and tar. If the fast pyrolysis products are not sufficiently diluted with relatively inert gases such as nitrogen, steam, or carbon dioxide, the formation of oxygen containing water-soluble tars appears to be favored rather than the gases. By very rapidly cooling the pyrolysis products, hydrocarbon compounds similar in product distribution to that obtained by cracking naphtha to produce ethylene are obtained in addition to oxides of carbon. Very little water is formed in this fast, high temperature reaction, which minimizes the problems involved with the production of contaminated water requiring extensive waste-water treatment. At these conditions, the pyrolysis reaction is extremely fast, which allows very short residence times of 0.050 to 0.150 seconds to pyrolyze powdered feedstocks (Diebold, Smith 1979). To fast pyrolyze larger feedstock particles to these desirable products apparently requires very high heat transfer rates, extended solids residence time, and very short gaseous residence time. If the pyrolysis products are allowed to stay several seconds in the reactor, the unsaturated hydrocarbons react with each other to form desirable aromatic liquids such as benzene, toluene, xylene, etc. as well as undesirable polyaromatic tars. The unsaturated hydrocarbons and aromatic liquids can be purified using established technology to produce petrochemical feedstocks, or they can be used to produce high octane gasoline. Because of their use as gasoline blending stocks, the value of the aromatic liquids is about 150% that of unleaded gasolines. Due to the high hydrocarbon content and the use of indirect heating, fast pyrolysis produces a "medium Btu" gas (500 Btu/ft^3 or more) which may find applications where low Btu gas is less desirable for combustion, as in stationary engines. One to two percent by weight char and small amounts of tars are formed as byproducts, which can be recycled or burned in the pyrolysis furnace.

The processing of the gaseous products of fast pyrolysis has received limited development. A straightforward application of the hot carbonate process, now in widespread use in the petrochemical industry, has been shown at the bench scale to very selectively remove carbon dioxide from compressed fast-pyrolysis gases. The use of absorption of ethylene and other unsaturated hydrocarbons into a hydrocarbon liquid has been partially demonstrated for fast pyrolysis gases, although some undesirable coabsorption of carbon monoxide was encountered which could probably be eliminated with additional effort to produce a concentrated unsaturated hydrocarbon gas stream.

These purified gases would process in a manner identical to gases now obtained by the petrochemical industry from the cracking of naphtha to produce ethylene. Consequently the biomass-fast-pyrolysis gases can be considered to be an alternate petrochemical feedstock.

The other approach to the use of biomass-fast-pyrolysis gases is to react the unsaturated hydrocarbons portion (ethylene) in the raw gases to create high value liquid products. This can be done with or without catalyst. Without catalysts, the ethylene will react to form low boiling, high octane aromatic liquids at a few atmospheres pressure and temperatures over $600^\circ C$ (Davidson 1918). With catalysts, the reaction proceeds at lower temperatures and can react some of the carbon monoxide for form n-propanal along with high cetane straight chain hydrocarbons (Kuester 1979).

These latter products of pyrolysis are the ones that could approach the compactness of gasoline as a vehicle fuel.

Acid Hydrolysis. For converting wood chips or agricultural wastes to fermentable sugars, acid hydrolysis is the only process which is in operation at full commercial scale. There are 10 plants in operation in the Soviet Union, and they are about 40 years old. They were designed by the German Schoeller. The process is far from optimal. Hot 0.5% H_2SO_4 percolates through a bed of wood chips to produce a sugar solution and a residue. Byproducts furfural, methanol and organic acids are flashed off. Lime is used for neutralization, producing calcium sulfate. A 3-6% reducing sugar solution is produced. About 60% of the original reducing sugars are recovered. The dilute sugar solution and the low yields are disadvantages. So too are the slow rate of production and the high reactor capital cost.

In Europe and the Soviet Union, chief interest is to produce single cell protein from sugars. For ethanol production a rough estimate of cost of production is $1.60/gal. The yield of alcohol depends on the substrate and the efficiency of fermentation. A yield of about 50 gals of 95% ethanol from oven dried ton of wood is claimed.

There is research work in the U.S. to improve efficiency of hydrolysis, by a low residence-time system to give better yeilds. This includes work on hemicelluloses. These processes still have to be scaled up.

Although acid hydrolysis offers an option for producing ethanol out of the vast quantities of woody biomass (an option that vitiates the necessity of ever producing methanol-compatible automobiles), the inefficiency of the process is a major barrier to be overcome. The problem of making methanol compat-

ible auto engines is likely easier to solve (such a factory for auto engines is now under construction in Brazil).

Biomass Utilization

Having estimated the biomass resource potential, it is necessary to construct an estimate of the amount of biomass which can be used, given different assumptions about policies designed to assist industry. Plainly, no estimate such as this can be precise since unforseen market conditions will dictate the actual mix.

Two scenarios for biomass utilization are presented. The first assumes that biomass is used for energy only up to the point where production costs are at or below competitive fuel and resource costs (Table 2.33). The second (Table 2.7) assumes a program wherein the government acts agressively to encourage the use of biomass: This scenario is probably close to the maximum total sustainable biomass production achievable through intensive management of both croplands and forests. It does not, however, assume the use of novel crops or arrays of algae ponds. In both tables ranges are shown which reflect uncertainty about the supplies, limitations on the use of the products, and the costs of competing products. The estimated costs of various biofuel products are also in both tables. To achieve the levels of biomass penetration shown in these tables no significant advances in technology are assumed; all technologies contributing to the projected supplies are now either commercial or in the prototype stage. However, in the enhanced case the optimistic assumption has been made that some advanced conversion technologies (such as acid hydrolysis) will come into widespread use before 2000.

Tables 2.33 and 2.7 reflect a judgment that biomass will be used primarily where self-generated sources are available--the food processing industries, forest product industries and agriculture. The most extensively used biomass source will be wood (at least 3 and perhaps as much as 9 quads will be used) primarily in the forest products industry, industries next-to-the-woods, the residential sector, and to lesser extent the transportation sector (in the form of methanol).

It should be understood that the biomass utilization projections in Tables 2.33 and 2.7 are speculative, given the great uncertainties associated with biomass supply, accessibility, conversion cost, etc. The ranges of projected use reflect these uncertainties. In most cases estimates were made with knowledge of the accessibility of the biomass resource, the costs of conversion with commercial or near-commercial technologies, and the anticipated costs of competing sources of energy. These estimates have not been formulated with precision but should provide an adequate basis for the construction of public policy.

Use Sectors	Ethanol[n]	Methanol[b,m]	Methane[c,q]	Direct Combustion /Gasification[l]	Total
			Primary Supply		
Transportation	.1-.25	0-2[o]	0	0	.25-2.4
Food Processing	0	0	0-.05	0-.1	0-.15
Forest Products	0	0	0	2.1-2.8	2.1-2.8
Chemicals	0	0	0	0	0
Residential	0	0	0	.5-1.0	.5-1.0
Agriculture	0-.2[i]	0	0-.1	0-.1[h]	0-.4
Other	0	0	0	0-1.0[g]	0-1.5
Total	.25-.4	0-2	0-.15	2.6-5.0	2.85-5.75[k]

Cost per Million Btu (1979$):

Ethanol: $12.40 - $18.40[r] ; 12.40

Methanol: $8.00 - $20.90[s]

Methane: $5.00 - $15.00[t]

Direct Combustion/Gasification: $3.00 - $6.00[u]

Table 2.33b BASE CASE FOR ENERGY DISPLACED BY BIOMASS IN 2000[a] (Quads)

Use Sectors	Ethanol[n]	Methanol[b,m]	Methane[c,q]	Direct Combustion /Gasification[l]	Total[k]
			End Use Energy Displaced (Oil Equivalent)		
Transportation	.076-.32	1.86[o]	0	0	.08-1.8
Food Processing	0	0	0-.05	0-.09	.05-.09
Forest Products	0	0	0	1.87-2.5	1.87-2.5
Chemicals	0	0	0	0	0
Residential	0	0	0	.4-.8[p]	.4-.8
Agriculture	0-.1	0	0-.1	0-.09	0-.29
Other	0	0	0	0-.9	0-.9
Total	.076-.32	0-1.86	0-.15	2.3-4.4	2.4-5.1[k]

Cost per Million Btu (1979$):

Ethanol: $12.40 - 18.40

Methanol: $8.00 - $20.90[s]

Methane: $5.00 - $15.00

Direct Combustion/Gasification: $3.00 - $6.00[u]

See notes to Tables 2.33a and 2.33b at the end of this section

Biomass in Transportation

In the base case shown in Table 2.33, 1-3 billion gallons of ethanol (.1-.25 quads) are produced from starch and sugar crops. Recent studies indicate that above this level of ethanol production (from starch and sugar crops) the cost or availability of food or feed could be jeopardized (see policy section) (Tyner 1980). Assuming, optimistically, that advanced conversion technologies could come into widespread use in 2000, as much as 2.4 quads (almost 28 billion gallons of ethanol) could be produced from starch and sugar crops and lignocellulose materials other than wood. This high target assumes that the entire 5 quads of non-wood lignocellulose shown in Table 2.6 is used to make ethanol, which seems unlikely. Achieving this high level of production will require massive government subsidies, availability of accessible and economically attractive feedstocks, and advanced conversion technologies which may not become economically viable before 2000. The 5 quads of lignocellulosic material could also be used to produce methanol.

Methanol, on the other hand, could be the most widely used liquid fuel derived from biomass in the year 2000. Like ethanol, methanol can be used as an octane booster as well as a stand-alone fuel. Assuming a 35% conversion efficiency 25-30 billion gallons of methanol could be produced annually using 4 quads of primary wood input, if efficiencies of 50-60% are achieved 36-43 billion gallons could be produced from the same input. The actual production of methanol depends heavily on the output of the forest product industry, how much wood the forest product industry and other consuming sectors will use for energy, the creation of a wood fuel supply infrastructure, and the commercialization of economically attractive technologies. Methanol production shown in Tables 2.33 and 2.7 has been determined by taking the total amount of wood available for conversion to energy and reducing it by the wood fuel demands of the industrial and residential sectors. It is likely that significant amounts of fuel wood will be used in the forest product industry and other consuming sectors having easy access to wood resources; liquid fuels are assumed to be supplied from the remainder. The ranges shown indicate that uncertainty about the competing uses of fuel wood and the feasibility of wood-to-methanol technologies.

Biomass in the Food Processing Industry

In the analysis of industrial energy demand exhibited earlier, it was estimated that by 2000 the food processing industry would require 0.8 quads of energy used as process and space heat and another 0.3 quad as shaft power. Biomass is expected to be used to provide heat via direct combustion and low- and intermediate-Btu gasification. Up to 10% of the industry's needs could be supplied by biomass in 2000. A portion of the mechanical drive could also be supplied with cogeneration. The amount used will depend on non-energy markets for the waste, the availability of nearby wastes and crop residues, and as whether biomass supplies are available when they are needed.

Biomass in the Forest Product Industry

About 75% of the fuel consumed by the forest product industry in 1978 was used in the pulp and paper industry and 25% was used to make wood-products (API 1979). The industry now uses wood to supply nearly half of its fuel needs, generating about 1.3 quads from wood wastes (Glidden 1980). Trends over the past ten years indicate that the industry is likely to continue replacing fossil fuels with wood and reduce its energy requirements per unit of product.

In the base case shown in Table 2.33 the forest product industry is assumed to be 70%--80% self sufficient. Direct combustion will be the most common conversion technology with on-site gasification increasing in importance by the mid 1980's. In the "enhanced case," shown on Table 2.7, the industry would be almost entirely energy self-sufficient. The forest product industry, with its experience in wood management and existing supply infrastructure, is more likely to obtain the wood energy supplies it requires than any other industry.

The wood fuel not used by the forest product industry could most readily be used by other industries located near forests, or in the residential sector. It is assumed that if these needs were saturated the manufacture of methanol for transportation could be increased to the highest levels shown in Table 2.7.

Biomass in the Chemical Industry

The chemical sector may provide a key market for biomass products. However, because of the uncertainties associated with future bioconversion technologies (those producing ethylene, for example) and limitted ethanol and methanol markets (methanol is produced from natural gas and is relatively inexpensive) biomass is not shown as penetrating this sector measureably. Rapidly increasing natural gas prices and technological advances could change this appreciably.

Wood Use in Residences

Residential wood stoves now consume between .25 and .5 quads of wood energy annually (excluding wood burned in fireplaces) (WEI 1979) (Glidden 1980). Following current trends, this could increase to 2 quads by the year 2000. If the residential conservation targets outlined in the chapter on building are met, residential wood use will only total 0.5 quads.

Biomass Energy for Agricultural Industries

Approximately .45 quad of natural gas is presently used to produce ammonia for fertilizers (Reed 1976). Ammonia can also be produced from biomass using gasification or anaerobic digestion technologies. The agricultural sector also uses .13 quad of energy for grain and crop drying, some of which could be supplied by biomass combustion or gasification. Approximately 0.2 quad/year of ethanol could be used to operate stationary mechanical equipment on farms. Methane can also be produced in small rural systems using anaerobic digestion although the total amount of energy produced will probably be limited to a few tenths of a quad/year.

Other Uses of Biomass

Very little wood is used as fuel outside the forest products industry. Many industries located close to large forest areas, however, can almost certainly make use of wood as an energy source. Examples range from shoe factories in New Hampshire to textile industries in Georgia. These industries could use as much as 2 quads of wood per year by the year 2000 (Gidden 1980).

Utilities may also use biomass to generate electricity although the total use of biomass for these purposes is likely to be relatively small. Some biomass may be burned to generate electricity directly in small steam-electric generators and co-generation systems, but this application is likely to be limited by the attractiveness of high E/S ratio cogeneration based on conventional fuels.This is due to the relatively low cost of competing fuels and the cost of coal and other competing utility fuels and the relatively slow growth in electric demand expected in many areas with access to large tracts of forests. Alcohol fuels on the other hand can be used to provide peaking power in diesels and gas turbines and may well be competitive with alternative sources of peaking energy by the year 2000 (Adelman, Pefley 1979).

Policy for Biomass

The analysis presented in the previous section indicated that biomass may displace 2.4-5 quads of energy in the U.S. by the year 2000 without any additional federal programs. Given appropriate federal, state, and local programs, it appears possible to increase this displacement to 4.8 to 9 quads.

Over 80% of the total biomass energy supplied is likely to be derived from wood or other forms of cellulose. Exploiting this enormous potential is not without risks. Poor forest management can result in the destruction of forests and irreparable damage from erosion. Even careful forest management can change a wilderness into disciplined arrays of highly productive trees resembling corn crops. Large scale use of agricultural products as sources of energy could lead to increases in national and world food prices. Residues from biomass processing can pollute streams and rivers if not properly controlled.

National programs in biomass must, therefore, be constructed around two basic themes: encouraging and developing efficient new techniques for growing, harvesting, and converting biomass, and; ensuring that increases in national use of biomass result in the development of a truely renewable resource and one which does not inflict large costs on other parts of the national economy. Given a well managed national program, biomass may provide more than half of the renewable energy used by the nation during the next twenty years.

Wood Energy

Wood represents the largest current and potential source of renewable energy (excluding hydropower). Wood currently supplies about 2 percent of U.S. primary energy needs. By the year 2000, wood energy is expected to suply 3.0 to 4.0 quads of energy with no additional federal initiatives. Between 5.0 and 9.0 quads of renewable wood energy could be produced without serious enviromental or economic impacts, if aggressive programs are launched to develop this resource.

Significant increases in the use of wood energy and sound forest management techniques can--and must--occur simultaneously. Clearly, a program to ensure proper management of forests should be the centerpiece of national, state and local efforts to develop wood energy resources. Federal forest policy should be designed to increase industry efficiency and, on the demand side, allow maximum market freedom to determine the mix of energy and conventional forest products derived from wood resources.

Wood fuel use can leverage investments in forest management by providing a relatively short-term return on forest management investments (most forest management investments not linked to producing wood fuel do not yield returns until timber is harvested), thereby increasing the productivity and availability of all forest products.

This will not occur without sound forest management and harvesting practices. Lacking this, a sizable expansion in the use of wood energy could cause serious environmental and economic damage. For example, a sudden increase in the demand for fuel wood in a region could stimulate private woodlot owners to cut portions of their timber holdings in the interest of receiving near-term profits off land holdings which do not normally yield high returns. Any policy which encourages such action compromises both the renewable and multiple-use potential of forest lands. This is a significant problem. In many areas of the country, forest product industries rely on private non-industrial lands as a supply buffer and if this is not available because of extensive harvesting of wood for fuel, the cost of other wood products could be significantly inflated.

It is, therefore, apparent that Federal incentives to stimulate the development of a significant fuel wood industry must be carefully constructed to ensure proper forest management. The following analysis will suggest techniques for promoting such management using policies which rely principally on marketplace mechanisms.

The programs will have several specific objectives:

o developing and communicating techniques for optimizing the production of wood for energy and conventional wood products in each region in a way that will lead to profitable businesses in the near-term as well as the long run;

o promoting the development of new technologies in wood harvesting, transportation and conversion;

o ensuring that the many small industries likely to grow up around wood energy are not inadvertently disadvantaged by regulatory programs or existing tax policy.

The generic problems faced by companies interested in investing in new technologies was reviewed in Section 2 of this chapter. That section contained a number of specific proposals for ensuring that small and large industries have access to capital for retooling and new construction, for promoting investments in innovative new ventures, and for ensuring adequate support for corporate reserch. These measures would all apply to the industries needed to promote the use of biomass and they will not be reviewed in this section. This discussion will concentrate exclusively on programs with unique and specific applications in biomass.

Forest Management. Policies to encourage proper managment of forest lands must be the centerpiece of a national effort to increase wood production. Failure in this regard could result in both reduced productivity of forest products and significant environmental damage.

As Table 2.6 and 2.34 indicate, the potential production of wood from the nation's forests, given an intensive program to improve forest productivity, is significant. To achieve the 9 quads of wood energy use stipulated in Table 2.7 will require an unprecidented investment in forestry. Without such an investment, however, the nation runs a serious risk of irretrievably damaging precious woodlands, continuing to rely on high levels of wood imports, and failing to utilize an important domestic resource.

The federal government can encourage proper forest management through the provision of incentives which, additionally, would enhance (rather than distort) traditional markets for wood products. For example, incentives to encourage commerial thinning, stand conversions or reforestation can increase forest productivity, increase the availability and decrease the cost of low-quality biomass (e.g., wood residues). Wood fuel supply infrastructures can be expected to develop; fuelwood costs will decrease as a consequence, and wood energy economics will become increasingly attractive.

There are two distinct resource bases for wood produces: (a) private industrial and nonindustrial forests (common in the East), and (b) national forests (most prevalent in the West).

Table 2.34 POTENTIAL WOOD AVAILABILITY BY FOREST REGION

Area Forest Region (million acres)	Commercial Forestland[a] Owned	Potential Wood Percent Federally Uses[b] (Quads/yr)	For Energy and Nonenergy
South	188.4	7.6	3.0-6.0
North	170.8	6.6	2.3-4.6
Pacific Coast	70.8	50.8	1.4-2.8
Rocky Mountains	57.8	66.1	0.6-1.3
Total	487.7[c]	20.4	7.3-14.6[c]

[a]Commercial forests are those that have good productive potential and have not been set aside as wilderness areas, parks, or land reserves. About two-thirds of the forestland in the United States is classified as commercial.

[b]Assuming 40 percent of the total growth potential (18 to 36 Quads/yr) is accessible (See "Forestry Under Biomass Resource Base" in vol. II). Note that relative productivity factors as follows are assumed: Pacific coast = 1. South = 0.78, North = 0.66, Rocky Mountains = 0.58. These are calculated from the weighted average productivity potentials for the various commercial forestlands, using data from USDA Forest Statistics, 1977.

[c]Sums may not agree due to roundoff error.

SOURCES: Office of Technology Assessment and U.S. Department of Agriculture, Forest Statistics of the U.S. 1977.

In general, separate forest management programs and policies will be required for each, as desbribed below.

Private Forest Lands.
Sixty percent of all commercial forests belong to private, non-industrial owners. In the East, nearly three-fourths of the commercial forests are in this ownership class. As old growth timber inventories in the West continue to decline, and the price of oil continues to rise, private, non-industrial forest lands can be expected to play an increasingly important role in wood energy development and forest product production generally.

Unfortunately, neither market forces nor past federal policies have stimulated the four million individuals who own these woodlands to maximize the potential for their forests through proper management. A recent industry report revealed that, while five million acres of forest lands are being harvested annually, only 2.2 million acres are being regenerated through planting or purposeful site preparation. An even lower ratio--one out of nine harvested acres--is being purposefully regenerated on private, non-industrial forest lands (NFPA 1980). Industry estimates that, in the 25 major timber-growing states, almost 80 million acres of private, non-industrial forest lands need silvicultural treatments to boost productivity. The industry estimates that a $6 billion investment over the next 50 years in increased forest productivity would increase timber growth by six billion cubic board feet per year (NFPA 1980).

Technical Assistance Programs. Most small, non-industrial forest landowners lack the training and capital resources necessary to maximize the wood fuel and forest products potential of these lands. (Capital assistance programs will be discussed in a latter section.) Nor does the average forest landowner possess the marketing and forest management skills, and resources necessry to develop a wood fuel supply infrastructure.

The importance of providing technical and cost-sharing assistance to small, non-industrial forest landowners cannot be overstated. Recent increases in the demand and the corresponding price for residential fuelwood in the Northeast has already created a situation where forest landowners are frequently cutting the wrong trees in the interest of making a quick profit on the sale of cordwood (Boffinger 1980). Instead of increasing forest productivity, wood energy use in this case is actualy contributing to its decline.

The New England Fuelwood Pilot project (recently initiated by the USDA's Agricultural Stabilization and Conservation Service) is an excellent model program which provides small, non-industrial forest landowners with technical assistance and cost-sharing opportunities to improve the management--and profitability--of their forest lands. Landowners receive both technical and financial assistance in the marking of trees for cutting, evaluating forest stands, the construction of access roads, etc. Cooperative forest management is also encouraged, where appropriate, to realize the economies-of-scale, maximize labor and gain access to capital resources.

The program appears to have been strikingly successful. With a net expenditure of $1.4 million, the program has led directly to the management of about 20,000 acres. It is estimated that the program increased the average value of the land by about $78 per acre, and $118 per acre for land where roads were built. Construction of roads subsidized by the program led to the construction of roads serving an additional 18,500 acres. The program has generated enough residues so that 20 barrels of #2 fuel oil have been displaced for every acre of forestland treated; a total of about 400,000 barrels of #2 fuel oil have been displaced.

Clearly, the New England Fuelwood Project should be expanded to other regions and more aggresively funded if the potential for fuelwood resources is this country is to be realized. At a funding level of about $9 million (perhaps in part suported by DOE) the program would have national impacts.

Other existing programs could enhance forest management and wood energy development. These included programs in USDA' Forest Incentives Program (FIP) and the Agricultural Conservation Program (ACP). Neither of these programs attempts to encourage comprehensive forest management nationwide, or wood energy development per se. But clearly, the base established by these programs should be expanded and more aggressively funded to encourage sound forest management practices. These two programs could be improved by:

o review of a current limitation of the ACP program which disallows selling stand treatment residues as wood fuel;

o establish a cost-sharing program (along the lines of USDA's Forest Incentives Program) to subsidize forest stand treatments which generate economic wood fuel;

o increase funding for the Agricultural Conservation Program devoted for energy-related initiatives, perhaps with DOE funds;

o more aggressive actions to encourage stand conversions in the FIP program;

o increase funding to state foresters by lifting the "1 percent limitation" clause of the ACP program;

o make timber marking elligible for funding under the FIP program;

o provide technical assistance for low-income and small forest landowners.

Other ways to establish technical assistance programs, making use of existing statutory authority, would be to:

o Provide funds for forestry assistance and related programs authorized in the Cooperative Forestry Assistance Program of 1978;

o Provide funds for the education and technical assistance programs authorized in Sec. 255 of the Energy Security Act of 1980;

o Provide funds for the Model Demonstration Biomass Energy Facilities authorized in Sec. 251 of the Energy Security of 1980.

Capital Assistance Programs. Three capital-related programs face private, non-industrial landowners who attempt to invest in improved forest management:

1) the high cost of, and limited access to, capital resources;

2) the uncertainties surrounding forest production costs and potential returns;

3) cash-flow difficulties resulting from the periodic harvesting of forest stands (USDA 1978).

Tax credits or loans to offset the costs of reforestation or other cultural treatments could minimize the three capital related problems cited above. The following initiatives could also be useful:

o Provide direct and/or guaranteed, interest-subsidized loans to finance reforestation and stand treatments (on the model of USDA's Soil Bank). Special payment plans could be adopted to stimulate stand treatments, and timed with anticipated investment returns;

o Establish a program to assist individuals who can demonstrate that the collection of estate taxes would force the owner to clear timber prematurely;

o Increase the capital resurces of landowners by providing an investment tax credit to apply to a percentage of their forest management capitalization costs. An accompanying provision for 1-3 year amortization of capitalized costs would permit the rapid recapture of front-end management costs (rather than until wood is actually harvested);

o Eliminate current estate tax provisions which penalize non-industrial landowners (i.e., heirs) when they retain professional forestry assistance.

The Role of States in Providing Technical and Capital Assistance Programs. Since forest management techniques must be adapted to local conditions, it only stands to reason that the States should play a prominent role in forest management and wood energy development. In fact, twenty five states are expected to produce major amounts of wood for energy by the year 2000; the majority of wood fuel harvested in these states will be from private nonindustrial lands.

State forestry offices can assess and monitor forest inventories, overssee or regulate harvests on private lands, and provide technical and financial assistance to private, nonindustrial land-owners. At present, however, most state forestry programs devote most of their resources to forest fire prevention. The role of the states could be expanded with federal financing of forest management activities on a formula and/or cost-sharing basis. The Cooperative Forestry Assistance Act of 1978 provides a good structure for a cooperative federal-state forest managment effort.

Because of their sensitivity to local conditions, state forestry offices may also be able to play an ideal role in ensuring that fuelwood is harvested without major

economic or environmental problems. Federal technical or capital assistance funds (or even fire prevention funds) could be awarded to the states, contingent upon the successful implementation of forest management programs.

National Forest Lands

The federal government is directly responsible for the management of the National Forest System, which comprises 18 percent of total U.S. commercial forest land. The preponderance of national forest lands are located in the Northwest. In 1976 16 percent of total roundwood supplies and 20 percent of total sawtimber supplies came from the National Forests (RPA 1980). In 1978, nearly $900 million in timber was removed from the national forests (Stat Abst 1979).

Because of the scale of resources harvested from the national forests and the federal government's ability to ensure their proper management, these forest lands can play an important role in wood energy development. Important initiatives designed to stimulate wood energy development and forest management on National Forest lands include:

Logging Residue Utilization. Timber harvests in the National Forests--particularly those in the Pacific Northwest--typically gnerate large amounts of logging residue. About 2 quads of wood in the form of residues are burned (broadcast burned) or returned to the soil in the forests each year (OTA 1980). (The U.S. Forest Service subsidizes a portion of the broadcast burnings.) These residues, which represent a potential energy feedstock, now instead pose serious forest fire hazards and impede regeneration in many cases. Uncertain economics, lack of roads, lack of markets, and USFS policy currently impede the use of these residues as an energy or forest product feedstock.

USFS could encourage the use of logging residues by establishing 'concentration yards'. Residues would be hauled to a common site near potential markets to be sorted and sold. Instead of creating environmental hazards, these residues could be used in pulp, particle board or fuel markets. The revenues generated from a properly managed concentration yard could then be used to pay for the costs of harvesting and transporting the residues, allowing the yard to become economically self-sustaining. This concept is similar to practices used in the Scandianvian countries for many years.

The Forest Service has already established some concentration yards. A significant expansion of this effort is appropriate. To accomplish this, the USFS must allocate start-up funds from existing operating budgets, receive pass-through funds from DOE, or receive additional funding from Congress.

The Forest Service could also assume a position of leadership by demonstrating the use of integrated harvesting techniques. Systems which harvest and deliver logging residues, as well as the merchantiable wood can decrease the cost of wood fuel, increase the speed of reforestation, and increase the utilization of forest resources.

Forest Management Planning. The Forest Service does not yet treat wood energy development as an integral element of its management policies. Wood energy production should be integrated into existing residue utilization and reforestation programs, particularly in the Renewable Resource Planning Act activities.

Resource Assessment. Lack of reliable information on wood resource availability is a barrier to wood energy development. There is little data on regional and site-specific wood supply. Most of the forest inventory evaluations conducted by the U.S. Forest Service have made aritrary assumptions about the utilization of the forests and have focussed on determining "merchantible" wood availability. The agency has virtually ignored evaluating the production and availability of relatively low quality wood suitable for use as fuel. In fact, USFS inventory and potential growth estimates are at odds with recently completed assessments (OTA 1980). The Forest Service should immediately step up its survey efforts to comperhensively determine the supplies of various qualities of wood-biomass on both public and private lands.

Encouraging Wood Energy Use Through Direct Actions.

Harvesting

The evolution of a sizable wood energy supply industry in the U.S. will depend in large measure on the development of inexpensive harvesting equipment. In many areas of the country there are large quantities of low-quality wood which could be used as fuel if a means were available for harvesting and moving it at low cost. Current efforts to develop equipment suitable for harvesting small, low-grade and inaccessible wood must be strengthened. Integrated harvesting systems which can harvest fuelwood as well as conventional forest products must be developed (see R&D section in the following).

Small businesses and cooperatives will play an important role in future forest management and wood fuel harvesting. Their involvment, however, will be constrained by the high cost of equipment and by a lack of technical and business expertise. Therefore,in addition to R&D, it is also important to establish capital assistance programs as discussed earlier in this chapter. The Farmers Home Administration and the Small Business Administration could play key roles in these programs. It is important that the programs provide capital to harvesting systems which can remove fuelwood as well as conventional forest prod-

ucts from the woods so that truly integrated harvesting systems will evolve.

Conversion Equipment

Small industries with capital formation problems perceive the capital costs of wood-fueled systems as a major drawback. Moreover, as conventional fuel prices rise these industries will be even less able to convert to wood fired capacity because liquidity will be tighter. Consequently, capital assistance programs as discussed earlier in this chapter are needed. DOE should consider passing through a portion of its funds to SBA to establish loan programs to help small industries convert to wood.

Funding for feasibility studies and cooperative agreements for wood utilization can also be made available through the Alternative Fuels Production Program (authorized in P.L. 96-126). Based on analysis presented in Appendix D, it could be expected that $3.2 million in private funds would be invested for each dollar the federal government invests in this program. A total federal outlay of $1 million (administered through the Alternative Fuel Production Program) could result in the production of the equivalent of 15,000 barrels of oil per day by 1995.

Any tax placed on industrial fuels, of course, would provide an extremely powerful incentive for firms to convert to biomass energy use.

Energy From Agriculture

Agricultural crops, crop residues, animal manure, and food processing wastes all posses significant energy-producing potential. They can be used on farms to provide heat from grain drying or fuel for agricultural equipment. They can also be converted to fuel for off-farm uses, most notably as alcohols for use as transportation fuels.

Today American agriculture is dedicated almost exclusively to the production of food and feed, two essential commodities. It is intuitively obvious that any policy which could re-direct some part of the nation's agricultural resources to the production of energy bears careful scrutiny. The variations of climate, international and domestic market conditions, pestilence and disease, alternative or evolving agricultural markets, land productivity and the enviromental and economic costs of bringing new land into production all make it difficult to predict the extent to which agriculturaly produced biomass can increase the domestic production of energy without disrupting traditional agricultural markets.

Alcohol Fuels. Few alternative sources of energy have received such windspread public support as gasohol (defined as 1 part ethanol, 9 parts gasoline). Gasohol enjoys one of the largest subsides ever implemented to bring a new energy resource on line. Yet no other energy alternative is perhaps so widely misunderstood.

It has been suggested, for example, that because alcohol production from corn uses only the starch and preserves the protein for use as animal feed, large amonts of corn can be used for alcohol production without jeopardizing food/feed availability. Although it is true that alcohol production uses only the starch, this fact does not minimize the "food versue fuel" tradeoff. Corn has a high value as an animal feed in part because of its high starch content; sixty one percent of our corn is fed to animals in the U.S., and thirty one percent is exported--mostly for use as animal feed. Hence, use of large amounts of corn for alcohol will undoubtedly increase the price of corn for animal feed and, consequently, the prices of animal food products. The criticl, and unresolved issue, is the exact amount of alcohol which can be produced without significant adverse impacts on food prices. Many studies indicate that at the point that ethanol production from starch and sugar crops approaches a few billion gallons annually, food and feed prices will be inflated (OTA 1980).

It has also been suggested that the nation should set a goal of producing 10 billion gallons of ethanol annually by 1985. Producing this much ethanol from the agricultural sector is technically feasibile, but it is highly unlikely that the ambitious goal can be reached without massive federal subsidies and dramatically inflating the costs of food and feed. Economic ethanol production technologies designed to use non-food feedstocks may not become commercially available in time to precede increased food and feed prices (OTA 1980).

In addition, it has been suggested that it is possible to alter the crop mix (e.g., use greater amounts of sugar beets or forage beets) in order to achieve high yields of food and fuel. Although these concepts have considerable merit and deserve investigation, the implication and consequences of changing the crop mix are hightly uncertain at this time (because of the uncertainties of crop productivity and environmental impacts).

It may be possible to develop improved species and techniques which make more complete use of a plant's value as an animal feed and as an energy resource. Such improvements could change the estimates (on the maximum practicle level of ethanol production) made in previous analyses. For purposes of making national policy, however, the operative assumption should be that the final word on this issue is not in.

The foreign policy and international economic consequences of significantly expanded corn-based ethanol production have not been adequately considered. Most of our corn exports have been to OECD member countries and the Soviet Union. A unilateral decision by the U.S. to use substantial amounts of corn for alcohol could increase the cost of corn imports for OECD countries, and could be viewed as violating the spirit of cooperation on energy matters among IEA countries. Also, substantial increases in corn prices could lead to increases in the price of wheat, a food grain, which the world's poor are dependent upon. In general, the supply and demand responses to changes in corn prices and agricultural exports is volatile and uncertain.

Current Alcohol Subsidies Should be Restructed

Subsidies currently available for ethanol include exemption of the 4¢ federal excist tax on gasohol (40¢ per gallon subsidy on gasohol since gasohol is defined as 10 percent alcohol), additional 10 pecent investment tax credit for plants and equipment, and low interest or guaranteed loans for plant construction. In addition, over two dozen states have exempted gasoline from all or a portion of their gasoline taxes (worth as much as $30 per barrel). The most important of these is the federal gasoline excise tax exemption.

Forgiving the 4¢/gallon federal gasoline tax on gasohol mixtures is the equivalent of a $20 per barrel tax on crude oil. It would be far preferable to tax fossil fuel directly. While the argument is explained in greater detail in the chapter on transportation, it should be clear that a $20/barrel tax on oil or on gasoline would have as great an effect on the markets for gasohol as the current program, but would have the great advantge of allowing the market to choose between investments in transportation efficiency and investments in various synthetic fuels.

Some form of differential subsidies may be desirable however, even if a fuel tax becomes the centerpiece of policy in this area. For example, it may be desirable to give some subsidy to alcohol fuels derived from biological feedstocks because they have less environmental impact than alcohol produced from coal. (Coal production leads to environmental difficulties during mining and contributes to world burdens of CO_2.

In addition, the current excise tax exemption may result in higher than desired farm prices and uncomfortably low food/feed reserves. There is no limit on how much alcohol can receive the subsidy under current law, although research done to date indicates that major feed/food price increases could result from use of more than 10 to 15 percent of the corn crop for alcohol production (Tyner 1980). Also, the subsidy is not linked in any way to grain or oil prices, and if oil prices rise faster than corn prices, on average, the tax exemption becomes a large windfall subsidy.

Because alcohol fuels policy is intimately related to both agriculture and energy policy, it is recommended that future alcohol policy be coordinated more closely with existing policy in USDA and DOE. For example, farm program management decisions in USDA should incorporate projected demands for crops used for alcohol. The crop reserve programs will need to be reconsidered and perhaps expanded to include an energy reserve as well as the food/feed reserve. Also, alcohol production plans will need to consider projected feed/food demands.

If the current structure of the alcohol incentives is to continue, it would be desirable to develop greater flexibility in administering them. An interagency committee should be given responsibility for monitoring the effects of alcohol fuels production on food and feed prices and exports, and providing an 'early warning' of any potential supply or price problems. Authority should be provided to the task force to react appropriately to any market imbalance by: increasing food/feed reserves, altering regulation of cropland, restricting permits for alcohol production, and even altering or removing subsidies given to the production of alcohols.

Methanol Production Should Be Given Equal Emphasis with Ethanol Production

The principal objective of the alcohol fuels program should be to stimulate the most economic production of alcohols capable of displacing liquid fuels. Current policies do not let reason and sound market principles determine the type or extent of alcohol fuel utilization.

Methanol production for use directly as a fuel or as a blended fuel in proportions less than 10 percent (most methanol blends will be less than 10 percent) is not elligible for the gasohol excise tax exemption, but instead receives a 40¢ per gallon tax credit. Since this tax credit is taxable the effective subsidy for methanol is, at most, 22¢ per gallon, whereas the ethanol subsidy is 40¢ per gallon. Clearly, current tax law favors the production of ethanol over methanol.

In terms of research and development, the federal government should give equal emphasis to ethanol and methanol production (see R&D section following). In fact, in order to fascilitate equitable treatment, minimize duplication and ensure proper technological administration, ethanol and methanol R&D should be administered under the same program, preferably DOE's Biomass Energy System Program.

In addition, the comprehensive biomass development plan which USDA and DOE are required to prepare under Sec. 211 of the Energy Security Act of 1980 should provide a detailed strategy of the federal role in methanol and ethanol development.

Using Oil or Natural Gas as a Distillery Boiler Fuel Should be Strongly Discouraged by the Federal Government.

If the policy objective is net premium fuels production, oil and natural gas should not be used in producing alcohol. Yet, the Energy Security Act of 1980 provides that certain sugar refineries producing alcohol, or certain existing distilleries producing fuel grade alcohol be allowed to use natural gas on a priority basis (and not subject to curtailment). This provision will probably have the effect of promoting the displacement of gasoline with natural gas and/or increasing premium fuel usage with gasohol production. This is an inappropriate policy which should be changed, post haste.

General Agriculture Initiatives.

USDA's Extension and Outreach Organizations Should Begin to Aggressively Promote On-farm Energy Technologies

The Department of Agriculture is uniquely equipped to implement a technical assistance program directed toward on-farm energy production technologies. The USDA has a long record in the area of agriculture technology transfer emanating from its strong research support system and the extension service arm, the Cooperative Extension Service (CES). A most pressing need is to coordinate its research efforts with DOE and the private sector.

A committee comprised of USDA Extension directors, DOE program managers, and industry representatives should be established to identify on-farm energy producing technologies that are ready for application. In addition, USDA expertise should be utilized by DOE research managers to identify barriers to technology implementation, in order to mitigate design problems before technologies are offered for introduction.

The committee should ensure the distribution of technical information concerning all aspects of the application of on-farm energy technologies, and provide that information in a useful format to be distributed to extension agents. In addition, regional training programs should be developed to disseminate information about technology cost, availability, etc. to the local agents.

The technical assistance available from the Agricultural Stabilization and Conservation Service should also be utilized to provide direct technical assistance by trained personnel for the installation and operation of on-farm technologies. The existing ASCS is comprised of specialists in engineering and environmental areas who can provide technical assistance to agricultural producers. With a minimum of effort these technicians could be trained in the nuances of construction and operation of on-farm energy producing technologies, such as anaerobic digestion and on-farm distillation. Research priorities are discussed at the end of this section.

The Farmers Home Administration Should Take a More Active role in Energy Development

The Farmers Home Administration has historically provided funds for onfarm improvements. With appropriate adjustment of the FmHA Charter or reinterpretation of the existing charter, the FmHA could act to insure that adequate funding is available for on-farm energy producing technologies. Many financial institutions will consider on-farm energy producing technologies with considerable caution, and such loans and loan guarantees will be necessary.

The Rural Electrification Administration Should Take a More Aggressive Role in Renewable Energy Development

REA cooperatives typically have limited funds to invest in energy conservation or renewable energy production. However, REA's loan programs can provide funding for the purchase and installation of conservation and renewable energy systems in rural homes and farm related buildings, and play an important role in biomass development.

Research and Technology Development Priorities in Biomass

Federal work recommended in this area is examined in two major categories: reseach and development appropriate for federal contracting with industrial research laboratories, universities, and national laboratories; and technology development activities appropriate for industrial liaison programs.

The research work is divided into three basic categories: the development of biomass supplies with improved forestry and agricultural practices; development of improved techniques for converting harvested biomass (e.g., wood) into a useful energy product (e.g., high temperature steam or methanol); and more basic work designed to encourage fundamental innovations. The funding recommended for each category is shown as a percentage of the total funds recommended for research. Percentages in parenthesis indicate the recommended allocation of funds within each major category.

Technology development work is divided into the following categories: equipment for harvesting biomass, equipment for converting biomass, and equipment for making better use of these conversion products.

Research. The areas listed below are those requiring reserch for the mid- and long-term. In many cases some research in the area is already underway. The budget allocations recommended assume that the total federal biomass R&D budget does not decrease appreciably.

Budget Allocation:

Conversion Development   30% of total

Gasification
Efficient conversion processes for   (40%)
producing methanol from a variety of
feedstocks, particularly small (less than
200 dry tons/day input) mass-produced units.

Ethanol from Cellulose
A variety of advanced processes to   (60%)
pretreat and convert lignocellulosic
materials to ethanol and bench-scale
processes.

Innovative and Exploratory Research   25% of total

Thermochemical
Research on the basic thermochemistry of   (25%)
biomass, including the use of pyrolysis
and advanced gas-phase chemistry to
produce high yields of fuels and
petrochemical substitutes from biomass.

Biotechnology
Advanced processes for producing fuels   (25%)
and petrochemical substitutes using new
types of bacteria and yeasts (e.g.,
fermentation, screening of novel organisms,
biophotolysis, etc.) and other techniques.

Advanced Plant Research
Basic research in photosynthesis and plant (25%)
growth with emphasis on nitrogen-fixing
plants, hormonal treatment, rubber plants,
and hydrocarbon plants.

Other Peripheral Research
Support of basic research in high-risk and (25%)
(potentially) high-payback areas which may be
enormous value, but whose outcome cannot be
confidently predicted.

Supply Development   45% of total

In general, supply development research should be directly
managed by USDA.

Wood Supply
(a) Development of improved forest   (20%)
    productivity through better selection
    of species and better forest manage-
    ment strategies.

(b) Comprehensive surveys of forests   (10%)
    to estimate current and future wood
    resources on a regional basis.

(c) Analysis of the impact of large   (10%)
    increases in forest management on
    water quality, nutrient cycles, and
    other environmental aspects.

Crop Development
(a) Development of a variety of plant   (25%)
    species specialized for fuel and

chemical feedstocks and species
capable of yielding both food and
fuel. High yield perennial and
annual crops which can be grown on
marginal lands need to be developed.)

(b) Development of more sophisticated   (15%)
    techniques for crop-switching, food
    and fuel co-production, and evaluate
    the interaction between food, feed,
    and fuel production.

(c) Assessment of the short- and long-term (10%)
    environmental impacts of extensive biomass
    development on the agricultural sector.

Aquatics
Basic and applied research into the   (10%)
production of salt water and freshwater
aquatic plants to produce chemical feedstocks,
as well as food and fuel. (This should be
complete before any large scale
demonstration efforts are undertaken.)

Technology Development. Work in this area involves
technology or product development and must be done
in close cooperation with industries developing proprietary products. (See section on industry.)

Budget Allocation:

Supply Development   30% of total

Wood Supply
Low-cost easy-to-use wood harvesting   (50%)
and transportation equipment which can
collect low-quality wood from small tracts
of forest lands, or which can be integrated
with conventional harvesting.

Crop Development
Low-cost biomass harvesting equipment   (50%
which can be integrated into conventional
agricultural harvesting practices.

Note that, in general, supply development work should
be administered by USDA.

Conversion Development   50% of total

(The largest fraction of the technology development
budget is allocated to this category, not necessarily
because it is the most important, but because of the
high costs of developing prototype systems.)

Wood Combustion
Development of efficient, low-polluting   (20%)
industrial/commercial technologies

Development of small wood stoves which   (15%)
are safe, inexpensive, have low emissions
and maintenance requirements, and
efficiencies well above 50%.

Gasification
Development of low- and intermediate Btu   (65%)
gasifiers (now in prototype stage) which
can accomodate a variety of feedstocks in
industrial and commercial applications.

Applications Development   20% of total

Use of Alcohol Fuels
(a) Examination of technical and legal   (20%)
    problems faced in distributing alcohols
    (methanol and ethanol) through conven-
    tional fuel systems.

(b) Development of low-cost additives   (20%)
    which prevent alcohol-gasoline blends
    from separating.

(c) Analysis of the costs of modifying   (10%)
    current and future transportation fleets
    to accept different blends of alcohol
    fuels.

(d) Examination of direct and indirect   (20%)
    environmental impacts of using alcohol
    fuels.

By-Product Utilization
(a) Examination of alternative strategies   (9%)
    to displace petroleum-derived chemicals
    with biomass.

(b) Development of wood ash disposal   (1%)
    techniques

Use of On-Site Produced Gas
Development of efficient use of biogas   (20%)
produced from thermochemical processes
for on-site direct heating and cogeneration.

DIRECT SOLAR INDUSTRIAL PROCESS HEAT

Heat for industrial processes applications can be
generated directly from solar radiation using a variety
of mechanical devices a variety of devices for
providing such direct solar heat have been proposed
and tested. Simple covered and salt gradient ponds,
flat plate collectors, similar to those which provide
space heat and hot water to buildings, can provide hot
water and hot air. Evacuated tube collectors, and
small concentrators can provide low pressure steam;
more elaborate (and more expensive) systems which

track the sun can provide temperatures of 1000°F or more. Figure 2.11 indicates the range of temperatures which can be produced from the major class of collectors.

As with many solar technologies, the attractiveness of these devices depends critically on whether costs can be reduced. For example, evacuated tube and concentrating collectors penetration are expected to have a significant only when installed system prices can be reduced to less than half that of the current generation of experimental units, the cheapest of which cost approximately $40-$50 per square foot. Contributions above about one Quad of energy annually are likely if costs can be reduced to a third of current system costs. There are sound reasons for believing that such cost reductions are possible, but sustaining a major development program is plainly necessary to achieve them.

The use of direct solar industrial heating systems will also depend on the temperature of the heat required, the amount of solar radiation received in the area, and the availability of land. It is expected, for example, that solar systems will be able to supply a larger fraction of low temperature industrial heating requirements than high temperature requirements. The total contribution of the technology also depends on the extent to which efficiency technologies, including industrial cogeneration, capture the markets which appear to be most easily compatible with the direct solar systems. Large land areas may be available near plants in the rapidly expanding industrial areas of the sunbelt, but northern industries--particularly those in crowded urban areas, will face greater difficulties. The constraint would be eased significantly if techniques can be developed for moving energy inexpensively from remote collector fields to factories; techniques for doing this also deserve major research attention.

It must be emphasized that a precise anaysis of this potential market would require a detailed analysis of the merits of different combinations of direct solar and energy efficiency technologies placed in competition with conventional fuels which may be available to industry during the next twenty years. The analysis of future demands for industrial process energy reported earlier did not examine many of the industries which may make the best markets for direct solar heat in detail. The earlier analysis directed most of its attention to the primary materials industries because they represented such a large fraction of the nation's industrial energy consumption. Solar equipment, however, may be most compatable with such applications as food processing, mining, textiles and a variety of other industrial classes. It must also be recognized that separate projections have not been made for each temperature range; no account was taken of the fact that greater savings are likely to occur in low temperature processes than in high

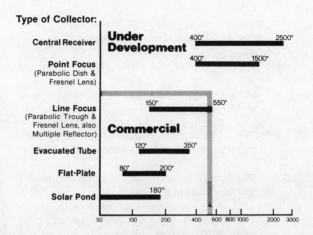

Figure 2.11. **Operating Temperature (°F)**

temperature processes. Low temperature heat can be supplied through a variety of heat recovery and control improvements. Industrial cogeneration was not considered in the demand reductions cited and would also reduce demands for industrial heat. These effects make it difficult to make confident statements about the total potential markets for direct solar equipment. It is unlikely that the demands cited in Table 2.8 have overstated the market, however, since these figures are based on aggressive implementation of conservation, and do not consider competition between solar and conservation technology.

The cost of energy delivered by a direct solar thermal system depends on the amount of sunlight available in a region, the percent which is direct as apposed to difuse, the correlation of the availability of this solar energy with the demands of the industry, the efficiency with which the system collects sunlight and converts it to a heated fluid, the total installed cost of the system, its expected lifetime, operation and maintence costs, and a number of other details depending on the integration of the solar equipment with each the industrial system. In any well designed system, it will be necessary to integrate the design of the solar equipment with the overall engineering of the industrial process to ensure that the device is economically optimum. Since it will not be possible in this study to do this for the enormous variety of industrial facilities in a large number of climates, it will be necessary to select a relatively small number of surrogate systems for the purposes of this analysis.

The analysis will proceed in the following steps:

(1) The total potential market for solar direct heat will be determined by examining solar

compatability with different temperature ranges in each major U.S. industry.

(2) The fraction of the total demand which can be economically supplied by the solar equipment and associated storage systems will be determined.

(3) The amount of sunlight available in each major industrial region will be computed.

(4) The potential future cost of solar equipment will be estimated.

(5) The value of solar equipment in different parts of the nation will be estimated on the basis of future fuel costs.

(6) The constraints due to land requirements will be estimated.

(7) The total amount of energy provided by direct solar equipment will be estimated by examining the previous items in light of the economic expectations of several different classes of potential system owners. This calculation is performed in each region for both high and low temperature (less than 212°F) systems and the results combined.

(8) An analysis will also be presented which shows the impact of different assumptions about growth in industrial energy efficiency in areas related to solar energy.

The Potential Market

Table 2.8 summarizes the judgements made about the compatability of different kinds of industrial systems with direct solar equipment. This was done by estimating markets in each temperature range in each industry type using the data in Table 2.35. These analyses have been combined with statistics on industrial shifts taken from a recent survey (Insights West 1980) to estimate the fraction of industrial demand which could be supplied with solar equipment in different regions. The result clearly varies with climate, the cost of storage, and overall system costs but the optimum solar fraction appears to be in the range of 42-55% in most parts of the nation of interest to direct solar markets. The calculations in this report will assume that, on average, a solar direct heating system for industry will supply 50% of the industry's annual process heat demands. It was also assumed that all of the direct heating needs of the paper industry and 25% of the heating needs of the food processing industry could be met more economically using biomass available onsite. The table also reflects a

judgment that direct solar equipement will have more difficulty finding markets in industrial applications which require high temperatures. A significant amount of development must be done to develop systems capable of providing energy at temperatures higher than 1100°F. The net effect of these assumptions is a judgment that the potential market for direct solar equipment will be limited to about 50% of the total process energy used in the year 2000.

The Solar Fraction

It is seldom economically attractive to build a solar unit large enough to supply 100% of an industry's process heat requirements. Some source of backup heat is typically used during periods when solar heat is not available directly or from storage.

Determining the optimum size of solar equipment requires an analysis of the correlation of energy demands and energy supplies which is specific to each plant. One critical factor is the number of shifts operated. Solar units are likely to be more compatible with one-shift operations where demands occur during the day. Many of the small one-shift plants, however, do not operate on week-ends and thus the solar heat produced during this time will be wasted if it is not stored. The fraction of the demand which can be supplied economically by solar units will plainly also depend on the local climate.

In the absence of a detailed analysis of optimum system designs for each industrial class in category and climate, this study will use a simple national average extrapolating from the results of the few detailed analyses available (Dickenson 1978, OTA 1978, Hock 1980), and the results of computer simulations (e.g. BALDR).

Available Sunlight

The effective cost of energy delivered from industrial solar heating units depends on the amount of sunlight available at major U.S. industrial sites. This analysis can be performed by examining Figures 2.12 and 2.13. Figure 2.12 shows the annual distribution of "direct normal" sunlight in the U.S. Direct normal solar energy is the solar energy which can be focused using concentrating collectors such as troughs heliostats and dishes; it is the energy which is received from the sun's disc without scattering. This information can be used to compute the annual output of different classes of solar collectors located anywhere in the U.S. Figure 2.13 indicates the total energy received on a horizontal surface. This energy includes direct normal sunlight and sunlight which has been scattered from clouds or turbidity in the atmosphere. Figure 2.13 can be used to determine the energy available from a solar pond or flat-plate.

Table 2.35 PROCESS TEMPERATURES IN MAJOR INDUSTRIES
(Percent of Total Demand in Each Temperature Range Used in Each Industry Class)

Process Temperature (°F)	Food	Paper	Chemicals	Petroleum	Stone/glass	Primary Metals	Other
0–212	16	21	8	0	2	16	36
212–350	11	33	29	5	1	1	20
350–550	5	18	33	16	0	11	17
550–1100	2	0	8	71	0	0	18
1100–2000	0	0	0	46	54	0	0
2000+	0	0	0	0	34	66	0

Source: Intertechnology 1977, May 1980, Brown 1980, Ketels 1979.

Figure 2.12.

Manufacturing Value Added Density with Average Daily Direct Normal Solar Radiation (KJ/m²)

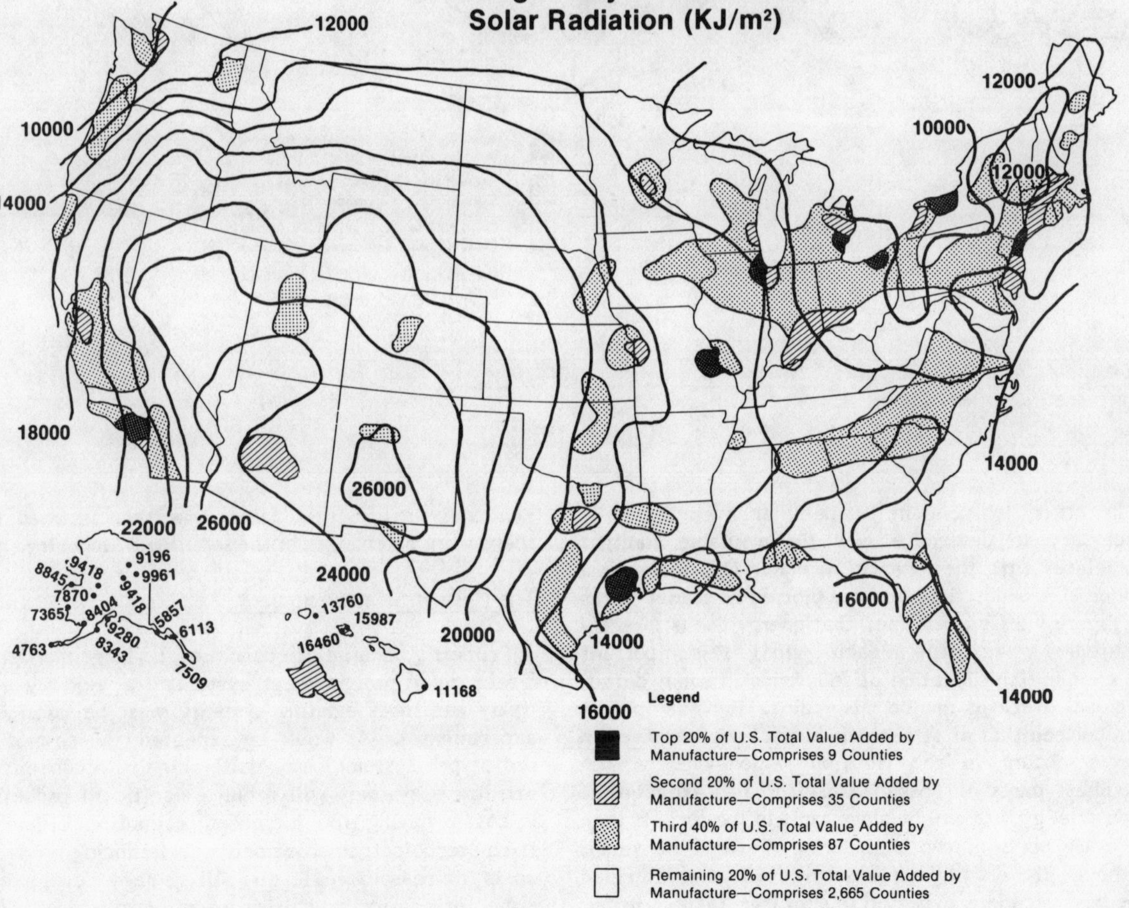

Legend

■ Top 20% of U.S. Total Value Added by Manufacture—Comprises 9 Counties

▨ Second 20% of U.S. Total Value Added by Manufacture—Comprises 35 Counties

▧ Third 40% of U.S. Total Value Added by Manufacture—Comprises 87 Counties

□ Remaining 20% of U.S. Total Value Added by Manufacture—Comprises 2,665 Counties

Figure 2.13.

Manufacturing Value Added Density with Average Daily Total Radiation on Horizontal Surface (KJ/m²)

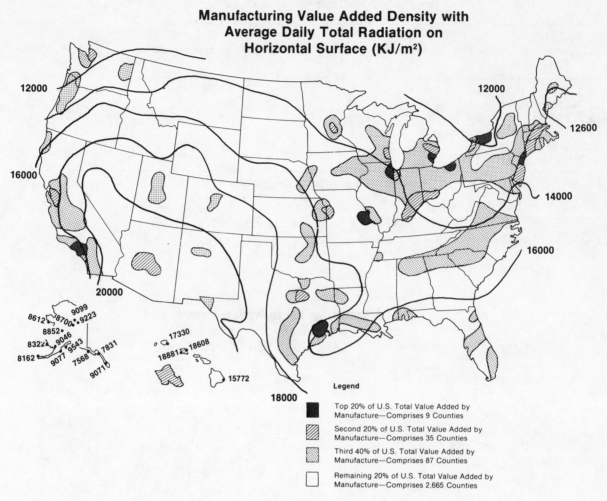

Legend

■ Top 20% of U.S. Total Value Added by Manufacture—Comprises 9 Counties

▨ Second 20% of U.S. Total Value Added by Manufacture—Comprises 35 Counties

▦ Third 40% of U.S. Total Value Added by Manufacture—Comprises 87 Counties

□ Remaining 20% of U.S. Total Value Added by Manufacture—Comprises 2,665 Counties

In order to compute a national average, it is necessary to determine how the available sunlight correlates with the location of major U.S. industries. Major U.S. industrial sites are plotted as shaded areas in Figure 4.6 (it is assumed that energy use is directly related to valve added in each region). It is important to examine the location of industries in some detail; regional averages can be misleading. For example, it can be seen that most of the industrial heat required in Texas occurs in the Houston region--one of the cloudiest parts of Texas. Most forcasts expect that industrial growth will be more rapid in the sunbelt than in other parts of the U.S. during the next 20 years, although the dramatic shift south which characterized the past decade is expected to slow (see Table 2.36).

The analysis presented later, however, assumed that there were no changes in the location of industry.

The Cost of Solar Equipment

Primarily because of economics, no large market for direct solar process heat systems for industry exist today and most existing systems must be considered experimental. As would be expected, the cost of the prototype systems are quite high. Economically attractive systems will require significant reductions in cost resulting from increased production volume and from technical improvements. Reducing collector costs to reasonable levels will require developing a collector design which requires significantly less

Table 2.36 REGIONAL BREAKDOWN OF INDUSTRIAL ENERGY USE[a]

Census Region	Fraction of Total U.S. Use (Quads- 10^{15} Btu)	
	1974	2000[b]
New England	2.4%	2.4%
Middle Atlantic	13.2%	12.8%
South Atlantic	11.8%	10.2%
East North Central	21.9%	20.2%
East South Central	7.4%	9.7%
West North Central	4.4%	4.0%
West South Central	27.5%	28.5%
Mountain	2.8%	2.8%
Pacific	8.6%	9.4%

[a]Includes mining, construction, and manufacturing.

[b]Year 2000 estimates were prepared from numbers used by EIA as background data for their annual report to Congress, 1979.

Source: Energy Consumption Data Base, Volume I, Summary Document: FEA.

weight per unit of collector area than current equipment. The development and testing of very low cost collector designs should be at the center of any private or public program to develop markets for direct solar process heat.

It will also be necessary to develop techniques for reducing system costs not including collectors. Installation, controls, storage, piping, adjustments needed in the plant receiving the energy, and other costs represent a large fraction of the total price of solar heating systems even at todays relatively high system costs.

A major study is presently underway to collect engineering analysis which can be used to estimate the potential for reducing the cost of collectors used to supply industrial heat. The results are summarized in Table 2.37.

The Cost of Delivered Solar Energy

The cost of the energy delivered from a solar installation can be calculated given the annual useful ouput of the collector (system efficiency and insolation), the installed cost of the system, operating costs, and the effective cost of capital. The economics of the systems will be computed from two perspectives: (1) that of a private corporation expecting a simple five-year payback on all capital invested in the system, and (2) that of a privately owned electric or gas utility. In all cases, the annual cost of maintaining systems is assumed to be 2% of the initial installed cost; these costs can be combined with capital costs which are also directly proportional to the first costs of the system. The effective capital costs (including operations and maintenance costs)

used in the analysis are 22% for the case of industrial ownership and 14% for the case of utility ownership. (E.g. the effective capital costs of a dollar invested by a utility is 14¢/year.)

The information provided in Figures 2.11 through 2.13 and Table 2.37 can be combined with these capital costs to compute the effective cost of solar energy in each area where there is a major concentration of industries in the U.S. This regional analysis was performed separately for low temperature (less than 212°F) systems and higher temperature systems. A region-by-region analysis has been performed and the results displayed in Table 2.38. This table indicates the fraction of the potential market for solar industrial heat which is economically attractive if competing fuel prices exceed the costs shown in the table. Land limitations and limitations of storage will further reduce the total contribution of solar industrial heat.

The Value of Solar Energy

The fraction of industrial heat which can be supplied by solar equipment can be computed from the information presented previously given an estimate of the value of the energy delivered. The value of the fuel displaced by the solar equipment also depends on the region examined since the types of fuel used and the cost of fuels vary with location. The fuels used to supply industrial heat in each major U.S. census region are shown in Table 2.39. The future fuel prices examined follow the assumptions used throughout this report. Low estimates assume that world oil prices reach 40 dollars per barrel in 2000 and natural gas prices reach $5.9/million Btu. High estimates assume that world oil prices reach $60 per barrel in the year 2000 and gas prices reach $9.7 per million Btu. If it is assumed that the mix of fuels used in each region remains the same through the year 2000 (see Table 2.39), these fuel prices can be used to calculate value for the solar energy which can be supplied by industrial systems.

If coal or biomass can substitute for oil or gas, the value of the energy displaced may be much lower than the values used in the following analysis. The analysis leading up to Table 2.8 however, removed many industries which are most likely to shift to solid fuels. It has been assumed that the remaining industries would compare solar energy with the price of oil and gas. This may be an optimistic assumption.

Land Use Constraints

The fraction of an industry's process energy demands which can be met with solar equipment also depends on the land available for collectors. Factory roof tops can be used, but in some cases the cost of reinforcing roofs to make them able to support collectors adds

Table 2.37 POTENTIAL FUTURE COSTS OF SOLAR INDUSTRIAL HEAT SYSTEMS (1978 $)

Confidence Level	Concentrating Collector		Flat Plate	
	System Cost ($/ft^2)	Efficiency (%)	System Cost ($/ft^2)	Efficiency (%)
1. Conservative	19	50	10	35
2. Expected	15	50	8	40
3. Optimistic	11	60	5	40

Source: Solar cost goals, 1980.

Table 2.38 FRACTION OF PROCESS HEAT DEMAND WHICH CAN BE MET AT DIFFERENT COSTS
(in percent)

Fuel Cost ($/MBtu)	Industrial Financing			Utility Financing		
	Optimistic	System Costs Expected	Pessimistic	Optimistic	System Costs Expected	Pessimistic
6.2	36	3	2	93	40	14
8.1	71	21	3	97	67	41
9.9	97	42	19	97	92	46

Note: The fraction is the ratio of the energy which can be produced at the indicated price in the industrial energy markets judged to be compatible with direct solar heat to the total compatible market (see Table 4.1). The Table is the result of an examination of high and low temperature systems operating in each major industrial region.

Table 2.39 REGIONAL INDUSTRIAL FUEL MIX (1975)
(%)

	NE	MA	SA	ENC	WNC	ESC	WSC	M	P
Distillate (#2)	8.3	8.3	8.9	5.2	4.2	6.2	2.0	11.5	5.4
Residual (#6)	60.8	29.0	24.0	10.4	6.0	7.6	2.8	10.5	9.4
Natural Gas	30.9	62.7	67.1	84.4	89.8	86.2	95.2	78.0	85.2

Source: FEDS Statistical Summary, EIA, Feb. 1978.

prohibitively to the overall system cost. This problem may become less serious if it is possible to develop very low cost collectors which would probably be much lighter than the current generation of collectors.

However, factory roof areas, even if completely covered by solar collectors, would in most cases supply only a small fraction of the industrial process heat demand. Therefore, it is necessary to examine land availability at industrial sites in order to get a better estimate of the potential contribution of solar IPH. Unfortunately, very little such information is presently available. The little data there is indicates that land availability may be a significant factor constraining the potential use of solar IPH, although most estimates of the potential for solar IPH to date have not taken this factor into account.

An analysis of the distribution of industrial process heat demands by the size of the industrial plant indicates that in a moderately sunny climate, a solar unit large enough to supply half of the energy requirements of for the average industrial plant in the region would require more than 140 acres of land.

While new industrial plant sites in the south or southwest might have that much land available to them for collector arrays, almost all of the older, more densely packed industries of the eastern and midwestern industrial states most likely do not. As can be seen from Table 2.36, the mid-atlantic and east north central regions represent a significant percentage of industrial energy consumption. The relatively high growth rates of industry in sunnier and less congested areas will of course, increase the liklihood that the plant can be located close to an area adequate to support the needed collectors.

A recent study done for the Gas Research Institute (GRI, 1980) collected data on land and roof availability, and total energy use for 200 different industries in all nine census regions. Less than 40 percent of the industries surveyed would be able to supply 100 percent or more of their total plant demand from solar thermal energy if all available land and roof were covered with solar collectors.

In light of the large regional diversity and uncertainty associated with this issue, and considering the lack of sufficient data necessary for a more careful analysis, the analysis conducted here will examine land constraints which reduce the potential solar contribution by 25-50%. The 25% reduction is probably optimistic, and is contingent upon significant increases in the net efficiency of energy used in industrial process applications. The 50% reduction is more likely if energy efficiency improvement targets are not reached.

Any more accurate assessment of land constraints would have to rely on a plant by plant investigation.

It must be recognized, however, that surveys typically examine only the land currently owned by factories. They do not examine the availability of land which might be purchased by the company if it has an interest in solar equipment.

The land available for such purchases would increase significantly if it proves possible to transmit energy from solar collector fields to a factory at low cost. Energy can be transmitted in the form of steam or pressurized hot water or a chemical technique can be used to convert the solar heat to a fluid or gas which

can be piped to the industrial site. Studies conducted by the General Electric Company and Dow Chemicals have shown that it may be possible to transmit energy in the form of high pressure steam (650°F, 2200 psia) for distances up to 20 miles. Further analysis of the cost, safety, and feasability of energy transport systems compatible with solar heating systems of varying temperatures are needed. (GE, ; Dow,)

The Total Potential Contribution of Solar Process Heat Systems

Estimates of the expected cost of solar ollectors (shown in Table 2.37), the resulting cost of solar energy (shown in Table 2.38), the value of solar energy (computed from the fuel mixes shown in Table 2.39), limitations placed on the use of solar energy imposed by compatibility with different types of industry (Table 2.8), the limitations of storage, and the limitations imposed by land availability can be combined to produce estimates of the total potential contribution of direct solar industrial process heat systems. These estimates are shown for a in Table 2.9. It can be seen that the outcome is extremely sensitive to the assumptions made about fuel prices and collector costs. The advantage of achieving the "optimistic" solar system prices is, however, much reduced if fuel prices incrase rapidly. Low system costs have a pronounced effect if fuel prices are low.

It should also be recognized that the solar contributions shown in Table 2.9 assumes full exploitation of the potential for increasing industrial efficency described in earlier sections. No attempt has been made to optimize solar and efficiency investments. It is reasonable to expect that many efficiency technologies assumed will be less expensive than the solar heating systems examined here. If not, the total contribution of solar could be larger than the contribution shown in Table 2.9. The current policies and new policies columns represent different levels of efficiency improvements as discussed earlier in this Chapter.

Since the range shown in Table 2.9 is so large, an effort was made to compute an "expected" value for the potential contribution of direct solar equipment. This was done by assigning a probability to each possible choice shown on Table 2.9 (except fuel price and finance option). The results of the analysis, together with the probabilities selected, is shown in Table 2.10.

Policy Implications

It is clear from the preceeding discussion that federal programs in direct solar industrial process heating systems should place major emphasis on research designed to greatly reduce the costs of solar collectors, storage equipment, and systems designed to transport thermal energy. This will require a substantial work in lightweight materials, durable reflecting surfaces, designs which withstand wind loadings wihtout large use of materials, development of receiver surfaces which can withstand repeated temperature fluctuations, and systems designed for quick installation. An analysis of problems encountered in recent demonstration projects indicated that the following problems also need attention (Kutscher, 1980):

o degradation of absorber surfaces and glazings;

o improved system designs taking careful account of parasitic power losses in pumps and other equipment and losses in piping.

Work on the development of prototype systems with a real prospect of providing thermal energy at attractive costs should clearly preceed any widespread demonstration of technology.

The economic attractiveness of solar equipment will plainly also be improved if they can be financed by organizations able to accept relatively long paybacks. Table 4.11 indicates the increased use in solar thermal energy which would result if utilities financed the systems instead of industrial investors. The Table indicates that utility financing could increase the solar contribution in industry dramatically in all but the one case which assumes the highest oil price and the lowest solar equipment cost. Techniques for encouraging more venturesome industrial investments and utility investments are discussed in later sections of this analysis.

Notes to Table 2.33a and 2.33b

[a]This scenario assumes biofuel exploitation up to the point where production costs are at or below competitive fossil fuel costs and competitive resource uses. This limits supplies of methanol to that produced for less than $10.50/MBtu. In all likelihood, an escalation in the energy value of wood would lead to a similar increase in the fiber value. Many of the substitutes for wood are petroleum derivatives, so that an increase in energy prices will raise the value of wood both as a fuel and a structural material.

[b]Includes 0.34 quad of municipal solid waste--approximately the amount available at prices up to $2/MBtu. Category indicates quads of feedstock input--mostly wood. Significant quantities of agriculturally-based lignocellulose could also contribute to this category.

[c]At high fertilizer prices the energy contribution from manures as fertilizer may exceed that of biogas (used to produce ammonia as fertilizer). At the same time, development of hybrid crops that use ammonia much more efficiently could reduce the fertilizer demand for ammonia (now 0.45 quad per year).

[d]This methanol could be used in turbine or fuel cell applications.

[e]High range (from OTA 1980) assumes feedstock costs as high as $1.26/gallon ethanol for sugar cane. Costs could be higher if feedstock production costs (on potential cropland, for example) are higher.

[f]Higher methanol cost assumes wood feedstock costs as high as $80.00/dry ton.

[g]Assumes the use of wood-fired boilers and gasifiers in the non-forest product industry sector.

[h]Grain and crop drying.

[i]Ethanol categories not necessarily additive. Table indicates quads of ethanol (not quads of feedstocks).

[j]This methanol could be used in turbine, fuel cell, or chemical applications. This methanol is not necessarily additive with other methanol categories because of restrictions on the wood resource base.

[k]Subtotals are not additive but are reduced by those resource categories (e.g., wood) which are not additive.

[l]In this category, about .89 quads of fuel (oil equivalent) is displaced per quad of feedstock input, assuming an average direct combustion/gasification conversion efficiency of 76% replacing oil burned at 85% efficiency (Glidden 1980).

[m]Under ideal conditions, it is assumed that for every Btu of wood (primary supply) used, approximately .93 Btu of oil equivalent can be displaced. This conversion efficiency will be achieved if future methanol plants achieve between 60 and 70% conversion efficiencies, and if future automobiles are engineered to exploit methanol's unique characteristics (by utilizing automotive engines which disassociate methanol into hydrogen and carbon monoxide, and by increasing compression ratios) thereby increasing end-use efficiency by 20 to 35%. (Willimas 1980) (Reed 1981).

[n]For this ethanol category, between 6.5 and 11 billion gallons of gasoline equivalent are displaced for every quad of ethanol produced. For simplicity, it is assumed that most of the ethanol is produced from grain grown on potential cropland, and that the distilleries are coal- or biomass-fired. The range of displacement is a function of whether or not the ethanol is used in a blend or stand-alone fuel, or as an octane booster. Future octane booster markets are expected to be very limited (OTA 1980).

[o]Methanol categories are not additive, dependent upon wood availability.

[p].8 quads of oil are assumed to be displaced for every quad of wood input in the residential sector (see Buildings Chapter).

[q]Quads of primary methane are shown in terms of product. It is assumed that using biogas in on-farm applications has the same process efficiency of conventional fuels.

[r]See Appendix for description of estimates for ethanol costs. This range only represents ethanol fermentation costs.

[s]See Appendix for description of cost estimates for methanol.

[t]See Appendix for description of cost estimates (OTA 1980).

[u]See Appendix for description of cost estimates (OTA 1980).

CHAPTER 3
TRANSPORTATION

Section One
Overview

Most Americans discovered the "energy crisis" in 1973 when they found that their routine trips to the local filling station had become long and trying waits for a suddenly scarce commodity. Tempers were lost and there was an eager search for explanations. The explanations proved to be unpalatable. The nation was forced to discover that it depended on foreign sources for a substance that formed an indispensable part of the national economy and had penetrated its personal life in everything from commuting to work to driving a Chevy to a levee. The energy required to move people represented more than 80% of the annual U.S. production of crude oil; the energy required for all forms of transportation required 12% more oil than the U.S. produced. The real, although seldom articulated, basis for the anger was the question of whether the nation's transportation system could survive without radical changes and sacrifices.

The analysis in this chapter results in relatively good news. The study concludes that it should be possible to sustain national transportation habits almost unchanged over the next two decades if the nation is able to make sensible use of technologies which are already well understood. (Results are shown in Table 3.1) In fact, it shows that with properly designed policies, it is possible that the total demand for energy in transportation can be 15 to 35% less than the U.S. used in 1977, while transportation services provided for each person (both personal transportation and freight) are increased significantly. The savings, of course, could be much larger if the transportation needs for each person do not increase as rapidly as assumed (a future which may be very plausible given that Americans are already among the world's leading travelers). Significantly, these savings do not require any major change in the kinds of vehicles now in use or in the way they are used. Public transportation, for

example, is not assumed to play a very large role; telecommunications, which may revolutionize transportation in this century and save large amounts of energy, is not credited with any savings in Table 3.1. The only "mode shift" contemplated is a transfer of some long-distance freight from truck to rail. The private automobile is expected to dominate personal travel through the indefinite future, although it is likely that the typical vehicle will decrease somewhat in size. In all sectors, the critical assumption has been that the nation will find ways to increase the efficiency of vehicles rather than sacrificing freedom of mobility. As a bonus, improved vehicle efficiency will, with few exceptions, translate directly into improved environmental quality. Increasing efficiency may provide an attractive way of improving national productivity and reducing harmful emissions simultaneously--there are few such opportunities.

TRANSPORTATION FROM RENEWABLE RESOURCES

A successful program for improving efficiency will provide more time to perfect the technology of alternative fuels and allow industry, and the nation as a whole, more time to assess the alternatives. The first, and in the short term the most important transportation fuels derived from renewable resources will be alcohols (primarily methanol) derived from wood, agricultural crops, and other sources. Without government interference methanol is likely to be the dominant alcohol fuel in 2000 because it can be produced from a greater variety of feedstocks (including coal) and because it is likely to be cheaper. Given analysis provided elsewhere in this report, it seems alcohols manufactured with high efficiency

Table 3.1. ENERGY USE IN TRANSPORTATION
(Quads)

		1977	2000 No New Policy	2000 New Program Effects	2000 New Programs With 50¢/gal. Gasoline Tax	New Programs
	Personal Transport					
A.	Automobiles and Light Trucks (personal use)	10.0	7.5	6.2/4.3/3.5	5.9/4.1/3.3	R&D, fuel taxes or "gas guzzler tax", and vehicle labeling programs achieving 30/55/75 mpg in 1995.
B.	Automobiles and Light Trucks (fleets)	2.9	2.7	2.3/1.5/1.2	2.1/1.4/1.1	R&D, fuel taxes or "gas guzzler tax", and vehicle labeling programs achieving 30/55/75 mpg in 1995.
C.	Commercial Airlines	1.4	1.8	1.6	1.34	R&D, increased load factors, efficient air-traffic control.
D.	General Aviation	0.15	0.37	0.37	0.30	R&D, equitable user fees.
E.	Military Aviation	0.50	0.60	0.60	0.60	
F.	Buses	0.13	0.20	0.20	0.20	Mass transit programs, increased efficiency.
G.	Rail	0.06	0.09	0.06	0.09	Mass transit programs, increased efficiency.
	SUBTOTAL	(15.14)	(13.26)	(7.6-11.4)	(6.9-10.5)	
Freight*						
A.	Trucks	2.0	3.6-5.4	2.1-2.5	2.1-2.5	R&D, deregulation, equitable user fees, deregulation, piggy-back useage resulting in increased efficiency.
B.	Rail	0.55	1.0-0.7	1.1-1.0	1.1-1.0	R&D, deregulation, equitable user fees, deregultion, piggy-back useage resulting in increased efficiency.
C.	Water	1.1	1.5	1.5	1.5	
D.	Air	0.09	0.30	0.20	0.20	R&D
E.	Pipeline	0.60	0.80	0.80	0.80	
	SUBTOTAL*	(4.34)	(7.2-8.7)	(5.7-6.0)	(5.7-6.0)	
	TOTAL DEMAND	19.5	20.5-22.0	13.3-17.4	12.6-16.5	

*Effect of fuel taxes not explicity considered in these categories.

from biomass could provide up to 35-45% of the nation's transportation needs on a sustainable basis, given a successful program to improve transportation efficiency. This results in part from a judgment that some of the nation's biomass resources would be used directly by the wood products industries and other companies using wood or other biomass material. It would be theoretically possible to supply 100% of the nation's transportation needs from alcohols if all sustainable yields of biomass from conventional sources were used to meet national transportation demands although it is extremely unlikely that this would be an economically preferred way to use biomass resources. The biomass fraction of transport energy could also be larger if national demand for transportation services grows more slowly than has been assumed.

By the turn of the century, it should be possible to generate electricity from renewable resources (wind or photovoltaics for example) which could also provide transportation services from renewable resources at competitive prices. A brief survey of the alternatives indicates that telecommunications, electric rail, and electric automobiles appear to offer the greatest promise for using electricity in transportation. Of these, only telecommunications and possibly electric

rail are likely to result in any significant energy savings; electric road vehicles may use somewhat more primary energy than internal combustion engines to deliver the same transportation service. The advantage of an electric vehicle, of course, is that electricity can be generated from a variety of sources of power--including solar, coal and nuclear energy. It may eventually also be possible to produce hydrogen and other transportable fuels directly from solar energy.

All of the transportation alternatives examined here must be compared with the possibility of producing synthetic liquids from coal and other fossil sources. A detailed analysis of the feasibility, costs, and environmental problems associated with producing a significant volume of fuel from coal is beyond the scope of this study. Instead, it will simply be assumed that synthetic fuels can be produced in significant quantities by the early 1990's for the equivalent of $40 per barrel of crude oil. This is roughly equivalent to gasoline costing $1.50 per gallon (1980 dollars). These prices are felt by some to be an optimistic assumption (DOE/PE-0021) because they do not adequately consider either the real environmental costs or the technical risks associated with synthetic oil. For comparison, an alternative analysis will be presented in which it is assumed that synthetic crude oil can be produced for only $60 per barrel ($2.10 per gallon of gasoline in 1980 dollars).

The programs and policies examined in this chapter are not inconsistent with the development of a large synthetic fuel industry. Efficiency will minimize the risk of failing to meet technical production goals, minimize the penalty associated with failing to achieve forecast prices, reduce the net demand for capital in the synthetic fuel industry (capital demands which may be astronomical if no attempt is made to improve the efficiency of consumption), minimize the environmental penalty associated with massive production, and extend the useful life of finite fossil resources.

The future envisioned in this analysis is plainly not a confident forecast; many plausible events could dramatically alter the outcome; skillful authors of science fiction may be far better able to imagine the future of national transportation. The rather conventional assessment presented here is intended to be an extrapolation of technologies which are now available. It is entirely possible that technologies not imagined in this study will radically transform the future of American transportation.

Conservatism is necessary, however, because of the enormous national investment in the current transportation system. The nation made vast investments in existing highways, rail networks, and facilities for manufacturing vehicles suited for these roadbeds. Between 16 and 18% of the U.S. gross national product is associated with transportation (Booze, Allen and Hamilton, Inc. 1977), and one job in six in the American labor force is directly involved in transportation (National Transportation Statistics 1979). Moreover, current transportation systems have deeply affected the physical structure of modern American society as well as social convention in everything from sexual mores to funerals. The automobile has shaped modern cities, creating suburban environments where personal vehicles are essential for the most basic functions--shopping, commuting to work or school, visiting friends. Inexpensive transportation has greatly increased opportunities for selecting housing and employment.

Any change in the transportation habits of Americans must, therefore, be treated with considerable caution. There can be little doubt, however, that major changes of some kind are necessary. The most dramatic change will be a reconstruction of the nation's automobile industry. A recent examination of the future of the industry concluded that the changed market for cars will require "the automobile industry to replace virtually its entire range of models, engines, transmissions, and components" (DOT 1981). This is no mean task realizing that three of the nations ten largest industries manufactured automobiles in 1979 (Fortune 1979). It is estimated that the industry will need to invest $70 billion in new plant and equipment during the next five years--something which will require the companies to double historic investment rates. This would be the largest capital project ever undertaken by private investors (DOT 1981). Finding the money to invest for such a heroic project would not be easy even in the best of times; it plainly will not be easy given that the two largest automobile companies in the U.S. together lost more than $2 billion a year in the past two years. The losses resulted both from an overall slowdown in the economy and from the fact that efficient imported cars began to capture a large fraction of U.S. automobile markets. Imports were 15.2% of sales in the fourth quarter of 1978 and 28.2% in the third quarter of 1980 (DOT 1981). Because of their size, trouble for the nation's automobile industry means trouble for the national economy. The downturn in sales by U.S. manufacturers was accompanied by the indefinite loss of 250,000 jobs and a trade deficit of $12.6 billion (DOT 1981).

The changes being forced on the nation's transportation systems must be managed in a way that can ensure the long-term health of the nation's automotive industries. The next few years will be crucial for the critical industry. The transition should leave the industry with the capability to manufacture cars which will have national, and possibly world markets for many years. A mistake in determining the direction of the new retooling transition would be disastrous since it would be difficult to undertake another massive reconstruction program before the end of the decade.

With wise investment, however, the transition could leave the nation with a revitalized industrial base. Japanese manufacturers have been able to underprice equivalent American cars in part because their factories use energy, labor, and materials more efficiently. This advantage can be eliminated, or possibly reversed, with a cogent rebuilding program.

Good management, and effective federal programs, should provide the nation with an automotive industry which can save energy with efficient manufacturing processes, and by producing efficient vehicles. There is probably no single element of industrial reconstruction that will be as important to the nation's energy future during the next decade. Given skillful use of technologies already in-hand, it should be possible to maintain the freedom of mobility which low-cost transportation has provided for many years. Plainly this is a freedom which most Americans passionately cherish. The central object of this analysis is to ensure that this freedom is preserved.

PUBLIC POLICY ALTERNATIVES

The Federal Government has had a dramatic role in shaping the structure of the Nation's transportation system. For example, the western railroads were heavily subsidized in the 19th century only to be burdened in recent years with a baffling array of regulations and taxes. The national highway system represents a staggering investment funded almost entirely from federal and state user taxes and public revenues. Between 1955 and 1975 approximately 1% of the gross national product was invested in building and maintaining roads. While the original intent was to have users bear the full costs of highway construction, nearly a quarter of the costs are now derived from general revenues. Petroleum price controls imposed in the 1970's maintained the price of petroleum products at artificially low levels, and gave badly distorted economic signals.

This history makes it impossible for the government to escape some responsibility for improving the national transportation system. Indeed, many of the policies examined here are designed to alleviate market imperfections caused by federal policies.

An examination of Table 3.1 makes it clear that public policy should be concentrated in three areas: improving the efficiency of automobiles and trucks; slowing or revising the shift from rail to truck, and; improving the performance of aircraft. Programs are summarized in four areas: tax and regulatory policies; technical assistance and grant programs; research and develoment programs, and; programs to increase the efficiency of federal vehicles.

Tax and Regulatory Policies

Reform of the confused regulatory and tax policies which govern transportation offers one of the most important means to increase energy efficiency. These policies are reviewed in detail in later sections. The most important are:

o Fuel economy standards could be extended at higher efficiencies or, if allowed to expire in 1985, replaced by a 'gas guzzler' and a fuel tax or petroleum import tax keyed to ambitious fuel economy goals. A properly designed tax can be imposed without regulatory complexity to: allow the market to reflect the true national cost of petroleum fuels, lower the risk borne by automotive manufacturers investing in efficient vehicle production facilities, stimulate synthetic fuels develoment, and encourage productive and efficient use of petroleum. The revenues accrued from such a tax should be rebated directly or indirectly to the consumer to: alleviate any economic inequities caused by the tax, assist consumers in purchasing efficient vehicles, support the domestic automotive manufacturers' retooling efforts, support synthetic fuels development, or other similar purposes. Care must be taken,

however, to introduce the tax gradually over time so that consumers and manufacturers can make a graceful transition.

o The efficiency of commercial air travel is already improving rapidly because of programs undertaken by the industry. The federal government can help by: (a) Improving air traffic control and operation systems. (The Federal Aviation Administration and the Civil Aeronautics Board should be given clear authority to implement air traffic control procedures which derive energy as well as safety benefits. The airline industry argues that air traffic control delays in 1979 wasted about 784 million gallons of fuel and cost the airlines about $1 billion.) (b) Taking steps to alleviate peak period conjestion at major airports through technical assistance programs or other means. (c) Review aviation user fees to ensure they are levied equitably.

o The national highway financing system could be restructured to encourage more efficient use of capital and energy. Heavy trucks, in particular, do not appear to be paying for their share of highway construction and maintenance costs. Taxes or fees to equitably cover maintenance and construction costs could be tied to a truck's "axle-weight" and distance travelled.

o Some restructuring of the highway financing is essential; current funding methods are no longer adequate. Highway maintenance and construction costs have skyrocketed in the last few years and many highways, particularly those built early in the interstate program, are beginning to require major maintenance. The real value of gasoline taxes has fallen steadily (a decline exacerbated by the recent decline in fuel consumption. The federal highway program will spend more money than it takes in for the first time in 15 years.

o Deregulation of rail and truck freight and implementation of an equitable highway financing system will make it easier for the rail industry to raise capital for needed system improvements, but will not necessarily provide the rail industry with enough capital to rebuild the railbeds. Federally sponsored capital assistance programs might be needed if an examination of the effects of deregulation indicate that such support is needed.

o Tax and regulatory policies which discourage van and car pooling, probably the most efficient form of commuting, should be removed. Non-business owners or operators of van pools hould be afforded the same federal tax advantages as other businesses. States should refrain from restricting or prohibiting fares collected by pool operators (17 states have already moved to 'deregulate' van pools.) Local zoning codes and parking regulations should be designed to encourage van pooling and ridesharing. High occupancy vehicle express lanes can also stimulate van and car pooling.

o If the 55 mph speed limit were enforced, national consumption of gasoline would be reduced by approximately 4%.

Technical Assistance and Grant Programs

The Federal Government can streamline the transition to an efficient transportation system with information, labeling and technical assistance programs.

o Technical assistance and information dissemination programs can assist states and local governments in encouraging car owners to maintain their vehicles properly. The efficiency of existing vehicles can be improved 5 to 15% if their owners make greater use of high efficiency radial tires, improved lubricants, and pay more attention to vehicle maintenance.

o The Environmental Protection Agency can incorporate new mileage testing procedures which accurately predict on-the-road driving performance. The new test results should be published along with the standard EPA test results.

o Tire and oil labeling programs can be developed so that consumers can accurately judge the effectiveness of various products. Some radials can decrease rolling resistance by 43% and some oils can increase automobile efficiency by 6%, but there is great variance among products and little information is presently available.

o A thorough review of federally sponsored mass transit policies is needed to ensure that the objectives of the programs are adequately understood and that funds are expended to meet these objectives efficiently. Current funding systems need to be restructured so that they are flexible and so that subsidies match the needs of cities. At present it

appears that a shift to most types of mass transit systems (with the exception of van and car pools) will not result in major energy savings; the objectives of the mass transit programs should be to assure the continued vitality of the cities. Improved use of existing systems can, of course, lead to energy savings. A variety of strategies can be pursued by cities to improve ridership by improving management and reducing travel time, thereby reducing the need for federal assistance.

o The Federal Government can encourage truck-to-rail transfers, or 'piggyback' applications through deregulation by providing assistance in locating and acquiring point-of-transfer sites, and by ensuring that trucks pay their fair share of road costs.

Research and Development

A careful review must be made of research priorities to examine both technologies designed to improve the efficiency with which energy is used in transport and the technologies by which energy can be produced in a form compatible with transportation demands. Unlike many European nations and Japan, the U.S. has rarely given automobile manufacturers direct support for automotive research and development. In fact, the U.S. manufacturers have traditionally been cool to any offers of support. The recent financial reverses of the U.S. auto-makers makes it appropriate to reexamine this issue. Research facilities are often the first victims of financial reverses and the federal government may have a special role in ensuring that automotive R&D facilities are adequately funded while the industry regains it profitability. Automotive research is as important to the nation's energy future as any other single national research project. Specific techniques which might be used to encourage and supplement corporate investment in research and new ventures are discussed in the chapter on industry.

Priorities in national transportation research should include:

o Develoment of a reasonably priced 60-80 mpg 4-passenger vehicle capable of providing passenger safety, low emmissions, and acceptable performance. Research should concentrate on improving internal combustion engines (such as 3 cylinder direct injected turbocharged diesels or direct injection stratified charge gasoline engines). Long-range research on stirling engines and other advanced engine cycles, transmissions, oils, body aerodynamics, and tires should also be supported.

o Develement of trucks 30% more fuel efficient than new trucks now being sold.

o Development of engines for automobiles and light trucks capable of achieving high efficiency with alcohol fuels.

o Development of advanced commercial aircraft capable of achieving efficiencies after 1990 30-40% above current designs.

o Design of electric storage technologies for electric vehicles capable of long life-times and low weight per unit of energy stored. (Research efforts should preceed demonstration programs).

o Research designed to reduce the emmissions from alcohol and diesel fueled internal combustion engines.

Federal Fleet

The Federal Government should take a position of leadership in its use of transportation fuels by:

o Instituting a comprehensive fuel economy maintenance program for all federal vehicles.

o Utilizing efficient oils and tires for all federal vehicles.

o Implementing van and car pooling programs at federal installations.

o Establishing programs, dovetailed to those already legislated in the Energy Security Act of 1980, to require the future use of ethanol and methanol in federal vehicles.

o Encouraging the use of teleconferencing and telecommunications wherever possible.

The energy savings which the above-mentioned policies can effect are summarized in Table 3.1.

One critical element in any national policy for ensuring the transition to a sustainable national transportation system involves finding adequate capital for the domestic automobile industry. Programs designed to increase investment in efficient new manufacturing facilities and to ensure that adequate funding is available for corporate research are discussed in detail in the chapter on industry. Since most of the programs discussed in that chapter would apply directly to the automobile industry, the analysis will not be repeated here.

FUTURE TRANSPORTATION DEMAND

Trends in Transportation

The energy demanded for transportation in the U.S. depends both on the amount of travel and freight hauling that the nation requires and the efficiency of the vehicles used. Efficiency, of course, can be improved by shifting from one class of vehicle to another, ensuring better use of existing vehicle classes, (such as by reducing the time trucks travel with no freight), as well as by increasing the performance of each vehicle type.

Passenger Travel

Since passenger travel currently consumes over 70% of the energy used in transportation (TEDB-4, p. 1-11), it will be necessary to pay particularly close attention to the nation's habits of personal travel. For the last thirty years, the nation has developed a society in which large amounts of personal travel are indispensable, and Americans have shown a continued willingness to invest significant amounts of their income in travel. The average American (man, woman, and child) travels about 11,500 miles each year (TEDB-4, p. xii), mostly in cars and light trucks. This means that each person spends more than an hour a day in some sort of vehicle. The fraction of each person's disposable income invested in travel has actually increased in recent years, and is now slightly more than 14%. One particularly astonishing feature of American travel patterns in the increasing use of pick-ups and other light trucks for personal travel. In 1977 nearly 16% of the personal vehicles were light trucks, and 13% of the people carried in personal vehicles rode in light trucks (TECDB-4, p. 1-24). Most of these vehicles were used almost exclusively for personal transport, carrying no cargo. It is interesting to note that the manufacturers of light trucks complained when EPA suggested that the performance of light trucks be computed under the assumption that some weight was carried in the cargo compartment; the manufacturers argued that cargo was transported so infrequently that tests should be made assuming that the vehicles contained only passengers (Von Hippel 1980).

Given that Americans are already spending a significant portion of their time in personal vehicles, how much more time are they willing to spend if their incomes increase? A careful examiniation of national driving habits presented in an appendix to this chapter indicates that people in upper income groups use relatively less of their discretionary income on automobiles than persons with low incomes. As a result, the amount of gasoline consumed by automobiles in the U.S. has not increased as rapidly as personal income (TECDB-4, p. 5-33).

In assessing past trends in driving habits, it is important to observe that the cost of operating an automobile in the U.S. remained virtually unchanged for a decade (TECDB-4, p. 5-56). Figure 3.1 illustrates the unexpected fact that gasoline prices, adjusted for inflation, did not rise above 1960 levels until late in 1978. The dramatic price increases of the past year, however, will plainly have an effect, but the nature of this effect is difficult to forecast. There is little firm evidence on which to base forecasts; Americans have never experienced such a dramatic increase in energy prices. It must be remembered, however, that in spite of recent price increases, gasoline expenses still represent only about a quarter of the total cost of operating an automobile (TECDB-4 p. 2-49). A number of studies have attempted to relate the behavior of American drivers to changes in gasoline prices. It is not surprising that their results vary considerably. A number of these studies conclude, however, that at least in the short term, a doubling of energy prices would decrease mileage driven by approximately 13%. This seems to be a suspiciously small change, but it has been used in the analysis of this discussion. (For present purposes, it has been assumed that the public would react to the "effective price" of fuel that is, the price of fuel required to drive a given distance. As a result, it is assumed that if both energy prices and fuel economy double, there would be no net change in the number of miles each person would drive in a year). One reason for believing that the estimate used here is extremely conservative can be seen in Figure 3.2, which illustrates the relationship between gasoline prices and driving habits overseas.

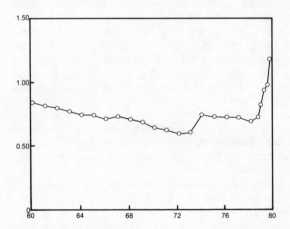

Figure 3.1. **History of Gasoline Prices in the U.S. (1979 $/gal of Unleaded Regular)**

It seems sensible to conclude that the amount of driving done in the future will depend on the efficiency of the automobile fleet, the income of the drivers, and the price of gasoline. Using the assumptions just discussed, the amount of driving which can be

expected in the U.S. by the year 2000 is shown in Table 3.2 for two different assumptions about the price of fuel and three different assumptions about the average fuel economy of the cars on the road in the year 2000. (The results presented in Table 3.2 can also be read simply as the assumptions made by this study about future driving habits). The analysis used throughout this report assumes directly or indirectly that if gasoline prices reach $1.50 per gallon in the year 2000, each person will drive 32% to 52% more than he does today, depending on the efficiency of vehicles.

While these forecasts imply that the absolute link between average national income and miles driven will not apply by the year 2000, they do imply a significant increase in the amount of time each person spends in a car each day will increase from an hour a day to 1-1/2 hours per day, but somehow this seems intuitively unlikely. The actual future of automobile travel will obviously depend heavily on such things as the growth of densely populated urban centers and other factors outside the scope of the study.

Predicting demands for air travel presents a different set of problems. There is no doubt that people enjoy using discretionary income to fly. In fact, air travel has increased much faster than income. Between 1970 and 1978 the number of passenger miles flown increased at nearly 7% per year-roughly 3-1/2% faster that the gross natinal product (TECDB-4, p. 5-33.) It is safe to assume that people will be anxious to spend some part of any increased person income on the speed, comfort, and convenience of air travel.

With the deregulation of the airlines and a near doubling of fuel prices, trends have not been followed for the past two years. A number of techniques have been used to predict the future of air travel, but none shows much sophistication. For the purposes of this study, future air travel was estimated by assuming simply that the total revenues earned by the industry will increase in the future at the same rate as during the period 1972-1978. These were relatively good years for the industry and revenues grew by 3-1/2% each year (in constant dollars). The miles traveled in aircraft can then be computed from estimates of the cost of fuel and the efficiency of aircraft. The results are summarized in Table 3.3.

Freight

As in the case of automobile travel, Americans spend far more on freight transport than Europeans or Japanese. Shipments of fresh California strawberries to the east coast and tractors from Detroit to Dallas have become a part of the basic fabric of the national economy. Recent trends, shown in Figure 3.3, indicate that freight movement continues to increase, although at a rate which is slightly slower than the growth in

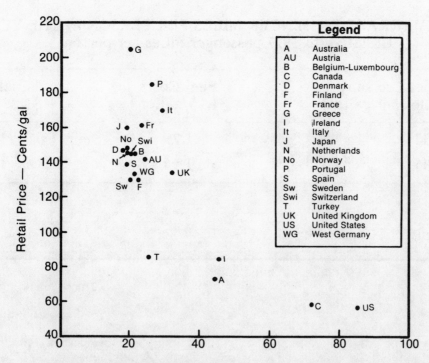

Figure 3.2. **Ratio of Transportation Motor Gasoline Demand to Real GDP (Thousand Btu/1975 US $)**

Cross-country Comparison of Motor Gasoline Demand in 1975.

Source: Data are from the **Economic Report of the President** (Washington, D.C.: Government Printing Office, 1978). The conversions to 1975 U.S. dollars are apparently based on

Source: Data are from the **Economic Report of the President** (Washington, D.C.: Government Printing Office, 1978). The conversions to 1975 U.S. dollars are apparently based on the then prevailing exchange rates, not purchasing power priorities as used elsewhere in this chapter. Price is not the only explanation for the consumption differences; for example, the highest consumption is in countries with longest travel distances. Calculation performed in **Energy and the Economy**

Table 3.2.   MILES TRAVELED PER PERSON IN AUTOMOBILES AND LIGHT TRUCKS IN THE YEAR 2000[b]

1995 Fleet Average fuel economy	Gasoline Costs $2.10/gallon in 2000[a]			Gasoline Costs $1.50/gallon in 2000[a]		
	Poor and Middle Income[c]	Upper Middle and Upper Income	All Income Groups	Poor and Middle Income	Upper Middle and Upper Income[c]	All Income Groups
35	1.45	1.09	1.23	1.55	1.16	1.32
55	1.58	1.19	1.35	1.68	1.27	1.43
75	1.68	1.26	1.43	1.78	1.34	1.52

[a]In <u>1980</u> dollars.

[b]Expressed as a ratio of miles per person in the year 2000 to miles per person in 1977.

[c]Persons in groups with a mean household income of $8000/year or less.

Table 3.3. AIRCRAFT PASSENGER MILES PER PERSON IN THE YEAR 2000
(Relative to 1977 passenger miles per person)

Average Fuel Economy (Passenger miles per gallon)	Fuel Costs $1.40/gallon[a]	Fuel Costs $1.88/gallon[a]
30	1.79	1.55
35	1.94	1.72

[a]1980 dollars.

See Appendix B

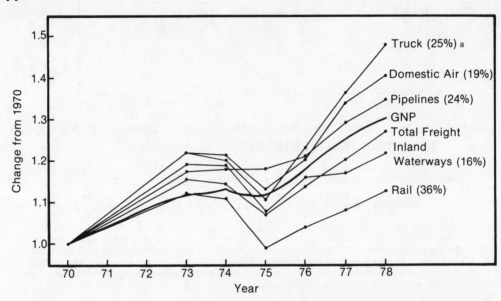

Figure 3.3. **Trends in U.S. Freight Traffic (Ton-Miles Compared With 1970)**

Source: TECDB - 4, p. 5-37

a Percentage of all freight carried in 1978

the economy as a whole. It appears that some of the growth in the economy is shifting from basic materials, requiring the movement of large amounts of metal and other materials to services and finishing processes. (This trend is discussed in greater length in the chapter on industrial energy consumption.) Curiously the Nation's demand for freight transportation is growing more rapidly than demands for basic materials. This is due, at least in part, to the fact that the length of an average shipment has increased. Truck hauls increased in length by over 10% between 1960 and 1978 and the length of rail hauls increased by 1/3 (TECDB-4, p. 5-38). It is likely that national demand for freight shipments will eventually saturate and follow the growth in national demand for materials. No saturation has been assumed in this analysis, however, and freight demand is probably overestimated as a result.

Following the trends of the past decade, it will be assumed that the overall growth in rail and truck freight combined (60% of all domestic freight) will be precisely equal to the rate of increase in the nation's gross national product for the next 20 years. This means that by the end of the century, the amount of freight carried to support each American would increase by 53%. The energy needed to deliver this increased freight service will depend most critically on increases in the efficiency of trucks. A discussion presented later in this chapter indicates that it should be possible to increase the energy efficiency of trucks at least 30% by the end of the century.

The energy used in land freight will also depend on the fraction of this freight carried by trucks. Freight hauling in trucks has increased rapidly during the past decade, largely at the expense of rail freight. This has occured in spite of the fact that railroads are nearly four times as fuel-efficient as trucks in hauling freight between cities (see Table 3.4). If diesel fuel prices double during the next twenty years, however, the price of hauling freight by truck will increase by 8-16% (depending on the future fuel-efficiency of trucks).

It is interesting to observe that in 1980 the relative growth rates of rail to truck dis not follow established trends; rail freight remained almost unchanged between 1979 and 1980 while truck freight actually decreased. The change was apparently due in part to increased grain shipments by rail, while a decline in sales of merchandise resulting from the 1980 recession hurt the trucking industry. It is also likely, however, that the change in the relative growth rates resulted

from increasing fuel prices. So many factors influence the choice between truck and rail, however, that is difficult to determine the role played by increased fuel prices.

Air freight is assumed to increase in the future as rapidly as it has during the past decade--6.2% per year. (The energy now used in air freight is small, but given the relative inefficiency of this kind of shipment and its rapid rate of growth, it is plainly an area to watch.) Water and pipeline shipments are assumed to increase approximately 1/2% per year more slowly than the GNP since coal, petroleum, and other energy products represent a large fraction of the total shipments in these carriers (63.5% in the case of water shipments and 100% in the case of pipelines).

Forecasting Energy Efficiency

Table 3.5 reviews the performance characteristics assumed in this report for the future efficiency of automobiles, commercial aircraft, and trucks. Table 3.6 presents alternative energy sources for transportation.

Significant improvements in automobile efficiency are possible given successful implementation of existing technology and some change in the size of vehicles on the road. Recent increases in the price of gasoline have accelerated U.S. sales of smaller cars, bringing the U.S. into closer conformity with Europe--

Table 3.4. ENERGY USE IN FREIGHT TRANSPORT
(1977)

	Gallons of Fuel Required to Haul One Ton of Freight One Hundred Miles[a]	Fuel Costs as a Fraction of Total Operating Costs	Share of Intercity Freight Market (Percent)[b]
Trucks	2.0 (diesel)	0.312	24.1
Rail	0.48 (diesel)	0.092	35.6
Inland Water	0.32 (diesel)	not available	16.1
Aircraft -- freight only -- freight carried with passengers[c]	20 (kerosene) 2.4 (kerosene)	0.388 ----	0.19 ----
Pipeline	----	0.01	23.7

Source: TECDB-4, p. 1-28, 5-23

[a]Diesel fuel contains 138,500 BTU/gallon, kerosene 135,000 BTU/gallon.

[b]Percent of total ton miles of freight.

[c]Additional fuel required to carry cargo in a passenger flight which would have been flown even if no freight were carried.

Table 3.5. THE EFFICIENCY OF VEHICLES USED IN U.S. TRANSPORTATION[a]

			2000	
		1977	No New Programs	New Programs to Accelerate Growth In Efficiency
1.	Automobiles (miles per gallon)	13.9	27.5	34.4-70.8
2.	Light Trucks for Personal Use (miles per gallon)	10.0	20.0	
3.	Commercial Aircraft (passenger miles per gallon)	21.2	30	40
4.	Medium and heavy trucks (vehicle miles per gallon)	8.3	9.5	10.8

[a]It has been assumed that car efficiencies will increase steadily between 1985 and 1995--many older model cars will be on the road in the year 2000, and as a result the average performance of all cars driven in the year 2000 will be somewhat lower than the performance of new cars sold in 1995.

Table 3.6. ALTERNATIVE SOURCES OF ENERGY FOR TRANSPORTATION IN THE YEAR 2000

		Potential Petroleum Displacement (Quads)	Primary Energy Consumed For this Displacement (Quads)	Source of Primary Energy From Renewable Resources
1.	Methanol and Ethanol[a]	.4-5.5	.3-6.4	wood, grasses, crops
2.	Electric Automobiles[b]	0.3-0.6	0.6-1.0	hydroelectirc, wind photovoltaics, etc.
3.	Electric Rail[c]	0.16	0.15	hydroelectirc, wind photovoltaics, etc.

[a]See biomass discussion in industrial chapter.

[b]Assumes that 30% of commuting is done in electric vehicles by the 2000 (assuming that about 1/3 of miles traveled in automobiles is used for commuting in the year 2000 following current patterns of use). Range of petroleum displaced assumes that the electric vehicles replace either 35 mpg commuter cars or 75 mpg commute cars. Range of primary energy consumption required to operate electric vehicles assumes either 0.3 kWh/mile or 0.5 kW/mile for commercial electric vehicle efficiencies.

[c]Assumes electrification of 9000 miles of track; up to 40,000 miles may eventually be eligible for conversion.

where a 6-passenger vehicle is a novelty. In 1980, more than 40% of the vehicles sold in the U.S. may be 4-passenger vehicles--up from 29,4% in 1978. It is interesting to note that if all of the vehicles bought in the U.S. in the first part of 1980 were the most efficient vehicles of their class on the market, the fuel economy of the new U.S. fleet would have been nearly 37 mpg. (The vehicle "classes" were used in this calculation: 2-, 4-, and 5/6-passenger cars).

Given improved technology, based on direct injection stratified charge engine or a direct injection turbocharged diesel and a modest shift toward smaller vehicles, it is clearly possible for the national automobile fleet to achieve 50-70 mpg by the year 2000 at a cost less than the equivalent of $1.50/gallon (Von Hipple 1981, TRW 1980, CBO 1980, Shackson 1980, Grey and Von Hipple 1981).

The improvements in air transport efficiency result primarily from greater use of the kinds of wide-bodied and higher efficiency jets that the major aircraft industries plan to introduce during the next decade, high load factors, and better air traffic control. The Boeing 757 and 767, and the new L-1011 will be able to improve on the 25 passenger-miles per gallon achieved by the current fleet (ATA 1980). Other savings might result after 2000 from the use of advanced propfan engines on planes designed for short haul, weight reduction with advanced materials, and other advanced technologies.

In comparing the efficiency of aircraft with that of automobiles in Table 3.5, it is important to remember that the average automobile carries 1.9 persons per trip (TECDB-4 1-26). As a result, automobiles achieved an average of 26 passenger miles per gallon in 1977 and would achieve 52 if the 1985 CAFE standards are met. This is better than aircraft, but is not as much better as might be expected.

The 30% improvement in truck efficiencies assumed in Table 3.5 results from assumed use of radial tires (where appropriate), air deflecters (both on top of the truck's cab and under the cab), and improvements in the performance of truck engines, transmissions, and auxilliaries. Greater improvements are possible, but a conservative estimate has been used in the analysis.

The efficiencies summarized in Table 3.5 were combined with the estimates of transportation activity in the year 2000 to produce the estimate of total energy demands displayed in Table 3.1. Several features of this table deserve careful attention. First, it is intersting to notice that the energy needs of the U.S. transportation sector may remain virtually unchanged even given conservative assumptions about the introduction of new policies or technology. The energy used to move people in cars, aircraft, buses, and rail may actually decrease even though the table assumes no growth in the fraction of people carried by mass transit, van pools or, car pools. These savings are offset by a large increase in the energy used by trucks; the forecast shown in the "no new policies" column reflects simply an extrapolation of 1970-1978 trends. The increase in truck use assumed this category is somewhat suspect, given the significant increases in the price of fuel likely to occur over the next twenty years.

Table 3.1 also illustrates the impacts of different fuel costs on vehicle efficiency. The savings are shown for two assumptions about the price of gasoline in the year 2000: one assumptions is roughly equivalent to a crude oil price of $40 per barrel ($1978); the other is equivalent to crude oil costing $60 per barrel. The difference between the two can be read to indicate the sensitivity of U.S. energy demand to any of three separate uncertainties:

-- it can reflect optimisn or pessimism about the future of world oil prices (will world oil prices rise to $60/barrel or stabilize around $40/barrel?),
-- it can reflect optimism or pessimism about the future cost of synthetic oil made from coal or biomass, (can synethic crude oil be produced from coal at $40/barrel?), or
-- it can indicate the effect of increasing gasoline prices by approximately 50 cents per gallon using a fuel tax.

The impact of these fuel price increases has been calculated explicitly only for travel in automobiles and passenger aircraft. As expected, the effect of higher fuel prices (or of higher fuel taxes) is greatest when fuel efficiencies are relatively low. High fuel economy plainly makes the nation much more resilient to fluctuation in energy prices.

The "business as usual" projection shown in Table 3.1 shows a range of possible consumption in rail and truck freight. The high end of this range assumes no increase in the performance of trucks and a continuation of the trend of the 1970's that saw the truck share of combined rail and truck freight increasing by 1% per year. The low end of the range assumes that this rate is slowed to 1/2% per year and that truck efficiencies are increased by 15%. Given the policies suggested, it should be possible to keep the fraction of truck/rail freight carried by rail at about 60%. The consumption shown in Table 3.1 for the (new programs) case assumes that the share carried by rail increases slightly since it is assumed that 40% of the freight carried more than 200 miles by trucks are shifted to rail. (Approximately 40% of truck freight is currently carried more than 200 miles.)

The savings shown in Table 3.1 do not include any contribution from alternative energy sources. Two primary sources of alternative fuels for transportation will be reviewed in some detail in this chapter--a shift to methanol (and, to a more limited extent, ethanol) derived from biological feedstocks, and a shift to electricity (see Table 7). The characteristics of synthetic fuels other than mrthanol and ethanol that can be derived from other renewable sources are so poorly understood at present that no attempt was made to estimate their potential in quantitative terms.

Section Two

Increasing Energy Efficiency in the Near Term

While it is clearly critical to begin work on technologies that can improve the performance of vehicles, most of the petroleum consumed by automobiles and light trucks during the next 5 years will be consumed by vehicles now on the road; and therefore, it will also be important to develop a strategy for reducing the consumption of the current fleet. Such a strategy could become critically important if the nation is suddenly confronted with a collapse of international oil markets during the next few years. Four major categories of opportunities are described in this section.

First, the performance of existing vehicles can be improved by 5-15% if their owners make greater use of high efficiency radial tires now on the market, replace their motor oil with improved graphite additives or other efficient lubricants offered by a number of manufacturers, and pay more attention to vehicle maintenance and tire pressures. Federal programs designed to encourage the use of these cost-effective measures would be relatively painless, since they involve no change in existing driving habits.

A second category of programs examined here involves encouraging greater use
of van pools and car pools. Such programs save consumers money, but can decrease convenience or comfort. Many individuals and companies, however, have found van pooling attractive, and the nation clearly benefits from the energy savings, reduced highway congestion, and cleaner air.

A third category of programs involves the use of gasoline taxes, parking taxes, and speed limits to reduce the demand for automobile transportation services. These approaches are unquestionably the least attractive of the alternatives examined and they violate the fundamental premise of this study--to find ways of increasing the efficiency of the nation's transportation system without asking consumers to sacrifice. In the long-term, of course, fuel taxes will encourage the development of efficient vehicles which will allow travel at pre-tax costs (although with somewhat reduced luxury). In the short-term, however, people can only react to taxes by driving less. The tax option is considered here for a number of reasons-- reasons which are fully consistent with the premise that the programs would only address real defects in the market.

Finally, there can be no doubt about the government's responsibility to ensure the efficiency of the existing fleet of federal vehicles; special attention will be given opportunities for ensuring that the government assumes a position of leadership in transportation efficiency.

OPPORTUNITIES FOR RETROFITTING EXISTING VEHICLES

Several opportunities for reducing the energy consumed in existing vehicles will be examined, and federal programs suggested for encouraging them.

Automobiles and Light Trucks

Tune-ups and tire pressure checks

Many vehicles on the road today are not achieving good fuel economy because they are not properly maintained. Problems typically include:

- poor engine condition (spark plug wires, spark plugs, air filter, spark timing, carburetor idle, air-fuel ratio),

- sub-optimally inflated tires,

- misaligned front wheels.

All of these problems can be corrected if owners take more care in vehicle maintenance and are better informed about the energy implications of this maintenance. Fuel economy inspections could provide owners with the information they need to make informed decisions.

Tune-Ups. The annual safety and emissions inspection programs now required in many parts of the U.S. provide a solid basis for any initiatives designed to improve existing vehicle fuel economy. Vehicle safety inspections have been in existence for many years. Twenty-seven states now require some form of periodic vehicle safety inspection and five areas in the country now have annual vehicle emissions inspection and maintenance programs. By 1983, many more such programs are expected to be operational. The safety and emissions programs are designed to tell owners and mechanics which vehicles need safety or emissions-related maintenance, to give advice on what type of maintenance is needed, to motivate owners to seek the maintenance, and to verify that the maintenance has been performed. These programs could easily be expanded to provide fuel economy check-ups. Existing and planned safety and emissions inspections could also offer fuel economy inspections for about 40% of the national fleet at low cost and with little lead time. One or more procedures specifically designed to evaluate fuel economy could be added to the list of items currently being inspected. If new check-up programs are started for the 60% of the fleet not otherwise subject to an inspection program, along with the congressionally-mandated emissions inspection programs thath start operation by 1983, the incremental negative impacts on owners, state governments, and the repair industry could be minimal.

It is important to recognize that an effective tune-up which both improves engine efficiency and reduces emissions requires a number of different adjustments. Some of the problems associated with improper engine adjustments for new cars may be eliminated during the next few years, however, as sealed, tamper-proof engine components are introduced. A tamper-proof choke, vacuum break and idle mixture setting will be required by 1981. Inspections will still be essential, however, to check for spark timing adjustment and equipment failures. It must be emphasized, however, that the bulk of the gasoline consumed during the next 5-10 years will be consumed by vehicles now on the road, which are not equiped with the tamper-proof units just described.

Underinflated Tires. If all tire pressures were increased to and maintained at the vehicle manufac-turer's recommended pressure, about 28 pounds per square inch (psi), the average fuel economy of the vehicles involved would improve two percent. If, as is technically feasibile, tire pressures were increased to the tire manufacturer's maximum inflation pressure, about 32 psi, the average fuel economy would improve a total of four percent. Other studies show that the rolling resistance can be reduced by as much as 15% by increasing the pressure of a tire from 25 to 40 psi (Corporate Tech. Planning 1978).

Maintaining a given tire pressure requires periodic checking and reinflation, since slow leaks reduce the pressure gradually. The average speed of this pressure loss and the corresponding check interval needed to maintain a desired pressure are not known at present, although field surveillance is in progress to address this question. Improper tire pressures result at least in part from the fact that drivers are not aware of low tire pressure. To achieve the full four percent improvement in fuel economy from the maximum inflation pressure would require that vehicle owners and servicers have easy access to air hoses, be aware of the fuel savings from higher inflation pressures, and check and inflate tires regularly. National Highway Traffic Safety Administration is developing a device which could sense tire pressures automatically and give a warning when pressures are too low.

Until reliable data are obtained on the rate of pressure loss, the fuel savings which would result from maintaining tire pressures at the manufacturer's recommended maximum pressure is estimated by EPA to be about two percent. Care must be taken before higher tire pressures can be widely recommended. Under current practice many manufacturers recommend a different pressure for front and back tires. Inflating all tires to a higher pressure than recommended could affect handling and performance as well as driving comfort in some vehicles. The U.S. Postal Service began using tire inflation pressures 5 psi higher than the vehicle manufacturer recommended pressure in 1974 (Bolger 1980). The Tennessee Valley Authority has inflated tires to the tire manufacturer maximum recommended pressure for many years (Rozek 1980). The long experience of these two agencies indicates that high inflation pressures do not have any adverse safety impact but the issue needs careful examination.

Fuel-Efficient Oils

New motor oils presently on the market using graphite and other substances can increase the efficiency of an automobile by 2-6% (Goodwin 1978, Wadley 1978, Bowman 1978, Reister 1979, Bennington 1975). It may be possible to increase vehicle performance by as much as 12% by using synthetic oils and anti-friction additives in both engines and transmissions (South Coast Technology 1978, VonHipple 1981).

A federal program could begin immediately to conduct tests to determine the impact of different oils on vehicle performance and provide information manufacturers for labels. The American Society of Testing and Materials is already helping to develop testing procedures. The government could also request new legislation requiring that all oils be labeled as to efficiency. A more powerful technique for encouraging use of fuel-efficient oils would be to place an excise tax on inefficient oils to increase their cost by an amount equivalent to the cost of the extra gasoline consumed by a vehicle using the lower grade oil. This tax could be substantial. If 8 quarts of oil were pruchased every 10,000 miles, the improved oil would be worth over $4.75/quart more than standard oils, given current gasoline prices.

Improved Tires

Below 40 mph, the majority of the energy used in a typical automobile is used to overcome the rolling resistance of tires (above 40 mph air resistance becomes the dominant factor) (Gleming 1974). Numerous tests have shown that automobiles equipped with radial tires are able to travel approximately 5% further per gallon of gas than vehicles equipped with standard bias ply tires (Thompson 1977, Klamp 1977, GM 1978). Also, the longer life of radials and their inherent safety advantages are becoming widely recognized in the marketplace. About 95% of all new cars sold in the United States are equipped with radials, 50% of replacement tires sold are radials (Uniroyal 1980), and more than half the vehicle miles traveled in automobiles are now driven on radials. By 1984 80% of all cars on the road may use radials (Uniroyal 1980). It is less well known, however, that there are significant differences among the radials now on the market. The best radials now on the market, such as those on the 1980 GM "X-cars," can increase the efficiency of vehicles 3-4% more than the average radial sold (GM 1978). General Motors estimates that in comparison with the standard bias-ply tires, the following improvements in rolling resistance have been achieved with tires in use or planned for use in their new cars (Stofflet, 1980):

> 20% for baseline original equipment radial (introduced in 1975)
> 24% for lightweight original equipment radial (introduced in 1977)
> 35% for "low rolling resistance" radial (introduced in 1980-1982)
> 43% for "low rolling resistance" radial (introduced in 1980-1982) used at higher pressure.

A federal program designed to encourage the use of the most efficient available radial tires would follow the same basic steps recommended for the development of efficient oils: a tire testing procedure, voluntary labeling, and eventually a law requiring labeling or, perhaps, an "inefficiency excise tax." It is important that consumers be warned that all tires should be replaced at the same time if a shift to radials is contemplated. Vehicles can be difficult to handle if driven with a mixture of tires.

EPA has already initiated work to develop a voluntary tire labeling program (Costle 1980). EPA and the Society of Automotive Engineers began to develop acceptable rolling resistance tests for tires over two years ago. Tests have been developed that would cost a tire manufacturer approximately $250 per tire per test. Since about 100 tests could be required to label a tire design for rolling resistance, some small tire manufacturers could find the test requirements burdensome. In addition, some firms not now manufacturing radial tires could be seriously hurt by the suggested program.

The transition to radials may be slowed in the short term by the current ill health of the automobile industry. In 1980 the nation's capacity to produce radial tires exceed demand by 33 million units per year. It is estimated that the industry will need to invest $400-500 million to complete the conversion to radial tires, but poor 1980 sales have discouraged investment.

Other Retrofits

A number of other devices have been suggested for improving the performance of the existing fleet. EPA and other agencies should be encouraged to pursue and test any innovations that appear to have merit. (One of the more promising recent suggestions is a technique for deactivating four of the cylinders on an eight-cylinder engine). In particular, a vehicle retrofit program should be developed for all federal vehicles, including all reasonable retrofit procedures (including oils and tires) as a part of routine maintenance. The program could also be made available to states and private institutions interested in offering efficiency services or developing efficiency retrofit programs for their own fleets.

Heavy Trucks

The trucking industry has begun to react to increasing fuel prices as evidenced by a growing interest in the subject in recent trade journals and periodicals. Significant savings are possible both from improved oils, the use of turbochargers and more efficient tires, and air deflectors installed on the top of truck cabs and under vehicles.

Tires present a particularly important opportunity for near-term improvements since they account for 65-85% of the total drag force on a large truck moving at 30 mph and for 35-50% of the drag on a truck moving at 55 mph. Radial tires now on the market can reduce fuel use by 0-8% (Hurter 1975) in urban driving and 4-14% in intercity driving (Goodyear 1975, Gleming 1975 pp. 13-14). Advanced radials that can be expected in the market during the next two to three years can save an additional 2-3% on urban driving and 4-5% in intercity trips (Knight). Curiously, radials have not captured a large fraction of the truck tire market, representing only 8% of all truck tires now on the highway. Acceptance is greater among large fleet operators who use radials on approximately 35% of their vehicles. (Modern Tire Dealer 1980, Diesel Equipment Superintendent 1979).

One calculation (Grey 1980) linked to truck-driving patterns indicates that radials could save $4400 over the 500,000-mile life of an intercity transit truck or $80 over 250,000 miles of service in a local truck ($10,400 and $1,300 if advanced radials are used).

Policies for encouraging the use of radial tires in the truck industry are identical to those proposed for encouraging better automobile tires. An acceptable test of the rolling resistance of passenger car tires has already been developed by EPA working in coordination with the Society of Automotive Engineers and is readily adaptable to truck tires (Thompson, 1978; SAE, 1979). It should be recognized that radial tires may not be preferred in all truck applications. Radial tires are more vulnerable to sidewall damage than standard bias ply tires, and therefore may not be preferred in situations where the truck will frequently be used in off-road applications or in areas where the tires would frequently be scuffed by a curb. It must also be recognized that radials will not significantly improve the fuel economy of empty trucks. The significance of radials will therefore increase if truck loads factors can be improved.

Apart from the labeling of tires and fuel taxes, few federal programs are likely to succeed in significantly influencing the efficiency of trucks in the near term. The federally sponsored Voluntary Truck and Bus Fuel Economy Improvement Program (GAO, 1980), which provides information to the trucking industry, has already made a valuable contribution. In addition, several large trucking fleets have used driver training and peformance-related bonuses to reduce fuel consumption; others have installed governors or tachographs to monitor and train drivers in efficient vehicle operation.

VAN AND CAR POOLS

Inefficient use of the private automobile for commuting presents significant opportunities for reducing the use of gasoline. Approximately one third of U.S. gasoline consumption is used in commuting (DOT 1972 p. 4). Three quarters of all private cars used for commuting are occupied only by the driver, and the average number of people per commuter car is 1.4 (Nationwide Personal Transportation Survey 1970).

Over a third of commuters now drive more than a 20-mile round trip; 30% commute 10-29 miles each way (SRI 1978). In spite of the expense and time required for such commuting, the public seems firmly committed to this style of transit: an FEA survey found that 80% of the people interviewed would prefer to sacrifice pleasure driving rather than turn to public transit for commuting (Grey 1977). There is a clear need to provide commuter technologies that can provide an acceptable alternative.

One such alternative is the use of van pools for persons commuting over 10 miles each way and an improved intracity bus and rail system for short commutes. Measured in terms of people moved per gallon, van pools are much more efficient than commuter automobiles, although the advantage can be greatly reduced if more efficient cars are introduced in the market.

Major corporations and government facilities offer the greatest potential for van pools. A surprisingly large fraction of the population is employed in such facilities; the largest 500 industries, for example, employ 18% of the American work force (DOC 1978). The potential of van pools has been demonstrated by several large firms. The 3M Company in St. Paul, Minnesota, for example, began a van pooling program in 1973 and now provides transportation for 10% of its staff. Other companies have followed suit. Employer-owned van pools currently number 300, and while the number of other types of van pools which exist has not been determined there are indications that individually owned and operated van pools outnumber all van pools in organized programs. The industry manufacturing vans suitable for pooling has adequate capacity to meet any expected future demand for vans needed for commuting.

Several federal programs designed to stimulate the use of van pools are already in effect:

o   The Federal Highway Aid Act of 1976 provides grants for: (1) Systems for locating prospective riders and informing them of pooling opportunities; (2) Designation of highway lanes as preferential for pools; (3) Designation of existing publicly-owned facilities for preferential parking $7.5 million was authorized for this program, with no one grant to exceed $1 million. According to data from the Federal Highway Administration, such grants have enabled more than a dozen large cities to offer special freeway laws for busses, car pools and van pools. All but four states now have one or more ride-sharing agencies.

o   The Department of Energy is currently providing grants to states to implement energy conservation plans under the provisions of Title III of the Energy Policy and Conservation

Act of 1975. In order to be eligible for these funds, each plan must include five specific program activities, some of which relate to transportation. The promotion of mass transit and ridesharing is one of the areas specified as a required energy conservation measure.

o Under the Clean Air Act, as amended August 1977, state and local governments are revising their State Implementation Plans to include all reasonably available control measures needed to attain air quality standards by 1982, or in some cases, by 1987. Local agencies have been designated in each area, and the Environmental Protection Agency, in cooperation with the Department of Transportation, has published transportation-air quality planning guidelines, as well as information documents; and is also providing planning funds to these lead agencies to support the analysis, adoption and implementaion of transportation measures, including ridesharing. All areas with serious air quality problems are required to consider van pooling programs for inclusion in a package of comprehensive measures needed to attain air quality standards.

Ironically, in spite of these federal programs to promote van and car pools government regulations continue to present several barriers to the use of van pools.

First, many states regulate van pools by restricting or even prohibiting fares collected by pool operators. At least 17 states have passed new legislation to exempt van pools from some kinds of regulation, but van pools should be completely deregulated to allow the market to work properly. Van pools are currently growing fastest in those states which have deregulated them.

Second, the federal government should consider some form of tax relief for van pool owners. Under present law, employers or businesses can take a 10% investment tax credit on van pool vans with an estimated life of three or more years. Other van pool owners can only take the credit on vans with a life or seven or more years. All van pool owners should be eligible for the credit for vans with a three-year life. It might also be useful to exempt the small amount of profit van pool owners/operators receive from income taxes.

Third, where local zoning standards and parking restrictions interfere or discourage ridesharing, van pools as well as car pools should be exempted. (See discussion of urban transportation.)

Finally, the government can encourage van pooling or car pooling through information dissemination programs and the implementation of van pooling at government installations.

FUEL TAXES

There are compelling reasons to increase the existing federal tax on transportation fuels or possibly to replace it with a tax on petroleum imports. Inflation has eroded the purchasing power of the revenues obtained from gasoline taxes to the point where user fees are not able to cover the expense of improving and maintaining the highway system. The purchasing power of fuel taxes has fallen by a factor of nearly two since 1960; 23% of the cost of improving the national road system must now be paid from general revenues (MVMA 1979 p. 89). Maintenance costs, however, are expected to rise dramatically during the next decade as the national highway system begins to reach its design life of 30 years (Jennings 1979).

The most fundamental reason for taxing fuel, of course, is based on the arguments presented in the first chapter of this report: existing fuel prices do not adequately reflect the national cost of fuel consumption. A case was made in that chapter that a tax on the order of $20/barrel (equivalent to 43¢ a gallon of gasoline) could be justified to encourage the market to react more rationally to the real cost of gasoline consumption.

Most European nations have chosen to use high fuel taxes as a primary instrument in encouraging fuel conservation. In January, 1980, for example, gasoline taxes were $1.62 per gallon in France and $1.82 per gallon in Italy. These astonishingly high taxes were first implemented in Europe at a time when petroleum shortages threatened post-war recovery; they were seen as a luxury tax which could safely be imposed on the privileged few who could afford to drive (Heaton 1980 p. I-1-9). Their survival in an era where middle-class drivers dominate the roads, however, has clearly had the effect of encouraging fuel economy. They have also been a lucrative source of revenue.

Fuel taxes have a number of clear advantages in transportation policy. They impose few regulatory complexities on customers or manufacturers--the rules of the game are extremely clear. Marketplace freedom is maintained since there would be no direct barrier to the purchase of an inefficient vehicle--simply an added cost in operating one. They can be implemented quickly and can take immediate effect. A clear federal fuel tax policy could be an important part of a national program designed to provide a stable future market for efficient vehicles as well as alternative fuels. This could lower the risk borne by manufacturers when they invest in facilities for manufacturing efficient cars. Care must be taken, however, to introduce the tax gradually--in a way that would allow consumers and manufacturers to make a graceful transition. The tax would also be inequitable unless some kind of a grant or rebate program.

An improperly designed fuel tax can inflict considerable damage on the economy. There do not, however, appear to be any difficulties which could not be avoided. If receipts received from the tax were returned to the economy by reducing taxes on income, the result could well be an increase in national productivity and a decline in inflation.

A second objection to fuel taxes is that they would fall more heavily on low-income groups, farmers, and other persons living in areas with relatively low population density. Thus, a tax would transfer some wealth out of states with high gasoline consumption where consumer demands is elastic to states with relatively low gasoline consumption where demand is less elastic. Rural dwellers in these states may not have the option of cutting back fuel consumption without making major sacrifices, and may not be able to afford newer, more fuel efficient vehicles.

In addition, a tax might price oil products well out of the range of low income groups. While the poor do have a lower than average level of auto ownership, a regressive tax on oil would raise the price of all products to some extent. The issue of income equity deserves special attention. If, however, the tax has been designed to make prices reflect actual national costs of road construction, road maintenance, and the full cost of importing petroleum to burn as fuel, it is clear without such a tax low density areas are presently being subsidized. The tax would simply remove the subsidy.

Finally, if fuel taxes are imposed to raise revenue beyond that needed to build and maintain highways, care must be taken to ensure that the taxes do not distort the market place in a way that encourages greater use of aircraft and other means of transport which are relatively less fuel-efficient that road vehicles. It would be necessary to accompany any highway fuel tax with a tax on aviation fuel.

Many of the inequities just described can be eliminated with a properly designed tax rebate program. Some of the revenues can, for example, be rebated directly (Williams, 1980) or indirectly to low income groups with reductions in social security taxes. Williams estimates that a $2 per gallon gasoline tax fully rebated to each adult would result in a net benefit to low income groups as well as a 10% near-term reduction in gasoline consumption. Another proposal is to use the revenues from a fuel tax to provide direct grants for research and development in efficient vehicles, to provide support for manufacturers investing in efficient vehicles, or for prospective purchasers of efficient vehicles. The relative merits of these aproaches are very complex, and much more analysis is needed to sort them out.

Two alternative techniques for taxing transportation--a ton-mileage tax for trucks and a modification of the "gas guzzler" tax levied on the sale of inefficient vehicles--are discussed in a later section of this chapter. It must be recognized, however, that although such taxes can serve specialized functions, none can substitute adequately for a fuel tax that can work directly through the marketplace. All other taxes run a risk of artificially distorting market choices or evasion. A gas tax would apply uniformly to all vehicles. It would tax on-road mileage directly without regulatory complexity.

REGULATORY AND PROCEDURAL PROGRAMS

The laws and regulations now governing automobiles and automotive fuels could be used more effectively to encourage fuel efficiency. Three specific examples will be examined: better use of national speed limits; development of an accurate label indicating the fuel efficiency of new cars; and careful deregulation of gasoline.

A regulatory technique for reducing gasoline consumption in the near future would be to take steps to ensure strict compliance with the national 55 mph speed limit. Most automobiles operate at peak efficiency at speeds of 30-40 mph and become progressively less efficient at higher speeds. At 80 mph a car is typically 40% below its efficiency at 35 mph (EPA 80). A careful state-by-state analysis of compliance with the 55 mph speed limit indicates widespread violations. If the speed limit were enforced, national consumption of gasoline would be reduced by approximately 4% (DOT 1977). As with the gasoline tax, however, some of these savings could be offset if the prospect of longer automobile trips increased the demand for air travel United Technologies 1978). Care should be exercised in designing any such program.

All vehicles sold in the U.S. must have a label indicating the vehicle's performance driven on an EPA "city cycle." These labels have proven to be effective tools for marketing efficiency vehicles. Unfortunately, the tests used to determine a vehicles performance tend to overstate the efficiency of the car, particularly smaller cars. The reasons for this are discussed in a later section of this report. The tests now used for the labels are linked to tests which determine whether a manufacturer is in complience with the CAFE standards; these tests are prescribed by law. A test should be developed which provides the consumer with more accurate information about the performance his vehicle is likely to achieve on the road.

FEDERAL FLEETS

Because the federal fleet is such a small fraction of the entire national vehicle fleet, the fuel savings from a strong conservation effort for federal vehicles is not very large. However, federal fleet conservation measures perform three very important functions: (1) they reinforce the public's perception that the federal government is sincere about using energy efficiently; (2) they provide concrete examples of specific conservation measures for the rest of society to consider, and hopefully, emulate; and (3) They can test techniques designed to improve vehicle efficiencies. The federal government could usefully implement the following conservation measures:

o   Utilization of fuel-efficient engine oils in all federal vehicles.

o   Utilization of low-energy dissipation tires on all federal vehicles.

o   A more careful and comprehensive fuel economy maintenance program for all federal vehicles, with particular emphasis on tire pressure, front wheel alignment.

o   Consideration of engine cylinder deactivation for all federal 8-cylinder engines, (if EPA certification is possible).

o   An aggressive program to shift the federal fleet to alcohol fuels and more efficient vehicles (39 mpg for all federal automobiles purchased in 1984; 50 mpg in 1988).

The Energy Security Act of 1980 requires the federal government to "use gasohol in any motor vehicle capable of using gasohol which it owns or leases. Exceptions apply where gasohol is not available in reasonable quantities or at reasonable prices or where the President finds an exemption is necessary to protect the national security" (Congressional Record (6/19/80 p. S7406). The Act also requires the DOE and DOT to prepare a study examining the wisdom of requiring that "any new motor vehicle be capable of operation on gasohol or on pure alcohol."

Section Three

Increasing Energy Efficiency in the Mid and Long Term

The previous discussion was limited to opportunities for saving energy in transportation during the next five years, placing major emphasis on making better use of the existing fleet of vehicles. Savings on the order of 5-10% seem achievable without major changes in transportation habits, and savings of 15-20% may be possible if technical fixes are accompanied by a new gasoline tax. Without further action, however, near-term savings could be eliminated by a continuous growth in demand for transportation services. The following discussion proposes a series of programs that can drastically reduce demand for transportation energy during the next 20 years, primarily through efficiency improvements.

IMPROVING THE FUEL ECONOMY OF THE NEXT GENERATION OF VEHICLES

Automobiles and Light Trucks

The efficiency of the nation's future automobile fleet is likely to depend more on consumer tastes than it does on new technology. In many cases, the technology required for high fuel economy is relatively well understood--it has not been converted into a commercial product simply because of a lack of demand. The research conducted for this study suggests, that the automobiles sold in the U.S. in the mid-to late 1990's may average more than 50 mpg if consumers are willing to accept relatively some change in vehicles size and performance. Average efficiencies could exceed 70 mpg if the public chose to purchase significantly smaller vehicles (for example if 20% of the cars sold are 2-passenger commuter cars "runabouts", or if any of several promising new technologies find their way into the marketplace. The analysis presented here will provide an assessment of possible changes in the market for new types of cars, a review of the technology available to meet these demands, a discussion of the costs and economic merits of achieving different levels of fuel economy, and a survey of programs that can be used to encourage greater vehicle efficiency.

The domestic auto industry has already made improvements in the economy of its fleet in response to higher gas prices and government regulations. 1980 was a transitional year. It saw the introduction of the GM "X-cars" and many small light trucks. These changes, however, are only the harbingers of a major series of new vehicles that the industry plans to introduce: The 1981 Ford World car (Escort and Lynx) is rated at 30 mpg (EPA); Some Ford 1/2 ton pickups in 1981 are rated at 21 mpg (EPA city). GM is scheduled to introduce the efficient J-car in mid 1981 and later the P-car; GM expects to have a diesel engine available for the Chevette in April 1981; Cadillac now offers a variable displacement V-8, and may introduce a 3-cylinder "commuter car". On-road milage may be significantly less than the EPA rating for reasons discussed later. These developments strongly suggest that significant improvements will be made in the efficiency of U.S. vehicles over the next few years.

The Future Automobile Market

Consumers face a complex choice in selecting a new automobile. While a number of factors enter the decision, comfort performance, cost, fuel efficiency, and safety play a major role in the decision from the consumer's point of view. From a national perspective, it would be necessary to add a test of environmental impact (involving exhaust emmissions and overall vehicle noise) to the list. Both taste and ratiocination come into play, therefore, when vehicles choices are made. Given the significant increase in gasoline prices over the past two years and the clear expectation that prices will continue to increase, there is no reason to believe that the mix of vehicles now on the road accurately reflects the kinds of choices consumers will make during the next few years. Thus, rising gasoline prices may significantly affect the types of vehicle design in the future. Many of the decisions are likely to be carefully reviewed in the near future as gasoline prices rise. Tastes are also changing as a result of increased urban congestion, new knowledge about vehicle safety, and social stigma associated with owners of "gas hogs."

There is, in fact, ample evidence that American tastes in cars have already begun to change dramatically. Table 3.7 compares 1980 car sales in the U.S. with sales in 1978. The shift toward 4-passenger vehicles during the two-year period is breathtaking given the glacial speed with which such changes usually occur. It is interesting to observe that while 21 mpg was the average of automobiles sold during the first three months of 1980, the average would have been 36.6 mpg if consumers had chosen the most efficient 2, 4 and 6 passenger vehicles available for sale during this period. (The average would have been 37.7 mpg if light trucks were excluded.). The trend shown in Table 3.7 is supported by projections made by the General Motors Corporation shown in Figure 3.4.B. In essence, the nation's largest automobile manufacturer has prepared a elegy for the V-8 engine, once the standard of the industry, and confidently forecasted it will be replaced by 4-6 cylinder engines and diesels by 1986. What this means, of course, is that U.S. cars will become much more like the vehicles driven everywhere else in the world. Whereas, in 1970, new U.S. cars engines were nearly five times as large as those of European and Japanese cars sold during the same year, Figure 3.4.A. indicates that the gap has been closing rapidly since 1976.

How far will styles change? Will the U.S. follow the trends of the car markets in Europe and Japan where 4-passenger cars dominate the roads and there is growing interest in 2-passenger "mini's"? In assessing pos

Table 3.7. AUTOMOBILE SALES IN THE U.S.

Vehicle Class (Cars Only)	1978 Actual Sales	1980 Projected Sales	3 Months 1980 Actual Sales
2-passenger*	1.7%	2.0%	2.0%
4-passenger	29.4%	36.7%	44.0%
5/6-passenger	68.9%	61.3%	54.0%
(Includes Light Trucks)			
2-passenger*	1.3%	1.6%	(1.6%)
4-passenger	22.3%	28.8%	35.4%
5/6-passenger and light trucks	76.4%	69.7%	63.0%

*Dominated now by the Corvette.

sible changes in the U.S. market, it is useful to begin by examining the actual needs of U.S. automobile drivers. The Personal Transportation Survey, summarized in Table 3.8, makes it clear that there is a large mismatch between the vehicle requirements for most trips and the vehicle actually used. For example, 78% of all trips taken in cars could have been taken in 2-passenger cars, but fewer than 2% of the trips were actually made in such cars; 94% of all trips could have been made in 4-passenger cars. Moreover, 92% of all commuting trips (typically taken in a family's "second car") could have been made in a 2-passenger car; 98% of all commutes could have been made in a 4-passenger car.

Table 3.8. USE OF VEHICLES IN THE FLEET (percent)

Number of Occupants	To & From Work	All Purposes
1	74.5	50.9
1-2	92.1	78.2
1-3	96.2	88.1
1-4	97.8	93.8
1-5	98.8	96.7
1-6	99.2	98.2
All	100.0	100.0

Source: Grey and Von Hipple 1981.

The average size of vehicles on the road is likely to decrease as the cost of fuel makes it increasingly expensive to drive a vehicle with much more capacity than is needed. For example, most "live alone" adults could drive four passenger, or even a two passenger car without much loss in convenience. The same is true for drivers of the second or third car owned by family; principle use of these vehicles is for commuting, for shopping trips requiring room only for one to two persons, or for a son or daughter using the vehicle for recreation or travel to school. In 1977 53.6% of all households owned at least two vehicles, 11% owned three or more vehicles, and 17.7% owned at least one truck or van.

More detailed information about the principle drivers of vehicles purchased in 1979 is provided by a recent poll taken by Newsweek Incorporated (Cheslow 1980). This information, which is summarized in Table 3.9, provides a basis for projecting future trends in vehicle sizes since census data can be used to determine how the population of each category of "principal driver" will change by the year 2000. This forecast is also shown in Table 3.9. The statistics shown for "2-passenger" vehicles are a little misleading since they refer only to 2-seater sports cars.

Table 3.9. PATTERNS OF CAR BUYING IN 1979

Principal Driver	Percentage of all Registrations by Owner Type in 1979			Percent of All Registrations	
	2-passenger	4-passenger	5/6 passenger	1979	2000
Male Head	1	34.5	65.5	46.9	43
Wife of Male Head	0.4	33.5	66.5	22.3	20
Female Head	0.2	40.0	60.0	9.9	13
Live Alone Male	.4	37.5	62.5	7.5	12
Son	.2	61.1	38.9	4.0	2.6
Daughter	.1	60.3	39.7	3.8	2.4
Other	.1	50.5	49.5	1.0	1.0
AVERAGE OF ALL DRIVERS (1979)	2.4	37.9	59.7		

Source: Cheslow, 1980

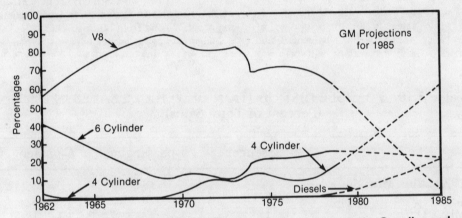

Figure 3.4.A **The Rise and Fall of the V8 Mix of 4, 6, and 8 Cylinder Gasoline and Diesel Engines in New U.S. Automobiles**

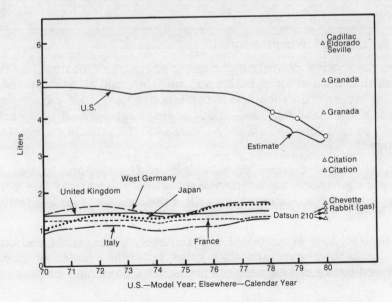

U.S.—Model Year; Elsewhere—Calendar Year

Figure 3.4.B **New Fleet Automobile Engine Displacements in the U.S. and Some Other Major Automobile Producing Nations (U.S.—Average; Elsewhere—Median)**

It is critical to remember, however, that the data shown in this table was gathered before the rapid increases in gasoline prices late in 1979. As a result, the figures may significantly underestimate the shift of market preferences toward smaller vehicles. The figures can, however, be used to construct projections of car sales in the year 2000. Four such projections are shown in Table 3.9. The first simply assumes that sales will follow 1979 patterns; the only factor assumed to change is family size. (The prediction is that there will be be fewer children in the driving ages and more people living alone.) Using this approach, the projections show that 4-passenger vehicles would actually capture a slightly smaller fraction of the market in the year 2000 than they do today! This scenario, of course, assumes that there has been no response to higher fuel prices. Three hypothetical consumer reactions to higher fuel prices are also illustrated in Table 3.10. In the extreme case, it is assumed that all live alone, and that males and females, wives, sons, and daughters purchase only 2-passenger cars while sales to all other classes of owners are evenly divided between 4-passenger and 5/6-passenger cars. These projections will be used

later to estimate overall efficiency of the U.S. vehicle fleet in the year 2000.

Purchasing decisions are based on luggage carrying capacity and comfort as well passenger capacity. It may well be that the trunk space and ultra-smooth rides now available in U.S. vehicles exceed what people will be willing to buy if they are aware of how much they are paying for the privilege. A large luggage compartment, for example, may not be required in a family's second car. Driving a pickup with an enormous load carrying capacity that is only used infrequently may become a luxury of bygone days. Many people may find it cheaper to rent trucks or large cars in the instances when they are actually needed.

Size may also contribute to a feeling of interior quiet, and luxury, and smoothness of ride; but customers will plainly need to ask penetrating questions about the real cost of incremental value in these areas. Their response is likely to be a strong function of the information they receive from industry marketing programs and from other sources.

Judgments about proper engine size can be expected to charge in the next 20 years. Most vehicles pur

Table 3.10. POSSIBLE DISTRIBUTION OF VEHICLE SALES IN 2000
(Percent of Total Sales)

		2-passenger	4-passenger	6-passenger
1.	1979 Sales Mix[a]	2	38	60
2.	Mix #1[b]	18	32	50
3.	Mix #2[c]	22	42	36
4.	Mix #3[d]	44	28	28

[a] Adjusted for changes in demography.

[b] This is the "extensive downsizing" case specified by Cheslow. In this case he assumes that all live alone males and females, and all sons and daughters own nothing larger than a compact car (counted as a 5/6 passenger car in the tables shown above), and that approximately half of each class "downsizes" by one vehicle category--he uses eight vehicle categories in the 5/6 passenger.

[c] Assumes that half of the live alone males and females and half of the sons and daughters purchase 2-passenger cars while the rest of the persons in this class drive 4-passenger cars. All other owner classes follow 1979 patterns.

[d] Assumes that all live alone males and females, wives, sons, and daughters purchase 2-passenger cars and that sales to other classes of owners are evenly divided between 4-passenger cars and 5/6 passenger cars.

chased today have engines larger than needed for expected road conditions; fuel economy suffers greatly as a result. Large engines add to vehicle weight, increase the overall friction losses in the engine and drive-train, and seldom operate at full capacity (most engines are relatively inefficient when operated at a fraction of their capacity). An assessment of the kind of engine required for operating a typical vehicle on the U.S. highway system established the following minimum levels of needed performance:

o An ability to maintain 55 mph while driving up a 5% grade with a full load of passengers (96% of all highway driving is on grades less than 5%). This specification requires 30 to 50 hp depending on the size and weight of the vehicle.

o An ability to go from 0 to 50 mph on the flat in 20 seconds or less. Some low powered cars now in widespread use go from 0 to 50 mph in 15 sec. Vehicles capable of 0-50 mph in 13-15 sec can be constituted which have fuel efficiencies about 10% lower than those calculated in this analysis (Grey and Von Hipple 1981)

o An ability to operate all conventional accessories now available in vehicles (air conditioners, window wipers, headlights, radios, etc.). This specification requires an additional 4 to 5 hp in four- to six-passenger vehicles.

o Some larger vehicles are used to haul goods or pull trailers. It is assumed that one third of six-passenger vehicles and light trucks require an additional 30 hp for this purpose and one fourth of four-passenger vehicles require an additional 20 hp. The power requirements of vehicles designed to satisfy these requirements are summarized in Table 3.11.

A 3000-lb 5-6 passenger vehicle capable of meeting all four criteria should only require a 50-hp engine; the typical engine for such a vehicle in the U.S. today is nearly twice as large.

Vehicle Technology

What kind of fuel efficiency can be achieved by vehicles capable of satisfying the four criteria just explained? The answer depends on technology available in each of the following categories:

Vehicle Weight. The performance and efficiency of a vehicle is strongly influenced by its weight, and even when the vehicle is fully loaded, the bulk of its weight is in its basic vehicle structure. Although some six-passenger automobiles now on the market weigh over two tons, careful vehicle design and choice of materials (following European standards of space and weight efficiency) can lead to substantial weight reductions. Ford recently exhibited a 2500-lb, six-passenger vehicle constructed from advanced materials that would have weighed 4000 to 5000 lb if constructed from conventional materials. Other designers have exhibited foam-filled metal structures that may increase safety while reducing weight. (Advanced technology materials may reduce the weight of six-passenger vehicles to 1750 lb and four-passenger vehicles to 1400 lb.) Weight is also a function of vehicle design. The use of front wheel drive, for example, can save weight and add to interior and trunk space. If advanced materials were combined with efficient vehicle design, further weight reductions would be possible. It must be recognized, of course, that some of the advanced materials now under examination are expensive and their economic attractiveness remains to be determined. Moreover, use of new materials can result in major re-designs and possible increase in the volume required for the new part.

Table 3.11. POTENTIAL VEHICLE EFFICIENCIES

Vehicle Class	Vehicle Test Weight (lb)[a]	Cruise 55 mph (hp)[a]	Extra hp Required to Maintain 55 mph Up a 5% Grade	Accessory Load (hp)	Utility Factor (hp)[d]	Acceleration Power (hp)	Average Engine Size Needed		Projected mpg
							hp	Cylinders	
2 Passenger[b]	2250/1500/1050	14.8/9/7.2	--/11/7.7	3	0	--/16.9/12.2	--/23/18	--/2/2	--/110/140
4 Passenger	2000/2000/1400	13.0[c]/9.8/7.8	--/14.7/10.3	4	5	--/21.4/15.4	--/32.5/27.1	--/3/3	--/78/93
5-6 Passenger & Light Truck	2500/2500/1750	13.5[c]/10.1/8.1	--/18.3/12.8	5	10	--/25.9/18.6	--/43.4/35.9	--/4/3	--/58/70

[a]Lowest Class Value Now/Today's Prototype Technology/Advanced Technology.

[b]Dominated now by the Corvette.

[c]Second lowest; several vehicles exhibited similar results.

[d]For trailer towing and hauling of goods; for the 5-6 passenger and light truck class, 30 hp per vehicle for utility, but assuming only one-third of the vehicles elect this utility; for the 4-passenger class, 20 hp per vehicle for utility use, but assuming one-fourth of the vehicles elect this utility; and for the 2-passenger class, no utility power assumed.

[e]Sufficient to accelerate from 0 to 50 mph in 20 seconds, includes power needed to accelerate the vehicle mass and to overcome frictional resistance (assumed 0.4 of total 55 mph road load), and assumes maximum power continuously available (CVT with 90% efficiency).

[f]Cruise plus grade plus accessories plus utility - generally acceleration is not limiting.

NOTE: The "advanced technology" cases are shown here only to indicate technical possibilities for a future generation of vehicles; they have not been used in the analysis used to derive possible goals for U.S. fuel consumption in the year 2000.

The trend toward lower weight, more efficient cars has already begun. The weight of the average automobile manufactured in the U.S. has been reduced by 800 lbs since 1975. It is interesting to observe that the weight savings possible in smaller cars can reduce the energy needed to manufacture them. The effect is significant. In 1978 the average new car weighed about 3600 lbs (TECDB-4, p. 2-8, 2-33) and its manufacture required the energy equivalent of approximately 1000 gallons of oil. (CBO, 1977, p. 66). If the average weight all new vehicles sold were 2000 lbs the nation would save approximately 0.5 quads per year in manufacturing energy.

Movement resistance. The energy delivered to a vehicle's wheels is mostly lost pushing aside air and overcoming the rolling resistance of tires. The potential of efficient tires is discussed earlier. Air resistance can also be reduced significantly with careful vehicle design although some design work needs to be undertaken to ensure that the market will accept the new body shapes required. Based on European experience, it seems that a producible advanced vehicle can be designed with a 30% reduction in air resistance without unduly sacrificing stylistic flexibility (Seiffert 1980 p. 4).

Power train. A vehicle's power train (primarily a combination of engine and transmission) contains all equipment needed to convert the energy in fuel to energy delivered at the car's wheels.

There are a number of "Engine and Transmission" opportunities for improving the performance of engines and transmissions. The bulk of the analysis which follows will be based on the performance of small three- and four-cylinder turbocharged diesels since these appear to be the most promising engines for the next generation of efficient vehicles, but the performance of "standard" gasoline engines (i.e., spark ignition Otto cycles) can also be significantly improved.

Diesel engines, which are typically 20% more efficient than "standard" engines of equivalent power (see Barry, et al., 1977; VonHippel, 1981), can be improved approximately 10% by using the technique of direct fuel injection now used in most heavy trucks (Shackson, 1980, p. 33). Performance can be further improved with a turbocharger that can provide extra power when it is needed for passing, climbing steep hills, or other purposes. This means that the basic engine can be smaller and designed so that it operates at its highest efficiency during typical driving conditions when extra power is not needed.

The 1980 VW Rabbit diesel (a four-passenger car) gets about 45 mpg. When the Department of Transportation converted this vehicle to use a turbocharger, it was tested at 60 mpg (Barth and Kranig, 1979; VonHippel, 1981). Volkswagen has tested prototypes of new vehicles of this class and test vehicles may achieve 80 mpg (Seiffert, 1980; DOT, 1979). Much more testing, of course, is needed to prove that a marketable vehicle with such characteristics can be produced.

Efficient, two- to four-cylinder stratified-charge gasoline engines operated at relatively low RPM are under development at Ford, Texaco, MAN, and elsewhere. The Ford PROgrammed COmbustion (PROCO) engine is claimed to have 20% greater fuel economy than standard engines (although research has apparently been halted). The MAN system has been run at a test track on approximately 90-octane gasoline with the fuel economy of a diesel (25% better than a gasoline engine). These engines can tolerate a wide variety of fuels, although the adaptability of the PROCO is apparently not as great as the Texaco TCCS design. General Motors has developed an electronic device cable of detecting the onset of "knocking" and automatically change engine operating characteristics to prevent "knocking". Such devices can increase the life of efficient high compression engines. They have been available on Buick's turbocharged engine since 1978 and three GM divisions now have such a device in production.

Advanced engine cycles that are also under development could exceed the performance of the VW power train just described. The Stirling cycle offers potential 10% to 40% improvements in efficiency over existing gasoline engines. Although the Stirling offers enormous potential advantages--efficiency, quiet performance, low emissions, and a complete multifuel capability--a number of engineering problems must be overcome before an affordable, reliable engine can be marketed. Electric motors powered by batteries instead of fuel may also power cars in the future. They are examined in the section of this chapter dealing with alternative fuels for transportation.

The conventional automatic transmissions used in most automobiles today consume about 25% of the fuel burned in vehicles traveling 55 mph. In addition, there is an inherent inefficiency caused by an inability to provide a perfect match between engine performance and vehicle demand. A simple indicator can be used to tell a driver when to change gears.

None of the analysis which follows will be based on the shift gears to a clutch that automatically disengages the engine when the driver is coasting or idling (Seiffert 1980 p. 8). An exact match to driving conditions is possible with a variable mechanical transmission; the first practical transmission of such design will be marketed by DAF/BW Fiat next year. A major barrier to the use of these transmissions in the past was the difficulty of designing transmissions that could operate at 50 hp or more. The small power requirements anticipated in future vehicles can eliminate this obstacle. While these cycles offer interesting research opportunities, the performance estimates used for forcasting.

It must be recognized that some constraints must be imposed on a prototype design if it is converted into a production vehicle. Prototype designs must be modified to make them compatible with a variety of realistic road conditions and with practical mass-production equipment. In some cases, for example, it may not be possible for a production vehicle to achieve the precise tolerances used in producing the prototype designs. For this reason, Table 3.1 reflects a sensitivity analysis which covers a wide range of possible outcomes. The range chosen will be described in greater detail in the following sections.

Vehicle Efficiency

The technologies just described, combined in a vehicle designed to meet the performance requirements developed earlier are capable of delivering the performance shown in Table 3.11. These forecasts were developed using the following assumptions:

o   The "prototype" technology shown assumes vehicle weights for each class similar to that of the best vehicle now on the market. The "advanced technology" weights are today's weights reduced by 30%.

o   At 55 mph, aerodynamic load on the vehicle is about 50% of the total opposing force, tire energy dissipation is about 30% and drive train losses are about 20%. The "state-of-the-art" technology vehicle was assumed to have achieved a 50% reduction in rolling resistance and drivetrain losses. The "advanced technology" vehicles were assumed to have achieved an additional 30% reduction in aerodynamic drag.

o   The projected mpg values assumed the efficiencies provided by small turbocharged direct-injected diesel engines; PROCO engines can be used but will be about 10% less efficient.

o   The efficiencies projected in Table 3.11 could result from a number of combinations of assumptions about vehicle weight, air resistance, tire resistance, oils, engines, transmissions, etc. The analysis was based on the assumption that a 4 passenger vehicle with a 36 horse power could achieve an EPA composite fuel economy of 70 mpg; extrapolations to higher and lower horsepowers was by simple ratios (Grey 1980) (Seiffert 1980). A number of recent studies have indicated that vehicles capable of accelerating significantly more

rapidly than the vehicles assumed for the construction of Table 3.11 can achieve 60-70 mpg (TRW, 1980; Grey and Von Hipple, 1981). In fact, the calculation leading to an assessment of engine size shown in Table 3.11 usually results in an engine which is larger than the minimum engine size needed to accelerate from 0 to 50 mpg in 20 seconds. As a result, the vehicles assumed in the construction of Table 3.11 will be able to accelerate to 50 mph in less than 20 sec.

The potential vehicle efficiencies for each class of car shown in Table 3.11 can be combined with possible market demand for different classes of cars shown in Table 3.9 to produce an estimate of the average efficiency of new cars entering the market in the year 1995. The result of the calculation is shown in Table 3.12. The Table can be used to compute the net national demand for automotive fuel shown in Table 3.1.

The performance of the new vehicles entering the fleet in 1995 can be used to compute the average efficiency of vehicles on the road in the year 2000 using the methods described in Appendix A. Translating the statistics shown on Table 3.12 to a realistic estimate of on-road performance in the year 1995, however, requires resolving three problems:

-- an estimate must be made about the number of light trucks used for personal transportation in 1995 and about whether the performance of these vehicles will be significantly less than the performance possible for a 5-6 passenger automobile. Two cases are explored in Table 3.12: one in which light trucks are assumed to be included in the marekt for 5/6-passenger cars in the markets summarized in Table 3.9, and; one which is assumed that 10% of all new vehicles sold for personal transport in 1995 are light trucks achieving a fuel efficiency of 30 mpg. Over 15% of all personal vehicle miles traveled in 1977 were in light trucks (TECDB-4, p. 1-24), which are among the least efficient passenger vehicles now on the road.

-- an estimate must be made about the difference between on-road performance and the EPA rating of vehicles entering the fleet in 1995. Standard techniques would reduce the EPA fuel economies by 15% (Shackson, 1980; CBO, 1980). there is evidence, however, that the EPA tests may provide a much more accurate estimate of the on-road performance of the front-wheel drive diesels postulated in the designs assumed in constructing the estimates shwon in Table 3.12 (Hayden, 1979; Von Hippel, 1981).

Table 3.12. POTENTIAL AVERAGE EFFICIENCIES FOR CARS SOLD IN 2000
(miles per gallon)

	Sales Mix	Prototype Technology Including 10% Light Trucks	Prototype Technology	Advanced Technology
1.	1979 Mix	58	65	78
2.	Mix #1	61	69	84
3.	Mix #2	63	72	89
4.	Mix #3	69	80	98

Notes:

-- See Table 10 for assumptions about downsizing.

-- The second column applies to a case in which light trucks are included in the average as if they were 5/6 passenger vehicles and are assumed to achieve the same efficiency as 5/6 passenger vehicles.

-- The first column assumes that 10% of the vehicles included in the average are light trucks capable of achieving 30 mpg (a 50% improvement over the 1985 standards).

NOTE: The "advanced technology" cases are shown here only to indicate technical possibilities for a future generation of vehicles; they have not been used in the analysis used to derive possible goals for U.S. fuel consumption in the year 2000.

-- an estimate must be made about the fraction of the vehicles entering the market which will use diesels. The performance of the fleets shown in Table 3.12 would need to be reduced by 20% if direct injection stratified charge engines were used instead of diesels; 10% if half the vehicles sold were diesels and half were stratified charge.

If on-road mileage is 15% below EPA mileage, 10% of all vehicles are light trucks, and no diesels are used, the new fleet entering the market in 1995 would achieve approximately 41 mpg. As a result, a range of 35-75 mpg was used in Table 3.1 to explore the effect of improved automobile performance on national energy consumption.

The Costs of High Efficiency

The previous section established the technical credibility of achieving building cars with very high mileages. It is now relevant to inquire into the costs of doing so. Costs can be measured in a variety of ways. Four categories of possible costs will be reviewed here: the costs borne by the consumer, costs to the manufacturing industry, costs in vehicle safety, and costs of environmental impacts. The use of small cars would also reduce wear on the U.S. road system and reduce the overall demand for steel in the U.S.--

thereby contributing to the decline in national demand for a material which requires significant amounts of energy to manufacture.

Possible Consumer Costs. The cost of high-efficiency vehicles depends entirely upon whether the performance improvements are the result of changing the size (and possibly the performance) of the vehicles or the result of advanced technologies. One suspects that both will be involved to some extent. It is obvious that if fuel economies are achieved with smaller cars, there may be no measureable cost to the consumer--indeed the buyer may benefit both from a lower initial vehicle cost and from lower operating costs over the lifetime of the vehicle.

Calculating the cost of technical improvements that increase efficiency without altering any feature of vehicle performance is much more difficult. Few of the technologies have actually entered mass production and manufacturers are not typically willing to talk publicly about their own estimates of cost. Estimating life-cycle consumer costs also requires an analysis of the lifetime and maintenance costs of future car designs. It is clearly possible that poorly designed, lightweight cars will not last as long as current vehicles. Quantitative estimates in these areas are virtually impossible without extensive road testing.

The cost a consumer should be willing to pay for improved vehicle performance is summarized in

Table 3.13. This table indicates, for example, that a car buyer making a decision in 1995 should be willing to pay $860-$1400 more for a car that gets 59 mpg on the road than for a car with 36 mpg and $2500-$4200 more for a 36 mpg car than a 16.7 mpg car. The first increments of fuel efficiency are worth more since they save more gasoline.

A number of recent studies have attempted to quantify the costs of different levels of efficiency improvements. They include:

-- An analysis conducted for the U.S. Department of Energy which estimated that improving the average of all automobiles sold from the 1980 level of 21.3 mpg to the 1985 standard of 27.5 mpg would cost consumers approximately $460 per car.

-- An analysis conducted by the TRW corporation which concluded that the performance of cars could be increased from 27.5 mpg to about 60 mpg if the consumer were willing to pay $800 for the increased performance (Gorman, 1980).

-- An analysis conducted by the Congressional Budget Office which concluded that the efficiency of cars could be increased to 37-42 mpg for an incremental cost of $600-$650 above the cost of a 1980 vehicle (CBO 1980 p. xiv)

-- An analysis conducted by the Mellon Institute in cooperation with the Automobile U.S. Motor Vehicles Manufacturers Association which concluded that the average efficiency of automobiles could be increased from 16.7 mpg to 43 mpg for a total added cost of about $2600 (Shackson 1980). Performance improvements considered included improved materials, use of front-wheel drive, improved transmission, efficient accessories, efficient oils and tires, improved aerodynamics, diesels with turbochargers, and direct injection stratified charge engines.

Table 3.13.   THE PRICE CONSUMERS SHOULD BE WILLING TO PAY FOR INCREASED VEHICLE EFFICIENCY IN 1995

Increase in on-road mileage	10% Discount Rate		3% Discount Rate	
	40 $/BBL Oil in 2000	60 $/BBL Oil in 2000	40 $/BBL Oil in 2000	60 $/BBL Oil in 2000
16.7 - 36	2500	3300	3200	4200
36 -49.6	600	780	760	1000
36 - 59.4	860	1100	1100	1400

Allowed price (P) calculated as follows

$$P = \sum_t \frac{M(t)\, p(t)\, F(t)}{(1td)^t} \left(\frac{1}{E_1} - \frac{1}{E_2} \right)$$

where $M(e)$ = miles driven by a car that is t years old

$p(t)$ = probability that a car will survive to the year t

$F(t)$ = fuel price in year t ($/gal)

d = discount rate

E = fuel efficiency (mpg)

-- A reevaluation of the efficiency improvements possible from the technologies suggested by the Mellon report and a more carefully documented estimate of the cost of turbocharging, indicates that it should be possible to increase the on road fuel efficiency of gasoline automobiles from 17 mpg to 50 mpg for about $2263, and acheive 59.4 mpg in diesel vehicles for about the same cost. The total cost of improvements would be about half the total shown after the cost of the initial tooling is paid off (Von Hippel 1981).

Comparing these results (Table 3.14) with the allowed costs shown in Table 3.13, a convincing case can be made that consumers could benefit from improvements which result in vehicle efficiencies reaching 60 mpg. None of the analysis applies directly to the characteristics shown in Table 3.12 although the final study cited (Von Hippel 1981) evaluates a case which is roughly comparable to the "mix #1" case shown in Table 3.12 using prototype technology. As noted earlier, the "cost" of shifting to further downsizing is likely to be entirely in the form of costs in convenience and not in additional direct costs.

Possible Costs for the Manufacturers. Converting the nation's manufacturing industry to an industry capable of producing the kinds of vehicles required for improved performance will plainly be a massive undertaking (John, 1979; Business Week, 1980; Shackson, 1980; CBO, 1980). The current financial difficulties being experienced by large portions of the industry are evidence of the problems confronted in making rapid changes. Given adequate time and planning, however, the industry can clearly meet the challenge. Planning is critical. One of the reasons for the relative strength of the General Motors Corporation is that it made a major decision to develop small "world class" cars well before the market shifted. The company began a massive investment in downsizing in 1977 (Defiglio, p. 2).

In fact, A recent study completed for the Department of Transportation suggests that the industry may need to raise 70 billion dollars by 1985 (DOE, 1981, p. 64). This would include $9 billion to rebuild 40 domestic engine lines, $6 billion to rebuild 28 automatic transmission lines, and an astonishing $13 billion for new front-wheel-drive-stamping and assembly lines. Much of this, of course, would need to be invested even without a shift to higher efficiencies (DOT, 1981, p. 65.) The industry has spent an average of $7-9 billion on capital improvements annually for the past decade, on general product improvements not related to vehicle performance.

CBO estimates that shifting to a 40 mpg fleet would require the industry to invest $1-5.5 billion per year more than they would if no efficiency improvements

Table 3.14. THE COST OF EFFICIENCY IMPROVEMENTS IN AUTOMOBILES
(technology assumed available by 1995)

		miles per gallon of gasoline equivalent (on road)	incremental cost (1978 dollars)
1.	gasoline engines		
	a. downsizing, component weight reduction, new materials, front wheel drive, efficient accessories	27.4	1385
	b. improved aerodynamics	30.2	18
	c. reduced rolling resistance	33.3	40
	d. improved lubrication	36.0	20
	e. direct injection stratified charge engine with turbocharge	49.6 (58.4 EPA)	800-900
2.	diesel engines		
	a. downsizing, component weight reduction, new materials, front wheel drive, efficient accessories, improved aerodynamics, rolling resistance, and lubrication	36.0	1463
	b. direct injection diesel	51.4	600
	c. turbocharger	59.4 (69.5 EPA)	200-300

Source: Frank Von Hipple 1981 (This paper is based on analysis in Shackson 1980 corrected with a number of other references, many of which have been discussed in this text).

Notes: -- mileages are average on-road miles per gallon equivalent for all new cars entering the fleet. In 1979 the on-road average was 16.7 mpg. One gallon of diesel fule is assumed to contain the same primary energy equivalent as a gallon of gasoline since, although diesel contains 10% more energy per gallon, 10% more energy is needed at the refinery to produce a gallon of gasoline. (EPA 1976.) Both "gasoline" and "diesel" engines could be converted to use alcohol fuels.

-- it is assumed that EPA miles per gallon are 15% greater than the mileages shown. 59.1 mpg on road, therefore translates into 69.5 mpg EPA. EPA mileages should be compared with the statistics in Table 12.

-- the net on-road mileages shown here assume no change in size mix and litle change in overall vehicle performance (although some downsizing is assumed). The net improvements are calculated by simply multiplying the fractional efficiency improvements resulting from each measure following the technique used in Shackson 1981. For a more exact technique of combining vehicle improvements, see Grey and Von Hipple 1981.

-- incremental costs assume that tooling is paid for in 6 years but no cost reduction is passed to consumers after this five year period.

were undertaken for a total incremental investment of $10 - $27.5 billion (CBO, 1980). Raising the kind of capital needed to rebuild one of the nation's largest industries will plainly be an extraordinary task. The problem is made all the more difficult by the fact that the industry has in the past attempted to raise most of the capital it needed by issuing stock instead of going into debt. This low debt-to-equity position made it relatively easy for them to weather large fluctuations in automobile sales. If a large part of the proposed reconstruction program must be financed with debt, this flexibility would be lost at a time when it is most needed.

A program to convert the industry to manufacture an entirely new generation of vehicles may offer the opportunity to revitalize many obsolescent plants. It must be remembered, however, that the U.S. automobile industry is relatively old. Most of its production facilities date to the 1920's and 1930's. Revitalization seems essential if the industry is to survive the mounting competition from imports. By 1985 virtually every U.S. car design will have been changed. Given proper foresight, and an effective national program in the area, it seems likely that the industry can accommodate the improvements examined in this analysis in the ordinary cycle of model changes. It may be preferable for manufacturers to develop a clear concept of the vehicle that can be expected to find markets through the remainder of the century and to move directly to the production of such vehicles rather than to continue to invest in intermediate vehicles or incremental changes in previous designs which may rapidly become obsolete. The direct approach would appear to require less capital. If the nation moves with resolution toward the right kinds of vehicles, however, there is no reason why the U.S. should not only block the growth of imports but direct its production in ways which could lead to significant exports of U.S.-made vehicles. One of the largest problems with any change of the sort contemplated here is that manufacturers would be taking a risk--gambling on assumptions about future consumer preferences. A representative of General Motors corporation commented on these risks in a review of this analysis as follows:

o In June 1977, it appeared small car sales would pick up and we were about to introduce another model of the Chevette and the second shift was reinstated in our first Chevette plant. Near this same time, the decision was made to bring on a second Chevette plant for start of production March 1979. In August 1978 (start of production 1979 model year), sales of foreign-produced automobiles like the small domestics, were slower than expected. By December 1979, just before the Iranian situation, the day's supply of these types of cars rose to 154 for imports and 99 for Chevettes (a 60-day supply is considered average).

o Thus, the small cars were available, but the consumer was not buying; yet mid-size and large cars were in short supply. Our concern is that this could happen again only producing a much more severe condition if such drastic changes in lifestyle are regulated and the consumer refuses to change.

o As we stated earlier, there are so many external factors of unknown dimensions that could influence consumer purchase decisions that it becomes an exercise in pure speculation as to what level of fuel economy will be closest to what consumers demand. What the free market mix of cars will be several years in the future is highly uncertain, but the manufacturer must remain flexible and be able to respond to that market. These risks can plainly be minimized given an effective national program designed to assure markets for efficient vehicles (i.e., a fuel tax or CAFE).

Possible Costs in Safety. Without proper attention, some of the cars required to achieve the high fleet mileage averages called for in the performance goals (see Table 3.3) may be less safe than the heavier vehicles that they replace unless techniques are developed for ensuring the safety of small cars. The National Highway Traffic Safety Administration recently released results from their 35 mph barrier impact test. The thirty three vehicles tested included a full spectrum of vehicle sizes. Eight passed the test. The vehicles which passed were spread fairly evenly over the range of cars tested, with no correlation to vehicles weight or size.

Indeed, public concern about light vehicle safety is widely enough shared so that it may be slowing the current market shift (Schneider, 1980). Paradoxically, although most Americans refuse to use their automobile seat belts, many of them justify buying cars larger and heavier than they need because they believe that larger cars are safer (Schnedier, 1980, DOT). It is certainly true today that the occupants of the small cars are much more likely to be killed than the occupants of large cars (South Coast Technology, 1978) (see Figure 3.5) although smaller cars seem to have a lower accident rate (Lee 1980). (The net impact of small cars on safety will also depend on how they are used. For example, most very small cars are likely to be used primarily for commuting at relatively low speeds).

Alarm generated from viewing Figure 3.5 should, however, be conditioned by the fact that during the period represented by the statistics the percentage of small cars on the road was small and to some extent, the crashing of a small car with a large car or truck may have accounted for the higher death rates in the small cars as compared to the large cars, rather than any inherently poorer barrier crash protection capability of the small car versus the large car. Another consideration is that during this period, smaller cars were likely driven to a larger extent by younger drivers who traditionally are involved in more fatal accidents.

The National Highway Traffic Safety Administration (NHTSA) has over the past few years consistently pushed for safety-related design improvements in

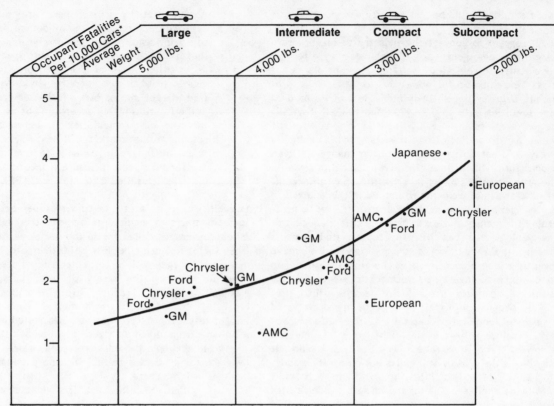

* The fatality rate of a car on the road depends on how, when and where the car is driven, as well as on the safety of the car itself.

Occupant death rate of cars in the U.S. during calendar years 1976-78 by manufacturer (or area of origin for imported cars) and weight size class for the 1974-77 model years.

SOURCE: U.S. Department of Transportation, National Highway Traffic Safety Administration, **The Car Book,** 1980, p. 19.

Figure 3.5. Fatality Rates as a Function of Car Weight

small cars. Some recent progress in this area is indicated by the fact that, of the four model year 1980 U.S. cars NHTSA determined provide a high level of protection for seatbelted passengers in 35-mph front and rear crashes into massive barriers (the Federal Motor Vehicle Safety Standard requirement is such protection at 30 mph), two were subcompacts (the Plymouth Horizon and the redesigned Ford Mustang) and another was one of GM's radically downsized X-body cars. It must be recognized, of course, that tests using dummies cannot provide confident predictions of injuries that might result from actual accidents and it would be a mistake to assume too much on the basis of these high speed tests. Some analysists have argued that the tests are so misleading that they should be discontinued (Stofflet, 1981).

NHTSA has a demonstration program to show how much automobile safety can be improved. In this program the agency has modified a VW Rabbit diesel to obtain a composite EPA fuel economy of 60 mpg <u>and</u> to provide occupant protection in 40 mph frontal crashes,

(Volkswagen of America, 1979) and has developed another light research safety vehicle which provides protection to front seat occupants in frontal crashes at speeds up to 50 mph as well as substantially enhanced protection in crashes of other types. Whether the even lighter vehicles described in Table 3.11 for 2-passenger cars could be brought up to these safety standards remains to be established, however.

Plainly, however, a program to increase the average efficiency of the U.S. automobile fleet must be accompanied by an ambitious set of safety standards reinforced with a program of research and testing. Ironically one of the most serious safety problems associated with small cars may be that the low cost transportation that they offer may result in more travel and consequently in more accidents. The net impact of improved fuel economy on safety, however, remains unresolved.

<u>Possible Environmental Impact Costs.</u> In most cases, reducing the gasoline consumed per passenger-mile

will also reduce vehicle emissions. Improved electronic controls designed to increase vehicle fuel economy can be designed to ensure a balance between performance and emissions for each point in the driving cycle. The major environmental problem associated with higher fuel economy is likely to result from increased use of diesel engines. The remainder of this discussion will be devoted to this problem.

Diesel engines are today about 20-30% more fuel-efficient than spark-ignition engines (Barry, et al., 1977), but without emission controls they emit about one hundred times the weight of small particles produced by comparable gasoline engines, and the particles carry absorbed organics of as yet unknown potency (South Coast Technology, 1978). The results of studies in this area are controversial, however, and more research is in order. The automotive industry has shown considerable concern over the ability of diesel vehicles to comply with future particulate and nitrous oxide emissions standards particularly if diesel emissions are proven to be highly carcenogenic and stringent standards imposed.

There is no fundamental reason to believe that such emissions are unmanageable. The particulate emissions from the current VW Rabbit diesel, for example, are approximately three times lower than those from conventional-size cars, and apparently when this engine is modified to reduce its fuel consumption, its particulate emissions are also reduced (Kelderman, 1980; VW, 1979). EPA tests of 1979 VW Rabbits indicate that the automobile, tested at 2250 pounds, met 1985 particulate emissions standards and 1981 NOx standards. Furthermore, EPA is testing experimental exhaust gas treatment devices that are expected to make possible a reduction of diesel particulate emissions to one third of their uncontrolled level (Kelderman, 1980). It appears possible that EPA 1985 automotive standards of 0.2 gms/mi particulate emission can be met with small diesels.

Assuming that diesel carcinogens have the same potency as those associated with coke oven emissions, an EPA task force estimated in 1979 that if 25 pecent of the U.S. light vehicle fleet were diesels emitting 1.08 grams of particles per mile, an additional 600 cancer deaths per year might ultimately result (South Coast Technology, 1978). Perhaps the most important concern about emissions which remains poorly understood is the extent to which diesel exhaust is carcinogenic. The extractable organic fraction of diesel particulate has been shown to be mutagenic. Even though the 1985 particulate emission standard of 0.2 gpm is thought by many to be a fairly stringent standard, it will still permit light-duty diesels to emit approximately 20 times more particulate matter (by mass) than that emitted by today's unleaded gasoline-fueled light duty vehicles.

The preceding discussion has established two things: first that it is technically possible for the U.S.

automobile fleet to average 50-80 mpg by the year 2000, and second that a plausable case can be made for shifting to such a fleet on economic grounds. The problems of vehicle safety, while critical, appear resolvable, and the efficient fleet could improve environmental quality. Research needs to Continue.

Policy Options

In all likelihood, the sales of efficient cars will grow rapidly in the absence of any new federal programs; it is possible that the market will begin to demand 50-80 mpg vehicles in the near future. Certainly a case can be made for such a demand given the actual pattern of use of cars now on the road. Given this situation, what should be the objectives of federal policy?

The objectives in general should guarantee energy sufficiency by promoting efficiency with a minimum amount of federal intervention. First, the suggested policies will be premised on an assumption that even with deregulation market prices will not reflect the full cost which the nation pays for liquid fuels. Why is this bad? The government can move to ensure that the full price is recognized by the market. Secondly, efficient market decisions may not emerge quickly enough to reduce a dangerous national dependence on imported fuels. There is enormous inertia both in opinions and in capital investment which can delay a move toward economically efficient choices. The government can move to grease the wheels. Finally, the existing automobile industry is in a perilous condition brought about in part by enormous uncertainties about future markets. Federal programs can help to establish greater predictability in the markets and can help ease some of the problems which have prevented the industry from investing more heavily in research and new equipment.

Four generic types of federal programs are possible to meet these objectives: taxes that have the effect of raising the price of fuels, inefficient vehicles, or parking; labels and other information programs which permit consumers to make more informed choices; regulations which would require the development of more efficient vehicles; and direct incentives for manufacturers. In addition, a program should be developed to ensure that all vehicles purchased for federal use be the most efficient, cost-effective vehicles available from U.S. manufacturers.

Taxes, Regulations and Labels.   Existing national policy on automobile efficiency is based on a combination of mandated Corporate Average Fuel Economy (CAFE) calculated for each major manufacturer, together with a sales tax on inefficient vehicles, the so-called "gas guzzler tax" (PL 91-190). The programs are implemented as shown in Table 3.15 and 3.16.

Taxes, labels, and regulations all are directed to encourage the market to move rapidly toward effi

Table 3.15.   PASSENGER CAR AND LIGHT TRUCK CORPORATE AVERAGE FUEL
ECONOMY STANDARDS
(mpg)

| Year | Passenger Cars | Light Trucks[a] | | Limited Product Light Truck[b] |
		Two-Wheel Drive	Four-Wheel Drive	
1978	18.0			
1979	19.0	17.2[c]	15.8[c]	
1980	20.0	16.0	14.0	
1981	22.0	16.7	15.0	14.0
1982	24.0	18.0	16.0	14.5
1983	26.0	19.5	17.5	
1984	27.0	20.3	18.5	
1985	27.5[d]	21.6	19.0	

[a]Less than or equal to 8,500 lb.

[b]Based on International Harvester medium-duty truck engines.

[c]Less than or equal to 6,000 lb.

[d]Secretary of the Department of Transportation may amend the fuel economy standard for mode year 1985, or for any subsequent model year, to a level which he determines is the maximum feasible average fuel economy level for such model year, except that any amendment which has the effect of increasing an average fuel economy standard to a level in excess of 27.5 mpg, or of decreasing any such standard to a level below 26.0 mpg, should be submitted to the Congress in accordance with Section 551 of the Energy Policy and Conservation Act and shall not take effect if either House of the Congress disapproves such amendment in accordance with the procedures specified in such section.

Source: TECDB-4, p 2-25.

Table 3.16.   GAS GUZZLER TAX ON NEW CARS FAILING THE MEET CORPORATE AVERAGE
FUEL ECONOMY STANDARDS, AS ESTABLISHED IN THE NATIONAL ENERGY ACT
(in Dollars per Car)

Vehicle Fuel Economy		1980 20.0	1981 22.0	1982 24.0	1983 26.0	1984 27.0	1984 27.5	1986 27.5
		Corporate Average fuel Economy Standard						
22.5 mpg				0			0	0
21.5-22.5				0			0	500
	21.0 mpg	0	0		0		0	
20.5-21.5				0		0		650
	20.0-21.0	0	0		0		500	
19.5-20.5				0		0		850
	19.0-20.0	0	0		0		600	
18.5-19.5				0		450		1050
	18.0-19.0	0	0		350		800	
17.5-18.5				200		600		1300
	17.0-18.0	0	0		500		1000	
16.5-17.5				350		750		1500
	16.0-17.0	0	200		650		1200	
15.5-16.5				450		950		1850
	15.0-16.0	0	350		800		1500	
14.5-15.5				600		1150		2250
	14.0-15.0	200	450		1000		1800	
13.5-14.5				750		1450		2700
	13.0-14.0	300	550		1250		2200	
12.5-13.5				950		1750		3200
	13.0	550	650		1550		2650	
12.5				1200		2150		3850

Source: Department of Energy, Office of Public Affairs, "Department of Energy Information," Washington, D.C., October 20, 1978.

ciency goals judged to be in the national interest. A successful program in this area could work to the benefit of manufacturers by helping to clarify the direction of future markets. They could also help the industry if the policies resulted in a more rapid turn-over of existing vehicles or encouraged the purchase of more cars. Caution is needed, of course, since a poorly designed program could change the market so rapidly that the industry would be unable to move gracefully to accommodate it. Policies could also have the effect of encouraging present car-owners to hold onto their existing vehicles rather than to purchase a new one.

Direct taxes on gasoline and diesel fuels may offer the most powerful single tool for promoting market decisions toward increased efficiency. As previous sections pointed out direct taxes on gasoline have the advantage of offering the greatest freedom for individual consumer decisions; moreover, they are the hardest taxes to evade but require virtually no bureaucracy to enforce. Such taxes operate equally well in encouraging use of alternate fuels and increased efficiency. Their fault, of course, lies in the fact their implications may not be clearly visible to a consumer at the critical moment when he is considering which vehicle to purchase (a time when a gleam of chrome may overwhelm reason).

If use of fuel taxes to promote a sustainable energy future proves undesirable for political or other reasons, an alternative would be to tax vehicles on the basis of inefficiency. The tax could be imposed when the vehicle is purchased--following the model of the existing "gas guzzler" tax (see Table 3.16) or, it could be imposed annually when the vehicle is licensed. Several European nations have imposed high annual automobile licensing fees that depend on the vehicle size or weight. The annual tax offers a number of advantages: First, unlike a large tax on the initial purchase price, the tax would not discourage the purchase of new cars. Inefficient vehicles now on the road would be taxed and presumably become less desirable than newer, efficient cars. Second, the vehicle owner would be reminded of his vehicle's efficiency on an annual basis, whereas a sales tax would be buried in monthly payments. Care should be taken in the design of all such taxes to ensure that they are introduced at a rate which will allow the market to react advantageously to the changed economic environment.

The design of a "gas guzzler tax" requires resolution of two issues: how large should the tax be? and what goal should be used for vehicle efficiencies?

The first question must be clearly tied to the objective of the tax. If the tax is to be designed to make the purchaser pay the full cost of the energy used by the vehicle he purchases, then the tax can be clearly linked to an estimate of the extent to which the market price of fuel falls below its full society cost.

Table 3.17 illustrates the kinds of taxes which would be appropriate if it is assumed that the economic, national security, and environmental cost of burning a barrel of oil is $20 (equivalent to a tax of about 42¢ per gallon.) If the tax is keyed to a 60 mpg goal, for example, it indicates that a 40 mpg vehicle should be taxed $320. This is calculated on the assumption that the average vehicle travels about 90,000 miles during its life (based on of data in Difiglio, 1979). It it is argued that national cost of oil is $40 above market prices, the tax would be double that shown. The tax could, of course, simply be used as a convenient tool for raising federal or state revenue.

Table 3.17. VEHICLE TAXES EQUIVALENT TO A $20 PER BARREL OIL TAX

Efficiency of Vehicle Purchased (mpg)	Taxes Imposed on Vehicle, assuming as Goals of 60 and 80 mpg	
	60 mpg	80 mpg
80	$ 0	$ 0
60	$ 0	$ 160
40	$ 320	$ 485
20	$ 1290	$ 1450
10	$ 3230	$ 3390

Problems encountered in finding acceptable techniques for measuring the performance of vehicles in a way that will accurately predict their mileage are important for developing tax policies. The reminder that the published mileage may not agree with actual mileage has now become a familiar footnote to automobile advertisements. The problem, however, is serious since the tests now used to ensure compliance with the CAFE program were not designed to produce accurate estimate of "on-road" mileage; On road mileages of most cars is lower than the EPA rating. The discrepancy appears to be greater for small vehicles. Table 3.18 compares actual performance with the performance measured using EPA's combined city-highway driving cycle. The discrepancy results from the fact that conditions of the EPA test do not match real driving conditions: air temperatures, loads carried, terrain, wind loadings, tire/road interactions, etc. A variety of administrative and equipment factors also contribute to the inaccuracy of the tests.

Table 3.18.   MEASURED PERFORMANCE
VS. ON-ROAD PERFORMANCE

Combined Cycle (mpg)		City Cycle (mpg)	
EPA	On-Road	EPA	On-Road
25.5	20.3	10	11.6
27.0	21.2	15	14.5
28.0	21.8	20	17.05
		25	19.06

Source: McNutt 1979

It may be desirable for the EPA to develop new tests or "correction factors", which allow consumers to accurately predict real driving performance. Although CAFE compliance cannot be based on these tests they should nevertheless be published along with the standard EPA test results.

Direct Incentives to Manufacturers.   Unlike many European nations and Japan, the U.S. has never given automobile manufacturers direct support for automotive research and development.   In fact, the U.S. manufacturers have traditionally been cool to any offers of support.   Present circumstances may have changed some minds, so it may be appropriate for the federal government to explore ways of using federal funds to ensure an adequate development program in the U.S. industries. If the industry's reaction is favorable, a more detailed proposal could be developed for allowing the use of public funds for the development of critical automotive components. (Ford 1980.)   One evil of financial hard times is that funding for advanced ventures is often sacrificed so that monies can be available to meet near-term emergencies; thus, research and development funds are often absorbed to address more immediate concerns.

The Federal Fleet.   Although statistics on the subject are not good, it appears that very few trips in federal passenger vehicles involve more than 4 passengers; most trips only carry 1 or two passengers. Virtually no trips in federal cars require substantial cargo space. Over eighty percent of federal vehicles have gross vehicle weight ratings less than 8500 lbs. Federal postal vans, pickup trucks and other cargo vehicles, therefore, are roughly comparable to passenger vehicles and could be included in the overall federal fleet efficiency. Postal vehicles require cargo space but these vehicles need only carry a single rider and can use additional space for cargo.   In effect all federal vehicle travel could be done in two-passenger or four-passenger vehicles.

A program to improve the efficiency of the federal fleet could play a major role in creating a market for new, efficient vehicles.   It would, moreover, be necessary in a program designed to encourage private investment in similar vehicles. Executive order 12003 (dated 20 July 1977) established a set of mandatory standards for all passenger automobiles acquired by the federal government after FY78.   The standard requires that the federal fleet achieve an average of 24 mpg in 1980 and 31.5 mpg in 1985. There is no separate standard for trucks.   Standards for light trucks and vans do not differ from the standards set for commercial trucks. This executive order should be changed in several ways:

o   All light trucks and vans purchased by the government should be included in the fleet average; and

o   The fleet average should be set at 35 mpg in 1985 and 50 mpg in 1988.

The federal government should also develop model standard programs for state, county, and municipal governments and work to encourage the adoption of these programs through state energy offices, energy extension services, and other organizations. Standby legislation requiring state compliance may be needed.

Standards applying only to federal vehicles would affect less than one tenth of one percent of all passenger vehicles in the U.S. and approximately one percent of all freight vehicles.   The standards can be expected to have an impact that exceeds the energy they save directly, however, since they would serve notice that the federal government takes transportation efficiency seriously.

The proposed federal standards would save on the order of 0.017 quads per year by 1992 if it is assumed that, without the standards the average economy of federal automobiles will follow current federal guidelines and that the on-road performance of federal vans and light trucks would reach 20 mpg in 1985. Federal fleets now consume approximately 0.032 quads a year and roughly 20% of the fleet is replaced each year. The standards could be implemented rapidly because a variety of vehicles are now on the market which exceed 35 mpg (EPA).   These include the Dodge Colt and the Plymouth Champ.

Heavy and Medium Trucks

The trucking industry has already improved vehicle performance in response to the rising price of fuels. Efficiency measures include turbocharging heavy trucks, developing more flexible engines which require fewer gears, installing superior radial tires; and increasing the fraction of efficient deisel engines entering the new truck fleet. Nevertheless there are still major opportunities for improving the fuel economy of future vehicles.

For example, diesel engine efficiency can be increased by using more of the energy currently rejected as waste heat. One proposal is to build diesel engines out of materials which need not be water cooled - the "adiabetic diesel." The resulting efficiency increase is estimated at 10% (Julsing, 1979). The use of the pressure in the exhaust from a diesel turbocharger to drive a second turbine directly coupled into the engine crankshaft via a speed reduction gear train that is "turbocompounding" has been found to give a 5% fuel economy improvement (Automobile News, 1980). Finally the use of heat in diesel exhaust to vaporize freon which then drives a turbine coupled to the engine crankshaft has been found to reduce fuel consumption by 15% (Sorge, 1980). The efficiency of the nation's fleeted Medium Trucks can be improved by shifting to diesels. In 1978, nearly 80% of heavy trucks used diesels but only 40% of medium trucks used them (EEA, 1980).

Other possible improvements in the heavy truck drive trains can be achieved with microcomputer controlled automatic transmissions that can automatically put the truck in the most efficient gear. A prototype device is now being tested in a Class 7 truck, and the transmissions may be available on trucks sold in the late 1980's. They would be particularly attractive for heavy vehicles used for stop-and-go driving in urban areas, where frequent gear shifts are required. Savings of 14-20% are possible in such driving cycles.

It should also be possible to increase truck efficiencies about 7% by paying more attention to the energy used for truck accessories such as fans and pumps. Most of these devices are connected to the engine and are turned at all times, even though they may actually be required only about 5% of the time. Relatively straightforward clutch-actuated fans controlled by a thermostat can achieve significant savings.

After engine improvements, the major targets for fuel economy improvements in heavy trucks are reduction in aerodynamic drag and tire rolling resistance. The aerodynamic shaping of trucks is extremely primitive today--even with the now-common deflector that has been added on to the cab roofs of many long distance tractors. NASA has shown that in an aerodynamically shaped tractor with flexible sheet connections to the trailer, the fuel consumption of a tractor-trailer combination can be reduced by 20-25% (Steers, 1977).

There are also major opportunities to reduce the rolling resistance losses of truck tires. Only 30% of new heavy trucks were equipped with radial tires in the 18-month period ending in mid-1979, despite the associated 5-10% fuel savings (GAO, 1980). An estimated additional 4-5% savings could be achieved by using radial tires made with lower loss materials (Knight, 1979). Further savings of 5-10% could be realized if some double tires on heavy trucks were replaced by "wide-singles", thus halving the losses in the tire sidewalls (Giles, 1979).

As in the case of automobiles, truck efficiencies can be improved by the weight savings resulting from increased use of aluminum and advanced composite materials. A significant increase in the use of composites cannot be expected until after 1985.

All of these savings are of course not additive. In Table 3.1, a 30% improvement in average truck efficiency is estimated for the year 2000. (DOT, 1976; DOE, 1975) This must be considered a conservative estimate of the savings possible.

Air Travel and Air Freight

Background

Air travel and freight have grown faster than any other form of transportation during the past decade because of speed and convenience. Passenger miles increased by 7.1% per year between 1965 and 1977, and air freight increased 6.2% per year during the same period. Over 60% of the passenger miles on aircraft in 1977 were traveled on trips to visit friends or relatives or for other personal reasons (TECDB-3, p. 133).

Surprisingly, commercial air travel, which consumes about 90% of aviation fuel, is at present approximately as efficient as automobile travel, measured in terms of passenger miles per gallon (Bond, 1980).

The deregulation of the aircraft industry coupled with higher fuel prices have already led to significant improvements in the energy efficiency of the aircrft industry. In 1973 fuel costs were approximately 25% of the direct operating costs of a Boeing 727. By 1975, fuel costs were nearly 40% of total costs (Klineberg, 1978). Although revenue passenger miles increased some 57% in the 1973-1979 period, fuel consumption increased only 5%. The energy use per passenger mile for domestic aircraft has been reduced by about 5% per year since 1972 (TECDB-3, p. 107). These savings have been achieved by increasing the number of passengers on each plane, by more carefully operating the plane (reducing cruising speeds, reduced holding, more efficient climb rates, and better maintenance), and by increased use of more efficient wide-bodied jets and modern high-bypass turbofan engines (Rose, 1979). Load factors (the fraction of available seats filled) in 1979 were 63.1%, up from 55.9% in 1977 (Thomas, 1980), and overall fuel efficiency increased from 17.5 passenger-miles per gallon in 1973 to 25 in 1979 (ATA, 1980). The aircraft industry would probably have improved upon this record of increasing energy efficiency if fuel prices had not been regulated at artificially low levels.

The next generation of aircraft (Boeing 757 and 767, DC9-80, advanced L-1011) will be substantially more efficient than the current generation of aircraft. When new wide-bodied jets begin to dominate the commercial air industry fuel efficiency can be expected to rise to around 45 seat-miles per gallon. This average is slightly less than what would be achieved if all aircraft were to have the same efficiency of wide-bodied jets in use in 1976 (Rose, 1979) (Swihart, 1979). The technologies that will probably have the greatest energy impact in this century include advanced gas turbine engines, low-energy taxiing, advanced lightweight materials, and better active controls.

Changes in flying or operating procedures (altering descent procedures or climb-out techniques) can decrease fuel consumption significantly in the near term without technical advances. In addition, air traffic control delays in 1979 wasted about 784 million gallons of fuel and cost the airlines about $1 billion (ATA, 1980). These delays were, to a large extent, caused by peak-hour congestion at airports, however, a problem created by consumer demand and not easily alleviated.

It will be possible to use advanced technologies to improve on the generation of aircraft just described, which will be introduced in the 1980's. The rate at which the next generations of aircraft will be introduced into the fleet will be limited by the fact that the average age of an aircraft now in the U.S. is approximately 8 years, and the average aircraft is expected to last about 18 years in domestic U.S. service. It takes about six years to take a concept from an R&D stage to a manufactured aircraft.

Technical Potential

Designing an improved aircraft requires optimizing a number of criteria, including performance, safety, handling, and fuel efficiency. Complex trade-offs are often necessary. For example, attempting to improve the efficiency of an airfoil could increase the aircraft's weight and decrease the aircraft's overall fuel efficiency. No accurate estimate of the actual fuel savings that could result from new technologies can be made until the are integrated into aircraft designs.

The high-bypass ratio gas turbine engines now used on aircraft are attractive both because of their high efficiency and their low noise levels. Improvements on present designs are likely to continue throughout the next decade. The generation of powerplants introduced in the 1990's may be able to reduce cruise fuel consumption between 15% and 20% with three-shaft engines or geared fans with lower specific thrust (Pope, 1978). NASA's Energy-Efficient Engine program predicts that turbofan engine improvements might reduce the fuel use per unit of weight by up to 12% (Klineberg, 1978).

Another class of powerplants still in the research stage are prop-fans. These propeller type powerplants may be 10% to 15% more efficient than an advanced turbo-fan, particularly in medium-range flights at somewhat reduced speeds (Pope, 1978; Klineberg, 1978; Rettie, 1975). The advantages of these designs may be offset to some extent by interference drag or high operating costs incurred from propeller and gear-box maintenance (Nored, 1978).

Fuel could also be saved by using power wheel systems to move aircraft on the ground, rather than the aircraft's engines. This concept could also decrease air and noise pollution in congested airports.

Use of lighter, stronger materials can also reduce aircraft weight and save energy. For instance, certain carbon-fiber epoxy composites are about 5 times stronger and 3 times stiffer per unit weight than the aluminum now used. Fuel savings could range from 10% to 20% (Linn 1976).

Recent advances in the development of high-integrity flight control systems are enabling a number of traditional constraints in the design of aircraft to be relaxed. Aircraft are traditionally designed so that when they are perturbed from steady flight (by turbulence, for example), the aerodynamic forces on the aircraft tend to bring it back to equilibrium without intervention by the pilot. This leads to design constraints that reduce an aircraft's efficiency. In addition, aircraft are typically designed to accept certain kinds of manuvering loads in such a way that overall aircraft efficiency is sacrificed. Both these conventional design problems can be alleviated with the use of active control technology. The Lockheed Tristar already uses active control technologies. When incorporated into the design of new aircraft, active controls may well lead to reductions in fuel consumption of the order of 10-15% (POPE 1978).

With a continuation of historic trends in fuel economy (25 passenger-miles per gallon, up from 17.3 PM/gal in 1973), by the year 2000 average fuel efficiency would reach 28.8 PM/gallon. But if research and development efforts prove fruitful, average load factors reach about 70% and if the policies described in the following are successful, it is reasonable to assume that the average fuel efficiency can reach about 35 PM per gallon.

It can be assumed that general aviation will follow trends similar to commercial aviation, particularly if equitable air service user fees are imposed and fuel prices rise substantially. Military aircraft may become more efficient in the future for strategic reasons, but determining the extent of their development is not within the scope of this study. No efficiency improvement has been assumed.

Policy

Policy options for improving aircraft energy efficiency include:

o Increased funding for research on advanced aircraft designs.

o Incremental or phased-in fuel taxes to raise the price of aviation fuel to its true marginal or social cost.

o Taxes or user fees on general aviation to ensure that general aviation pays its fair share of federal airport and airway systems costs (Goldschmidt, 1980).

o Improve air traffic control and operation systems. The FAA and the CAB should be given the clear authority to implement air traffic control procedures which derive energy as well as safety benefits.

o To the extent practicable, airlines and/or airports should take steps to alleviate peak period conjestion at major air carrier airports through pricing policies or public relations campaigns.

Shipping

In 1977 the U.S. international and domestic shipping industry purchased 1.1 quads of petroleum (1.1 quads annually) (TECDB-3 pp. 2-13, Booze Allen and Hamilton 1977). Used in computing official estimates of U.S. energy demand, these figures significantly understate the vulnerability of U.S. ship transportation to declining world petroleum supplies. Only the petroleum products purchased in the United States were counted. Eighty-one percent of the fuel used for foreign shipping and 73% of all fuels used for U.S. Commerce were purchased abroad in 1974. Assuming that the ratio of foreign to domestic fuel purchased had remained constant, U.S. shipping required 4.1 quads of petroleum in 1977, about twice as much energy as was consumed by combination (large) trucks (TECDB, p. 2-13).

U.S. opportunities for reducing the energy required in shipping are constrained because 43% of the total shipping energy required by the United States is consumed in ships flying a foreign flag. One opportunity for reducing the oil required in shipping, of course, is to reduce the nation's demand for imported fuels, since about half of the energy required in both domestic and international shipping to the U.S. is used to transport petroleum or petroleum products. Beyond this information, the current literature is surprisingly sparse. There may be opportunities for shifting some ships to direct use of coal. Improvements in hull design may

also be able to increase efficiency. A number of recent studies have also indicated that it may well be possible to use contemporary technology to design advanced sailing ships and sail assisted ships that can compete favorably with conventional transport craft. These ships would be designed with highly automated rigging and would use computers, satellite navigation, and advanced weather forecasting techniques to chart economical and safe routes across the ocean. (This may represent the most attractive use of solar energy in marine transport.)

Since the problems confronted in ship transportation are international, it would be desirable to undertake a study under the auspices of OECD (and other international organizations with a membership likely to be interested in ship efficiency) that would examine the extent of western dependence on petroleum for national and international ocean commerce, the potential for major changes in the character of international shipping (e.g., from an increasing commerce in coal), and the potential of new marine technologies. Joint research and development projects may be natural outgrowths of such an analysis.

Telecommunications

Future electronic communications may have a greater impact on national transporation patterns than any other technology likely to be introduced during the next two decades. Teleconferences can replace business trips; high speed data links, mailings and couriers; and home computers may replace some shopping trips. A home computer can provide a source of news, communicate with bank accounts, place orders in local stores, and countless services limited only by the imagination and courage of the forcaster. Like electric vehicles, electronic communication replaces conventional transportation energy with electricity. In addition, telecommunications can indirectly conserve energy by decreasing the use of a natural resource such as wood (by decreasing paper consumption).

In spite of its potential impact, it is virtually impossible to make any quantitative estimate of the net impact of electronic communications. Recent studies have indicated that up to 20-35% of business trips would not be needed if adequate videoconference facilities were available (Kollen, 1975; Tyler, 1977) (or 7-12% of all personal air travel). Sophisticated videoconference facilities have been available from AT&T for nearly a decade (Ryding, 1980), and a new system scheduled for operating in January 1981 will be introduced by a consortium between COMSAT and IBM (Satellite Business Systems, 1980). The economic advantages of teleconferences and telecommunication systems are increasing with rising fuel prices.

Modern electronic systems involving video, sound, and data transmission, have also been proposed as a mechanism for reducing energy used in commuting. It may be possible to decentralize large corporate work forces (such as insurance or engineering companies) by developing a series of regional centers located in a way that would reduce commuting time. These centers could communicate electronicly, and achieve a great saving in energy. In some cases, commuting may not be needed at all if a coaxial cable can bring multi-channel communication links directly to individual homes. Cables now being installed to carry television signals provide an adequate basis for many channels of communication.

SHIFTING TO EFFICIENT TRANSPORTATION MODES: MASS TRANSIT AND RAIL FREIGHT

Urban Transportation

The energy efficiency of several kinds urban passenger transportation shown in Table 3.19. This table makes the complexity of any decision about personal transportation very apparent. Plainly, the van-pooling options (or car-pooling) discussed in the earlier section are the preferred mode of urban transportation under any of the circumstances examined. An urban bus system also appears to be preferred to efficient cars. Buses offer distinct advantages over heavy rail systems, since they are not tied to fixed guideways and are able to adjust routes to meet changing traffic and demand patterns.

New heavy rail systems do not appear to be particularly attractive from the perspective of energy conservation. It is possible that efficient automobiles may consume less energy per passenger mile even if they are driven on urban driving cycles with an average of 1.4 persons per vehicle. There are, however, many reasons to support a rapid rail system which are not tied directly to energy. Rail systems (like bus systems) can reduce congestion and pollution in urban areas and may lead to more attractive urban designs in the future but trains tend to be much faster than buses and thereby more attractive to commuters, most of whom value time more than cost in comparing commuting systems. With careful planning a rail system can reduce urban sprawl, and thereby indirectly encourage urban designs which do not depend on large amounts of transportation energy.

Truck and Rail Freight

Railroads are 3-4 times as efficient in hauling freight between cities as trucks, but the railroads share of this traffic has declined dramatically during the last two decades (see Table 3.20). In 1959 railroads carried 56% of all intercity freight and trucks 16%; in 1980 rail carried about 36% of this traffic and trucks 25%. The decline in rail traffic market share has resulted in part from a program of federal subsidies and regulations which have become outdated. Artificially low

Table 3.19. EFFICIENCY OF PERSONAL TRANSPORTATION

Transportation	Btu/Passenger Mile	
	Direct Fuel Use	Direct & Indirect Energy
Automobiles[a]		
14 mpg (on road)	4250 (7970)	(10,270)
40 mpg (EPA)	1490 (2790)	(4,782)[g]
60 mpg (EPA)	0990 (1860)	(3,852)[g]
80 mpg (EPA)	0745 (1400)	(3,392)[f,g]
Advanced Electric Vehicles[b]		
Traveling at 25 mph	1350-3000 (2025-5700)	
Traveling at 35 mpg	1614-3200 (2421-4800)	
Motorcycles (current average)[c]	2270	
Vanpools (8 riders)[d]	1200	2060
Bus[c]		
Transit	2960	3420
Intercity	1010	
Rail		
Commuter	3500	5895
Transit	3030	
Air[c]		
Local	8320	
Domestic Truck	6620	
Possible Future	4500	

[a]It is assumed that automobiles carry an average of 2.1 passengers. The numbers in parentheses indicate the mileage expected in commuting. In this mode it is expected that the average vehicle will contain 1.4 passengers and that vehicle efficiency will be 20% lower than the EPA composite fuel economy value.

[b]TEDB (4027). Range indicates different degrees of technological optimism; current vehicles are near the upper end of the range. Numbers in parenthesis assume 1.4 persons per vehicles; others assume 2.1 persons.

[c]TEDB (2-28).

[d]Assumes 14 mpg for the van and that van miles are increased by 10% to account for pickup and delivery of 8 persons.

[e]Includes energy required to manufacture the vehicle and guideway construction. (CBO, 1977). Note parallel with assumptions in TEDB (e.g., passengers per van may differ slightly from column 1.

[f]High efficiency autos may be more energy intensive than indicated due to the use of advanced materials.

[g]Assumes minor credit for energy saved in manufacturing light-weight automobiles.

fuel prices, massive federal subsidies to highway construction, a tax system which failed to make trucks pay their fair share of highway construction and maintenance, and regulations which made it impossible for railroads funds to improve their roadbeds and equipment have combined to accelerate the decline of the rail industry. The association of American Railroads estimates that the total capital needs for the industry during the next 10 years will be close to $50 billion. It will be difficult for them to raise this capital since net profits per ton mile have decreased by more than a factor of three since 1950 and the average return on equity invested in the rail industry is now about 3%.

It is plainly necessary to review the programs now governing investments in rail and truck freight to ensure that trucks are not inequitably subsidized and that the rail industry is able to raise enough capital to meet its urgent needs. A program which seems to offer the most promise for removing subsidies to trucks would be a direct tax, linked to maximum axle weight and truck miles traveled, which is adequate to pay for all highway costs traceable to truck traffic. A fuel tax

Table 3.20. MARKET SHARES OF TRUCK AND RAIL FREIGHT

Year	Total Rail and Truck Freight (billions of ton-miles)	Fraction Carried by Rail	Fraction Carried By Truck
1960	855	0.67	0.33
1970	1160	0.65	0.35
1973	1363	0.63	0.37
1074	1340	0.63	0.37
1975	1213	0.62	0.38
1976	1243	0.61	0.39
1977	1373	0.60	0.40
1978	1478	0.59	0.41
1979	1548	0.59	0.41
1980*	1490	0.62	0.38

*Preliminary

Source: TECDB-4 p.

would also ensure that trucks paid full "social cost" of their fuel use. Given recent actions which deregulate many aspects of the truck and rail industry, the rail industry may be able to adjust its rates to raise enough capital to compete with the trucking industry if inequitable subsidies are removed. The new rules allow the rail industry to raise up to 160% of their variable costs with ICC intervention (higher rates can be approved but only after the ICC approves).

Revenues derived from these higher rates may well be adequate to support the rail industry's needs, but the situation must be watched carefully to see whether any direct federal assistance is needed to rectify years of inequitable treatment. One specific program examined here involves direct support for municipal facilities designed to shift freight from trucks to rail for long-hauls.

Taxing Trucks to Pay for the Highways

The techniques now used to raise funds for federal highways do not send the proper signals to the market for two reasons: first, highway users do not pay the full cost of building and maintaining the national highway system and second user, payments are not charged in a way that taxes users in proportion to their share of highway costs. Excise taxes, for example, do not reflect miles traveled. Taxes imposed on highway users in the form of excise taxes, user taxes, registration fees, licensing fees etc. only covered about 65% of the cost of building and maintaining the federal highway system; 23% came from general revenues and 12% from investment income and borrowing.

Perhaps the most critical question is whether trucks are paying their share. There are two parts of the question: (1) allocating the costs of building thicker, highway surfaces and stronger bridges to support heavy trucks, and (2) allocating the costs of maintaining the highway system. The Federal Highway Administration is conducting an extensive examination of this issue and is expected to report its results in 1982. Meanwhile, however, several states have already begun to take action on the basis of existing information.

It is generally agreed, for example, that trucks are disproportionately responsible for a majority of highway maintenance costs. One study showed, for example, that an 80,000-lb truck inflicts as much damage on a road as 6,000-10,000 passenger automobile (Jennings, 1979). The discrepency is likely to increase as cars become lighter and trucks heavier.

The point at issue is whether trucks pay user fees commensurate with the highway costs traceable to trucks. This question is now being studied by the Federal Highway Administration. (A report is expected to be completed early in 1982). Several state and national studies, however, indicate that heavy trucks do not pay a fair share of taxes to offset the fixed and variable costs for which they are responsible. An Urban Institute Study in 1977 found that as of 1975, heavy trucks were making just enough payments to offset direct use-related pavement deterioration costs, and thus were making no contribution to fixed costs (Urban Institute, 1977). According to an Arkansas State Highway and Transportation Department study, considering road damage on a proportional basis, cars pay 1160 times in taxes and fees as much as a 73,280-lb truck (Gray, 1980). According to a 1978 Georgia study, four-axle and five-axle truck only paid 51% of their share of highway costs. (CSRC, 1980). Oregon, California, and Florida have reached similar conclusions (Gray, 1980; Oregon, 1979). A Bureau of Public Roads Study in 1970 found that combination freight trucks paid 32% less than their fair share of construction costs.

One way to ensure that highway users are forced to pay the full cost of the service provided by the national highway system would be to tax trucks on the basis of the maximum weight per axle and the number of miles traveled each year. A schedule of taxes designed to raise a billion dollars in annual revenue is exhibited in Table 3.21. The rates can be scaled to raise the amount of revenue needed to cover the trucking industry's share of highway costs. Some provisions must be found, of course, to ensure that mileage is accurately reported each year. There are precedents for such a system. Oregon, Colorado, New York, and Ohio all have truck-ton/mile tax systems in place.

Truck-to-Rail Transfer

The technique of "piggybacking" (or transferring a containerized cargo unit or trailer onto a flatcar) combines the resource and energy benefits of rail with the flexibility, convenience, and delivery time of trucks. Piggybacking is particularly attractive for intercity freight, carried over 200 miles which is not sensitive to changes in time of delivery. Its use has grown every year since 1975. A result study showed that piggyback was found to be twice as efficient as intercity trucking. Piggyback service takes longer than direct truck service due to terminal delays and delivery and pickup times will continue to limit its use. In the long run, higher prices will increase the relative fuel effectiveness of rail freight. Tentative results of one study indicate that as fuel prices reach about $2.00 per gallon, as much as 50% of freight trips over 200 miles might transfer to rail (GM 1980).

It appears that the majority of the factories and warehouses in the U.S. built since 1960 do not have rail access. Federal assistance in locating and acquiring yard sights could be beneficial. It might also be appropriate for the federal government to sponsor demonstration programs to encourage truck-to-rail transfer; determining new strategies to decrease point-of-transfer and related shipping delays is important (OM 1980). (See Table 3.5).

Table 3.21. HYPOTHETICAL TARIFF SCHEDULE ON HEAVY TRUCKS

Maximum Single Axle Loading (Pounds per Axle)	Tariff in Costs per Ton-Mile Required to Raise $1 billion
Less than 10,000	.18
10,000 - 12,000	.5
12,000 - 14,000	1.0
14,000 - 16,000	1.75
16,000 - 18,000	3.0

Section Four

Transportation Energy
from Renewable Resources

Alcohols made from wood, crops, and organic wastes will undoubtedly be the most important renewable resource available for use in transportation during the next few decades. Methanol is likely to be the dominent synthetic fuel in the future because it can be manufactured inexpensively from a variety of renewable and fossil energy sources. In the relatively near future it may also be possible, to use electricity generated by wind machines, photovoltaic arrays, or other solar electric generating equipment to operate electric vehicles, electric rail, and telecommunication devices. Longer term possibiltites include the use of hydrogen derived from electrolysis of water or directly from a photochemical systems, and direct photochemical production of fuels other than hydrogen. It is also important to acknowledge that there are many traditional ways of using renewable resources for transportation which may offer advantages in the future, including the use of modern sailing ships and increased use of pedestrian walkways and bicycles.

Given the relative importance of various renewable fuels, the bulk of the analysis below will be devoted to alcohol fuels,while electric vehicles, hydrogen and other technologies will be given secondary emphasis.

ALCOHOL FUELS

Most of the solar energy used in transportation in the next two decades is likely to result from the use of alcohol fuels made from wood, crops, and organic wastes. Ethanol (the alcohol in beverages) and methanol (or "wood alcohol") will undoubtedly be the primary solar alcohol fuels. There is some possibility that other alcohol mixtures including butanol and other higher alcohols, will find applications. The costs and resources of alcohols from biological sources are discussed in detail in the chapter on industrial energy

use. The present discussion will mainly examine how the fuels can be used in automobile and truck engines.

It is apparent from previous chapters that potential supplies of methanol are significantly larger than supplies of ethanol since methanol can be obtained from wood, other biomass feedstocks, peat, coal, and natural gas at a lower price. In addition, methanol can be derived from coal or biomass using a process that is simpler than the one needed to convert coal to gasoline. In fact, methanol derived from biomass or coal will almost certainly be the least expensive "synthetic" fuel. The analysis provided in the Chapter dealing with industry indicated that methanol from biomass can be competitive with gasoline manufactured from oil at today's price of oil imports unique characteristics. With improvements in biomass-to-methanol conversion technology that exploit the advantages of biomass feed stock, methanol from biomass may also be competitive with methanol produced from coal (see Appendix in Industrial Chapter).

Whether the savings in methanol production compensates for the cost of converting the present distribution and retailing systems to methanol remains an open issue, but it seems unlikely that the nation would find any long-term benefits in sustaining production of a synthetic gasoline that is significantly less efficient, both in terms of production and end-use efficiencies. The attractiveness of a methanol fuel, of course, would significantly increase if the industry determined that vehicle manufacturers were interested in designing fleets of vehicles and.distribution systems able to accomodate methanol fuels. Interest in such conversion, in turn, would be influenced by the availability of reliable sources.

Alcohol has two significant advantages as a motor fuel. Pure alcohol has a high octane rating so that the engines designed to operate on pure alcohol can have higher compresson ratios and thermal efficiencies than conventional gasoline engines which often use lead or other toxic compounds to boost octane. Alcohol also burns in leaner mixtures than gasoline. This can improve efficiency and lead to reductions in net vehicle emissions. Disadvantages arise because the specific energy is lower which means that a larger gas tank is required.

Alcohol Fuel Blends

Both methanol and ethanol mixed with gasoline or diesel fuel can be used in vehicles. A 10% mixture of ethanol with gasoline (or about a 7% mixture of methanol with gasoline) will result in a fuel about 3 octane points higher than the gasoline used in the mixture. The octane-boosting property of alcohols offers several advantages. First, since low-octane gasoline requires less energy to manufacture per unit of energy delivered to a vehicle than high-octane fuels, energy savings can accrue by blending alcohols with gasoline; the savings will depend in complex ways on the mix of products produced by refineries and the characteristics of the crude oil (or synthetic oil used as a feedstock). Second, substituting alcohol for octane boosters such as tetra-ethyl lead may decrease certain kinds of vehicle emissions and reduce pollution. In fact, some refiners are now using oxynol, a mixture of tertiary butyl alcohol with methanol, as an octane booster. The total market for octane boosters is, of course, a small fraction of total national gasoline production.

Use of "gasohol," or 10% mixtures of ethanol with gasoline, has increased rapidly in the past year, although 1981 consumption will be only a few hundred million gallons. Gasohol can be used in most existing automobiles with minor changes in mileage and performance, although some early model cars will experience some difficulties when driven on ethanol (e.g. stalling or poor performance). Most new cars sold in the U.S., however, are warranted for gasohol use.

Methanol mixtures are in some ways less attractive than ethanol mixtures. The mixtures are more sensitive to water than ethanol blends (small amounts of water added to a gasoline/alcohol mixture can cause the alcohol and gasoline to separate.) There are technical solutions to this problem such as the use of selective solvents, but these solvents may be expensive.

Methanol mixtures can cause relatively high evaporative emissions (i.e., vapors released by fuel in a tank even when the vehicle is not being used) and must have special equipment to trap these vapors. Gases from methanol/gasoline mixtures may also lead to vapor locks.

.Another serious problem with methanol mixtures is that methanol attacks many materials used in automotive fuel system components. Some vehicles now on the road could not tolerate methanol mixtures without serious damage. Retrofits making vehicles compatible with methanol blends could cost between $20 and $145 per car. Retrofits making the vehicles compatible with 100% methanol could cost between $300 and $400 per car (Bolt 1979).

Engines that run on alcohol blends have different exhaust emissions than engines run of straight gasoline. Alcohol changes the vapor pressure of the gasoline with the result that the evaporative emissions are increased. This problem may be overcome by altering the chemistry of the gasoline intended for alcohol blending, or auto manufacturers may certify their systems on alcohol blends to insure that emissions are kept at an acceptable level. Although the emissions of alcohol-fueled cars are known with some certainty, the long-range environmental effects of alcohol use has not been examined in detail.

Engines can be designed to operate on alcohol-gasoline mixtures where the alcohol fraction is higher than 10%. For example, engines with advanced closed-loop feedback controls or most 1981 vehicles can tolerate 10-20% alcohol without loss of performance and without retuning. As indicated earlier, it is likely that efficient, new automobile engines will include a range of different stratified charge engines. Although these engines are designed only to burn gasoline, it is expected that they may be readily compatible with various alcohol-gasoline mixtures. It may be somewhat easier to use alcohol mixtures in a stratified charge engine because of its burning characteristics.

Diesel engines can also be run on 30-40% ethanol and 20-30% methanol blends, although there are fuel separation problems and the injector timing has to be advanced to correct for the altered heat content of the fuel. 30-40% blends have been handled by aspirating the fuel into the induction system through a separate tank. An advantage of alcohol-diesel fuel blends is that particulate emission rates are substantially reduced. However, diesels may not be the ideal application for alcohols as they do not take advantage of the high-octane properties of alcohols.

Pure Alcohol Fuels

General Motors, Ford, Fiat, and Volkswagen currently have vehicles for sale in Brazil which are designed to operate on 100% ethanol. The Bank of America, headquartered in San Francisco, California, is engaged in a program to convert a significant portion of its automobile fleet to allow use of 100% methanol instead of gasoline. The Tennessee Valley Authority may have similar plans.

The use of pure alcohol has a number of advantages. Engines may be able to achieve 10-20% greater efficiency with pure alcohol rather than conventional fuels, primarily because the alcohols can be used at a much higher compression ratio. By the late 1980's it may be possible to develop an engine that can increase the efficiency of contemporary engines by 20% to 30% by using engine exhaust heat to disassociate methanol into hydrogen and carbon monoxide before it enters the engine (Pefley 1979).

The use of pure alcohol fuels results in different vehicle emissions than those associated with conventional fuels. In general, engines burning alcohol can be tuned to emit less carbon monoxide. Alcohol fuels can also eliminate contaminants resulting from lead and other chemicals used as octane boosters. The chief emission problem associated with alcohol results primarily from aldehydes, such as formaldehyde, which could create significant health risks and aggravate smog. Depending on engine design, pure alcohol could increase or decrease emissions of nitrogen oxides and unburnt fuel. It has been found, however, that oxidation catalysts or other emission control systems can be used to successfully control the environmental difficulties associated with the use of alcohol fuels. Development of emission control devices for alcohol engines should be given high priority in the national research agenda. As mentioned before, the effects of alcohol fuel emissions on the environment is largely unknown.

A difficulty associated with the use of 100% alcohol fuels is that alcohols make the vehicle difficult to start when it is cold. A variety of schemes have been proposed for resolving this problem including additives, induction heaters, and small tanks of starting fuels. There are some indications that engines like the PROCO design will be relatively easy to start as the fuel is squirted directly into the cylinder at a spark plug.

Multi-fuel engines offer a possible use for alcohol fuels. Research has indicated multi-fuel capability in diesel engines equipped with a "glow-plug" surface ignition device. Both pure methanol and ethanol can be burned at the compression ratios of today's diesels (Nagalingam 1980). Future stratified charge engines also show potential for a multi-fuel capacity, although little published work has been noted in this area. The lean combustion characteristics of stratified charge engines could integrate well with the leaning properties of ethanol and methanol. Fuel volatility problems and the varying combustion air requirements of different fuels present difficulties for the design of carbureted or naturally aspirated multi-fueled engines; stratified charge engines are more readily adaptable to a multi-fuel capacity due to their great flexibility in air/fuel ratios.

A major obstacle to any conversion to alcohol fuels is the need to develop a distribution system. The retail dealers distributing transportation fuels have only recently been compelled to shift to a parallel system for delivering lead-free gasoline, and it is unlikely that they would enthusiastically embrace a proposal for still another mandated shift. A conversion to

alcohol could be disruptive, requiring new, specially lined storage tanks and new fuel pumps. Programs for introducing alcohol may need to consider the average life span of wholesale and retail gas station equipment. Complete conversion to alcohol fuel compatible systems in 20 years, however, does not seem to be an inappropriate goal.

Alcohol Fuel Policies

Programs designed to encourage the production of alcohol fuels from biomass are discussed in the chapter on industry. They include policies designed to encourage industrial research, development, and investment in producing biomass energy supplies including specialized policies designed to improve forest management and to support research on advanced techniques for manufacturing alcohols. The following discussion will be limited to programs whose objective is the development of a market for alcohol fuels in transportation; programs for using alcohol fuels in federal vehicles, programs for encouraging the use of alcohols in captive fleets, and research and development priorities will be suggested. As the earlier discussions have emphasized, however, fuel taxes would undoubtedly be the most effective policy for encouraging the use of alternative fuels in transportation.

Federal Fleets

Nearly half a million vehicles are owned and operated by the federal government. Federal, state, county, and municipal government vehicles total over two million. Some fleets are maintained and fueled at a central service area and provide an excellent opportunity for establishing test fleets or markets for alcohol fuels. Such a program should dovetail that legislated in the Energy Security Act of 1980 and should include the following:

o A study designing a program to make federal maintenance facilities compatible with alcohol fuels, including methanol fuels. The study should establish precise timetables and goals.

o An executive directive requiring that ten percent of the motor fuel dispensed by the federal government after 1984 should be an alcohol. (This does not mean that all the fuel dispensed should be 10% alcohol.) Methanol should be used whenever practicable.

o An executive directive requiring that all federal vehicles purchased after 1985 should be equipped with multifuel engines or with engines capable of using methanol or ethanol mixtures.

Captive Fleets

Over 13% of the new cars sold in 1977 were sold to fleets, and the percentage has been growing steadily for the past decade. Private businesses operate 46% of these vehicles; others are operated by utilities, police forces, taxi companies, etc. Corporations with large car fleets are in a good position to convert to alcohol fuels since they typically maintain close control over maintenance and fueling. The government can promote programs in which fleet owners could follow the federal lead in converting to efficient ethanol and methanol compatible vehicles. The programs could consist of the following:

o   Detailed guidance and information on fleet conversion based on federal experience and studies. Both technical and procedural information should be provided, and the information should include lists of suppliers of critical materials and fuels.

o   Standby authority, through legislation, to require compliance with these standards should a national liquid fuel crisis develop.

Research and Development

There are a number of uncertainties associated with using ethanol or methanol in blends or as pure fuels which need to be investigated as soon as possible. Research should continue to identify: (a) economic and institutionally acceptable methods of distributing alcohol through conventional fuel systems; (b) low-cost additives to prevent alcohol-gasoline blends from separating; (c) the methods and costs of modifying the transportation fleet to accept various blends of alcohol fuels; (d) macro as well as micro environmental impacts of using alcohol fuels. This research should be a part of the DOE study and plans mandated in the Energy Security Act, and should give equal emphasis to methanol as well as ethanol.

TRANSPORTATION SYSTEMS USING ELECTRICITY

Electricity can be used to substitute for the fuels now used in transportation in three ways: (1) electric telecommunication systems can be substituted for business travel, paper transfers, and possibly even for some commuting and shopping trips (discussed previously), (2) electric highway vehicles, probably used primarily for commuting, can replace conventional automobiles, and (3) electric rail systems can be used instead of conventional diesel run trains. Since alcohol fuels derived from biomass sources are unlikely to be able to provide more than about 30% of the total national demand for transportation fuel even with sucessful programs for encouraging transportation efficiency,

electric systems may provide the greatest promise for operating a transportation system based on renewable resources in the 21st century. (Coal, of course, can also be used to supply methanol but environmental constraints may also place limits on its use.) Electric systems enjoy the considerable advantage of being able to operate from a variety of renewable energy sources, including hydroelectric, geothermal, wind, and photovoltaic equipment.

Electric Passenger Vehicles

There is growing commercial interest in electric road vehicles and the DOE has funded a major research program in the area. Several commercial vehicles are now available, and GM recently announced a breakthrough in the development of low-cost nickel-zinc batteries that would accelerate the introduction of their electric cars. They could have an electric car in production by 1985 if market conditions are right.

Apart from their ability to use a variety of fuels, however, electric passenger vehicles may not be attractive substitutes for efficient vehicles with internal combustion engines. Compared with an alcohol vehicle with similar performance, electric vehicles are likely to cost more to purchase and operate and provide less range. The electrics may even be less fuel-efficient (measured in terms of the distance traveled per unit of primary energy consumed. The relative environmental impacts of electric and methanol vehicles are likely to be roughly proportional to the relative amounts of primary energy consumed. Electric vehicles, of course, have the advantage of shifting pollution from the area where the vehicle is used to areas where powerplants are located. Emission control could be easier since emissions are concentrated at a relatively small number of sites.

Direct comparisons of the costs of owning and operating electric and internal-combustion vehicles and calculations of their overall efficiencies are difficult to make because data available on the two types of vehicles are seldom comparable. Vehicles with similar interior space, similar use of auxiliary equipment (lights, heaters, etc), similar driving comfort, and similar acceleration over identical driving cycles should be compared. No such analysis has been performed. In fact, electric vehicles are frequently compared with gasoline vehicles capable of far superior performance. Existing data will, however, allow some important conclusions to be drawn.

The State-of-the-Art

With existing technology, typical electric vehicles must carry over 1000 pounds of lead acid batteries to achieve a 50-mile range. This added weight necessarily limits vehicle space, reduces vehicle efficiency and, above all, adds to the cost. Gasoline contains nearly 90 times more stored energy per pound. Liquid-fueled vehicles will, therefore, always offer much greater flexibility. Improved lead acid batteries, nickle-zinc batteries, and other advanced electric storage systems may be able to reduce that weight by factors of 2-9 during the next few years. This would mean that liquid fuels still could carry 10-45 times as much energy per pound. Electric vehicles currently cost significantly more than standard cars, and a recent EPRI analysis (EPRI, 1979) estimated that electric vehicles would probably cost more than comparable fuel-burning vehicles even if 100,000 electric vehicles were produced a year. The battery alone could make electric vehicles 20-40% more expensive than fuel powered-vehicles that were otherwise similar. A manufacturing study done for the advanced "Electric Test Vehicle #1 (ETV-1) estimated that at a production level of 300,000 per year, the electric vehicles would cost $2,872 more than a similar vehicle with an internal combustion engine (DOE/CS/51294-01). $1,470 of this cost is for the batteries. The comparison is biased in favor of electric vahicles, since the conventional vehicles used in the comparison (a plymouth Horizon TC-3) would have better performance than the ETV-1.

Table 3.22 summarizes the vehicle efficiencies of a variety of state-of-the-art test vehicles. Electric vehicle performance is calculated assuming an optimum charging rate. Most charging equipment now used, however, is quite inefficient when used to provide power to batteries which are only partially discharged. While an ideal charge can be 87% efficient, a typical charge may be less than 60% effi

cient. More research work is needed to develop better controls for charging units which can optimize the charging rate for different requirements (Barber 1981). The most efficient current commercially available 4-passenger electric vehicle can be expected to obtain approximately 1.6 miles per kwh of electricity from the wall-plug in actual commuting use (Modern 1981, Woa. 1981). The most efficient electric vehicle tested to-date is the ETV-1 built for DOE by General Electric on a Chrysler Body. Tests at the Jet Propulsion Laboratory found that the ETV-1 could achieve 2.9 miles per kwh delivered from the batteries over the SAE urban driving cycle (SAE 227aC), and 4.8 miles per kwh at a steady speed of 55 mph (Barber, 1981). The efficiency on an EPA combined cycle is then 3.0 miles per kwh assuming an ideal charging efficiency (87%).

The overall efficiency of an electric vehicle can be compared with a methanol vehicle by examining how far each could drive given the same amount of primary energy. If a million BTUs of coal or wood is burned in a modern electric generating station approximately 100 kwh can be delivered to a residence or office. If this electricity is used to power a current generation electric vehicle it could travel about 160 miles. A car similar to the ETV-1 could travel 300 miles.

If the same million BTUs of wood or coal is converted to methanol with an energy efficiency of 57%, about 10 gallons of methanol could be produced. A current generation diesel volkswagen Rabbit capable of 36-42 miles per gallon (EPA city cycle) should be able to travel 190-220 miles on 10 gallons of methanol. Previous sections of this chapter have indi

Table 3.22. ELECTRIC VEHICLE EFFICIENCY TEST RESULTS

Vehicle	Test Weight (kg)	Fuel Economy Driving Cycle B[1] (mi/kWh except as noted)	Fuel Economy Driving Cycle C[2] (mi/kWh except as note	Acceleration 0-20 (seconds)	Acceleration 0-30 (seconds)
Elcar (2-passenger)	653	2.86		8	
Waterman Renualt 5 (4-passenger)	1362	2.70		9.3	34
Renault 5	1025	23.1-23.8			
Waterman DAF (4-passenger)	1365	1.96-2.50		14	29
EPC Hummingbird (Converted VW Thing)	1463	1.20		9.4	22
Ripp-Electric (Converted Datsun 1200)	1622	2.46-2.76[+]	2.29-2.78[+]	8.2	14.5
ETV-1*	1678		2.42-2.49		
EVA Contractor (Converted Renault 12)	1700	1.21-1.30	.93-1.33[+]	8.8	15.4
DJ-SE Electruck (Converted Jeep)	1950	1.19-1.25		8.7	23.4
DJ-5	1500	14.3-14.4 mpg			
Fiat Van	1986	.84		9	20
EVA Pacer (Converted AMC Pacer)	2086	1.43	1.15-1.18	8	14
AMC Pacer	1787	12.8-14.7 mpg.	15.8-16.3 mpg.		
Battronic Minivan	2860	.91	.77	8	11
VW Transporter (Microbus)	3075	1.06-1.11	.96	7	14
VW Transporter**	?	1.25			
Gasoline VW Transporter	2100	15.9-16.2 mpg	17.2-17.4 mpg		

Sources: State-of-the-Art Individual Electric and Hybrid Vehicle Test Reports. HCP.M1011-03/1,2. NASA, July 1978.
        Except: *Tom Barber, JPB, personal communication, January 1981.
              **Jerry Mader, EPRI personal communication, January 1981: Based on 100,000 miles in-ues at TVA.

[+]Low end of range is without regenerative breaking high end of range is with regerative breaking.

[1]SAE J227a Test Schedule B. 19s acceleration to 20 mph, 19s cruise at 20 mph, 4s cost, 5s break to stop, 25s idle.

[2]SAE J227a Test Schedule C. 18s acceleration to 30 mph, 20s cruise at 30 mph, 8s cost, 9s break to stop, 25s idle

cated that vehicles in the same size range as the Rabbit should be able to achieve 75 miles per gallon of gasoline. Such a vehicle modified to use methanol would be able to travel 400 miles on 10 gallons of methanol. If any of the techniques for boosting the performance of a methanol engine are developed, it may be possible to increase this distance by 20-25%.

A similar comparison can be made of the relative cost of operating electric and alcohol vehicles. For electrics, it will be necessary to compute the effective cost of the battery system, operating costs, and the energy costs from purchased utility electricity. For the internal combustion engine, maintenance costs and gasoline costs will be computed. Data collected by Booze-Allen & Hamilton on the DOE Electric and Hybrid Vehicle Demonstration Project indicates that current lead-acid battery packs will probably obtain about 300 cycles in actual vehicle use. These 15-kwh-capacity packs cost approximately $1200, and weigh about 1000 pounds. Although under ideal conditions of constant low power demand as much as 12 kwh could be recovered from these packs per cycle, Booze-Allen & Hamilton data indicates that only 8-10 kwh can actually be used before performance becomes unacceptable. Thus a current generation electric vehicle (obtaining 2 mi/kwh from the battery) with a 1000-pound battery pack would have a range of only 16-20 miles. Assuming 300 cycles are obtained, this translates to an operating cost of 20-25¢/mile for batteries alone.

The EVT-1 is equipped with 18 experimental EV2-13 batteries. This battery pack weighs about 480 kg and has a capacity of 18 kwh. The manufacturing study cited above estiamtes that this battery pack would cost $1,470. The largest uncertainty in calculating effective battery costs is estimating how many miles can be traveled before they must be replaced. The ETV-1 final report indicates that a range of 30,000-50,000 miles is expected (DOE/CS/51294-01, p. E-14). However, in actual use very high peak-power demands could accelerate battery degradation and reduce this range significantly.

If DOE's ambitious battery-cost and performance goals (cost $918, range 50,000 mi, based on $40/kwh (1977$) or $51/kwh (1980$) (Webster, Yoa 1980)) are met, effective battery costs could be as low as 1.8¢/ mile. Under more likely assumptions (cost $1,470, range 30,000 mi), battery costs would be 4.9¢/mile.

Analysis presented elsewhere in this study indicates that the marginal cost of electricity delivered to a residence in the U.S. now averages 7-8¢/kwh (5.5-6.5¢/ kwh during off-peak periods. At these prices, the electricity costs for operating a current-generation electric vehicle would average 4.4-5.0¢/mile (3.9-4.1¢/ mile off-peak). Energy costs for the ETV-1 would average 2.3-2.6¢/mile (1.8-2.1¢/mile off-peak) under the same assumptions.

Data on standard internal combustion vehicles indicate that maintenance and repair costs average about 4¢/mile over a vehicle's 10-year life (TECDB-4, 1980). It will be assumed that half of these costs are due to tires, brakes, and other things which would be roughly equal for both electric and conventional vehicles, the remainder being primarily engine maintenance. Actually, tire and brake costs may be higher for electrics because of higher vehicle weights. If we assume that electric vehicles have half the motor maintenance costs of internal combustion vehicles, we arrive at a 1¢/mile incremental maintenance cost for fuel-burning cars.

At diesel fuel costs of $1.30/gallon, current-generation Rabbits have fuel costs of 3.1-3.6¢/mile, 4.1-4.6¢/mile when the incremental maintenance costs are included. This compares with 23-30¢/mile for the current generation of electric vehicles. (Note that these costs exclude operating costs common to electric and internal combustion engines.) Advanced generation electrics are likely to have operating costs around 7.2-7.5¢/mile (6.7-7¢/mile off-peak). Gasoline (or methanol) would have to reach $2.25-$2.75/gallon of gasoline equivalent if advanced electric vehicles are to compete with V. W. Rabbit technology ($2.50/gallon if off-peak rates are available). Fuel prices would have to reach $4.25/gallon of gasoline equivalent for future electrics to be able to compete with a vehicle capable of 65 mpg of gasoline equivalent ($3.90 per gallon for off-peak charging). If all DOE goals for electric vehicles are met, electrics could be come competitive with the 65-mpg vehicle at gasoline prices of $2.20/gallon ($1.80/gallon charged off-peak). It must be recognized, of course, that both the Rabbit and the 65-mpg vehicle are likely to be more attractive than the electric vehicle because of performance, range, and other features. The energy comparison would also change sharply to favor internal combustion engines if the comparisons were made under the assumption that all vehicle accessories (heaters, etc.) are operating.

Electric vehicles may be more attractive in areas that have an over-capacity in baseload generating facilities, or if they can be charged with wind machines or photovoltaic devices located near residences. the chapter on buildings indicated that there are circumstances under which electricity from these sources may be less costly to a homeowner than electricity from conventional sources.

One difficulty with using photovoltaics for a commuting vehicle, of course, is that the vehicle would not be at the residence during the sunny part of the day. On the other hand, the value of electricity from the photovoltaics is greatest to the utility during the day. Thus, given that PURPA allows the homeowner to sell all electricity generated from a photovoltaic unit at a relatively high rate and purchase electricity at average rates, it may be attractive to sell electricity to the utility during the day (helping to meet peak demands) and purchase (baseload-generated) electricity from the utility at night to charge an electric vehicle.

Hybrid vehicles combine features of both internal combustion engines and electrics. Two alternative designs are under consideration: a system that couples both a fuel and electric engine directly to the drive train and a design that uses a fuel-powered engine to operate an electric generator that in turn operates an electric motor connected to the wheels. In both systems, a relatively small battery unit is used that allows the engine to operate under conditions that maximize its efficiency while the vehicle is operated to match the speed and power required by road conditions. Future electric hybrid vehicles may also use a fuel cell to generate electricity instead of an engine generator.

One application of a hybrid vehicle could be extremely important. The engine, in a hybrid vehicle, could be connected to a home not otherwise connected to an electric utility and provide backup electricity for a residential photovoltaic or residential wind system (Reuyl 1980). The only energy which would be delivered to the home for the outside would be the fuel carried in the vehicle's tank. Much more analysis is needed to determine whether this application can offset the much higher costs which will almost certainly be associated with hybrids.

Policy

It is clear that current electric vehicles are not cost effective, and that major cost reductions and advances in battery technology will be needed to make them attractive in common applications. Current Federal "demonstration" programs aimed at early commercialization of electric vehicles are likely to be counterproductive if they simply demonstrate known deficiencies. On the other hand, the potential efficiency of electric vehicles, and the opportunity to use a wide variety of primary energy sources, including renewables, does make them an attractive option for the future. Therefore, further research on batteries, chargers, and control systems is indicated.

Electric Rail

Electric rail systems provide a more familiar technique for using electricity to provide transportation services. Electric rail is now used extensively in Europe because of a post-war decision to avoid diesel fuel. European rail lines also are heavily used and therefore are able to make better use of the capital which must be invested in the overhead wires, switching, controls and other expensive devices which must be added to a rail line when it is converted to rail. Electric rail offers several distinct advantages

over conventional diesel electrics in addition to its ability to use a variety of fuels. Electric rail can require less maintenance and can provide more power on hills ince electric motors are better able to meet high demand for short periods (SRI, 1980).

Two recent studies have examined the prospects of converting parts of the U.S. rail system to electric rail (SRI, 1980; Spenny, 1980). The expenses can be formidable, since in many cases major construction work must be done. For example, it may be necessary to raise bridges and alter tunnels to provide enough room for overhead wires. Both studies, however, came to the conclusion that a major part of the current U.S. rail system could be profitably converted to electricity. The SRI study, for example, concluded that up to 9,000 miles of track could be converted to electricity over a 14-year period at acceptable rates of return and 40,000 mile systems have been studied (Spenny, 1980). Both studies concluded, however, that capital would not be forthcoming for such projects uness ICC regulations were changed to allow the railroads greater freedom in setting prices. Both studies used marginal electric costs which may be below actual marginal costs.

Electrification does not necessarily save much energy. For example, converting 9000 miles of track to electricity results in a net annual petroleum savings of 0.175 Quads but an additional 0.15 quads of annual energy must be used to generate electricity. If the petroleum were derived from a coal or a biomass resource, however, the 0.15 quads required to generate electricity would have displaced 0.27 quads of primary energy. A much more careful analysis of this subject is needed.

HYDROGEN

The last renewable fuel considered here is hydrogen, which is significantly different from methanol and ethanol. Not only is hydrogen the most plentiful element in the universe, it has the highest energy density per unit weight of any chemical fuel (56,000 Btu/lb), it is essentially non-polluting in combustion (except for the formation of oxides of nitrogen), and can serve in a variety of energy converters from internal combustion engines to fuel cells (Reilly, 1980).

In general, water electrolysis and coal gasification are the two major processes or hydrogen production. Other technologies in use today include steam reforming of napthas, and other petroleum fractions, as well as partial oxidation of petroleum.

Solar technologies can be used in hydrogen generation processes in several ways. Wind, hydropower, and photovoltaics can provide electricity for electrolysis applications. Enzymes can produce hydrogen in photobiological processes. Photochemical conversion of solar electricity, can be employed in a water decomposition process.

Biomass gasification can also produce hydrogen (see Industry chapter). An attractive application for biomass is to use it in coal/methanol synthesis. In the methanol synthesis process, after coal is oxidized, extra hydrogen must be added to the process to produce the desired carbon monoxide and hydrogen. While more coal can be used to generate the required hydrogen, environmental restrictions may make the use of biomass in this process an attractive option.

Storage of hydrogen may present a problem. It can be stored as a compressed gas, a liquid or absorbed in a powdered metal.

Hydrogen-powered vehicles have been operated by Daimler-Benz in West Germany, and a 20-vehicle demonstration project is being planned in West Germany (Gray, 1980). The utility of hydrogen fuels in an urban fleet is enhanced by the virtual absence of toxic engine emissions (Gray, 1980). An experimental vehicle which can run on either gasoline or hydrogen has also been demonstrated (Reilly, 1980).

Because of hydrogen's volatility, safety hazards present an obstacle for hydrogen-powered vehicles (especially for hydrogen in liquid form). In one comparison with eight other fuels, hydrogen was judged to present the greatest hazard in the areas of leakage, volatility, confined dissipation, and deflagration. In addition, liquid-hydrogen distribution and refueling may be very difficult to institutionalize (Reilly, 1980).

Compressed methane may also prove to be a valuable vehicle fuel if adequate supplies can be maintained at reasonable prices.

CHAPTER 4
UTILITIES

Section One

The Need for a New Perspective

The nation's electric and gas utilities have reached a point of major transition. They, perhaps more than any other industry, have fallen victim to the uncertainties and rapid changes in energy economics witnessed by the past decade. Growth in demand has failed to follow historic trends; in the case of electric utilities the cost of new plant construction has risen precipitously, and the time required to obtain licenses and construct new plants has increased unpredictably. These uncertainties have made planning much more difficult at a time when the cost of an error has vastly increased. Regulatory restrictions and other factors have contributed to the problem, often making it difficult for utilities to maintain acceptable net incomes. The market value of utility stocks has fallen below book value and bond ratings, carefully nurtured over decades, have eroded perilously. Orders for new electric generating plants have declined dramatically as the industry and its investors re-evaluate their options. While the origins of the problem are complex, a major cause has been a consistent underestimation of both the cost of energy from new sources and the ingenuity with which the market would react to higher prices and the anticipation of higher prices. The trends leading to the industry's present difficulties are not ephemeral. Previous chapters of this study, for example, indicate that the trend toward declining demand may be reinforced with an impressive array of technologies designed for increasing efficiency, and for using solar energy.

New technology, and a new economic environment, have created a situation in which a fundamental reappraisal of the utility industry is urgently needed. From the point of view of public policy, one of the fundamental issues is whether the new circumstances have altered the rationale for regulating the utility industry. Alfred Kahn noted over a decade ago that "technological progress--which may itself be stimulated by the monopoly pricing of the regulated service--

ends likewise to break down cartels. And it confronts regulatory commissions with the dilemma of seeing their charges lose business and the regulated price structure threatened with undermining--or trying to bring the new suppliers as well under the regulatory tent." (Kahn, 1970, Vol II, p. 30). Much of the following discussion will be directed at an analysis of these conflicting forces in light of the new energy realities. The discussion, which will pressure to expand utility monopolies and pressure to break those monopolies to allow greater competition, will focus on three central functions of regulation: establishing prices and ensuring equity and quality of service, regulating the "right of entry".

RECONCILING SUPPLIES WITH DEMAND

Previous sections of this study concluded that demand for electricity during the next two decades could actually decrease if programs designed to encourage cost-effective energy efficiency and the use of solar equipment prove to be successful. On the other hand, the nation's ability to supply electricity will increase even if no new plants are brought on line after 1985 and most oil and gas burning generators are replaced.

Table 4.1 illustrates the difficulties faced in anticipating future demand for electricity. Experts have been severely embarrassed in recent years and mistakes have been costly. Some of these costs are evident in examining the change in "gross peak margin", a measure of the extent to which peak capacity exceeds peak demand. In the 1960's, this margin was maintained at 15-20%--an amount most analyists would agree can provide adequate reliability for a typical utility system. In 1979, however, the gross peak margin reached 36.3%, nearly double what is required.

Figure 4.1 shows that while a number of plants will be completed during the next decade, completions will decline rapidly toward the end of the 1980's. Many

Table 4.1. RECENT TRENDS IN ELECTRICITY DEMAND

	Growth Rate in Energy Delivered (%)	Growth Rate in Peak Demand (%)	Gross Peak Margin (%)
1975	1.9	2.1	34.3
1975	6.7	4.0	34.5
1977	5.5	6.5	31.0
1978	3.4	3.3	34.0
1979	3.0	0.3	36.3
1980[a]	0.3	6.6[a]	32.6

Source: "30th Annual Electrical Industry Forecast" _Electrical World_. 15 September 1980; p. 66.

[a]_Electrical World_ attributes the high growth in 1980 peak demand to the unusually hot summer. Electrical World estimates that 2.6% "real" growth in peak demand occured in 1980.

coal and nuclear units originally scheduled for completion in the late 1980's have been cancelled or indefinitely delayed. The increase in construction of coal

plants projected for the late 1980's reflects Electrical World's forecast of a favorable regulatory environment and a rapid increase in the demand for electricity over the next decade. The dramatic reduction in "new starts" is clear evidence that the financial community is reserving judgment and looking for a more reliable guide to the future. Energy-related investment must continue in some area, but there is confusion about where it should begin; the industry appears to have paused for reflection.

Table 4.2 shows that the problem may become much worse. This table displays a number of different possibilities for future electric demands from utility-owned electric plants (electric generation for residences and industry is not included, even though these units may be owned by utilities under some circumstances). Estimates range from a growth of 3.3 percent per year (made by a major electric utility trade journal, Electrical World), to a negative 1 percent per year, (assum

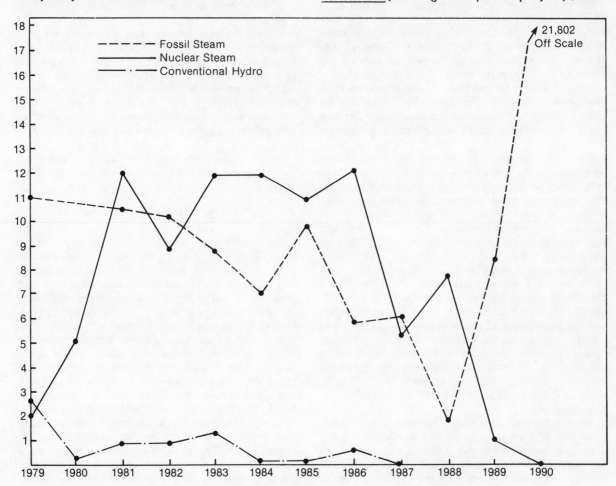

Figure 4.1.   **Baseload Plant Construction Forecast**
Source 31st Electrical World Forecast
Electrical World, 9/15/80, Page 67.

Table 4.2 PROJECTING THE DEMAND FOR ELECTRICITY FROM UTILITY PLANTS
(in billions of kWh)

	Residential	Commercial	Industrial & Agricultural	Transportation	Other	Total	1978-2000 Growth (percent Per year)
1978 demands[a]	679	481	782	0	76	2018	—
year 2000 demands							
(i) Electrical World base case[a]	1424	1100	1476	20	120	4140	3.3
(ii) "business as usual" case	1100	870	1020	20	140	3150	2.0
(iii) SERI "cost effective efficiency "investments"	625	475	900	65	140	2205	0.4
(iv) SERI "cost effective efficiency and on-site solar investments[d]	566	465	900	65	140	2136	0.2
(v) SERI "cost effective efficiency and on-site generation (industrial cogeneration) on-site wind and photovoltaic systems[e]	431	443	410-900[f]	65	140	1489-1979	-1.4 to -0.0

[a]Electrical World "31st Annual Electrical Industry Forecast" September 15, 1980 page 61.

[b]SERI/LBL base case as defined in the chapter on buildings.

[c]Oak Ridge National Laboratory 1980.

[d]Includes active and passive solar space heating and water heating and daylighting. It does not include photovoltaic systems on buildings or small wind machines.

[e]Does not include large wind, photovoltaic, solar panel, or OTEC electric generation or any hydroelectric power.

[f]Upper range assumes no growth in industrial generation after 1978. Lower range assumes that the steam load compatable with cogeneration consumed in the six largest steam consuming industries (in 1976) is supplied by cogenerators which average 200 kWh per million BTU of process steam delivered.

ing full implementation of cost-effective energy efficiency and on-site generating equipment discussed in earlier sections of this report). Electric demand would remain nearly constant if optimum investments in efficiency were made but no investments were made in onsite cogenerating equipment or solar energy. This demand forecast includes a rather aggressive program for increasing the use of electric rail and electric vehicles, even though the discussion in the chapter on transportation concluded that electric vehicles powered from conventional electric sources are probably not a profitable investment. It also reflects an assumption that there will be a rapid increase in U.S. production of aluminum from electrolysis (see the chapter on industry). It is quite possible that rising electricity prices in the U.S. will put U.S. aluminum producers at a disadvantage in international markets and that at least some of the increase in national demand for aluminum will be met from imports. It is interesting to observe that even the high electrical world forecast of electric demands shown in Table 4.2 is much lower than government and industry forecasts made in the mid-1970's.

The long-range forecasts shown in Table 4.2 are the sort that make utility planners reach for a strong drink. At a minimum, they seem to confirm the instinct of investors to be cautious and wary. A considerable risk must be attached to investments that depend on long-range forecasts for their success, since long-range analysis must based on probabilities, not on certainty.

Table 4.3 indicates two possible utility responses to a decline in demand. In the first it is assumed that all plants scheduled to begin operating by 1985 are completed but that no new plants are added between 1985 and 2000. In the second it is assumed that all plants scheduled to come on-line prior to 1990 are completed, but no plants are constructed between 1990 and 2000. In both cases it is assumed that all plants constructed before 1961 and all plants burning oil and gas constructed before 1970 are retired between 1985 and 2000. In this case 80 percent of all oil and gas facilities in operation in 1977 would be retired.

The results of the analysis illustrated in Tables 4.3 and 4.4 indicate that electricity utility equipment now operating or under advanced construction could sustain a growth in demand of 0.1-1.3 percent per year for 20 years, even if all obsolescent plants and most oil and gas facilities are retired to comply with the prohibitions against the burning of oil and gas in the Fuel Use

Table 4.3 ALTERNATIVE ELECTRICITY SUPPLY PROJECTIONS, 1978-2000

Date	Central Station		Industrial[h] $(10^9$kwh)	Total $(10^9$kwh)	Growth Rate 1978-2000 (%/yr)
	Capacity Factor	Production $(10^9$kwh)			
1978	45[a]	2232[b]	79[c]	2311	–
2000					
1985 cutoff [d]	45[e]	2276	80-570	2350-2830	0.1-1.0
	52[f]	2630	80-570	2710-3200	0.7-1.5
1990 cutoff [g]	45[e]	2570	80-570	2650-3140	0.6-1.4
	52[f]	2970	80-570	3050-3540	1.3-2.0

[a] The installed generating capacity in 1978 averaged 566 GW (see "1979 Annual Statistical Report," Electrical World, March 15, 1979), so that potential production (at 100% CF) was 4958×10^9 kwh. The capacity factor is CF = (2233/4958) x 100.

[b] Thirty First Annual Electrical Industry Forecast," Electrical World, 9/15/80.

[c] Energy Information Administration, Annual Report to Congress 1979, Volume Two.

[d] Assuming no new construction beyond plants scheduled to be brought on line before the end of 1985. See Table 4.4.

[e] Assuming the same average capacity factor as in 1978.

[f] Assuming the average capacity factor is restored to the level of the 1960's.

[g] Assuming no new construction beyond plants scheduled to be brought on line before the end of 1990. See Table 4.4.

[h] The lower value of industrial generation involves no growth beyond the 1978 level. The higher value is based on the assumption that cogeneration at an average rate of 200 kWh per million Btu of process steam is associated with that part of the 1976 level of the process load in 6 major steam using industries that is presently technically suitable for cogeneration (see chapter on industry).

Note: To avoid double-counting cogeneration, the demand growths shown as Table 4.2 should only be compared to the lower range of the supply growth rates shown on this table.

Act. A demand growth of 1-2 percent per year could be sustained given economic use of cogeneration in the six major steam-consuming industries. Comparing Tables 4.2 and 4.3, it is difficult to see how national supplies and demand for electricity can be made to match if the goals of programs designed to improve national energy efficiency are achieved. If demand grows as shown in the "SERI cost-effective efficiency investments" row of Table 4.2, utilities would, on a national basis, have too much capacity through the end of the century, unless all of the following occurred: (1) no capacity is brought on line after 1985; (2) all older fossil fuel plants are retired; (3) 80 percent of all oil and gas capacity is retired; (4) capacity factors are not improved through better load management, and (5) no cogeneration or solar electric capacity is added.

Electric utilities plainly face perplexing problems under these circumstances: should they proceed to construct new plants under the assumption that the projections of the on-site conservation and solar programs will not materialize and risk vast over-investments in generating capacity or should they radically revise their construction programs to reflect an assumption that demand growth will be substantially reduced, and risk reliance on relatively inefficient and expensive generating alternatives if the demand grows at close to historic rates? The question of strategy, therefore, now lies at the heart of utility

Table 4.4   ALTERNATIVE CENTRAL STATION GENERATING
CAPACITY PROJECTIONS FOR 2000 GW(E)

1985 cutoff[a]

1970 Generating Capacity[b]	+340
Net Additions, 1971-1978[c]	+240
Net Additions, 1979-1985[c]	+153
All Pre-1961 Vintage Fossil Steam Plants Retired After 1985[d]	-117
All 1961-1970 Vintage Oil and Gas Steam Plants Retired After 1985[e].	-39
	577

1990 cutoff[f]

1970 Generating Capacity[b]	+340
Net Additions, 1971-1978[c]	+240
Net Additions, 1979-1990[c]	+227
All Pre-1961 Fossil Steam Plants Retired After 1985[d]	-117
All 1961-1970 Oil and Gas Steam Plants Retired After 1985[e]	-39
	651

[a]Assumes no new plants beyond those scheduled to be brought on line by the end of 1985. Electrical World (9/15/80).

[b]Federal Power Commission, The 1970 Power Survey, Part I, p. I-18-3, December 1971.

[c]Electrical World (9/15/80).

[d]137 Gw(e) of fossil steam plants were built prior to 1961. From this, one must subtract the 10Gw of fossil steam capacity retired in the period 1971-1977 (Energy Information Administration, Steam-Electric Plant Construction Cost and Annual Production Expenses 1977, December 1978) and the roughly 10Gw of retired capacity implicit in the Electric World forecast (9/15/80).

[e]About 26% of the oil and gas fired capacity in seven states (which account for 73% of total U.S. oil and gas fired generating capacity) in 1977 was brought on line in the period 1961-1970. If it assumed that 26% of total U.S. oil and gas fired capacity in 1977 was brought on line in this period, then nationwide this would have been about 39 Gw(e). Also in these 7 states about 55% of oil and gas fired capacity came on line before 1961 and thus would be retired along with other pre-1961 vintage fossil steam plants. Thus 80% of total oil and gas capacity in 1977 would be retired by 2000.

[f]Assumes no new plants beyond those scheduled to be brought on line by the end of 1990.

planning and its resolution should be a central theme of analysis.

Table 4.3 indicates that cogenerating equipment can provide a kind of safety net for planning. The cogeneration potential of the six largest industrial consumers of steam would allow an increase in U.S. electrical output of nearly 0.8 percent per year for 20 years. More to the point, the existence of this potential would permit planners to take advantage of the possibility that demand growth can be small. The cogeneration potential can protect the industry against a forecast that is low by as much as 0.8% per year. The same argument could be made for cost-effective solar electric technologies. Cogeneration and solar equipment can be added relatively rapidly (many can be built in less than a year) and provide flexibility for dealing with unanticipated changes in demand. The cogeneration equipment could be powered by natural gas (if regulatory barriers are removed) or by biomass. The efficiency measures discussed in previous chapters can free enough natural gas from current gas customers to supply fuel for any increase in cogeneration that can reasonably be expected.

The fact that the analysis presented here deals only with national averages makes it somewhat artificial. Since growth rates vary significantly across the country, demand will almost certainly increase in some regions while other areas will face shortages. A promising approach to solving this problem is to reexamine the potential for improved transmission of power from one region to another.

The analysis has also assumed that 80 percent of the oil and gas capacity in the nation can be replaced by the year 2000. It will be relatively easy to achieve this reduction in regions where oil and gas plants have recently been converted from coal plants and can easily be converted back. In several states virtually all generating capacity results from oil and gas boilers that have never used coal. In these regions it will be extremely difficult to convert or replace 80 percent of current capacity by the year 2000 without a massive construction program. Furthermore, given the adequate supplies of gas likely to be available in the future (if only from gas "saved" in buildings) and marginal cost pricing, the Fuel Use Act may be unnecessary or counterproductive.

Electricity derived from renewable resources could also play an increasing role during the next two decades. Currently hydroelectric energy displaces the equivalent of 2.8 quads of primary fuel. Some increase in hydroelectric capacity can be expected in the future, although environmental considerations will preclude a large expansion. In the next few years wind machines should also begin to play a major role. Analysis presented elsewhere in this study indicates that the potential contribution of wind machines could reach 3 quads of energy displaced by the turn of the century. Photovoltaic systems, principally on-site systems located on buildings, could also contribute approximately 1 quad per year if expected cost reductions are achieved. There may also be some contribution from solar thermal, ocean thermal energy, and other advanced solar electric equipment.

It is difficult to estimate the impact of these solar devices, or of cogeneration equipment, without a careful analysis of national options in supply and demand. Such an analysis has not been conducted in the context of this study; plainly much more work in the area needs to be done. The result of an exact analysis has been approximated by assuming that electric utilities continue their expansion plans until 1985 but no plants are brought into operation after 1985 excpet as replacements for retired equipment.

Since this analysis suggests that total national demand for electricity in the year 2000 may not increase significantly, markets for solar equipment were defined to include only those markets in which it would be less expensive to replace existing utility generators with solar generating devices than it would be to replace them with a new coal-fired plant. In conducting this analysis it was assumed that no new generating plants of any kind would come into operation after 1985 except as replacements for oil and gas facilities or as replacements for obsolete equipment. This is a particularly severe test for solar equipment since the equipment would be looking for markets in a market that is not growing. Markets will, of course, be growing in some parts of the country even though national average demand may not increase significantly; markets for solar equipment may be underestimated in these regions.

The results of calculations conducted for this study are summarized in Table 4.5. It can be seen that hydroelectric and wind electric equipment are expected to dominate the market. Much more careful analysis must be conducted, however, before it can be determined whether the nation has a need for the levels of supply possible from these renewable electric generating systems or from any of the cogeneration systems shown in the bottom row in Table 4.2.

WHAT NOW?

The situation just described presents a serious challenge for the management of privately owned utilities and the public bodies that regulate them. There is so much uncertainty about the future, and so much diversity in the situations utilities face in different parts of the country, that it is not sensible to recommend any comprehensive "solution" to the problem. It is critical, however, that a regulatory environment be created in which a variety of solutions can be worked out in the marketplace. Existing regulation often has the effect of discouraging investments in energy efficiency and of providing so much protection for large utility plants, which involve large financial risks in times of demand uncertainty, that market signals that would discourage such investments are not able to work effectively. This has led to inefficient use of the capital available for national energy investments.

A program for undertaking a reform should proceed along three general lines: ensuring that utility prices are designed to send the proper signals to utility customers and utility stockholders; easing the regulatory and tax barriers that prevent regulated utilities from expanding their repertoire of investments to include some of the more profitable new energy-efficiency and solar-energy technologies; and allowing nonregulated companies greater access to markets historically reserved for regulated companies. The basic thrust of this effort would be to encourage greater opportunities for competition between regulated and nonregulated companies and between electric and gas utilities. Undertaking such reforms will require striking a delicate balance between the continuing need for some kind of regulation where existing companies still have the ability to exercise monopoly powers, and the efficiency which can be achieved through deregulation.

The federal role in facilitating the transition just discussed will necessarily be minimal since most regulatory authority remains at the state level. The federal government can, however, facilitate the process in a number of ways: it can make better use of recent

Table 4.5. SOLAR ELECTRIC GOALS FOR 2000
(quads of primary fuel displaced)

Hydroelectric	3.4-3.7
Existing Capacity	2.8[a]
New Capacity	.61-.9
Wind	1.0-4.0
Utility Applications	(0.5-3.4)[c]
Onsite Machines	(0.8-1.1)[d]
Photovoltaic	0.5-0.8
Utility Applications	e
Onsite Applications	(0.5-0.8)[f]
Solar Thermal	g
Central Receiver	g
Distributed Receiver	g
Ocean Thermal	h

[a]Energy production varies, depending upon annual rainfall.

[b]This is an estimate of the range of hydropower sites for which development is both economically and environmentally feasible. The hydropower technical potential has been estimated by as 5.4Q from large scale hydro (FERC) and small scale hydro (Corps of engineers).

[c]High estimate assumes that wind machines represent 20% of generating capacity in all regions except region 4.

[d]See the Chapter on Buildings.

[e]Although photovoltaic systems would be attractive in Regions 6 and 9 if the lower cost estimates are achieved, they do not, on average, compete with the other solar electric technologies or with new coal plants at initial penetration levels. Given the limitations of this analysis, it was not possible to estimate likely penetration levels.

[f]See the Chapter on Buildings. Penetration range depends upon projected photovoltaic costs.

[g]Although solar thermal systems would be attractive in Regions 6 and 9 if the cost estimates are achieved, they do not, on average, compete with wind and coal units. There are undoubtedly areas within Regions 6 and 9 for which adequate wind sites are not available and for which insolation is good. However, given the limitations of this analysis, it was not possible to estimate likely penetration levels.

[h]Although ocean thermal systems are likely to be attractive in the island and coastal areas of regions 6 and 9 if the estimated costs are achieved, they do not, on average, compete with wind and coal plants at initial penetration levels. Given the limitations of this analysis, it was not possible to estimate likely penetration levels.

legislation that acts to encourage competition in utility markets e.g., the Public Utility Regulatory Policy Act (PURPA) and the Energy Security Act (ESA); it can remove the remaining provisions in PURPA, several recent tax bills, and other legislation, which act to discourage utility investments in innovative energy systems; it can provide direct assistance to utilities and state regulatory authorities interested in designing improved techniques for assessing opportunities for future utility expansion; it can make better use of the regulatory authority it has over the federally regulated Federal Power Marketing Administrations (FPMA's) and the Tennessee Valley Authority to encourage investments in energy efficiency, solar equipment owned by industry and building owners, and larger solar electric systems; and it can support research in solar technologies designed to provide power directly to electric utilities.

The bulk of the discussion that follows will deal with electric utilities, but much of what will be said can be applied directly to gas utilities. One of the most important results of the savings that can be achieved through the programs described in earlier chapters is that a significant amount of gas can be freed from existing applications. Gas utilities will thereby enjoy continued supplies of gas at costs lower than could be expected if the gas were all produced from new sources. As a result, gas utilities can be expected to compete with electric utilities for some time. The competition could be expanded dramatically if gas-fired industrial cogeneration comes into widespread use.

Rates

One of the most critical products of utility planning will be an effective technique for designing rates. Existing utility rates send customers very misleading signals and have the effect of distorting decisions made about investments in energy efficiency and new energy supplies. They do not, for example, reflect the real "marginal cost" of supplying energy. Effective rates would examine the real cost of providing service as a function of customer demand patterns. Instead, rates are typically set on the basis of expected utility costs incurred by operating existing plants during a future "test period one or two years in the future." The average charges computed in this way are typically less than the cost of providing energy from new plants. As a result customers will undervalue investments which might have had the effect of deferring expensive new plants. In addition, use of the future test period means that utilities have little incentive to encourage efficiency would cause demand to fall below the demand expected during this period since their effective return would be reduced if this occurs. (Other pernicious effects of existing rate policies are discussed in later sections). Many rate studies have been initiated by utilities, some under the pressure of PURPA. There should be continued support for these studies and for experiments based on them. The experiments should include:

-- a detailed examination of the marginal cost of providing service to customers on the basis of their demand at different times of day, their demand in different seasons, and their willingness to accept lower reliability for lower rates.

-- an analysis of the utility of using "inclining block rates" in which a relatively low rate is charged for 'essential' utility services and the long-term marginal cost is charged for the last blocks of power purchased by customers.

-- an analysis of the potential for using "windfall profits" which may be earned by moving to marginal cost pricing to encourage utility research or to subsidize customer investments in efficiency or solar energy.

-- an analysis of techniques that can be used to provide relief for utilities that are now badly stressed by overinvestment in construction programs while not transferring so much of the risk of such investments away from utility stockholders that these stockholders are insensitive to the faults of utility management.

Encouraging Utility Investments

The Energy Security Act removes federal prohibitions against utility financing of investments in energy efficiency and solar energy equipment in buildings thereby leaving regulatory decisions about this issue in the hands of state utilities commissions. There are no federal prohibitions against utility investment in cogeneration and other generating equipment associated with an industrial site. A number of barriers to utility investment in such unconventional areas remain, however. The wisdom of allowing regulated utility monopolies to expand their investment horizons beyond conventional equipment has been heatedly debated. Utilities are in a unique position in the energy marketplace because they can construct exact comparisons between marginal investments in energy-saving equipment and investments in energy-generating equipment. Moreover, denying firms most experienced in providing energy services access to markets that may provide the most lucrative energy-related investments during the next few years, plainly creates an artificial tension. On the other hand, there are plainly risks inherent in letting a monopoly expand its reach into markets that can easily be exploited by nonregulated companies.

Protection against monopoly exploitation has been a critical element of recent laws allowing utilities access to new markets. The issues must be resolved on a state-by-state basis under the scrutiny of the Federal Trade Commission. Planning a proper role for utilities in this area should be a critical part of the model utility plans discussed earlier.

Specific actions that can remove some of the barriers to utility investment in energy efficiency include:

-- changing the rules governing tax credits in P.L. 96-223 and P.L. 95-816 to make them available to utilities, as well as nonregulated companies.

-- removing the provision in PURPA denying PURPA protection for plants in which a utility owns more than a 50 percent share.

-- allowing utilities to enjoy credits for investing in 'synthetic fuel production' even if utility equipment consumes all fuels produced.

-- examine the Public Utility Holding Company Act to remove historic barriers to innovative utility financing.

Encouraging Nonregulated Companies to Invest in Energy Production

PURPA moves boldly to remove cogenerators and small power producers using renewable resources from regulatory control. Since such equipment may represent some of the most attractive energy investments in the next few years, PURPA may already have deregulated an important area of utility growth. PURPA also provides nonregulated companies with a guarantee to fair prices for power offered for sale, protection against discriminatory rates charged by utilities for backup power, and guaranteed access to utility transmission lines (at a fair price). The greatest challenge of encouraging the expansion of nonregulated generating companies will be simply one of ensuring that PURPA is adequately enforced.

Small power producers can, however, benefit from other federal actions in a number of specialized areas. Examples include:

-- expanding section 402 of PURPA to provide funds for planning by small companies interested in developing small hydroelectric facilities or wind generating equipment.

-- funding regional wind and hydroelectric surveys following the excellent precedent set by the New England River Basin Commission in the case of hydroelectric power.

Utility Planning

Most of the measures just described must be managed entirely by states and their public utility commissions; the federal role will be minimal. The Federal government can, however, be of considerable assistance if it supports the development of plans and analytical tools. Such programs would cost relatively little.

The program could result in the development of the following kinds of products:

-- techniques for determining optimum utility investments in the absence of adequate data and sophisticated analytical techniques

-- a program for developing analytical techniques and information which will allow a more sophisticated evaluation of investment opportunities over the long-term

-- development of well documented "marginal costs" which can be used to set rates for selling utility power and purchasing utility power from nonregulated companies

-- development of innovative financial relationships involving various combinations of regulated utilities, unregulated utility subsidiaries, utility customers, equipment vendors, and third parties interested in selling energy services

The planning exercises would be a logical extension of the Residential Energy Efficiency Program (REEP) currently funded. The results of the planning experiments could provide a valuable opportunity for utilities directly involved in the program to develop strategies for the future. The program would be almost as valuable to nonparticipating utilities since they would benefit from the methods developed and the experience of the experimental programs.

In addition to the model utility plans suggested, the federal government should sponsor a new review of the national transmission system. Much has changed since the thorough analysis of the subject reported in the National Power Survey and elsewhere. The nation is now facing a situation in which some regions will have excess capacity while others are short. It will also face a situation in which renewable solar resources may be available far from areas where they are needed. These and other factors need to be included in a new assessment of national strategies of energy transmission.

Federal Power Marketing Administrations

The federal government operates the world's largest transmission system in its five regional power marketing administrations. These companies can exercise enormous influence over decisions made by the small utilities they serve; this leverage should be used to encourage efficient investments in energy efficiency and solar energy. In particular, the FPMA's could:

-- use the leverage inherent in offering contracts for low cost hydroelectric power available through the FPMA's to encourage their customers to reduce demand

-- provide financing for solar and efficiency investments in client utilities by using some of the techniques suggested in the Pacific Northwest Electric Power Planning and Conservation Act.

-- developing better techniques for dispatching hydroelectric resources to reflect the needs of solar equipment installed in their region.

-- promote the use of wind power by expanding their hegemony to include the transmission of power generated by wind machines as well as power generated from hydroelectric resources. With amendments to their enabling legislation they could directly finance the construction of wind machines.

Section Two
Regulation and Deregulation

BACKGROUND

The analysis that follows will review a number of opportunities for changing the structure of utility regulation, or for making better use of recent changes in regulatory authority, all of which work in the general direction of encouraging greater competition in areas now dominated by regulated monopolies. The increase in competition is justified on the grounds that new technologies for reducing demand for energy generated in centralized production facilities have become increasingly attractive. Economic efficiency demands that price structures and regulations on the "right of entry" into the energy business be altered to ensure that new investments in energy properly reflect these new opportunities. Electric and gas utilities required nearly a quarter of all investments in new manufacturing plant and equipment in 1977, and it is reasonable to ask whether funds would have been invested in this way given a more competitive environment.

Increasing competition will require, three kinds of actions: (1) Regulators should make proper use of short-run and long-run marginal costs for rates making an allowance for depreciation that properly covers risks inherent in construction during periods of rapid technical change (Kahn, 1970, Vol I, p. 118). Such policies should make it difficult to argue that real opportunities exist for "cream skimming" on the part of non-regulated companies, since long-run marginal costs are higher than average costs in most regions. (Kahn, 1970, Vol II, p. 224). (2) Utilities should be allowed to invest in energy efficiency in buildings and industry when they, or their regulatory commissions, judge that such investments would be practical and profitable for both utility stockholders and rate-payers. It will be shown that it is possible to develop rates by which all utility customers could benefit under a plan in which utilities invest in on-site units up to the point where marginal costs of savings equal the marginal avoided costs. All of this should be done under careful scrutiny to ensure that utility investments are not in restraint of trade. Several recent studies have shown that a well designed program for promoting equitable utility investments in efficiency and solar equipment on their customers' premises can result in higher earnings per share for stockholders and lower overall rates for utility customers. (Praul and Marcus, 1980; Kahn, 1980; Wiley, 1978). (3) Nonregulated companies should be encouraged to invest in cogenerators and power producers of all kinds by making effective use of the provisions of the Public Utilities Regulatory Policy Act of 1978. These actions can be encouraged by providing states and local utilities with adequate funds for a series of trial projects. Assistance would be valuable in developing data, planning tools, and implementing experiments. A small number of utilities should be selected for an initial test. Details of such a program will be discussed in a later section.

It must be recognized, of course, that economic efficiency is not the only objective of regulation. Regulated utilities are expected to provide a minimum level of service to a broad variety of customers. Their prices may also be regulated to adjust to other social objectives such as environmental quality and national security, which cannot be reflected in ordinary market economics. (These noneconomic objectives will remain in spite of the changed environment and must temper any movement toward deregulation of power services.) This means, among other things, that utilities may always need to maintain capacity in excess of minimum demands so that they can provide a minimum level of reliability; an unregulated firm might operate closer to the margin.

Regulation of utilities began in 1907 when Wisconsin and New York established commissions to prevent unregulated monopolies from exploiting a turbulent and growing market for electric services (Kahn, 1970, Vol. II, p. 119). Regulation was codified on a federal level

334

with the Federal Power Act of 1935, the Public Utilities Holding Company Act of 1935, and the Natural Gas Act of 1938. In the years between 1892 and 1907, the year Edison opened his Pearl Street Station in New York, many regions attempted to encourage competition between holders of franchises for generating electric power. Instead of leading to reduced prices, however, the programs often led to poor service and increasing manipulation of markets by a small number of firms able to hold, or manipulate all franchises in a region. (Kahn 1970, Vol. II, p. 118). This led naturally to the concept that utilities were "natural monopolies" and should be regulated as such.

(1) It was argued that utility generation, transmission, and distribution equipment enjoyed a natural "economy of scale" and that competition would result in inefficient duplication of resources and use of relatively small equipment that would not minimize costs. (Garfield, 1964; Scherer, 1970), Kahn argues that "the critical and--if properly defined--all-embracing characteristic of natural monopoly is an inherent tendency to decreasing unit costs over the entire extent of the market. This is only when the economies achievable by a larger output are internal to the individual firm. . ." (Kahn, 1970; Vol. II, p. 119). The problem becomes one of determining whether declining costs require a single firm or whether market conditions would have led to declining costs without a monopoly. Declining costs do "not necessarily" (Kahn 1970, Vol. I, p. 12) indicate the presence of a natural monopoly. In the case of electric sales, for example, there are clear efficiencies in having a large transmission system since such a system can lead to a diversity of demand, which irons out sudden peak demand from individual customers and allows a greater choice of generating facilities. Kahn concludes, however, that "the generation of electric power is not necessarily a natural monopoly at all." (Kahn, Vol. II, p. 124). Monopoly ownership of generation requires establishing both that economies of scale exist in generation and that these economies could only be captured by granting monopoly powers to a single company.

(2) A monopoly can lead to efficiency in a situation in which economies can only be achieved with enormous investments in capital-intensive plants. It can be argued that such facilities need regulatory protection because private investors might otherwise not undertake the risk of building them. Large capital investments mean that shareholders are vulnerable in uncertain market conditions since their large fixed costs can not be adjusted to meet fluctuating demand. The problem, of course, is whether the regulatory protection granted such large plants has the effect of giving them an artificial advantage over smaller, less risky investments.

(3) A monopoly can lead to an accumulation of resources needed for research on large projects (Kahn, 1970, Vol. II, p. 114), and regulatory control can encourage a company to undertake risks on new technologies that might not otherwise have been explored. In practice, of course, utilities have relied on direct federal support for most major innovations. The tendency toward risk-taking has often been countered by the reluctance of regulatory commissions to allow the companies to take risks that might affect the public interest. Moreover, regulatory control of rates may have had the effect of encouraging utilities to maintain unamortized generating plants when they might have been replaced by new technologies in an unregulated market (Kahn Vol I p 119.)

Regulation of utilities amounts to a compact in which a regulating body seeks to replace ordinary market forces by a "balancing of the investor and the consumer interests." The shareholder is allowed a return on equity "commensurate with returns on investments in other enterprises having corresponding risks" while the customer is assured of "just and equitable rates" (Federal Power Commission vs. Hope Natural Gas, 1944). As could be expected, it has often been difficult to get all sides to agree that a proper balance has been struck, but, prior to 1973, litigation led to a series of acceptable compromises. Agreement was relatively easy since the real cost of electricity declined continuously during the two decades preceding 1973; the advantages of maintaining a utility monopoly seemed to have been proven.

The situation was turned on its head in the early 1970's, and the events of succeeding years have called many of the fundamental reasons for utility regulation into question.

Does Utility Regulation Still Make Sense?

Undoubtedly the most significant change in utility economics is due to the fact that under present circumstances electricity from new plants typically costs more than the average cost of electricity from older plants. This, of course, meant that utilities could no longer argue that their monopolies guaranteed "decreasing unit costs." Table 4.6 provides a rough comparison of average and long-term marginal costs. In some cases the marginal costs are nearly twice the average cost, but the difference is significant in all cases. Differences of the magnitude seen here inevitably lead to questions about whether regulated utility management is the most efficient way to provide needed electricity services. Calculations presented elsewhere in this report show that it is reasonable to argue that in the future, investments in efficiency, cogeneration, and and on-site solar can provide incremental energy services at a lower cost than would be possible with a new utility plant. If this proves to be true, the concept of "natural monopoly" could be called into serious question. It is important to recognize, of course, that the real value of any innovative generating equipment can only be calculated if its operation is evaluated as a part of an integrated utility system charged with meeting varying customer demands with acceptable

levels of reliability. Efficiency investments always "supply" energy when it is needed, but small power producers may not and their value must be evaluated accordingly.

Rising marginal prices led to another fundamental change in utility economics. In the past it was argued that the regulated rate of return was actually higher than the cost of capital because of regulatory lags and other effects and that because of regulatory protection, utility investments were relatively risk-free. To the extent that a discrepancy existed between regulated rates of return and the cost of capital, utilities would be tempted to over-invest in capital-intensive technologies of all kinds (Averch and Johnson, 1962). This tendency could make utilities uninterested in rates that discourage use of energy during periods of peak demand and rates that offer reduced quality of service at reduced prices. (Kahn, 1970, Vol. II., p 49). With rising costs, however, regulatory delays have led

Table 4.6. MARGINAL AND AVERAGE ELECTRICITY PRICES IN CONSTANT 1978 DOLLARS (mills per kWh) (1980)

	Residential/Commercial		Industrial	
	Actual 1978 Price	Marginal[a] Cost	Actual 1978 Price	Marginal[a] Cost
New England	49.8	(60.1) 82.4	35.5	(46.8) 50.1
Mid Atlantic	54.8	(64.4) 90.1	31.4	(48.4) 53.3
South Atlantic	40.5	(60.4) 74.4	27.4	(47.9) 49.2
East North Central	43.4	(57.6) 75.8	27.6	(45.0) 46.4
East South Central	31.4	(57.0) 68.6	23.6	(44.9) 45.9
West North Central	39.6	(56.4) 71.0	27.7	(43.7) 45.8
West South Central	36.5	(52.3) 65.8	22.3	(41.3) 44.0
Mountain	36.9	(53.7) 70.4	19.5	(41.4) 44.0
Pacific	31.6	(60.2) 81.2	20.8	(47.6) 4.84
U.S. Average	40.3		25.9	

(a) = Numbers in parentheses exclude the marginal cost of distribution. Calculations based on EPRI Technical Assessment Guide.

to situations where returns on utility equity may actually be falling behind returns on other investments of similar risk, creating a situation where utilities find it difficult to raise capital for any investments. Regulatory delays have increased because public opposition to rate increases has turned routine regulatory actions into angry confrontations. All of these factors have significantly reduced the ability of utilities to embark on any new enterprises.

A related problem results from regulations that permit utilities to pass increasing fuel prices on to customers without going through ordinary rate-making proceedings. These "fuel adjustment" clauses have the effect of insulating utilities from fuel price increases, giving them little incentive to improve fuel efficiency except in areas where there is enough competition with natural gas or enough concern with consumer efficiency investments to make the utility concerned about whether demands will decline sharply if prices are increased. One symptom of the relative insensitivity of utilities to rising fuel prices can be seen in the relatively slow rate at which older fuel burning plants are retired (Smiley, 1980).

These effects are shown clearly in Table 4.7. The value of utility stocks, measured as a fraction of "book value," fell by nearly a factor of two during the decade following 1968. In a regulated utility, returns on equity are determined by regulatory agencies and the capital market typically compensates by bidding the value of the stock up or down. The high market value of utility stocks in the 1960's was a reaction to the relatively attractive returns on utility equity. The situation reversed in 1973, although pressures were clearly building before the oil crisis occurred in that year. Unfortunately, the fall in the value of utility stocks was accompanied by construction programs of staggering ambition. Table 4.7 also shows that the money utilities have tied up in construction projects represents a quarter of utility investments in operational electric plants. This ratio has increased by a factor of 2.5 since 1969. In some areas the ratio is even higher. The ratio of construction work in progress to total assets is 37 percent in California and will reach 50 percent by 1992 if current construction plans are continued, (Praul and Marcus, 1980, p. 35). One symptom of this over-building is that utility tax credits may actually exceed taxable income. PG&E in California, for example, carried forward $76 million in unused investment tax credits in 1979. (PG&E, 1980)

Regulation may also protect companies that have been poorly managed. For example, the terms of the Bankruptcy Act of 1978 appear to have the effect of requiring refinancing of all utility bonds if a utility goes into receivership. This would mean that long-term debt with relatively low interest rates would have to be replaced with new debt issues with much higher interest rates. The prospect of such an enormous increase in costs has been an effective barrier to actions on the part of of public utility commissions that might have the effect of forcing a utility into bankruptcy.

In the present economic environment, therefore, existing utility regulation may have the effect of discouraging on-site investments in efficient industrial and building equipment and on-site solar equipment by both the regulated utilities and their nonregulated customers. Regulations also have the effect of directly prohibiting utility investments in new and promising categories of technologies which could substitute more economically for conventional plants and, until recently, have prevented nonregulated investors from investing in new generating facilities.

Policy Response

The changed economics of utilities have already led to changes in utility regulations at the state and federal level. The programs adopted have both encouraged and discouraged utility investment in on-site equipment.

The Public Utilities Regulatory Policy Act (P.L. 95-617) and the Natural Gas Policy Act (95-621) deregulate important parts of the electric and gas utilities and encourage investments in generation and efficiency on the part of nonregulated companies. Both encourage the use of marginal cost pricing (although gas prices will not be deregulated until 1986), PURPA acts forcefully to protect private owners of small electric generating equipment from discriminatory practices by regulated electric utilities. If small generation devices fulfill their promise, PURPA may have already completely deregulated the most promising electric generating technologies likely to be available during the next few years. Other laws have the effect of preventing utilities from receiving certain classes of the credits which are made available to nonregulated businesses. PURPA, for example, does not cover cogenerators or small power producers in which a utility has more than a 50 percent interest. The tax credits available for certain categories of industrial investments under the terms of the Windfall Profits Tax Act of 1980 (P.L. 96-223) are not available to utilities. In some cases the regulatory disincentives to utility investments are so great that there have been proposals that nonregulated companies would finance an on-site investment, collect the allowed tax credits, and sell the completed project to a utility.

The Energy Security Act of 1980 takes the opposite approach by allowing utilities to expand their investments in areas that had been foreclosed by previous legislation. In fact, this act makes it possible for state regulatory bodies to require utilities to examine options for financing facilities on the customer's side of the meter.

If there is a common theme in these programs, it is to increase competition between regulated and unregu

338   UTILITIES

Table 4.7. STATISTICS OF INVESTOR OWNED ELECTRIC UTILITIES
1960-1978

Year	Average Operating Cost per Delivered Kilowatt Hour[1]	Value of Plant Under Construction Divided by the Value Operating Equipment	Rate of Return on Equity[3]	Ratio of Market of Equity to Book Value of Equity[4]
1960	1.38¢	—	10.6%	1.75
1961	1.38	—	10.6	2.15
1962	1.35	—	11.4	2.05
1963	1.33	—	11.4	2.22
1964	1.30	—	11.8	2.22
1965	1.27	—	12.2	2.31
1966	1.24	—	12.4	1.97
1967	1.24	—	12.4	1.86
1968	1.22	—	11.9	1.70
1969	1.22	0.10	11.7	1.56
1970	1.26	0.12	11.2	1.23
1971	1.34	0.15	11.0	1.27
1972	1.40	0.17	11.1	1.14
1973	1.49	0.19	10.8	0.99
1974	1.96	0.19	10.2	0.66
1975	2.30	0.20	10.5	0.68
1976	2.46	0.22	10.8	0.78
1977	2.77	0.23	10.8	0.86
1978	2.99	0.25	10.8	0.79

Prepared by Albert Smiley
Sources: 1. Edison Electric Institute, Statistical Year Book of the Electric Utility Industry, 1960-1978, Tables 20s, 52s.
2. Statistics of Privately Owned Electric Utilities 1979, p. 26. The ratio shown is the value of the electric utility plant construction work in progress to the value of electric utility plant in operating condition.
3. Edison Electric Institute, 1960-1978, Tables 51s, 52s.
4. Moody's Public Utility Manual, 1979, pp. a10, a12-a13.

lated companies in a variety of energy markets. Increased competition can lead to greater freedom in the marketplace and a better allocation of resources, but it is not clear how much competition can be allowed in an environment where there are still clear monopoly advantages.

As noted earlier, regulation has two major objectives: efficiency and equity. One of the challenges of effective regulation is striking a compromise when these two objectives are in conflict. (Cicchetti, 1977, p. 91; Kahn, 1970, p. 56). Both must be considered in any program to encourage greater competition in the utility industry. Without price regulation, for example, utilities might not have any incentive to provide

service to low income families with relatively small and uneven demands for electricity and the companies could take advantage of customers unable to reduce their demands. The issue of equity makes it unlikely that any acceptable technique can be found for deregulating the price of electricity charged by transmission and distribution companies even if it proves possible to move toward the deregulation of electric generation.

Other more dramatic approaches to electric utility deregulation have been proposed as alternatives to the current program of "muddling through." Demsetz and Hyman have suggested a detailed scheme in which the PUC would offer to buy blocks of incremental power from competitive bidders. Another suggestion would deregulate all generation of electric energy, leaving only electric transmission and distribution under regulatory control (Spann, 1976; Weiss, 1975). This could be done by prohibiting existing electric utilities from investing in new generating facilities or by seeking immediate divestiture of generating plants. Under these proposals the utility owning transmission and distribution facilities would purchase electricity offered for sale by nonregulated companies. The price could be established by competitive bidding and presumably would reflect the quality and reliability of the generating service offered. The price of power offered for sale by the transmission and distribution utility would be set by a state regulatory authority. Prices would be based on a revenue requirement that included the cost of purchased energy, an allowed cost for return on investments in transmission and distribution equipment, and operating costs. The rate schedule would be established using ordinary ratemaking procedures. This technique ensures that low-income groups would receive regulatory protection. One potential problem with any program that leads to a sudden or gradual separation of the "vertical integration" of existing generation, transmission, and distribution monopolies is that some regions might be left in a situation where power was produced from a small number of unregulated companies. In conventional coal-burning utilities the optimum plant size may be about 500 MW. Many areas would require only a few such plants (Loose and Flaim, 1980) Private generating companies could conspire to "fix" prices for inelastic demands in the area. Their ability to accomplish this would, of course, depend on the extent to which the region could purchase power transmitted from other regions, and on whether power could be generated economically from generating facilities accessable to a large variety of owners.

One major difficulty of separating generation from transmission involves control. Complex arrangements would need to be developed to ensure that private generating companies were able to provide ower when it was needed. It might be necessary for the transmission utility to insist on the right to dispatch all plants in the system.

Leonard Weiss has suggested a "second best" solution to the complete devestiture of generating and companies and transmission and distribution companies. He proposes, "(1) the elimination of private restrictions on sales to large industrial customers as well; (2) a general requirement of interconnection and wheeling (i.e., requiring utilities to transmit electricity generated by nonregulated companies at a fair price); (3) control of horizontal and vertical mergers; and (4) at least some divestiture of gas properties in connection with further mergers." (Weiss, 1975, p. 170) There is, of course, a structural problem in separating generation from transmission and distribution. Transmission companies would naturally fall into large multi-state units while distribution must remain a local industry. (Kahn, 1970,

Vol. II, p. 74) The issue of improving access to the nation's transmission systems is discussed at greater length in a later section.

Natural Gas

The Natural Gas Policy Act provides for the phased decontrol of the wellhead price of natural gas with full decontrol accomplished by 1986. State regulatory authorities will remain responsible for setting the selling price charged by distribution companies. Until decontrol, natural gas that is sold on the interstate market will be priced well below marginal gas costs. More rapid deregulation would result in a more efficient use of natural gas, but there are four principal objections to moving to immediately deregulate the price of gas;

o Customers need time to adjust their capital stock of energy using equipment to the new higher gas prices. If there is a sudden price shock this adjustment cannot be made in a smooth, efficient manner,

o Low-income gas customers may not have access to the capital required to convert to the new efficient on site equipment,

o Gas producers would receive an immediate "windfall profit" from the sudden increase in the value of their immediately available reserves,

o Producing and consuming states would not be treated equally.

If a program of deregulation is determined to be in the national interest, a few of the obstacles to rapid decontrol of natural gas could be removed if "windfall profits" resulting from rapid decontrol were taxed. The revenues could be used to help consumers weatherize their houses and buy more efficient on-site equipment to lessen the impact of fuel price increases. It would be economically inefficient to decontrol gas prices immediately even if a windfall profits fund were available since most customers could not react rapidly enough to avoid hardship inflicted by higher energy prices. Investments made in haste with insufficient information are likely to be relatively unproductive.

One interesting feature of the changed economics of on-site electric generation is that it may lead to greater competition between electric and gas utilities. Gas-burning cogeneration units may, for example, be able to provide electric power to buildings and industries at a price lower than electricity supplied by electric utilities.

PRICE DISTORTIONS

The Theory of Marginal Costs

Poorly designed rates are the most pernicious effect of utility regulation, discouraging adequate investment in energy efficiency and solar energy both by utility customers and by their customers. Rate structures often more accurately reflect a collection of historic quirks than a thoughtful program for promoting efficiency and lowering overall costs. The most deleterious result of this haphazard structure is that rates seldom reflect the actual cost of providing service but rather reflect an average cost based on a utility's need to meet all obligations and earn a "fair" rate of return on stockholder's equity during some "test period". If a constant rate is charged, customers will purchase too much energy during periods when a utility's marginal costs are high and too little when costs are low. This means that artificial pressures can develop for new construction. The problem can be even worse in jurisdictions where rates are based on hypothetical utility revenues during a future test period; this has the effect of giving a utility little incentive to reduce demand since their stockholder's returns depend on selling all electricity planned during the period. Rates that have the effect of placing all risk of construction on customers also have the effect of minimizing the risk of over-construction.

These problems can be alleviated if rates are designed to reflect more accurately the full marginal costs of producing energy from new sources, including the cost of building new plants or finding new sources of gas supply. As the later discussion will show, this is a complex problem. The cost of delivering electricity depends on the season, time of day and on the demand patterns of the customer. Costs are highest when the utility is operating close to its peak capacity and lowest when it is able to supply all needed service by operating its least expensive plants. Costs also depend on the quality of service required. If a customer is willing to reduce or eliminate demand during periods when utility costs are high, his marginal rates would presumably be lower than a customer unwilling to accept such restrictions. Costs will also depend on the facility with which customers are expected to change demand patterns in response to new rates. Marginal rates also depend on the characteristics of the utility; three types can be identified (Praul and Marcus 1980 p 166):

(1) utilities where variable operating costs (fuel, maintenance, depreciation, etc.) are higher than average energy costs. (These utilities have a clear incentive to invest in efficiency since average costs will decrease.) A utility based primarily on oil might fit this criteria.

(2) utilities where variable operating costs are less than average costs but where long-run marginal costs are higher than average costs. (In these cases efficiency will raise rates in the short term even though it may decrease rates in the long term).

(3) utilities where average costs are higher than variable operating costs or long-run marginal costs. This could occur when a utility is operating with a significant overcapacity. Any savings would only result in higher rates.

On average, U.S. utilities fall into the second category. The distortions inherent in existing electric rates are shown clearly in Table 4.6. Three kinds of prices are shown: (1) the average price actually paid by customers in 1980; (2) the cost of generating electricity for each type of customer assuming that all energy is generated from a combination of new coal-burning generators and new peaking units using oil or gas; and (3) the cost of electricity delivered to a customer, assuming that all equipment used to generate, transmit, and distribute the electricity was purchased at current prices. The analysis was based on the rather conservative projections of the cost of new generating facilities and future costs of coal published by the Electric Power Research Institute. Other projections estimate much higher future prices for new power plants and for future costs of coal. There can be little doubt, however, that customers are not able to make decisions based on the real cost of any incremental energy they consume. If prices are to be rationalized, two problems must be solved: (1) a technique must be devised for calculating the real marginal cost of consuming electricity and (2) an equitable technique must be found for sending proper signals through the rates.

The problem of rectifying gas prices is equally critical but the solution is somewhat more straightforward. The difference between the price paid for gas in areas where price is regulated and the marginal cost of gas from a variety of sources is shown in Table 4.8.

While there is general agreement that the current design of utility rates is not optimum, there is little agreement about how best to rectify the problem. There is disagreement both about how to compute the actual marginal cost of providing electric service and about how best to design rates to reflect these costs. The calculation is plainly difficult because it depends on assumptions about customer response to increased rates and on an assessment of distortions resulting from existing federal and state tax and regulatory programs. The following discussion will review the subject as follows: first it will review the theoretical basis for establishing marginal rates; second it will review practical difficulties which must be confronted in

Table 4.8   MARGINAL AND AVERAGE
PRICES FOR NATURAL GAS
($/MMBTU)

1.	Average wellhead value August 1980	1.50
2.	Average price charged to residential heating customers August 1980	4.13
3.	Marginal price for gas from Mexico and Canada delivered to residential heating customers January 1980	6.82*

Source: DOE Monthly Energy Review, December 1980, p. 84.
EIA Annual Report to the Congress, 1979, p. 67.

*assumes that the markup between the price at the border and the delivered residential price is the same as the difference between the U.S. wellhead price and the delivered price.

using marginal cost rates, and the results of some actual experiments; finally it will examine two special problems that must be confronted in setting marginal rates--allocating the cost of construction work in progress and allocating the full cost of externalities associated with consumption.

The role of the federal government in rate-setting will necessarily be minimal even through proper rates are critical to ensuring adequate investment in efficiency and in solar energy. The federal government can provide direct assistance in developing generic techniques for computing rates, assistance for utilities and state regulatory commissions interested in experimenting with innovative rates, and provide information about methods and empirical results to utilities and commissions not actively participating in the experiments. This activity should be coordinated with the utility planning initiatives discussed in detail later in this chapter. Existing electric rates, which include a charge for transmission and distribution, may have the same effect on consumer decisions as rate structures designed to charge marginal rates based on the full social cost of generation--including a 'Tax' on fuels.

Microeconomic Principles

Designing rates that adequately communicate the "opportunity cost" of using utility energy is one of the major challenges of regulation. Design of these rates has been explored at length by a number of authors, and no attempt will be made to reproduce these discussions in detail. (Kahn, 1970; Chicchetti, 1977; Turbey, 1973). In an ideal situation, prices should always be set at a point where they reflect the full "short-term marginal cost" of consumption. During periods when the utility has excess capacity, short-term marginal costs may reflect only the "variable costs" resulting from consumption of fuel, operating costs, and any depreciation that can be directly attributed to use. During periods when demand approaches a utility's maximum production capabilities, however, increased demand can lead directly to the need to construct a new plant; customers requiring energy during these periods should consequently be charged a rate which reflects the full cost of providing energy from a new plant. The calculation may not be quite this straightforward, since patterns of demand may well change in response to new price structures. An efficient rate must assess the ways demand can be expected to change in response to the rate, and it may therefore be desireable to charge some portion of the capital costs of new facilities to consumers not currently purchasing energy during periods of peak demand.

Assessing the price of electricity becomes very complex if the cost of new facilities differs significantly from the average cost of equipment now operating. New solar or nuclear plants, for example, may have total costs that exceed the average cost of existing facilities but their short-term variable costs may be much lower than current operating costs. Care must be taken to design rates that send consumers an appropriate signal through rates, maintain overall revenue requirements of the regulated utility, and ensure that utilities do not reap a "windfall profit" from higher rates. Rates that fail to communicate the full cost of new plants, however, will clearly not be in the best economic interest.

In practice there are many barriers to the use of marginal-cost pricing techniques following the idealized rules just discussed. The cost of designing such a structure, metering it, and managing it could be be prohibitive. Moreover, it would require a constantly shifting set of prices, making planning difficult for customers, and it would require extreme flexibility in regulatory procedures to allow all needed changes. Moreover, it should be apparent from the previous discussion that there is no single simple formula for computing marginal costs at any given time. The problem of allocating the capital costs of new plants is particularly difficult and requires some judgement about how rates should be fixed as a function of the "congestion" of demand. Some suggestions for approaching these problems in practical cases will be discussed in the following sections.

Social Marginal Costs

Another problem in designing marginal cost rates involves finding a way to adequately reflect costs not routinely reflected in economic analysis, such as environmental costs. The fact that utility prices are already regulated provides an opportunity for designing rates that take these full "social costs" into account. Market prices necessarily include all effects of federal tax policies and regulations but do not include costs related to national security, macroeconomic effects, or environmental liabilities. A calculation of marginal "social costs" should approach the problem in the following way:

o The calculation assumes that the utility pays no state, local, or federal taxes. From a national perspective taxes simply remove income from society at one point and return it at another. The resulting income transfers serve a number of objectives having little to do with energy policy; the tax structure, therefore, should not be included in an analysis of social costs.

o The calculation assumes that oil and gas are taxed at $20 per barrel to reflect the full costs borne by the nation when these fuels are used (Wood, 1979). These costs reflect national security costs and economic effects resulting from continued oil imports.

o The calculation assumes a "social discount rate" of 3 percent in constant dollars since a variety of authors have argued that this discount rate should be used to determine the time value of money to society. (In fact it is very close to the discount rate actually used by privately owned regulated utilities.)

It is interesting to notice that the effects just described tend to cancel one another. Removing taxes lowers the effective cost of capital, while higher fuel costs and relatively low discount rates increase the effective cost of electric energy.

Existing electric price structures have the curious effect of sending signals to residential customers about the economic merits of investments in on-site equipment similar to the signals that would be sent if rates were based on the "social cost" defined above. In an ideal economic system, customers should compare marginal investments in end-use equipment with the marginal cost of generating energy. The cost of distribution should be included in the calculation only to the extent that it can be shown that the on-site investment would reduce delivery costs. It is difficult to claim (for example) that the retrofit of existing buildings reduces distribution costs, since the lines, poles,

and transformers serving the residence would not be changed. Including delivery costs in the price of electricity, therefore, has the effect of making electric customers behave as if they were charged the full social cost of energy.

Implementing marginal costs rate

Generic Problems

The problem of establishing equitable rates would, of course, be eliminated if electric prices were simply deregulated. This seems to be impractical unless competitive alternatives to the existing utility monopoly were available to all customers.

A utility would always be able to use whatever remained of its monopoly position to raise prices for customers with the least flexibility to reduce demand. This could mean that prices would be relatively low for large industrial customers able to improve efficiency at relatively low cost or able to replace electric consumption with natural gas while prices would be high for the first block of electricity sold to residential customers since it would be impossible for most residential customers to reduce their demand below some minimum level. (Leonard, 1975).

Given the impracticality of price deregulation in electric utilities, the burden is placed squarely on the regulatory commissions. Allocating prices is controversial and complex and the problem is made more difficult by the fact that most analysis of marginal cost pricing examined cases where long-term marginal costs were lower than average costs (for example, see Kahn, 1970, Vol. II). In the few cases where attempts have been made to implement rates based on marginal prices, it has been acknowledged that "translation of the principle from theory into practice is an extremely difficult exercise" (45 Fed. Reg. 12, 225-226, quoted in Lock 1980). Much remains to be learned.

One particularly sensitive issue involves allocating costs fairly among different classes of customers. It appears, for example, that in some utilities residential customers are subsidizing industrial customers while in others the reverse is true. (DOE/OGC, 1980). This problem can only be overcome if utility commissions have the analytical tools to make a case, and the courage to undertake an initiative which could be very costly to some class of customer. The Department of Energy is authorized to intervene directly in rate cases to resolve this kind of issue, but it would probably be preferable for the federal government simply to make appropriate analytical tools available to the states and let the states fight the battles over details.

Another vexing problem results from the fact that a full shift to marginal-cost pricing could result in utility revenues that exceeded the amounts needed to cover regulated utility expenses. Three basic solutions can be imagined: The additional utility revenue could be transferred to federal or state governments through a "windfall profits" tax; the excess revenues could be returned to customers in the form of direct grants, which could be used to subsidize investments in on-site solar equipment and efficiency; or, the revenues could be returned to the economy by lowering rates for customers with relatively inelastic demands or with very low incomes. The last two techniques have been used by electric companies in Sweden and France (Mitchell, 1977).

The options of using separate rates for customers with inelastic demands and for low income families who need a certain block of electricity to maintain a comfortable life has a number attractions. It would have the effect of balancing revenues and benefits while still sending a clear signal of marginal costs to customers best able to react. The incremental energy sold to most customers could be priced at the margin even though total revenues received by the utility remained the same.

Most analysis in this area has examined cases where long-term marginal costs were lower than average costs (Kahn, 1970, Vol. II), and much more analysis is needed for the case where the reverse is true.

Construction Work in Progress and Tax Normalization

One special case of interest in establishing rates involves the regulatory treatment of "construction work in progress" or CWIP. The statistics exhibited in Table 4.7 indicate that a large fraction of the capital assets of U.S. electric utilities are currently tied up in construction projects that are not generating any electricity. Regulatory agencies must decide how utility investors should be allowed to earn a return on these investments. The problem has become critical since large new plants can be under construction for more than a decade.

The traditional technique for allowing a fair rate of return on capital used during construction is to capitalize these returns and add them to the value of the plant, on which the shareholders are allowed to earn a return once the plant begins operations. This policy means that the shareholders see only "paper profits" until the plant is complete. A growing share of the "profits" of many large utilities is now in the form of these deferred income allowances called "Allowance For Funds Used During Construction" or AFUDC.

The accounting procedure just described has the effect of insulating the utility's customers from the long-term marginal cost of electricity that must be charged for energy from the new plant until the new plant actually becomes "used and useful." This places a major burden of risk on the utility's stockholders who presumably suffer losses if the plant is not needed or if it unanticipated costs and delays are experienced in building it. The policy has the undesirable effect, however, of sending consumers the wrong signals. The

overall cost of providing energy services to a region over the long-term can be kept at a minimum if contemporary customers could be persuaded to invest in energy efficiency and solar energy up to a point where an incremental investment in such equipment would save energy at a cost greater than the cost of energy from the new generating plant. Consumers have no incentive to make such investments, however, if they are charged only short-term costs.

A policy that prevents utilities from charging their customers for the cost of construction can have a chilling effect on the ability of a utility to raise capital for large, lengthy construction projects. The risk of such large projects has increased as the demand for electricity during the projected lifetime of a new plant becomes increasingly uncertain. Risks are also increased by lengthy construction, since strikes, new regulatory policies, and inflation can cause unanticipated costs and delays. In an effort to make investments in large plants more attractive and to salvage construction projects that were running into trouble, forty states now allow utilities to pass on directly the costs of construction to their ratepayers. Adoption of this procedure seems to increase defined market-to-book values of utility stock and raising utility bond ratings (see Praul and Marcus, p. 80, Note 10 for a bibliography). Eight states have shifted to this policy since 1978 (Iwer, 1980).

The policy that allows utilities to charge customers for construction work in progress (CWIP) has the effect of moving near-term prices closer to long-term marginal costs. It also transfers some of the risk of new plants from the company's stockholders to its customers and from future rate-payers to contemporary rate-payers thereby encouraging overconstruction. (Praul and Marcus, p. 35) It can be argued that customers must assume the full burden of the risk in all cases, since utilities have never been allowed to go bankrupt. If a utility has constructed a plant that turns out to be a lemon or if it has grossly overinvested in new capacity, however, it is likely that the utility will have an extremely difficult time persuading a regulatory body to pass on the full costs to customers unless such costs are automatically passed along during the construction process. At a minimum, a request for a rate increase to cover a planning disaster is likely to be granted only after a protracted arguments. The CWIP program plainly reduces the impact of bad luck or bad management on stockholders. Ironically, the higher rates resulting from a CWIP program are likely to accelerate investments that have the effect of reducing demand and thereby increasing the risk of new plant construction.

The fact remains, however, that without a program like CWIP utilities may be unable to raise capital for any lengthy construction project. Arguments remain about whether the CWIP program distorts investment by giving a disproportionate incentive to large generating facilities. The only alternative to CWIP that is likely to improve the climate for large investments without a cost to the federal government would be to allow utilities to earn a higher return on equity, since the inability of utilities to raise capital for large construction projects without CWIP is due to the fact that the regulated returns allowed for utility equity are not adequate to compensate investors for the increasing risks of large construction. A higher return would pay investors for taking the risk and would have roughly the same effect on rate-payers as CWIP program. The market, of course, could react to this change by simply changing the price to book value of utility stock and current holders of utility stock would therefore obtain a windfall profit. The policy would, however, keep the risk of poor management squarely on the stockholders. It is not surprising to learn that formidible institutional obstacles must be overcome to implement such a policy. Several alternatives are possible: accelerated depreciation, depreciation on construction work in progress, increased the investment tax credit, and reduced regulatory delays (Iwer, 1980; GAO, 1980; Praul and Marcus, 1980).

Another accounting procedure that has many of the same effects as CWIP is called "normalized accounting". The technique is now allowed by most state regulatory agencies. Normalized accounting takes advantage of the fact that tax credits and accelerated depreciation give utilities relatively large tax breaks early in the life of a plant while these credits are much smaller later in the plant's life. Compared with "flow-through" accounting in which tax credits are passed directly to rate-payers, normalized accounting results in higher rates early in a plant's life but lower rates later in the plant's life. Instead of giving the customer the benefit of these tax credits in the year in which they are obtained by the utility, credits and deductions are spread evenly through the life of the plant. The rate-payers get the full benefit of the tax credits but at a later time; they have, in effect, given the utility an interest-free loan. During a rapid construction program, normalized accounting will result in rates that are closer to long-term marginal costs. If construction stops, normalized accounting could eventually lead to costs lower than those that would have to be charged by a "flow-through" utility operating under the same circumstances (Praul and Marcus, p. 50).

Experiments with marginal costs

Section 111 of PURPA required state Public Utility Commissions to consider several new kinds of rates. These include:

(1) <u>Cost of Service</u> rates in which a utility attempts to design rates that reflect the real marginal costs of each class of customer.

(2) <u>Inclining Block Rates</u> in which customers are not given a rate advantage for consuming large amounts of electricity.

(3) <u>Time-of-day rates</u> in which an attempt is made to charge customers for the full short-term marginal cost attributable to demand at each time of day.

(4) <u>Seasonal rates</u> which adjust prices to reflect marginal costs in each season.

(5) <u>Interruptable rates</u> which give consumers a lower rate if they are willing to accept interruptions in service when the utility is short of capacity.

(6) <u>Load management techniques</u> in which the utility offers customers equipment designed to move a customer's demand to times when marginal costs are low, if such equipment is proven to be cost effective.

Implementation of these standards is left to the states. Reaction has been mixed although many PUCs have moved to discourage declining block rates and encourage the use of seasonal or time-of-day rates (Feuerstein, 1979). The National Association of Regulatory Utility Commissioners recently surveyed 40 state commissions and the Tennessee Valley Authority, finding that 40 percent had adopted one of the six marginal rates (<u>Electrical Week</u>, 1980, p. 7). The California Public Utilities Commission has developed increasing block rates for residential gas and electric customers. An experiment with load management rates lowered peak demands by 20 percent in Davis, California and by 10 percent in Chico and Merced (Praul and Marcus, p. 93).

In addition to the rules in PURPA, Section 111, rules issued under Title II of PURPA require utilities to purchase energy from small power producers at a "full avoided cost" that includes the capital costs avoided if construction of a new plant can be deferred. Section 405 of the Energy Security Act requires the federal government to base all decisions about new federal buildings on long-term marginal costs.

UTILITY INVESTMENTS ON A CUSTOMER'S PREMISES

The preceding discussion examined ways of ensuring greater freedom for the market to choose among alternative investments affecting the supply and demand of electricity by setting proper rates.

This section will examine another approach to increasing investment freedom in the marketplace by exploring the advantages and problems created if existing regulated utility companies are allowed to invest in on-site generation and energy-efficiency technologies on a customer's premises. The next section will complete the picture by examining the merits of developing a market for electricity produced by non-regulated companies.

Previous chapters of this report have already explored a number of programs designed to encourage regulated utilities or unregulated subsidiaries of regulated utilities to invest on the customer's side of the utility meter. The buildings chapter examined techniques by which utilities could offer to audit and retrofit buildings. The chapter on industry explored the regulatory problems associated with utility investments in industrial cogeneration and other electric-generating equipment located in an industrial site. The details of these discussions will not be repeated here.

Advantages

There are a number of reasons to believe that utility investments in energy efficiency and on-site solar technology might be attractive:

o Stockholders and customers of utilities suffering financial hardship because of large construction projects could be hurt if utilities were prevented from investing in attractive alternatives. A utility confronted with only the alternative of constructing new large plants is likely to find that its obligations to provide reliable service requires a continuation of large investments.

o Without rates based on full marginal rates, only utilities are in a position to compare the marginal costs of adding a new source of energy with the cost of an investment which saves energy. All other investors must base decisions on average delivered energy costs which are lower than the costs avoided by delaying construction of a new plant.

o Utilities have access to capital not available to many classes of customers, and because of their protection as regulated monopolies, are able to accept investments with much longer payouts than conventional investors.

o Utilities have energy as their primary business, unlike most other classes of investors for whom an investment in energy-related equipment on their premises is seen to be at best only of secondary importance. This means that they can develop and maintain expertise in the design and operation of on-site energy technology. Moreover, it may be difficult for

companies other than utilities to attract talented management into activities associated with the design and operation of on-site generating equipment when this is not a major part of the industry's business, and opportunities for advancement in the company in the area are limited (Alley, 1980).

o There may be some economies of scale in offering retrofits for an entire community. Many buildings in a region are similar and a single audit may be sufficient to provide guidance for retrofits of many buildings of similar design and construction. Retrofit companies could be more efficient if they go from door to door with similar retrofits or are able to make bulk purchases of equipment.

o Utilities have an established mechanism for dealing with and billing virtually all energy consumers in a region.

o Utility economies of scale may facilitate use of cogeneration equipment serving several different customers, particularly systems involving centralized production of medium Btu gas.

Problems

It is clear, however, that utility investments are not a panacea, since a number of problems are associated with utility participation in their customer's investments:

o Such programs risk expansion of the hegemony of a monopoly unless care is taken in limiting the scope of its activities (Bossong, 1978). This could have the effect of imposing regulatory control on a market that otherwise might be competitive.

o In some areas utilities are simply not trusted by their customers, and a utility recommendation will be viewed with great suspicion.

o In some cases utility management may simply be unwilling to explore investment opportunities radically different from those on which the utility's future has been traditionally based.

o There will be areas where utilities have over-invested in conventional plants and equipment. In such cases the value of an on-site investment will be lower than if the on-site equipment displaces the need to build new capacity.

o Utility equity can be expensive, particularly in areas where utilities are in a poor financial condition.

Undoubtedly, the most serious problem confronted is one of whether utilities can use their monopoly powers to discourage competition in the manufacture, installation, or financing of efficiency and solar equipment (Smiley, 1980; Bossong, 1978; Satlow, 1981). Congressional concern over this possibility motivated the provisions in the National Energy Conservation Policy Act (P.L. 95-619 Section 216(a)) which prevented utilities from financing any residential investments of more than $30 or providing any equipment except load management devices, clock thermostats, and equipment for improving the efficiency of furnaces (P.L. 95-619 Section 216(b).) unless the utility already had such a program underway. This prohibition did not apply to utility investments in industrial equipment but Section 201 of the Public Utility Regulatory Policies Act of 1978 has the effect of discouraging utility investments by denying any facility in which a utility owns more than a 50-percent share status as a "qualifying facility". Equipment in which a utility owns more than a 50-percent share, therefore, can not claim certain exemptions from the sections of the Fuel Use Act prohibiting use of oil or gas as a fuel, and would be denied extra tax credits under the Crude Oil Windfall Profit Tax Act of 1980.

The Energy Security Act (ESA) lifted the Federal prohibition imposed by NECPA on utility investments in residences and gave the states the freedom to allow utilities to "treat the cost associated with a supply or installation or financing program in any way it deems appropriate, subject to approval of the State regulatory authority" (House Conference Report, 1980) This opens a variety of opportunities. Options include direct utility investment in equipment which would treat on-site investments as other utility capital investments are treated (e.g., "rate-basing"), or by treating the utility investment as a one-time "expense", charging all rate-payers for the subsidy in the year in which the grant was made. This second approach could be attractive in areas where a utility faced difficulties in raising capital, but it would deny utilities any of the advantages of profitable on-site investments. The preference of the utility's customers (both those that receive utility financed equipment and those that do not) will depend on the way in which they discount future costs, and on details of the utility's financial structure (Kahn, 1980).

Removing the prohibition on utility financing did not remove concerns about the potential liability of expanding monopoly power. NECPA (as amended by ESA) empowers the Secretary of Energy, acting with the consultation of the Federal Trade Commission, to terminate utility programs if it can be shown that they have "a substantial adverse effect upon competition"

or if they lead to "unfair, deceptive, or anti-competitive acts" (P.L. 96-294 Section 547). Nonregulated companies in a utility's service area are assured of protection under the Sherman Act (15 U.S.C. Section 1-7) and the Clayton Act (15 U.S.C. Sections 12-15).

State response to opportunities for utility investment in on-site equipment has been mixed. California, Illinois, Iowa, and Arkansas all have moved to allow utility financing. (Satlow, 1981). Several states have taken strong positions against expansion of a utility's hegemony in areas not related to solar energy or energy efficiency and these statutes may present barriers to utility sales. For example, Oklahoma has a direct prohibition against utility sales of gas and electric appliances. (Oklahoma Stat.). The New York Service Commission prevented AT&T from expanding its unregulated businesses into a venture involving waste disposal. (U.S. v W.E)

In Illinois, a private business won a case against Detroit Edison Company that prevented the utility from giving away light bulbs. The court ruled that "there is no logical inconsistency between requiring such a firm to meet regulatory criteria insofar as it is exercising its natural monopoly powers and also to comply with antitrust standards to the extent that it engages in business activity in competitive areas of the economy" (Cantor v. Detroit Edison Co.) Undoubtedly, there will be much more litigation before the proper degree of utility involvement in distributed energy systems is finally resolved.

A proposal

While no single solution can be proposed, the following points should be considered in the design of any program for encouraging utility investment:

o Utility investments on the customer's premises are justified up to the point where the cost of energy saved by an additional investment could be greater than the cost of producing an equivalent amount of energy from new generating equipment. (i.e., the "full avoided cost" as defined in PURPA.) The average cost of all energy saved by such an investment strategy will, of course, be considerably less than the cost the utility's customers would have borne if no on-site investment been made.

o Since society may attach a value to conservation and solar investments that is not measured in conventional utility economics (i.e., environmental quality, or reduced dependence on imported oil), programs can be justified that encourage utilities to invest in conservation and solar equipment beyond the point justified by the conventional costs avoided by such investments. This encouragement could be achieved either by taxing utility fuels or by providing subsidies for utility investments through the solar/conservation bank or some other mechanism.

o Ideally, each utility customer should be encouraged to invest in conservation and solar up to the point justified by the delivered (average) cost of electricity; the utility would invest an additional amount up to the point justified by its calculation of its marginal cost of energy from new sources; and the Federal government would invest an incremental amount to account for non-market values. A utility, however, could equally fund all investments up to its own marginal cost. The advantage of using a customer's equity is only that it reduces demands on utility capital and gives owners a greater stake in the investment.

o Electricity from new sources is more expensive than the average cost of energy from existing utility equipment except possibly in utilities highly dependent on oil and gas. Thus, all customers in a service area benefit from a conservation/solar investment which minimizes the need to build new facilities. As a result, all utility customers should pay for each investment since all would benefit (Kahn, 1980). If the costs of the investments are shared equally in the utility's service area, however, there will be an inequity in that customers owning the facilities in which the retrofit has occurred will receive greater benefits than those not receiving such investments. It could be argued that a customer refusing a utility investment should pay some penalty. There would be genuine inequities, however, for customers desiring a retrofit but unable to obtain one either because of the nature of the facility owned or because of the delays in obtaining the required serivce. There would also be an inequity for customers who had already invested in solar or conservation. Some utilities have reacted to this situation by investing their own funds only up to the point where there would be "no losers"; that is, up to the point where, if costs are equally shared, no customer would find rates higher than they would have been if no utility on-site investment had been undertaken. Such an investment strategy, however, would leave the service area far short of the on-site investments that would minimize the cost of providing service to all customers. This problem can be avoided by using a more creative rate schedules. For example, rate can be designed in

which neither participants nor nonparticipants would find their monthly bills increase more rapidly than they would have if the utility on-site investment program had not occurred. This still allows the utility to invest in each facility up to the point that would minimize total costs in the service area. (See Appendix D.) Timing, of course, is critical. Noanprticipants may pay more in the short-run but less in the long-run (Praul and Marcus, p. 69). The way they would value such investments, therefore, depends critically on the way they discount future costs.

o The supply curves developed in the chapter on buildings ranked investments in energy savings in order of cost. If a utility selects a set of investments (e.g., insulation, storm windows, solar waterheating) in which the last investment of the set saves energy at the marginal cost of electricity, all previous investments will have saved energy at a lower cost. Since all of the saved energy is worth the "marginal cost to the utility and its other customers, a decision must be made about how to allocate the savings which result from the inexpensive investments. Appendix D shows how to distribute these savings in a way that ensures justice for all utility customers. It is possible, however, to use some of the fund potentially available from investments in efficiency and small power porudcers to provide incentives for retrofit companies working in cooperation with utilities. The REEP program is designed to experiment with financing mechanisms in which a private company can "sell saved energy" to a utility at the utility's full avoided costs and profit from the difference betwen the actual cost of providing the savings and the marginal costs. Great care must be taken to design bidding procedures which prevent these companies from "cream skimming" when these experiments are undertaken. The calculations presented in Appendix D will also need to be revised in a way that ensures reasonable equity for all utility customers while leaving enough funds to encourage private retrofit companies to find profitable markets. If the REEP experiments are well constructed, it should be possible to encourage states and utilities to develop several creative solutions to these problems.

o There may be no reason for a utility to require a lien on a customer's home before making an investment made in a building; many customers, particularly elderly customers who own their homes, will be reluctant to encumber their homes with a lien. The lien may not be required as a protection against default since a default on a conservation/solar loan would be equivalent to default on a utility bill. Low-income customers should be protected from having their services terminated for nonpayment on loans in the same way that they are protected against termination of service for not paying routine bills. Regulations should place limits or foreclosure of loans similar to the limits placed on termination of service specified in PURPA Section 115(f).

o Care should be taken to prevent utilities from "cream-skimming," or making only the least costly and most profitable investments in each customer's facilities. In some pilot programs there are indications that utilities are investing in equipment which saves energy at an effective cost of 1-2 cents per kwh even though the marginal cost of electricity in the area is three or four times higher. While these cream-skimming investments are plainly tempting, in an ideal situation it would be desirable to encourage utilities to make the least costly investments first. In practice, however, major expenses are incurred in sending an audit team and a team of installers to each building and or to each industry. It may be difficult to return to each site several times. It may be possible to develop a schedule for retrofits after an initial thorough audit, which would allow the utility to undertake several retrofit phases, but it seems likely that the nuisance of returning to a number of sites repeatedly would outweigh the advantages. In most cases it would be preferable to encourage utility investments up to the margin in a single visit.

o Programs should be accompanied by rigorous techniques for evaluating their success. A fixed fraction of funds directed to the projects should be used to measure performance in quantitative terms and for making routine reports and recommendations for change. Programs should be flexible enough to react quickly to effects identified.

Utility investment in buildings and industries could also be encouraged by:

o Allowing utilities equitable access to tax credits under P.L. 96-223 and P.L. 95-618

o Changing the provisions of PURPA that deny protection to investments in which utilities

own more than a 50-percent share was modified to allow greater utility participation, (see chapters on industry)

o Making effective use of the RCS program, provisions in the solar and conservation that allow utilities to act as qualified lenders, and provisions of the Energy Security Act that allow utility investments in buildings (see buildings chapter)

o Amending Title II of P.L. 96-223 to allow utilities tax credits for producing synthetic fuels even if they consume all of these fuels themselves in a cogenerator. This would encourage utilities to produce a medium Btu gas from coal or biomass and use the fuel in their own generators or pipe the gas to cogeneration units located on other customer's sites.

The Public Utility Holding Company Act of 1935 (15 USCS 79 et. seq.) may also adversely affect utility investments in a customer's facility. First, the act prevents any utility holding company from making any investment which is not "reasonably incidental" or "economically necessary or appropriate" to its main line of business. No clear interpretation of this provision is available to determine the extent to which this restriction will limit on-site utility investments. Secondly, many utilities now not controlled by the provisions of the act may fall under its jurisdiction if they form subsidiaries to invest in on-site units. The act should be carefully reviewed and its impact clarified through new regulations or amendment.

INVESTMENTS IN GENERATION MADE BY NONREGULATED INVESTORS

The previous section examined techniques for encouraging greater competition in the energy industries by allowing regulated utilities access to a larger range of investment opportunities. This section will examine the reverse of this situation and explore programs in which nonregulated companies can invest in the generation of power. Improvements in the technology of small generating equipment and increasing costs of utility power have opened a surprising range of possibilities. Options include: industrial cogenerating units burning natural gas, coal, or biomass; cogeneration units in buidings; and a variety of solar electric devices including small wind machines and photovoltaic systems.

The market for this kind of equipment is difficult to forecast, given the enormous uncertainties about future demands for electricity. If the demand is growing slowly, the systems are likely to be attractive only if the owner of cost-effective generating equipment consumes most of the power generated. The uncertainties will affect the price which can be expected for any power offered for sale.

Allowing nonregulated companies to compete for utility markets has a number of clear advantages. Undoubtedly, the most important is that is allows competition in what has traditionally been a monopoly market. This could increase the rate at which innovations, particularly innovations in relatively small equipment, enter the market. Perhaps most importantly, however, competition can free some of the market from the inefficiencies of regulatory control.

Allowing competition in utility markets, however, is not without its drawbacks. For example:

o   Utilities are required to "furnish adequate and safe service" and "may not pick and choose, serving only the portions of the territory covered by their franchises which it is presently profitable for them to serve." (Or. Rev. Stat section 757.020 (2974) and New York & Queens Gas Co. v. McCall, 245 U.S. 345,351 (1918.) Care must be taken to ensure that competition does not deprive individuals and institutions of needed services.

o   Regulations requiring that utilities pay competitors a price for electricity equal to the costs avoided if the purchase prevents the construction of a new plant, have the effect of giving a windfall to owners of private generating equipment capable of producing at below the regulated rate. This advantage is compounded if they are guaranteed a rate over an extended period, leaving utility rate-payers vulnerable if the need for the power diminishes.

PURPA

Sections 202 through 210 of Public Utilities Regulatory Policy Act (PURPA) provide a powerful set of reforms that allow free market competition in the generation of electricity for owners of small generating equipment and cogenerators. This act removes most classes of small generators from regulation under a variety of federal statutes, provides that the owners of such units will not be charged discriminatory prices by regulated utilities, and ensures that the small producers will be able to sell electricity at a fair rate. In particular, the act exempts all cogenerators and electric generators (smaller than 30 Mw), using renewable resources or waste products for more than 75 percent of their primary energy from regulation under the Federal Power Act and the Public Utility Holding Companies Act and state laws. (18 C.F.R. Section 292.204(B). The Federal Energy Regulatory Commission (FERC) has taken a broad interpretation

of PURPA authority and acted to use it to "remove the disincentive of utility-type regulation." (45 Fed. Reg. 12,232-3) (Lock, 1980) PURPA also modifies section 211 of the Federal Power Act to ensure that small power producers and cogenerators have adequate access to utility transmission systems so that they can supply power to potential customers through utility lines. This section prevents utilities from charging cogenerators or small producers "unreasonable rate structure impediments, such as unreasonable hook-up charges and other discriminatory practices". (Joint Explanatory Statement) Perhaps most importantly, the PURPA requires utilities to purchase electricity from cogenerators and small power producers at their "full avoided cost;" utilities are required to publish a rate for purchase of electricity every two years starting November 1, 1980. Interestingly, the current rules allow a small power producer or cogenerator to purchase power at a utility's average rate and sell it at the 'full avoided cost', which may be much higher (although this provision is being challenged in court by the American Electric Power Service Corporation). (Lock 1980 p 724). (It should be noted, however, that if utility delivered costs were set properly, utility rates would always represent the full avoided cost at any given time; the only difference between buying and selling rates should involve treatment of transmission and distribution facilities.) In return for assurance of PURPA rates, the nonregulated generators must ensure that their equipment provides "reasonable standards to ensure system safety and reliability of interconnected operations" (45 Fed Reg 12,230 2/25/80) The major provisions of the PURPA are reviewed in Appendix E. This act moves so boldly to deregulate a promising new set of generating technologies that the major challenge for federal policy managers is simply to provide that the intent of the act is not frustrated by confusion or inertia on the part of state regulatory authorities.

The reaction of state regulatory authorities to PURPA has been mixed. Extremely favorable rates have been set for the purchase of electricity from small generators in a number of states. Vermont utilities now will pay 7.7¢/kwh for all electric energy and 8.2¢/kwh for firm power (New Hampshire 1980). Proposed rates for the PSE&G company in New Jersey are shown in Table 4.9.

As expected, the response has been rather unenthusiastic in states such as Illinois where some utilities have built too many conventional plants. In such states the costs that can be "avoided" by purchasing power from an unregulated investor are very small; indeed they may be negative. In February of 1981, a U.S. district judge in Mississipi declared Title I, III and Section 210 of PURPA unconstitutional on the grounds of states rights. At this time, the national implications of this decision are unclear.

Table 4.9.   PURPA SECTION 210, SCHEDULE 302(B) (1) ESTIMATES AVOIDED ENERGY COSTS
PUBLIC SERVICE ELECTRIC AND GAS COMPANY
(¢/kWh)

1979	1980	1981	1982	1983	1984
Summer (5/15-10/14)					
On-peak (8AM-10PM weekdays)	4.240	5.665	6.706	7.276	8.232	9.308
Intermediate (8AM-10PM Saturdays)	3.439	4.215	4.567	5.161	5.355	5.996
Off-Peak (All Other Hours)	2.205	2.994	3.160	3.361	3.340	4.059
Non-Summer (1/1-5/14, 10/15-12/31)					
On-peak (8AM-10PM weekdays)	4.088	5.843	6.604	7.400	8.263	9.546
Intermediate (8AM-10PM Saturdays)	3.367	4.344	4.826	5.463	5.441	6.434
Off-peak (All other hours)	2.511	3.751	3.937	4.414	4.503	5.147

Source: Gerald A. Calabrese, Board of Public Utilities, "In the Matter of the Federal Energy Regulatory Commission Order No. 69 Final Rule Regarding the Implementation of Section 210 of the
Public Utility Regulatory Policies Act of 1978. Docket No. RM79-55.

Note: Avoided energy cost is independent of the size of the sale. The above costs are based on PJM
running rates. While PJM-PSE&G billing rates may be a more accurate reflection of avoided
cost, projection of these rates are not available at this time.

It is not easy to determine whether PURPA has led
to increased interest in nonutility generators. Electricity generated by nonregulated industries has been
declining steadily for the past thirty years. Industrial
generation fell a surprising 9.6 percent in 1979 (Electrical World, 9/15/80, p. 63). This decline is probably
due mostly to the economic difficulties faced by industries that own their own generators.

Control problems

Other sections have examined the economic merits
of investments in small generating equipment. This
work will not be repeated here. One technical issue
which must be addressed here, however, involves the
control problems which may be created when small
units are connected to a large utility grid. In normal
circumstances all electric generating units are controlled from a central control facility. It is possible
that control will become more difficult if electricity is
sent into the system by generators which are not under
direct utility control. The exact control problem will
depend on the kind of generating facilities installed in
the system.

For example, the output of a large number of wind
systems is more likely to be correlated (all wind
machines are producing power or none are producing
power) than a diversified system containing wind systems, hydroelectric plants, cogenerators, and photovoltaic systems. Several recent studies conducted for
the Tennessee Valley Authority and other utilities have
shown, however, that a utility should not have signifi

cant control problems until these small generators are
able to supply roughly 20 percent of the total utility
demand (Hemphill, 1980). It may be possible to operate a system in which a much larger fraction of the
supply is generated from dispersed units, but this
would require developing more sophisticated control
techniques.

Programs Designed to Remove the Regulatory
Problems Faced by Small Power Producers

Full implementation of PURPA must stand at the
center of any program to encourage competition in
energy production by nonregulated firms. The program
could be improved if nonregulated producers had better access to transmission facilities and if the program
designed to provide funding for municipal utilities and
individuals considering investments in small hydroelectric facilities were expanded to include other generating technologies. Specific suggestions are as follows:

Improve Access to Transmission Facilities

o   PURPA simplified the wheeling regulations
    imposed by Section 211 of the Federal Power
    Act but wheeling can be difficult unless the
    facility protected by PURPA has the support
    of a wheeling utility or of a purchasing utility.
    Lock finds that "The qualifying facility may
    force wheeling under section 210 vis-a-vis the
    purchasing utility; the purchasing utility may
    force wheeling through an order under sec-

tion 211 vis-a-vis the transmitting utility. If neither cooperates, however, the qualifying facility is powerless". (Lock, p. 734) This problem should be remedied.

o Section 203 of PURPA should be revised to remove the restriction stating that power transmitted by a utility must not displace power that would otherwise have been purchased from the utility doing the transmitting. This restriction has proved to be a major practical barrier and distorts the market for energy from new sources.

Funds for Planning

Section 402 of PURPA provides loans for municipalities, electric cooperatives, industrial development agencies, electric cooperatives, nonprofit organizations, or "any other person" to conduct feasibility studies for small hydroelectric facilities and prepare applications for licensing. Funding is provided through 10-year loans given at favorable interest rates. The entire loan can be forgiven if the project is not completed. Approximately 33 loans have been made under the program to date; the value of loans has been in the range of $35-60,000. This program could be expanded to allow the use of these funds for any facility which are "qualifying facilities" as defined under PURPA section 210. The program could be used more effectively to encourage use of renewable resources if DOE would request appropriations for the Construction Loan Program authorized in the National Energy Act and section 402 of PURPA. This program might be ammended so that it would also apply to small wind and photovoltaic projects.

Specialized Programs for Small-Scale Hydroelectric and Wind Systems

Small Scale Hydroelectric Facilities

The nation needs an accurate inventory of its hydroelectric resources. Several studies are now underway but will apparently not provide enough detailed information about costs, potential, and environmental impacts of expanded use of hydroelectric resources to provide an adequate basis for making national policy. Both the Army Corps of Engineers and the WRPS are engaged in studies designed to determine the potential of small scale hydroelectric sites and the Corps is undertaking a massive analysis of all hydroelectric sites in the National Hydropower Study. An apparent problem with all of these studies is that they have not been able to produce information which allows a clear comparison between investments in hydroelectric systems and investments in competing energy systems. The studies typically also do not reflect the dramatic

changes which have occurred in energy economics in recent years.

The recent study of hydroelectric resources recently conducted by the New England River Basin Commission is an example of the kind of work needed nationally. The study concluded that the total hydroelectric resource of New England consisted of 1750 sites which could produce 1000 Mwe if fully exploited. It estimated that if facilities at these sites were financed at 14 percent they could produce electricity that could be sold for 4.5¢/kwh. Clearly existing PURPA rates would make a larger fraction of these sites attractive. The study found that if loans were available at a subsidized 7 percent rate, 50 percent of the potential would be developed. If further incentives reduced the effective interest to 3 percent (through investment credits for instance) and the power were purchased at 6.7¢/kwh, 80 percent of the potential would be developed. Methodology needs to be developed to assess the national potential in a similar fashion.

In addition to the programs described previously, small-scale hydroelectric facilities could be encouraged with the following actions:

o FERC should closely monitor its hydro license application process and take steps to insure that backlogs will not occur (GAO, 1980).

o FERC should revise its "generalized power values" used to qualify hydro facilities so that they more closely approximate actual marginal electric costs.

Wind Systems

The Wind Energy Act of 1980 authorizes an eight-year loan and grant program designed to facilitate the installation of at least 800 Mw of capacity from wind systems by FY 1988, of which at least 100 Mw are from small machines. The act also provides for federal procurement, research and development, contracts, cooperative agreements, a resource assessment program, and a wind data center. Loans for large and small machines will cover up to 75 percent of project purchase and installation costs at an interest rate equivalent to water development loans, currently about 8.5 percent. The loan term can be up to twenty years, but the aggregate capacity funded by any one loan may not exceed 320 Mw. Grants will cover up to 50 percent of project purchase and installation costs in the first six years of the program, and up to 25 percent in the last two years. Only large systems are eligible for grants; federal procurement will be the major assistance for small machine producers. The act provides funds to federal agencies to accelerate their purchase and installation of wind machines.

The program will cease to provide funding for large machines after FY 1988 and small machines after FY 1985, or earlier if there is prior determination that wind systems are competitive with conventional energy sources. The act authorizes $100 million for FY 1981. Funds for FY 1982 - FY 1988 will rise above that level and then drop back to $100 million. At least one fourth of the funds authorized in each of the first five years of the program must be directed toward small machines. The program, of course, is meaningless if not funded.

UTILITY PLANNING

The previous discussion has concentrated on techniques that can be used to encourage greater competition in utility markets. This increased competition places new and complex burdens on electric utilities and the state agencies that regulate them. In particular, the techniques that have been developed during the past few years for determining how utilities should invest in new generating equipment will need to be modified to reflect a new, and more turbulent environment. Improved planning techniques must be developed which explicitly consider a larger variety of investment opportunities, and which take better account of the possible reaction of utility customers to higher prices. Development of improved utility planning methods will be critical for implementing many of the programs discussed in previous sections. Planning will be needed, for example, to establish accurate marginal cost pricing, to determine effective 'buyback' rages under PURPA, and to determine economically optimum levels of utility investment in equipment on a customer's site.

The federal government plainly cannot, and should not, play a role in developing plans for utility investment. It can, however, provide many kinds of useful support. A useful precedent can be found in the work done by the Federal Power Commission (now the Federal Energy Regulatory Commission). The FPC developed a national transmission plan in response to its statutory mandate to "divide" the country into regional districts for the voluntary interconnection and coordination of facilities for the generation, transmission, and sale of electric energy . . .(and) to promote and encourage such regulatory coordination" (16 U.S.C. 824 (a) quoted in Kahn, 1970). The Commission worked successfully with the industry to develop such plans. Building on this successful enterprise, two new planning efforts are suggested: (i) development of improved techniques for determining how utilities should invest their capital under today's circumstances, and (ii) reviewing national transmission planning to determine whether changed circumstances call for fundamental revisions. In both cases, care should be taken to work with the industry and state regulatory commissions.

Improved Planning Methods

The Problem

The analytical tools available for determining how a utility should invest are deficient in a number of ways. The basic techniques needed to made decisions in an uncertain environment have not been developed and the use of sophisticated techniques is limited by the lacked crucial data.

Typically, current analytical techniques are also unable to calculate the real value of small generating facilities which can be constructed in a short time; almost none of the techniques now used is able to assess the value of solar electric generating equipment of different types. Highest priority must be given to the construction of conceptually simple methods which can improve on the analytical techniques now used to determine a utility's need for new capacity. While it is important that these computational problems be addressed, however, utility investment plans must clearly be developed before a satisfactory model can be perfected.

Most utility planners use models in which the growth in demand for electricity and the level of reliability expected by customers is specified. Demand is typically determined by examining the history of demand growth. In the more sophisticated models, the history of consumer behavior is examined to determine possible responses to future utility prices. Such approaches are deficient in two important respects. First, data used to predict consumer responses is usually derived from a period when energy prices were not a major concern. In the future, consumers can be expected to be much more concerned about energy prices and, possible more important, will have access to a new array of technologies which can reduce their demand. Moreover, customers will be increasingly pursued by entrepreneurs intent on selling energy saving equipment or ideas. Second, most models require the analyst to guess the future price of electricity deriving price on the basis of demands and supplies. This means that the computation can result in a situation where a utility's revenues do not equal its allowed costs. The calculations can be repeated until the correct result is reached, but the process is time consuming and unsatisfying. The Electric Power Researach Institute has supported the development of several models which attempt to resolve this problem. The "General Equilibrium Modeling System" (GEMS) is designed to consider both supply and demand simultaneously (Cazlet, 1977), and the model has been used by the Energy Information Agency to develop long-range national energy forecasts in a model called LEAP (EIA, 1979). An analytical technique has also been developed that treats demand as a statistical variable. This model allows a calculation of such things as the expected cost of electricity given different approaches to system expansion.

(Cazlet, 1978). Neither of these models is able to provide the industry with the kind of detailed information it has come to expect from models of this kind even though they are quite complex. Moreover, neither is able to incorporate a sophisticated appraisal of new on-site technologies, possible consumer behavior under new economic conditions, or expanded use of small generating equipment based on solar or other energy sources.

Solar equipment, of course, introduces an even more complex set of problems since solar energy may not be available when it is needed. An accurate appraisal of the value of solar equipment requires a detailed study of the correlation between solar supplies and energy demands, and the correlation between the wind and solar resources available in different geographical areas at the same time (SRI, 1978; Kahn, 1978) In the future, storage systems and demand management systems (possibly responding to time-of day or seasonal marginal rates can adjust the timing of both supplies and demand making the problem even more difficult. A simplified approach to some of these problems for a range of solar equipment will be examined in the last section of this chapter.

This discussion should not be read to imply that no attempt should be made to change existing utility planning techniques until a new technique is perfected or until accurate information is available about all important variables. It is clear that existing utility planning models are badly deficient in their treatment of efficiency investments and small power producers. Planning should proceed in two phases. Work should be undertaken to perfect sophisticated analytical techniques so that greater precision can be achieved in the long-run. A technique also must be developed for making decisions in the near-term, before the sophisticated techniques are perfected. It is likely that some extremely simple planning techniques can be developed that will give more accurate guidance than bulky computer models that consider a limited number of issues with great precision.

Model Utility Plans

The federal government can act to overcome some of the difficulties faced by utilities by funding the development of 10 model expansion plans and investment programs for investor-owned utilities. Such a policy could be used to support innovative utility companies and local regulatory bodies willing to experiment with new ideas and planning concepts and implement field tests. The development in these companies can serve as a model for the rest of the industry.

Funding for model utility plans could be based on competitive proposals submitted by privately owned electric or gas utilities acting in cooperation with a state PUC or EES. A requirement for selection would be that the utility whose service area was being con-

sidered must commit itself to making a major contribution to the study, providing significant staff time and funding to the project. An effort should be made to obtain a wide geographic dispersion of model utilities and to divide the awards between varying sized systems. It is possible that these studies could be made a part of the Residential Energy Efficiency Program (REEP).

Reassessing the Nation's Transmission Capabilities

Planning the need for new transmission capacity and for the improved use of existing transmission capacity presents specialized problems that require specific attention. One clear difference is that such plans must be developed on a regional, or possibly on a national basis and cannot be limited only to the problems of individual utilities.

The nation's transmission system has been examined at considerable length in the National Power Survey and subsequent documents, but new circumstances have created an environment in which the issue should be reopened. Some of these issues can be seen by enumerating the advantages of expanding the existing power grid:

o   During the next few years there will almost certainly be areas with surplus gas and electricity while other regions are experiencing shortages; transmission systems may be able to alleviate this problem.

o   Solar generating equipment may provide energy at low cost in areas where electricity is relatively inexpensive and therefore not be used effectively unless techniques are available for supplying wind-derived energy to other areas where electricity is more expensive but wind resources less attractive (Justus, 1977).

o   Solar energy, being inherently erratic, is best utilized if the generating equipment is integrated into the largest possible network of suppliers through some sort of grid arrangement. In a large network, local variations in direct solar and wind energy supplies can be averaged to a more constant load, the systems would have greater access to any storage units (i.e., pumped hydroelectric facilities) integrated into the utility network, and there would be more opportunity for dispatching energy in a way that could make maximum use of the solar energy that became available.

o   Improved transmission systems can open new markets to operators of cogenerators and non-regulated owners of energy systems operating from renewable resources. This would, for example, make it easier to find markets for energy generated by wind machines located in areas with relatively small electric demands.

o   Improved transmission will make better use of cogeneration since there may not be a good match between needs for process steam and needs for new electric generation.

o   In many cases municipal utilities can only have access to electricity from new generating sources (such as wind machines) if the utilities are adequately served with transmission facilities connecting them with attractive solar sites.

o   Transmission costs could increase if many on-site generators are connected--particularly if they both buy and sell electricity.

The advantages must be weighed against the problems that can result from too great a reliance on a large grid. Large, complex systems are vulnerable to failures (witness the recent blackouts in the Northeast). They are also vulnerable to attack by tee greater access to any storage units (i.e., pumped hydroelectric facilities) integrated into the utility network, and there would be more opportunity for dispatching energy in a way that could make maximum use of the solar energy that became available.

o   Improved transmission systems can open new markets to operators of cogenerators and nonregulated owners of energy systems operating from renewable resources. This would, for example, make it easier to find markets for energy generated by wind machines located in areas with relatively small electric demands.

o   Improved transmission will make better use of cogeneration since there may not be a good match between needs for process steam and needs for new electric generation.

o   In many cases municipal utilities can only have access to electricity from new generating sources (such as wind machines) if the utilities are adequately served with transmission facilities connecting them with attractive solar sites.

o   Transmission costs could increase if many on-site generators are connected--particularly if they both buy and sell electricity.

The advantages must be weighed against the problems that can result from too great a reliance on a large grid. Large, complex systems are vulnerable to

failures (witness the recent blackouts in the Northeast). They are also vulnerable to attack by terrorists or, in time of war, to foreign attack.

In addition to the technical issues involved in the assessment of the adequacy of the national transmission systems, DOE should also report on institutional arrangements which may hinder the adoption of lowest cost utility investment strategies. For example, private power pools in the Northeast and Mid-Atlantic states require that individual utilities each contribute an agreed-upon number of new generating plants to the pool's capacity reserves. These arrangements do not encourage utilities to satisfy demand and maintain reserves in other ways, i.e., financing of on-site equipment to reduce demand.

Other Planning Initiatives

In addition to the major utility planning efforts described above, the following initiatives should also be undertaken to provide support for the proposed planning system:

o   The proposed Energy Management Partnership Act contains provisions for energy planning on state and local levels. Care should be taken to integrate the planning contemplated under the terms of this act with the plans developed through the direct federal assistance proposed elsewhere in this catalogue. The EMPA could be used as a vehicle for funding local public utility commissions in their planning activities.

o   The proposed changes in rules governing the Technical Assistance Program of the Economic Development Agency (Federal Register, 5/7/80, p. 3032, proposed change to 13 CRF 307.3) would offer EDA funds to "establish and carry out effective economic development programs at local and multijurisdictional levels, and to provide a basis for improved coordination and continuity of Federal, State, and local economic development activities," which would include energy conservation planning. These funds, which should also be available for examining the potential for solar energy equipment, could also be used to provide coordinated planning of utility expansion in the regions. They could be essential if EMPA funding is not available.

o   Direct financing can be provided for state public utility commissions. Many of the programs suggested in this analysis, and many of the programs operating under existing statutes, require that sophisticated analytical work be conducted by the staff of local public utility commissions. (The federal government should be prepared to pay for the additional effort required.) While several states have large, and relatively expert staffs for dealing with the increasing complexity of utility regulation, many states do not have adequately funded commissions. Even the most ambitious states are feeling the impact of the increasing complexity of the work imposed on them. Section 207 of PURPA provides limited funding for state PUCs to work on specified PURPA regulations. $40 million was authorized in 1979 and again in 1980. Actual funds used in 1979 were about $10 million. Grants awarded to 46 states, and grants for 1980 are currently being reviewed. The actual procedures used by PUCs to comply with these titles are up to the states.

o   The federal government could (for utility planners and regulatory planners) sponsor a series of training workshops or possibly a program providing fellowships to universities offering relevant programs. These programs could be used to ensure that the participants have access to the most recent information and planning tools and that any group interested in using the analytical tools developed to assist utilities and commission staffs has an adequate opportunity to learn more about them.

o   A special fund should be established to train and provide information for personnel of municipal utilities and cooperatives. Regional and local planning should emphasize the use of municipal utilities where they exist and should consider the possibility of establishing a separate municipal financing authority in areas where municipal utilities do not presently exist. Planners should examine the possibility of using existing municipal agencies (i.e., water and sewer agencies) as administrative centers for undertaking on-site solar and conservation initiatives, as well as examining the option of establishing an entirely new organization. The Federal government could provide basic information useful for the construction of such plans, funding provided to support proposals for planning studies, and grants to support start-up costs.

Section Three
Federal Power Marketing Authorities

BACKGROUND

The Department of Energy administers five "Power Marketing Administrations" created to transmit electricity from federal hydroelectric projects built and operated by the Army Corps of Engineers or the Water and Power Resources Service (formerly the Bureau of Reclamation). The five agencies are the Western Area Power Administration (WAPA), the Alaska Power Administration (APA), the Bonneville Power Administration (BPA), the Southeastern Power Administration (SEPA), and the Southwestern Power Administration (SWPA). In addition, the Tennessee Valley Authority (TVA) operates an extensive system of transmission facilities for its hydroelectric projects. TVA is unique in many ways, however, not the least of which is that TVA can own and operate its own generating facilities. The areas served by power authorities are illustrated in Figure 4.2. These organizations market power produced in 121 federally-owned multiple purpose projects. In addition, the FPMA's operate and maintain approximately 30,100 miles of high voltage transmission lines. In 1978, the five FPMA's sold 125.1 billion kilowatt hours of electricity (DOE, 1980). Taken together, the FPMA's are the world's largest electric transmission network.

The FPMA's serve a variety of customers but current laws state that "preference in the sale of Federal power will be extended to public agencies, cooperatives, and organized Indian tribes." At present, the FPMA's serve 886 customers, of which 635 are public. With some relatively modest changes in their structure, the FPMA's could powerfully induce their public customers to maximize on-site energy efficiency and to consider a full spectrum of renewable resource technologies. Without FPMA assistance, for example, most of the FPMA's public customers would not have access to attractive sites for wind machines or centralized photovoltaic or solar/thermal sites.

THE ROLE OF THE FPMA'S IN ENCOURAGING EFFICIENCY AND USE OF RENEWABLE RESOURCES

The highly subsidized power provided by the FPMA's acts as disincentive for investments in on-site energy equipment. FPMA power rates are a bargain in comparison to alternative power rates. For example, according to EIA data, in 1977 the average electricity price for industry was 25 mills per kwh, while the average price of power (this is principally because of the low cost of the power produced at federal multi-use projects) marketed by PMA's was approximately 5 mills per kwh (EIA, 1980). The catch, of course, is that there is only a limited supply of low-cost power available, so that the lure of this cheap electricity can, if skillfully manipulated, provide a powerful incentive to encourage investments in efficiency or on-site generating equipment. Contracts between the PMA's and their customers provide one point of leverage. The PMA's could offer to renegotiate the term of their contracts and, as an example, offer to convert a contract to supply 100,000 kwh per year for five years into a ten-year contract for 50,000 per year. This would be an extremely attractive offer, particularly if it could be accompanied with information about and possible financing, of on-site technologies which could significantly reduce demands for PMA energy.

The leverage of the FPMA's is greatest in situations where demand is growing rapidly. The "boom towns" likely to grow up around coal mining in the West provide excellent targets of opportunity. Acting in cooperation with local utilities, the FPMA's can help to see that these new towns make the best use of energy efficiency and renewable energy technologies available on the site.

The FPMA's provide an excellent institutional mechanism for facilitating the introduction of generating equipment using renewable resources into electric

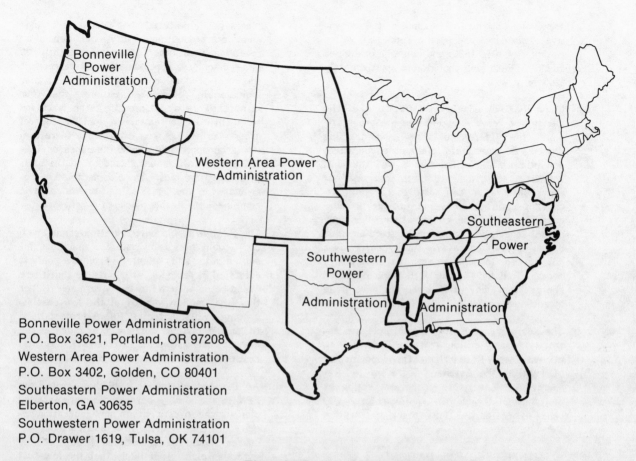

Bonneville Power Administration
P.O. Box 3621, Portland, OR 97208

Western Area Power Administration
P.O. Box 3402, Golden, CO 80401

Southeastern Power Administration
Elberton, GA 30635

Southwestern Power Administration
P.O. Drawer 1619, Tulsa, OK 74101

Figure 4.2. **Electric Power Marketing Areas**

utility systems, since FPMA's provide a basis for regional planning and have access to capital whose
effective cost is nearly half that available to investor-
owned utilities. Together with the Water and Power
Resource Service (WPRS) and the Army Corps of
Engineers, FMPAs were used effectively to promote
the development of national hydroelectric resources in
an earlier generation. Their mandate should be clearly
expanded to include all renewable resources. They are
in a good position investigate and encourage the possibilities inherent in renewable resources because their
enormous service areas allow them to access to many
desirable wind and direct solar sites.

The problems of energy storage and backup requirements sometimes associated with renewable energy
sources can be mitigated because the FPMA's are
uniquely equipped to offer backup power from their
large hydroelectric generating systems.

In the case of the renewable energy other than hydroelectric power, it would be expected that federal
purchases would serve as an example for private investors. The Federal government would absorb the initial risks.

It might also be desirable to allow the FPMA's to
purchase loans owned by their "preferred customers,"
and possibly by investor-owned utilities purchasing
power from the PMA's, to finance solar energy or
energy conservation. Under the terms of the "Pacific
Northwest Electric Power Planning and Conservation
Act," BPA is authorized to provide financial assistance
for an itemized list of conservation investments in
buildings and industries in its service area. BPA is
authorized to use its bonding authority to fund such investments with the provision that the obligations be
paid from net proceeds and not from general federal
revenues. One possible problem with this approach is
that recipients of FPMA funds may not be eligible for
an investment tax credit because of provisions of the
Windfall Profits Tax of 1980 (26 USCA 44C (c)(10)).

Other initiatives to encourage use of FPMA facilities should include the following;

o   Each of the five FPMA's should be required to
    develop plans for making maximizing energy
    efficiency and use of renewable energy re-
    sources in cooperation with public customers.

These plans should follow some of the innovative ideas developed by the Tennessee Valley Authority (DOE, 1980) and could be developed quickly. Each plan should cost approximately $0.5 million.

o   Using the knowledge gained through the efficiency and renewable resource use promoting plans, the FPMA's could revise contracting procedures with rural electric cooperatives and other public utilities in a way that will encourage these institutions to undertake energy conservation and solar energy projects. Current contracts may not provide an adequate incentive for conservation since they permit long-term guarantees of fixed amounts of power at fixed rates. Rural cooperatives not forced to supplement the inexpensive energy from the FPMA's with more expensive energy from conventional sources have little incentive to conserve.

o   The Secretary of Energy should delegate authority (under Section 4 of the Federal Nonnuclear Energy Research and Development Act of 1974) to the Assistant Secretary for Resource Applications to cooperate, acting by and through the power administrators, with other offices of the Department to carry out these programs.

OTHER RECOMMENDATIONS

It seems clear from the previous discussion that large and small hydroelectric plants and wind machines offer the most important opportunity for generating electricity used in electric utility grids during the next. The FPMA's could facilitate the construction of large wind machines with the following actions.

o   Each of the Agencies should be funded to identify the most promising wind sites in their regions. Particular emphasis should be placed on the region as shown to be most promising in the analysis presented in Tables 4.5 and 4.7.

o   In FY 1981, a goal should be established of identifying approximately 80 promising sites in the FPMA regions. Estimates from manufac-

turers have indicated that corporate investment in a production facility would depend on receiving firm orders for approximately 20 large machines, approximately 20 percent of the total production run of the pilot facility indicated in Table 4.8. Assuming that four manufacturers would produce large machines for utility use, a reasonable goal for federal orders during the next two years, therefore, would be approximately 80 machines or 160-200 MW. Procurement should be on the basis of competitive bids. The bidding could be coordinated by the WPRS so that orders could be combined into sizable blocks of purchases.

o   The FPMA's should determine the price which they could reasonably pay for machines installed in the sites selected. From the analysis exhibited in Table 4.6, it is possible that there would be no net cost to the government (other than the implicit subsidy of the low-cost financing available for all FPMA projects). If the initial systems cost twice as much as the highest costs estimated in Table 4.9, the total cost could be on the order of $140 million.

o   At the end of the first bidding cycle, the FPMA's should have developed a more detailed plan for the investments in new generation required in their service area and further purchases should be undertaken in accordance with this plan. It is likely that the low-cost capital available to the WPRS for funding wind projects will provide a subsidy adequate to support significant additional purchases. If price projections follow manufacturers current plans, the first round of FPMA purchases could encourage investments in production facilities which would result in sales prices low enough to encourage further investment by investor-owned utilities.

A similar program should be established for small wind machines (BPA, APA, WAPA), and residential photovoltaic systems (WAPA, SEPA, SWPA, TVA). The experience gained in constructing these programs could be used to design projects for other solar electric technologies which may be available later in the century.

Section Four

Assessment of Solar Electric Technologies

Given the enourmous uncertainties about the future cost and performance of solar equipment designed for generating electricity and the uncertainties which surround estimates of the future cost of energy from conventional sources, all estimates of the future demand for solar electric equipment must be considered highly speculative. If the nation is to make a serious effort to capture the potential of this equipment, however, a clear set of priorities and objectives must be developed The analysis which follows represents one attempt to cut through the complexity and arrive at such a set of goals. It must be recognized from the onset, however, that the analysis cannot be exact and it has been necessary to theorize about a number of variables that cannot be derived with precision. One major barrier to the construction of a set of priorities is the lack of consistency in the techniques used to estimate the cost and performance of solar equipment; each technology seems to have employed a different method and applies a different degree of scepticism to the cost estimates.

METHODOLOGY

Since it appears that demand for electricity is unlikely to increase significantly during the next two decades, the value of solar electric equipment has been determined on the basis of savings which result when it is used to reduce the power produced from conventional generating equipment; the mix of generating equipment used here as a baseline is the mix which would result if plants scheduled to be on line before 1985 are completed but no plants are added between 1985 and 2000. In this context, solar electric equipment derives its value primarily from the fact that it can save expensive fossil fuels and the cost of operating conventional equipment although it must be recognized that fuel can also be saved by constructing a new more efficient coal burning or nuclear plant.

The solar equipment, however, is preferred if the ratio of cost to value of the solar investment is lower than the cost to value of a coal or nuclear plant investment designed to achieve similar energy displacement. (No attempt has been made in this analysis to determine whether nuclear or coal plants are the least expensive option.) This method does not give the solar equipment explicit credit for reducing the demand for new generating capacity since it is assumed that the solar equipment will not displace new construction. In cases where demand is increasing, however, some credit can be given the solar equipment for the displacement of new capacity. This displacement depends critically on the location of the equipment and on the amount of solar equipment introduced. Table 4.10 summarizes the results of several studies that have examined the capacity value of wind equipment in some detail.

Details of the technique used to assess solar and coal costs and values are provided in Appendix B. This Appendix also summarizes the results of a number of different sensitivity studies. The major results of the analysis are summarized in Tables 4.11 and 4.12. The tables present the cost-to-value ratios of the addition of a coal plant and five different kinds of solar electric plants. Table 4.11 indicates the value of the first increment of new capacity and Table 4.12 indicates the values assuming that 20 percent of the capacity in each region is already being met with new equipment of the technology type under examination. In each case the range represents uncertainties about the future cost of the technology.

Two types of costs and values are computed. In one calculation, which is called the "market cost" case, it is assumed that the plants are financed with an annualized fixed charge rate equal to that paid by a typical investor-owned utility. The price of oil and gas in this case is assumed to be the marginal cost of oil and gas during the time in question. The second case, which is called the "social cost" case for convenience, is an at

Table 4.10. POTENTIAL WIND MACHINE PRODUCTION RATES

	Number of Machines	Capacity (MW$_e$ rated)
Small and intermediate systems		
Pilot Plants (1980-1983)	2000-6000	60
Initial Production (84-90)	21600-7200	216
at 100/month-300/month		
Mature Production (90-2000)	108,000-36,000	1080
at 300/month-900/month		
Total Small and Intermediate		(1356)
Mod-2 (2.5 MW$_e$ rated) class		
Pilot Plant (1980-1984)	100	250
Initial Production (85-88)	360	900
at 10/month		
Mature Production (1988-2000)	4320	10,800
at 30/month		
Total Mod-2 Class		(11,950)
Mod-5 (4.0 MW$_e$ rated) class		
Pilot Plant (1983-1987)	100	400
Initial Production (88-90)	360	1440
at 10/month		
Mature Production (1991-2000)	3240	12,960
at 30/month		
Total Mod-5 Class		(14,800)
TOTAL INSTALLED CAPACITY IN 2000		28,106 MW$_e$ = 1 Q at 35% cf

Assumptions:

Pilot plants for small machines operate for 3 years, large plants 4 years.

Decision to invest in a production facility must be made 3 years before initial production (allowing for one year of operating field experience); the production facility requires 2 years to complete.

After an initial production facility has operated for three years production is increased by a factor of three either by expanding the facility or by the addition of competing production firms.

The average capacity of small and intermediate systems is assumed to be 30 kW.

Production can yield 1.84 quads in the year 2000 if mature production reaches 60/month instead of the 30/month assumed in the table.

Conversion efficiency for electrical generation in 29.4% (11,604 Btu/kWh$_e$)

Table 4.11. COST/VALUE RATIOS FOR COAL AND SOLAR ELECTRIC TECHNOLOGIES--YEAR 2000
(Initial penetration level and 1985 generation mix)

DOE Region	Fuel Prices	Coal	Wind	Photovoltaics	Solar Thermal Central Receiver[a] STSAS	STSAN	Distributed Receiver	OTEC
1	M	0.48-0.52	0.33-0.42	0.80-1.64	*	*	*	*
	S	0.29-0.31	0.14-0.17	0.33-0.67	*	*	*	*
2	M	0.48-0.52	0.33-0.41	0.82-1.70	*	*	*	*
	S	0.29-0.31	0.14-0.17	0.34-0.69	*	*	*	*
3	M	0.60-0.66	0.63-0.80	1.14-2.35	*	*	*	*
	S	0.34-0.37	0.29-0.36	0.52-1.03	*	*	*	*
4	M	0.07-0.78	1.24-1.58	1.16-2.38	*	*	1.42-1.95	0.90-1.23
	S	0.39-0.42	0.58-0.73	0.54-1.08	*	*	0.70-0.97	0.46-0.60
5	M	0.85-0.95	0.81-1.03	1.90-3.93	*	*	*	*
	S	0.45-0.49	0.39-0.49	0.92-1.83	*	*	*	*
6	M	0.48-0.52	0.34-0.43	0.56-1.15	0.67	0.70	0.44-0.60	0.50-0.69
	S	0.29-0.31	0.14-0.17	0.23-0.47	0.30	0.31	0.20-0.27	0.23-0.30
7	M	0.63-0.69	0.52-0.66	1.12-2.31	*	*	*	*
	S	0.36-0.38	0.24-0.30	0.57-1.02	*	*	*	*
8	M	1.85-2.15	0.74-0.94	1.89-3.90	*	*	*	*
	S	0.89-1.00	0.39-0.49	1.00-1.99	*	*	*	*
9	M	0.48-0.52	0.34-0.43	0.54-1.12	0.74	0.77	0.53-0.72	0.56-0.76
	S	0.29-0.31	0.14-0.18	0.23-0.46	0.34	0.35	0.24-0.33	0.26-0.34
10	M	1.59-1.84	1.78-2.26	5.65-11.65	*	*	*	*
	S	0.77-0.87	0.83-1.03	2.61-5.21	*	*	*	*

[a]Cost/value ratios for repowered plants were not estimated bacause many existing oil- and gas-fired plants should be retired before the year 2000. Thus, repowering is not expected to be a major supply option in the year 2000.

Notes:

Symbols denote the following:
STSAS = solar thermal central receiver stand-alone plants with storage
STSAN = solar thermal central receiver stand-alone plants without storage
M = "market" cost case where the cost/value ratio is computed assuming that the cost of capital is equal to conventional private utility financing, and that fuels are priced at their marginal market value.
S = "social" cost case where the cost of capital is computed under the assumption that no taxes are paid and that the cost of oil and gas is the equivalent of $20 per barrel higher than in the market case above. (Coal and nuclear prices are the same as in the market case even though arguments can be made that their market prices do no adequately reflect the non-market cost associated with using these fuels.)
* = cost/value ratios were not estimated for these regions

The cost/value ratios for this table assume that the capacity of the solar or coal equipment added to the grid is a small fraction of the total system capacity. All systems are assumed to begin operation in the year 2000 and operate for 30 years. The calculation of value is based on a mix of conventional generating equipment which assumes no new capacity additions after 1985. See Appendix B for a detailed description of the methodology and assumptions about cost and performance of the equipment.

tempt to reflect the costs that society as a whole might attach to the investment in question. In this case the cost of capital is computed assuming that the utility pays no taxes, and the cost of oil and gas are increased by the equivalent of $20 per barrel to reflect costs not included in the market price of these products.

Table 4.12. COST/VALUE RATIOS FOR COAL AND SOLAR ELECTRIC TECHNOLOGIES--YEAR 2000
(20% penetration and 1985 generation mix)

| | | | | | Solar Thermal | | | |
| | | | | | Central Receiver[a] | | Distributed | |
DOE Region	Fuel Prices	Coal	Wind	Photovoltaics	STSAS	STSAN	Receiver	OTEC
1	M	0.57–0.62	0.38–0.48	0.85–1.76	*	*	*	*
	S	0.33–0.35	0.16–0.20	0.36–0.71	*	*	*	*
2	M	0.52–0.56	0.36–0.45	0.86–1.77	*	*	*	*
	S	0.31–0.33	0.15–0.19	0.36–0.72	*	*	*	*
3	M	**	0.73–0.92	1.24–2.56	*	*	*	*
	S	**	0.34–0.43	0.58–1.15	*	*	*	*
4	M	**	1.36–1.73	1.31–2.69	*	*	1.56–2.13	2.71.3.70
	S	**	0.65–0.82	0.63–1.25	*	*	0.79–1.08	1.70–2.23
5	M	**	0.97–1.23	2.08–4.29	*	*	*	*
	S	**	0.50–0.62	1.03–2.05	*	*	*	*
6	M	0.56–0.61	0.37–0.47	0.60–1.23	0.75	0.76	0.48–0.66	0.67–0.92
	S	0.33–0.35	0.16–0.20	0.25–0.51	0.34	0.34	0.22–0.30	0.32–0.42
7	M	**	0.63–0.79	1.23–2.55	*	*	*	*
	S	**	0.30–0.38	0.58–1.15	*	*	*	*
8	M	**	0.88–1.12	2.20–4.54	*	*	*	*
	S	**	0.51–0.64	1.25–2.48	*	*	*	*
9	M	0.64–0.71	0.40–0.51	0.61–1.26	0.88	0.87	0.59–0.81	0.89–1.21
	S	0.36–0.39	0.17–0.22	0.26–0.52	0.40	0.40	0.27–0.37	0.42–0.55
10	M	**	15.67–19.88	8.62–17.78	*	*	*	*
	S	**	9.01–11.27	4.24–8.45	*	*	*	*

[a]See footnote a in Table 11.

Notes:

Symbols denote the following:
STSAS = solar thermal central receiver stand-alone plants with storage
STSAN = solar thermal central receiver stand-alone plants without storage
M = "market" cost case where the cost/value ratio is computed assuming that the cost of capital is equal to conventional private utility financing, and that fuels are priced at their marginal market value.
S = "social" cost case where the cost of capital is computed under the assumption that no taxes are paid and that the cost of oil and gas is the equivalent of $20 per barrel higher than in the market case above. (Coal and nuclear prices are the same as in the market case even though arguments can be made that their market prices do no adequately reflect the non-market cost associated with using these fuels.)
* = cost/value ratios were not estimated for these regions
** = cost/value ratios are not .applicable for coal in this region at this penetration level because less than 20% of the energy produced is from oil and gas.

Assumptions are the same as those used in the preparation of Table 6 except that for this table, the cost/value ratio is computed assuming that 20% of the capacity in the region is new solar or coal units. The value computed is the value of one additional increment of new solar or coal capacity at 20% penetration.

Several features of the results deserve attention:

o Under the assumptions used here, wind machines are preferred to all other renewable electric sources shown on the table in all regions except region 4. This does not necessarily mean that wind machines will dominate all markets in these regions since the table represents average. values for large regions. There are undoubtedly areas within each region where no adequate wind sites can be found and where other solar technologies are particularly attractive. Unfortunately, there determinations can only be obtained with more detailed analysis than the one which could be conducted here.

o   The cost-to-value ratios of photovoltaic and solar thermal systems are below one in several regions, even after a 20 percent penetration has occurred, indicating that they would have economic merit. The ratio is seldom lower than that obtained by adding a new coal plant, however, indicating that without incentives the solar alternative would have difficulty competing.

o   The cost-to-value ratios for all renewable resource and new coal plant investments are high for all in region 10. This is because, under the assumptions of this analysis, most of the energy generated in Region 10 is derived from hydroelectric or nuclear generating facilities; the value of displacing fuel from a nuclear plant is very low. The analysis appears to have indicated that if demand growth in the area grows at a rate which does not require capacity beyond that indicated in Table 4.12, there would be no incentive to invest in renewable generating equipment in the region. This fact illustrates the difficulty of concentrating too narrowly on regional supplies and demands. Region 10 has a number of attractive wind sites, and in fact will be used as a test-site for the new Mod-2 wind machine. Wind energy available at low cost in Region 10 could have high value in neighboring regions if adequate provision is made for its transmission.

o   The value of photovoltaic equipment is extremely sensitive to the assumptions made about its future cost--particularly assumptions made about the cost of installing a utility-owned photovoltaic system. The costs and value of the average of the photovoltaic system are roughly comparable to those of a solar thermal repowering system, because both systems are assumed to have no storage. Low-cost storage equipment could improve the economic attractiveness of both technologies.

o   The photovoltaic systems had a cost to value ratio less than one in all regions except region 10. If social value analysis is employed and the lowest cost goals are met, even the most expensive photovoltaic system assumed had a cost value less than 1.1 in all regions save 5, 8, and 10. Using standard financing, however,

the more expensive photovoltaics did not have a cost-to-value ratio less than one in any regions, although the ratio is close to one in regions 6 and 9. The photovoltaic systems were preferred to coal systems only in these two regions and only if social costs were used to make the comparison. The coal and photovoltaic cost to value ratios are, however, extremely close in several other regions. Using this analysis, therefore, utility-owned photovoltaic systems installed in remote areas would be preferred to a new coal plant in most regions only if a penalty is charged for environmental costs of burning coal--a factor not included in the calculations.

o   Solar Thermal systems using dish collectors may compete with utility-owned photovoltaic equipment in some regions if the very ambitious cost goals of the dish program are met.

o   In most cases the cost-to-value ratio computed for the "social cost" case is roughly half of the cost value ratio computed for conventional financing. Utilities could be induced to make decisions similar to those preferred under the social cost analysis, therefore, if the effective capital costs of the solar were roughly half of what they are under the terms of conventional financing. Financing from federal or municipal bonds, provided by the federal power marketing authorities or a corporation of the type established under the terms of Title I of the Energy Security Act of 1980, could achieve this effect. (This issue will be examined with greater attention in the final draft.)

MAJOR CONSTRAINTS

Plainly there will be a number of barriers to the introduction of massive amounts of solar electric equipment. A brief review of some of the major points to be considered follows.

Environmental Constraints

o   One of the largest barriers to large-scale introduction of wind systems will be the location of a sufficient number of sites. If the nation expects to be able to generate 3 quads of energy from wind, it will be necessary to find sites for at least 21,000 large wind machines. Without better data, it is simply impossible to determine whether enough sites can be found. Surveys in California, however, have already identified sites for approximately 1500 large machines near the San Gorgonio pass (Lerner 1980). Such large-scale fields of wind machines might, however, be visually unattractive.

o Producing one quad of electricity from photovoltaic arrays will require on the order of 600-800 square miles of land (assuming that arrays cover 1/2 the land available). Large land areas are unlikely to be found in congested industrial areas in the northeast and north central regions. Similar constraints apply to solar thermal systems.

o Environmental constraints present the largest barrier to significant increases in hydroelectric production (see Appendix C).

System Stability

If solar electric systems, where production fluctuates with the weather and with the seasons, are integrated with a conventional electric generating systems, the utility may have difficulty adjusting conventional capacity to match supplies and demand in its service area. This difficulty is only likely to present a major problem in circumstances where the solar electric equipment represents as much as 20 percent of the installed capacity in a region. It is likely that improved control systems and short-term storage will be able to alleviate any problems encountered, but research at this point does not provide an adequate basis for confident statements. It is extremely important that field experiments be conducted to determine the exact nature of the problems encountered and to begin the search for solutions. Experience to date has been mixed. The 220 kW Mod-OA wind machine installed in Clayton, New Mexico, represents only about 15 percent of the installed capacity of the local utility and no stability problems have been encountered; "Spinning" reserve has gracefully covered wind fluctuations. On the other hand, the same machine installed on Block Island, Rhode Island, produces a large fraction of the total island demand on occasions. Problems have developed, largely because the diesel turbines on the island cannot be operated effectively below 40 percent of their rated power.

Construction Schedules

Adding a large amount of solar electric generating capacity during the next two decades will require the construction of a manufacturing capacity adequate to produce the needed equipment. This could be an enormous enterprise. A preliminary analysis of the requirements, however, does not indicate that any major barrier would be faced.

Table 4.10 indicates a construction schedule that could result in the production of 1 quad of wind equipment by the year 2000. If several competing firms manufactured Mod-2 or Mod-4 class machines simultaneously, it would not be difficult to install 3 quads of equipment by the year 2000. A production facility capable of producing 5,000 Mod-2 wind machines per year would require approximately 10 million square feet of factory space--approximately 1/3 the factory space now occupied by the Boeing Corporation (Lowe 1979). It has been estimated that a plant capable of manufcturing 30 machines per month would take 2-4 years to construct and would cost approximately $100 million.

It is likely that 0.5 quads of wind wnergy will be developed by the year 2000 with minimal government intervention. Southern California Edison is currently planning to install 55 MW of wind turbines by 1986 (Western Systems Coordinating Council, 1979, p. 37; Weir, 1980). Wind Farms Ltd. is planning to build an 80-Mw wind installation in Hawaii as early as 1984 (Solomon, 1980). Other utilities have also expressed a strong interest in wind turbines (Business Week, 1979; Smith, 1980).

An analysis similar to the one that produces these predictions on wind machines indicates that production of 1 quad of photovoltaic capacity would not require a prohibitive manufacturing schedule: if five plants per year begin operating each year between 1986 and the year 2000 and each plant is capable of producing 250 Mw/year, the goal can be met.

Materials

It does not appear that a shortage of materials will present a major barrier to large-scale production of solar electric equipment. For example, the Mod-4 wind machine is expected to weigh 880,000 lb and produce a peak output of 4 MW (General Electric 1979). Production of enough machines to generate 3 quads of energy by the year 2000 would require an amount of steel equal to approximately 7 percent of one year's output of the U.S. steel industry. (Metal Statistics 1979) Large-scale production of photovoltaic equipment based on silicon would not strain any raw material resource but would vastly exceed production of existing silicon purification plants.

CONSTRUCTION OF GOALS

The previous discussion has made it clear that uncertainties preclude any confident forecast of the solar electric capacity likely to be installed during the next twenty years. The penetration of solar electric generation depends on the success of research programs, resolution of the noneconomic problems identified in the previous section, growth in demand for electricity, and the success of policies designed to encourage solar investments. No analysis can provide exact prognostications at these areas.

The goals selected here will be based on assumptions that are presented here explicitly. A choice of goals is important to the construction of a federal program to since it permits an organized approach to the development of research priorities and priorities in programs, such as the FPMA work described earlier. The goals can also be used as the basis of an evaluation of the success of different implementation schemes. The results of a preliminary analysis of the information available have led to the solar electric goals summarized in Table 4.5.

Section Five
The Gas Industry

BACKGROUND

While the electric industry is at a crucial point, the gas industry is also at a crossroads. Conventional domestic production in the lower 48 states has been declining, although the production of some remains stable, or even increases, because of the price incentives provided by the Natural Gas Policy Act of 1978. At any rate, production will not be sufficient to supply the enormous demand for gas service that would exist if all oil users were free to switch to gas. A variety of more expensive gas sources--coal gasification, liquefied natural gas, Alaskan natural gas, Canadian and Mexican imports, and biomass-produced gas all could provide sizable increments of gas supplies in the next 10-20 years. In a completely free market, those supplies that could compete successfully by displacing oil (because of their cost advantage) would be developed. Federal control over gas prices, over the develoment of each of these alternatives, and over the use of gas by various consumers who are now using oil instead, results in a situation which is unusual in the U.S. economy since FUA and other current law has the effect of placng the decision about whether to develop these sources of gas in the hands of federal regulatory agencies, rather than in the marketplace during the next five years. If a rapid return to market economics in gas is not possible, DOE should move immediately to develop a clear and realistic policy toward the development and use of new gas resources.

RENEWABLE GAS SUPPLY

The procedures used to estimate the supply potential of gas produced from renewable resources involved estimating the supplies of biomass feedstocks that could be obtained given a range of feedstock prices and accounting for efficiency losses in conversion and processing to estimate the quantities and costs of gas production. Biomass feedstocks include wet biomass, such as animal manures and high-moisture crop residues and dry biomass, such as forest products, low-moisture crop residues, and municipal solid wastes. Conversion processes included biochemical processes (anaerobic digestion) for wet biomass and thermochemical processes (gasification) for dry biomass. The assessment generally assumed existing conversion processes and marginal improvements in the production of biomass feedstocks. (These are all discussed in greater detail in the chapter on industry.)

Between 0.2 and 4.0 quad of gas could be produced at gas prices ranging between \$5.9 and \$8.4/MBtu. Although the gas supply potential is large, it is probably unwise to establish specific goals. Biomass resources have many competing uses, and it would be best to allow the market to allocate them to their highest value applications.

This assessment also ignored the production yields that could be obtained from advanced biochemical processes. These processes might significantly reduce processing costs and might allow the production of a mix of products. Finally, this analysis ignored cashflow and financial constraints that might affect feedlot operators. Because the livestock industry is cyclical, feedstock supplies will vary from year to year. These annual variations will affect the optimal size of gas processing plants and may limit the total amount of gas that can be produced from manure.

In the future, it may be possible to generate gas, or substitutes for gas, from a variety of new technologies:

o hydrogen can be generated by the electrolysis of water using electricity derived from wind, photovoltaic, or other electric devices. (This technology has been proven but requires a reduction in the cost of solar electricity and improved electrolysis efficiencies.)

o hydrogen can be generated directly from sunlight using solid state electrodes or complex molecules. Experimental results are promising but measured efficiencies are low.

o hydrogen can be produced by bacteria or blue-green algae actng on a variety of organic materials.

o materials other than hydrogen are also suitable for storing the energy received from sunlight in the form of chemical energy. Active research is investigating a variety of such chemicals.

Policies to encourage the more efficient use of gas and the use of gas from renewable resources must:

o Rationalize gas pricing so that residential consumers do not find their price decreasing when oil prices increase; this will require amending the NGPA.

o Allow the use of gas in cogenerating facilities. (See industrial policy section)

o Permit the resale of gas saved by utilities and industry at the market clearing price to. This would accelerate deregulation of gas prices. (This is possibly allowed in experimental program in the Energy Security Act of 1980)

o Provide guidelines to assist gas utilities to invest wisely in energy conservation and solar energy on the customer's side of the meter and to sell saved gas at the clearing price. This can be done under the terms of the Energy Security Act of 1980. Specific recommendations could follow the previous discussion reviewing electric utility options.

Appendix
BASELINE ASSUMPTIONS

I
ALTERNATIVE GNP PROJECTIONS

GNP projections are made using the following expression:

$$G = \frac{P_I P_{NI} L}{P_I + (P_{NI} - P_I)(G_I/G)}$$

where

$$G = GNP = G_I + G_{NI}$$

G_I = industrial value added (gross produce originating)*

G_{NI} = nonindustrial value added (gross product originating)

L = employment (full time equivalent) = $L_I + L_{NI}$

L_I = industrial employment*

L_{NI} = non-industrial employment

$P_I = G_I/L_I$ = industrial productivity

$P_{NI} = G_{NI}/L_{NI}$ = nonindustrial productivity

The parameters P_I, P_{NI}, and (G_I/G) are given historical values from each of three periods (1953-1978, 1968-1978, and 1953-1968) and L is projected on the basis of expected demographic trends (based on an average of Census Series II and III projections).

---

*The industrial sector consists of agriculture, forestry, and fisheries; mining, construction; manufacturing.

369

Table 1.

	1953-1968	1968-1978	1973-1978	1953-1978	53-68	68-78	73-78	53-78
1. Gross National Product (G)	3.72		2.80	3.48	0.986	0.969	0.870	0.993
2. Industrial GPO(G_I)*	3.37	1.65	2.10	2.86	0.949	0.783	0.593	0.969
3. Industrial Productivity	2.69	1.25	1.55	2.07	0.988	0.876	0.756	0.977
4. Non-Industrial Productivity (P_{NI})	1.59	1.17	0.72	1.29	0.980	0.954	0.837	0.981
5. Total Productivity	2.02	1.18	0.98	1.57	0.990	0.939	0.796	0.981

 * GPO = Gross Product Originating, the value added measure used in the National Income and Product Accounts of the United States.

** Productivity is defined here as GPO per full time equivalent employee (FTEE), as FTEE is defined in the National Income and Product Accounts.

Table 2a. ALTERNATIVE GNP GROWTH RATE PROJECTIONS, 1978-2000
(percent/year)

Basis	4 UE*	5% UE*	6% UE*
10 year trend 1968-1978	2.21	2.16	2.11
10 year trend (no change in mix)	2.24	2.19	2.14
25 year trend (1953-1978)	2.53	2.48	2.43
25 year trend (no change in mix)	2.57	2.52	2.47
"Good times" trend (1953-1968)	2.93	2.88	2.83

Table 2b. ALTERNATIVE PRODUCTIVITY GROWTH RATES PROJECTIONS, 1978-2000

Basis	4% UE*
10 year trend (1968-1978)	1.17
10 year trend (no change in mix)	1.20
25 year trend (1953-1978)	1.48
25 year trend (no change in mix)	1.52
"Good Times" trend (1953-1968)	1.88

 * UE = unemployment. The alternative unemployment rates are reached by 1985 and sustained thereafter.

Table 3. GNP PROJECTIONS BASED ON 25 YEAR (1953-1978) TRENDS

	GNP(10^{12} 1972 \$)		
	4% UE by 1985	5% UE by 1985	6% UE by 1985
1978	1.399	1.399	1.399
1985	1.750	1.732	1.714
1990	1.973	1.952	1.931
2000	2.425	2.400	2.374
2010	2.976	2.945	2.914

	GNP Growth Rates (%/YEAR)		
	4% UE by 1985	5% UE by 1985	6% UE by 1985
1978-1985	3.25	3.10	2.94
1978-1990	2.91	2.81	2.72
1978-2000	2.53	2.48	2.43
1978-2010	2.39	2.35	2.32
1985-1990	2.43	2.42	2.43
1985-2000	2.20	2.20	2.20
1985-2010	2.15	2.15	2.15
1990-2000	2.08	2.09	2.09
1990-2010	2.08	2.08	2.08
2000-2010	2.07	2.07	2.07

V

Table 4. GNP PROJECTIONS BASED ON 10 YEAR (1968-1978) TRENDS

GNP(10^{12} 1972 \$)

	4% UE by 1985	5% UE by 1985	6% UE by 1985
1978	1.399	1.399	1.399
1985	1.709	1.691	1.673
1990	1.897	1.877	1.857
2000	2.261	2.237	2.213
2010	2.710	2.682	2.654

GNP Growth Rates (%/Year)

	4% UE by 1985	5% UE by 1985	6% UE by 1985
1978-1985	2.90	2.74	2.59
1978-1990	2.57	2.48	2.39
1978-2000	2.21	2.16	2.11
1978-2010	2.09	2.05	2.02
1985-1990	2.11	2.11	2.11
1985-2000	1.88	1.88	1.89
1985-2010	1.86	1.86	1.86
1990-2000	1.77	1.77	1.77
1990-2010	1.80	1.80	1.80
2000-2010	1.83	1.83	1.83

Table 5. GNP PROJECTIONS BASED ON "GOOD TIMES" TRENDS (1953-1968)

GNP(10^{12} 1972 \$)

	4% UE by 1985	5% UE by 1985	6% UE by 1985
1978	1.399	1.399	1.399
1985	1.799	1.780	1.762
1990	2.068	2.046	2.025
2000	2.643	2.615	2.588
2010	3.369	3.334	3.299

	4% UE by 1985	5% UE by 1985	6% UE by 1985
1978-1985	3.66	3.50	3.35
1978-1990	3.31	3.22	3.13
1978-2000	2.93	2.88	2.83
1978-2010	2.78	2.75	2.72
1985-1990	2.83	2.82	2.82
1985-2000	2.60	2.60	2.60
1985-2010	2.54	2.54	2.54
1990-2000	2.48	2.48	2.48
1990-2010	2.47	2.47	2.47
2000-2010	2.46	2.46	2.46

Table 6. ALTERNATIVE EMPLOYMENT PROJECTS

Employment* (millions)

	4% UE**	5% UE**	6% UE**
1978	83.13	83.13	83.13
1985	93.9	92.9	91.9
1990	98.0	97.0	96.0
2000	104.2	103.1	102.0
2010	111.0	109.8	108.7

Employment Growth Rates (%/year)

	4% UE**	5% UE**	6% UE**
1978-1985	1.76	1.60	1.44
1978-1990	1.38	1.29	1.21
1978-2000	1.03	0.98	0.93
1978-2010	0.91	0.87	0.84
1985-1990	0.86	0.87	0.88
1985-2000	0.70	0.70	0.70
1985-2010	0.67	0.67	0.67
1990-2000	0.61	0.61	0.61
1990-2010	0.62	0.62	0.62
2000-2010	0.63	0.63	0.64

* Employment is given in terms of full-time equivalent employees (FTEE), as defined in the National Income and Product Accounts.

** UE - unemployment. The alternative unemployment rates are reached by 1985 and sustained thereafter.

Table 7. HISTORICAL DATA (Industrial Sector)

	Industrial GPO* (billion 1972$)	Industrial Employment** (million)	Industrial Productivity (Thousand 1972$ per FTEE)
1953	239.5	22.559	10.617
54	229.6	21.225	10.817
55	250.0	21.783	11.477
56	254.2	22.139	11.482
57	254.5	22.039	11.548
58	241.1	20.425	11.804
59	261.2	21.271	12.280
1960	264.8	21.336	12.411
61	264.8	20.856	12.697
62	281.8	21.451	13.137
63	298.9	21.584	13.848
64	317.4	21.800	14.560
65	341.4	22.801	14.973
66	361.4	24.033	15.038
67	363.9	24.142	15.073
68	381.3	24.538	15.539
69	389.0	25.055	15.526
1970	371.0	24.097	15.396
71	376.0	23.331	16.116
72	401.1	24.003	16.710
73	426.4	25.437	16.762
74	402.6	25.409	15.845
75	382.9	23.221	16.489
76	412.7	24.185	17.064
77	440.6	25.114	17.544
78	460.3	26.387	17.444

*GPO = Gross product originating, the value added measure used in the
National Income and Product Accounts

**Full-time equivalent employees.

Table 8. HISTORICAL DATA (Non-Industry)

	Non-Industrial GPO* (billion 1972$)	Non-Industrial Employment** (million)	Non-Industrial Productivity (Thousand 1972$ per FTEE)
1953	382.3	32.009	11.944
54	384.1	31.518	12.187
55	404.8	32.175	12.581
56	414.6	33.094	12.528
57	426.4	33.623	12.682
58	438.4	33.417	13.119
59	459.2	34.071	13.478
1960	472.0	35.016	13.480
61	490.5	35.376	13.865
62	517.3	36.292	14.254
63	531.8	37.004	14.371
64	557.0	38.064	14.633
65	584.5	39.473	14.808
66	619.6	41.650	14.864
67	643.8	42.639	14.753
68	670.5	45.077	14.875
69	689.8	46.340	14.886
1970	704.3	46.754	15.064
71	731.5	47.238	15.485
72	770.0	48.345	15.927
73	808.6	50.047	16.157
74	815.2	51.067	15.963
75	819.4	51.153	16.019
76	860.3	52.539	16.375
77	899.9	54.355	16.556
78	938.9	56.738	16.548

Table 9. HISTORICAL DATA (Total Economy)

	GNP (billion 1972$)	L (million FTEE)	GNP/L (thousand 1972$ per FTEE)
1953	621.8	54.568	11.395
54	613.7	52.743	11.636
55	654.8	53.958	12.135
56	668.8	55.233	12.109
57	680.9	55.662	12.233
58	679.5	53.842	12.620
59	720.4	55.342	13.017
1960	736.8	56.352	13.075
61	755.3	56.232	13.432
62	799.1	57.743	13.839
63	830.7	58.588	14.179
64	874.4	59.864	14.606
65	925.9	62.274	14.868
66	981.0	65.683	14.935
67	1007.7	67.781	14.867
68	1051.8	69.615	15.109
69	1078.8	71.395	15.110
1970	1075.3	70.851	15.177
71	1107.5	70.569	15.694
72	1171.1	72.348	16.187
73	1235.0	75.484	16.361
74	1217.8	76.476	15.924
75	1202.3	74.374	16.166
76	1273.0	76.724	16.592
77	1340.5	79.469	16.868
78	1399.2	83.125	16.832

Table 10. HISTORICAL DEMOGRAPHIC DATA AND PROJECTIONS

	(a) P-16	(b) N.I.P	(c) b/a	(d) C.L.F.	(e) T.L.F.	(f) T.L.F.P.R.	(g) C.L.F.P.R.	(h) C.E.	(i) T.E.	(j) FTEE	(k) (j/i)
1960	109.141	106.745	.977	62.208	63.858	0.599	.592	58.918	60.568	48.527	.801
55	114.276	112.732	.986	65.023	68.072	0.604	.593	62.170	65.219	53.958	.827
1960	121.835	119.759	.983	69.628	72.142	0.602	.594	65.788	68.292	56.352	.825
61	123.404	121.343	.983	70.459	73.031	0.602	.593	65.746	68.318	56.232	.823
62	124.864	122.981	.985	70.614	0.597	.588	.588	66.702	69.530	57.743	.830
63	127.275	125.154	.984	71.833	74.571	0.596	.587	67.762	70.500	58.588	.831
64	129.427	127.224	.983	73.091	75.830	0.596	.587	69.305	72.044	59.864	.831
65	131.542	129.236	.982	74.455	77.178	0.597	.589	71.088	73.811	62.274	.844
66	133.651	131.180	.982	75.770	78.893	0.601	.592	72.895	76.018	65.683	.864
67	135.905	133.319	.981	77.347	80.793	0.606	.596	74.372	77.818	67.781	.871
68	138.171	135.562	.981	78.737	82.272	0.607	.596	75.920	79.455	69.615	.876
69	140.462	137.841	.981	80.734	84.239	0.611	.601	77.902	81.408	71.395	.877
1970	142.956	140.182	.981	82.715	85.903	0.613	.604	78.627	81.815	70.851	.866
71	145.435	142.596	.980	84.113	86.929	0.610	.602	79.120	81.937	70.569	.861
72	147.908	145.775	.986	86.542	88.991	0.610	.604	81.702	84.151	72.348	.858
73	150.489	148.263	.985	88.714	91.040	0.614	.608	84.409	86.735	75.373	.869
74	153.058	150.827	.985	91.011	93.240	0.618	.612	85.935	88.164	76.342	.866
75	155.724	153.449	.985	92.613	94.793	0.618	.612	84.783	86.963	74.374	.855
76	158.324	156.048	.986	94.773	96.917	0.621	.616	87.485	89.629	76.724	.856
77	161.047	158.559	.981	97.401	99.534	0.628	.623	90.546	92.679	79.469	.857
78	163.588	161.058	.985	100.420	102.537	0.637	.632	94.373	96.490	83.125	.861
79		163.620		102.908	104.996	0.642	.629	96.945	99.033		
1980	168.3	165.8	.985	105.3	107.4	0.648	.635				
1985	177.	174.3	.985	111.6	113.7	0.65	.64		109.2	93.	.86
1990	185.	182.2	.985	116.6	118.7	0.65	.64		114.0	98.0	.86
1995											
2000	197.	194.0	.985	124.2	126.3	0.65	.64		121.2	104.2	.86
2010	210.	206.9	.985	132.4	134.5	0.65	.64		129.1	111.0	.86

Notes to Table 10:

(a) P_ is the population aged 16 and over. The projections are based on the average of U.S. Census Series II and III.
(b) N.I.P._16 is the non-institutionalized population aged 16 and over.
(c) (b/a). The projections are based on a future N.I.P. of 98.5%
(d) C.L.F. is the civilian labor force, e.g., those civilians working or seeking work.
(e) T.L.F. is the total labor force (including military). The projections are based on military employment of 2.1 million.
(f) T.L.F.P.R. is the total labor force participation rate = (e)/(b).
(g) C.L.F.P.R. is the civilian labor force participation rate = (d)/(b). It is assumed to saturate at 0.64 in the projections. The participation rate can be expected to saturate because of the relative growth in the population aged 65 and over. In 2000 it is expected that there will be 32 million people aged 65 and over. Thus if the number of people aged 65 and over in the labor force is negligible, then about 65% of those in the age range 16-64 would be in the civilian labor force in 2000, with a 64% participation rate.
(h) C.E. is the civilian employment.
(i) T.E. = C.E. +2.1 million is the total employment. The projections are based on a 4% total unemployment rate for 1985 and beyond.
(j) F.T.E.E. stands for full-time equivalent employees, as defined in the National Income and Products Accounts.
(k) The projections are based on a FTEE/T.E. ration of 0.86.

BASELINE ASSUMPTIONS

II

CALCULATION OF LEVELIZED PETROLEUM PRODUCT AND NATURAL GAS
LONG RUN MARGINAL COSTS

The following formula should be used to calculate the levelized long run maginal costs of petroleum products and natural gas (PL) for a project started t years from 1980 and with a duration of n years. The formula is as follows:[1]

$$PL = Po + a(t + b_n)$$

where

PL = levelized product marginal cost
Po = estimated 1980 product marginal costs
a = annual product cost increase assuming linear escalation
b_n = arithmetic gradient factor for $i = .03$ (discount rate in real terms) and n number of years in the analysis.
t = date which project starts - 1980

Po, a, and the long run marginal cost in year 2000 (P*) are shown for various petroleum products and natural gas in Table 1. Both a marginal and a shadow cost are shown for each product. The shadow cost is representative of the social cost of each product.

Factors of b_n are shown in Table 2.

It is a simple matter to compute the levelized cost of a product from the above tables. For example, to consider a solar system competing with residential fuel oil which is to be constructed in 1985 and have a life time of 20 years: $t = 1985 - 1980 = 5$; $n = 20$; Po = \$9.2/MBTU (shadow price); $a = .135$; and $b_{20} = 8.523$.

Substituting:

$$PL = 9.2 + .135(5 + 8.523) = \$11/MMBTU$$

Table 3 provides the parametes for calculating average U.S. levelized electricity costs. Tables 4 and 5 provide parameters for calculating regional electricity prices. All of the numbers in these tables follow the approach outlined in Attachment A.

It should be remembered that in conducting analyses of the present value or annual costs of both solar and conventional systems, a 3% real discount rate should be used for consistency. Also, a price index of 118 should be used to convert from today's dollars to 1978 dollars.

The following conversion factors are used to convert petroleum product costs from \$/BBL to \$/MMBTU:

crude	1 BBL = 5.8 MMBTU
gasoline	1 BBL = 5.25 MMBTU
distillate	1 BBL = 5.825 MMBTU
residual	1 BBL = 6.29 MMBTU

---

[1] In general $b_n = \dfrac{1}{i} - \dfrac{n}{(1 + i)^n - 1}$

374

To calculate 1980 marginal costs the following esti-mates were divided by 1.18 to account for inflation: crude = $30/BBL, an approximation of the average January 1980, refiner acquisition cost for imported crude; natural gas = $2.15/MMBTU, the January 1980, FERC allowable price for new on-shore gas supplies. The appropriate markups for each product are added to obtain retail marginal costs.

In the case of natural gas two marginal costs are shown. The values in parentheses include the costs of transmission and distribution, whereas the other values represent wellhead prices. The latter should be used in determining the cost justifiable levels of conservation and solar investments in instances where there is a hookup to the gas system. The values in parentheses should be used for assessments of stand-alone systems.

Table 2.   ARITHMETIC GRADIENT FACTORS (A/G) FOR i = .03 AND n YEARS

n	b_n	n	b_n
2	0.493	15	6.450
3	0.980	20	8.523
4	1.463	25	10.477
5	1.941	30	12.314
10	4.256		

Table 1.   PARAMETERS FOR CALCULATING LEVELIZED PETROLEUM PRODUCT AND NATURAL GAS COSTS ($/MMBTU in 1978 $)

| | Po (January 1980) | | P* (2000) | | |
	Marginal	Shadow	Marginal	Shadow	a
Gasoline	7.2	10.6	10.2	13.6	.15
Diesel Fuel	5.3	8.7	8.0	11.4	.135
Residential Heating Oil (No. 2)	5.8	9.2	8.5	11.9	.135
Industrial Distillate	5.5	8.9	8.2	11.6	.135
Residual (No. 6)	4.4	7.8	6.9	10.3	.125
Natural Gas: Residential	1.8 (3.9)	5.2 (7.3)	5.0 (7.1)	8.4 (10.5)	.16
Natural Gas: Commercial	1.8 (3.0)	5.2 (6.4)	5.0 (6.3)	8.4 (9.7)	.16
Natural Gas: Industrial	1.8 (2.7)	5.2 (6.1)	5.0 (5.9)	8.4 (11.8)	.16

Notes: $a = \dfrac{P* - Po}{20}$

Table 1a. REGIONAL MARGINAL COST VARIATIONS (1978$) FOR 1980

	Coal Costs[a] ($/$10^6$ Btu)	Transmission and[b,c] Distribution Costs (mills/kwh)		Total Electricity Cost[c,d] (mills/kwh)	
		R/C	Industrial	R/C[e]	Industrial[f]
New England	1.60	9.7 (32.0)	7.9 (11.2)	60.1 (82.4)	46.8 (50.1)
Middle Atlantic	1.69	13.1 (38.8)	8.6 (13.5)	64.4 (90.1)	48.4 (53.3)
South Atlantic	1.70	9.0 (23.3)	7.9 (9.2)	60.4 (74.7)	47.9 (49.2)
East North Central	1.62	7.0 (25.2)	6.9 (7.3)	57.6 (75.8)	45.0 (46.4)
East South Central	1.70	5.6 (17.2)	4.9 (5.9)	57.0 (68.6)	44.9 (45.9)
West North Central	1.25	9.7 (24.3)	8.6 (10.7)	56.4 (71.0)	43.7 (45.8)
West South Central	1.30	6.1 (18.6)	5.6 (8.3)	52.3 (65.8)	41.3 (44.0)
Mountain	0.98	9.9 (26.6)	9.2 (12.2)	53.7 (70.4)	41.4 (44.4)
Pacific	1.42	11.7 (32.2)	10.6 (11.4)	60.2 (81.2)	47.6 (48.4)
U.S. Average	1.45	8.8 (25.4)	7.9 (9.4)	57.6 (74.2)	45.2 (46.7)

Notes:

(a) The replacement costs given for 1972 in M.L. Baughman and D.J. Bottaro, "Electric Power Transmission and Distribution Systems: Costs and Their Allocation," Center for Energy Studies Report, The University of Texas at Austin, July 1975, are converted to 1978 dollars by using as a deflator the Handy-Whitman Index of Public Utility Construction Costs. The values in Baughman and Bottaro are thus multiplied by 1.8.

(b) The first number of each entry involves only the transmission component of T&D while the number in parenthesis involves both transmission and distribution costs. The former number should be used in estimating the economically justified level of investment in a consevation or solar system when the alternative system is tied to the utility; the number in parenthesis should be used in assessing a stand-alone system.

(c) Transmission and distribution losses are assumed to be 9%.

(d) The generation component of the cost of delivered electricity (for 9% T&D losses) for R/C customers for a utility where 60% of the electricity is consumed by R/C customers and 40% by industrial customers (roughly the U.S. average for 1978) is estimated as follows:

$$(\text{Cost})_{R/C} = \frac{(\text{Cost})_{Ave} - 0.4 \,(\text{Cost})_{Ind.}}{0.6 \times 0.91}$$

It is assumed here (for simplicity) that industrial customers consumer only baseload electricity. Thus since (see Table 3)

$$(\text{Cost})_{Ave} = 0.55 \,(\text{Cost})_{Baseload} + 0.427 \,(\text{Cost})_{Cycling} + 0.023 \,(\text{Cost})_{Peaking}$$

we obtain

$$(\text{Cost})_{R/C} = 0.275 \,(\text{Cost})_{Baseload} + 0.782 \,(\text{Cost})_{Cycling} + 0.042 \,(\text{Cost})_{Peaking}$$

(e) Assuming (for simplicity) that industrial customers consume only baseload electricity.

Table 3.   PARAMETERS FOR CALCULATING U.S. AVERAGE ELECTRICITY LEVELIZED COSTS (Mills/Kwh)

	Po (1980)		P* (2000)		a
	Marginal		Marginal		
Residential/ Commercial	57.6	(74.2)	63.7	(80.3)	0.305
Industrial	45.2	(46.7)	49.9	(51.4)	0.235
Average*	52.6	(63.2)	58.2	(68.7)	0.280

*Assuming 60% Residential/Commercial and 40% Industrial.

Table 4. ECONOMICS OF POWER GENERATION IN 1980

	Baseload (Coal)	Cycling (Coal	Peaking Turbine (Resid)
Capital Cost (1978 $/kw)	750[a]	820[a]	210[e]
Plant Life (years)	30[a]	30[a]	20[e]
Annual Capital Charge Rate	0.106[b]	0.106[b]	0.120[b]
Average Annual Heat Rate (Btu/kWh)	9735[a]	10150[a]	14000[e]
Average Annual Capacity Factor (%)	65	45[d]	7[f]
Generation Costs (1978 mills/kWh)			
Capital	14.0	22.0	41.1
Fuel	14.1[c]	14.7[c]	82.6[g]
O&M	5.8[a]	6.9[a]	3.6[e]
TOTAL	33.9	43.6	127.3

Notes:

[a]Data obtained from Exhibit 8-4b in Technical Assessment guide, special report prepared by the Technical Assessment Group of the EPRI Planning Staff, Electric Power Research Institute Report No. OS-1201-SR, July, 1979. The coal plants are coal-steam units with wet lime/limestone FGD. The baseload plant is a high efficiency supercritical plant with a 3500-pis steam system. The cycling plant is a 4200-psi subcritical plant capable of cycling the minimal loss of efficiency. The coal costs are the middle of the range cited.

[b]The annual capital charge rate (with no subsidies and straight line depreciation) is

$$B_c = \frac{C(i,N)}{1-t} \qquad \frac{t}{NC(i,N)} = t_p + r_i - r_r$$

where $i = A_e r_e + (1-t) A_d r_d$ = constant dollar utility discount rate = 0.0344
$A_e (A_d)$ = fraction of investment from equity (debt) = 0.50(0.50)
t = combined federal/state income tax rate = 0.50
N = plant life
$r_e (r_d)$ = constant dollar annual rate of return on equity (debt) = 0.055 (0.028)
T_p = annual rate for ad valorem, property, and other taxes = 0.025
r_i = insurance rate = 0.0025
r_r = capital replacement rate = 0.0035

$$C(i,N) = \frac{i}{1 - (1 + i)^{-N}} = \text{capital recovery factor}$$

Thus, $B_c = 1.106$ for N = 30 years and $B_c = 0.120$ for N = 20 years.

[c]For a national average 30 year levelized coal marginal cost of $1.45/10^6$ Btu. See Table 4B.
[d]See not g, Table 3.
[e]Data from Exhibit 8-9b of Technical Assessment Guide (see note a).
[f]See note f, Table 3.
[g]The levelized cost or reside is $P_1 = 4.4 + .125 b_{30} = \$5.9/10^6$ Btu.

Table 5. ECONOMICS OF POWER GENERATION IN 2000[a]

	Baseload (Coal)[a]	Cycling (Coal	Peaking Turbine (Resid)
Capital Cost (end of year 1978$/kw)	750	820	210
Plant Life (years)	30	30	20
Annual Capital Charge Rate	0.106	0.106	0.120
Ave. Annual Heat Rate (Btu/kWh)	9735	10,150	14,000
Ave. Annual Capacity Factor (%)	65	45	7
Generation Costs (1978 mills/kWh)			
Capital	14.0	22.0	41.1
Fuel	18.4[b]	19.2[b]	117.6[c]
O&M	5.8	6.9	3.6
TOTAL	38.2	48.1	162.3

Notes:

[a] All entires are obtained as in Table 2a, expected where indicated otherwise.

[b] For a national average 30 year levelized coal marginal cost of $1.89/10^6 Btu. See Table 4b.

[c] The levelized cost of reside is
$$PL = 6.9 + .125b_{30} = \$8.4/10^6 \text{ Btu.}$$

Table 6. CHARACTERISTICS OF UTILITY [a]

Annual Load Factor (%)	59[a]
Reserve Margin (%)	25[b]
Distribution of Capacity (%)	
Baseload	40[c]
Cycling	45
Peaking	15[d]
TOTAL	100

Distribution of Electricity Production (%)	
Baseload	55.0[e]
Cycling	42.7[g]
Peaking	2.3[f]
TOTAL	100

[a] The synthetic model utility shown here is System A (a summer peaking system with loads evnly distributed throughout the service territory) as given in Synthetic Electric Utility Systems for Evaluating Advanced Technologies, EPRI Report E-285, February 1977. The load duration curves for spring/fall, summer, and winter are shown in Figures 4-24, 4-26, and 4-28.

[b] Recently (see Figure   ) the reserve margin has been much higher than this. We assume there is a 25% reserve margin for all kinds of generating capacity at the time of the summer peak.

[c] The winter (spring/fall) peak demand for System A is 0.78 times (0.73 times) the summer peak demand. Thus from inspection of Figures 4-24, 4-26, and 4-28 w see that the baseload demand is
summer baseload demand = 0.40 x summer peak
winter baseload demand = 0.78 x 0.55 x (summer peak) = 0.43 x (summer peak)
spring/fall baseload demand = 0.73 x 0.52 x (summer peak) = 0.38 x (summer peak)

Thus we assume that throughout the year

baseload demand   0.4 x (summer peak)

[d] We assume that peaking capacity accounts for 15% of peak demand in each season.

[e] At 65% capacity factor the fraction of all electricity produced by nuclear plants is

$$0.65 \times 0.40 \times \frac{1.25}{0.59} = 0.55$$

[f] If the peaking unit accounts for 15% of peak demand in any seanson the (from inspection of Figures 4-24, 4-26, and 4-28) the fraction of all the electricity produced by the peaking unit is approximately

$$1/2 \times 0.15 \quad (\frac{5}{12} \times 0.26) + (0.73 \times \frac{4}{12} \times 0.43)$$
$$\text{summer} \qquad\qquad \text{spring/fall}$$

$$+ (0.78 \times \frac{3}{12} \times 0.45 = 0.023$$
$$\text{winter}$$

The average capacity factor for the peaking unit is thus

$$\frac{0.023 \times 0.59}{0.15 \times 1.25} = 0.072$$

[g] The agerage capacity factor for the cycling unit is

$$\frac{0.427 \times 0.59}{0.45 \times 1.25} = 0.448$$

APPENDICES TO CHAPTERS ONE THROUGH FOUR

Appendix to Chapter One
BUILDINGS

APPENDIX A

ANALYSIS OF RESIDENTIAL INCENTIVES

Tables A.1 and A.3 provide a technique for comparing the impact and cost of several different strategies for subsidizing decentralized energy investments. The consumer costs computed represent the present value of all payments associated with the solar investment. This includes the down payment, annual mortgage payments, property taxes, insurance, and all tax credits and other subsidies. The federal costs comprise the value of all credits and subsidies, including the value of all taxes forgone because of tax deductions. It should be noted, however, that the calculation does <u>not</u> include any accounting of the additional federal revenues which would almost certainly result from an increase in solar investments. These unaccounted values include the additional taxes generated from the increase in solar and conservation scales (both indirectly and directly), potential deductions in health care resulting from improved environmental quality, credits resulting from the fact that investments in cost-effective solar equipment can reduce the inflationary pressures created by imported energy and other factors. The federal costs illustrated in the tables represent only the outlays directly associated with the installation.

Examining the first rows in Tables A.1 and A.3, it can be seen that if a unit costing $1000 is financed with a loan paying 10% interest covering 75% of the value of the house, the consumer cost would be $1055, if the consumer pays taxes at a marginal rate of 20%, $919 if the consumer pays taxes at a marginal rate of 40%, and $1192 if the consumer does not file an itemized return. The federal costs shown in this case represent only the value of the reduction available by deducting interest payments. It is interesting to note that for the high-income individual, the value of the

interest subsidies already available through the federal tax laws are the functional equivalent to an investment tax credit of 27% or equivalently an interest subsidy of approximately 4%. Higher income groups receive higher effective credits.

The statistics illustrated in the example just cited also demonstrate another basic point which will appear repeatedly in the following discussion. If the Federal Government and the consumer discount future expenses at the same rate, as assumed in the calculations exhibited here, the sum of federal costs and consumer costs is always the same. Any incentive which has a small impact on consumer costs would also be relatively inexpensive for the government. If the Federal Government's discount rate is less than that of the consumer, the present value of federal costs will be below the present value of the benefits to the consumer (in all the cases examined).

Tables A.2 and A.2a illustrate another technique for comparing consumer capital expenses. The costs illustrated in these tables represent the consumer and federal expenses which would result if the person purchasing a residential solar unit sells the house and solar equipment after owning them for seven years-- approximately the average length of time a house is owned by a single owner. In this calculation it is assumed that the solar equipment has lost 1/3 of its value (in constant dollars) by the time the unit is resold.

Tables A.1 and A.3 also illustrate consumer and federal costs for four types of subsidies: investment tax credits ranging from 22-30% available through the existing NEA for some types of solar investments, an interest subsidy program, in which the federal government gives grants to banks allowing a reduction in interest rates paid on loans for equipment (in the notation of the tables an "interest subsidy" of 6% means that the commercial interest rates are reduced by 6% e.g., from a commercial rate of 10% to a subsidized rate of 4%), and finally a "principal subsidy" in which a federal grant is made available to reduce the principal vale of loans made for solar equipment. Principal subsidies have much the same direct impact as interest subsidies. For example, a consumer's monthly payments will be the same for a 20-year loan with a 6% interest subsidy and a loan paying 10% interest with a 37.4% principal subsidy. It will be seen, however, that the principal subsidies offer significant advantages under some circumstances.

Investment tax credits clearly offer the advantage of being simple to administer; the tables indicate that they can have a significant impact on perceived capital costs. The use of the tax system to provide such credits, however, has a number of limitations.

The most glaring limitation involves the equity of the credits. It is difficult for low income purchasers to take advantage of them. In the first place, over half of the taxpayers in the United States do not have a tax liability large enough for them to take advantage of the maximum credit even by applying the credit to two successive years of taxation. Secondly, taking advantage of the credits requires an initial investment which must be tied up for a number of months while tax claims are processed. Persons owing less than $2200 in taxes must wait up to two years for the return of their capital.

Even persons in higher income groups are less likely to be lured into an investment with a credit which requires a significant outlay of cash, some of which may be repaid in a number of months, than by a credit which is available at the moment of purchase.

The Bank eliminates many of the problems encountered with the tax credits, but can create others. Use of the Bank working directly through existing lending institutions avoids complicating the tax laws, it provides a subsidy whose benefits are realized as soon as the system begins to operate, and the relatively low month payments it offers make life-cycle cost comparisons between solar and conventional systems much more straightforward. The Bank also has the effect of offering greater subsidies to low income individuals than to high income persons. (The high income individuals already enjoy a large effective interest subsidy, because they can deduct interest from taxable income.) Tables A.1 and A.3 indicate that a 6% interest subsidy reduces consumer costs by $281 (per $1000 of solar investment) for a person not itemizing on tax returns, $215 for a person paying taxes at a marginal rate of 20% and itemizing deductions, and $150 for a person paying taxes at a marginal rate of 40%. The Bank also offers the advantage of dealing with institutions which, it is hoped, will provide the bulk of the financing for solar investments after

federal grants are no longer needed. The value of doing this should not be underemphasized. One question which is seldom asked about new subsidy programs is, how they can be gracefully phased out, and the necessary activity picked up by private enterprise. A solar subsidy operating through commercial banks provides an obvious mechanism for accomplishing this transition. A final major advantage of the Bank is that the amount of subsidy given can be directly controlled through annual appropriations--all Bank funds would be on-budget.

The impact of the Bank on consumers, however, depends critically on how the grants are given. Table A.1, for example, shows that for a 20-year loan covering 75% of the value of the solar system, the Bank provides less credit to a consumer than the tax credits now available. This is due primarily to the fact that the Solar Bank, in this case, would provide assistance only for the fraction of the equipment financed through the Bank while the tax credit applies to the full investment. The situation is reversed if the Bank permits a loan value which covers 100% of the value of the property. In this case, the 6% interest subsidy provides lower consumer costs than a 30% investment tax credit. (It must be recognized that the 30% credit applies only to a relatively inexpensive solar installation.) Table A.1a illustrates the effect of a solar bank granting 30-year loans covering 100% of the value of the system. In this case, the Bank offers significantly more powerful incentives than the tax credits.

Another technique for increasing the leverage of the Bank, is to allow a consumer to apply both for an interest subsidy and an investment tax credit. The tables illustrate a case in which the consumer is permitted a 4% interest subsidy and a 22% investment tax credit. It can be seen that this provides a greater advantage to consumers than either the tax credits or the interest subsidies taken by themselves. This "double dipping" credit would, of course, require the consumer to apply twice for credits and a significant amount of additional paperwork could result. The benefits would plainly be more difficult to interpret than the impact of a single subsidy.

Tables A.2 and A.2a illustrate another potential difficulty with the Bank. The Bank's interest advantages could be enjoyed only for the period when the house was owned. (This is a significant disadvantage for high income purchasers not strongly leveraged by the Solar Bank in the first place.) The full value of the investment tax credit, of course, is enjoyed in the first year. Selling the house before the termination of the mortgage would require the consumer to forego the full value of the subsidy. The consumer would, however, still prefer a Solar Bank loan offering 100% financing and a 6% interest subsidy to a 22% investment tax credit.

The problem associated with early sale of solar equipment are entirely offset, of course, the principal subsidy approach, since the full value of the credit is received at the moment of purchase. (In this, it retains the psychological advantage of the Solar Bank appraoch over the investment tax credit.) The tables indicate that a 40% principal subsidy would be an extremely powerful incentive for all income groups. It should also be relatively easy to administer since there would be no need to take special care in adjusting interest rates so that banks could offer attractive loans to consumers without claiming undue profits. Any windfall from the program would go to consumers. In particular, the principal subsidy would eliminate the need to require banks to repay some part of a solar loan payment given to banks, if the subsidized loan notes issued by the bank were prepaid before the termination of the note. The requirement for bank repayment has the awkward feature of reducing the net federal outlays associated with a Bank without reducing the net appropriation required in the initial years of the Bank's life.

The tables just reviewed indicate how subsidies of varying depths can be achieved. How deep a subsidy is needed to stimulate purchases and overcome the market imperfections described at the beginning of this discussion? Clearly, there is no confident answer since consumer response to solar equipment will depend on a number of factors having little to do with rational economics. The skill with which solar equipment is marketed, consumer perceptions about the future of energy costs, local building covenants, and architectural tastes will have a major impact on purchasing decisions. Precise statements about the economic impacts of the subsidies are also difficult because of the great variation in the price and quality of equipment now available, because the price of energy from nonsolar sources in different parts of the United States can vary by a factor of five or more, and because the economics is very sensitive to local climatic conditions.

Table A.1.   TOTAL PRESENT VALUE OF CONSUMER AND FEDERAL PAYMENTS
FOR A SYSTEMS VALUE OF $1000
(20-year Loan Term)

Marginal Tax Rate of Consumer	20% ($12,000 annual income)		40% ($40,000 annual income)	
	Consumer Cost	Federal Cost	Consumer Cost	Federal Cost
No credits	1,055	136	919	273
22% ITC	855	336	729	473
30% ITC	782	409	646	546
6% IS	840	352	769	423
4% IS	906	285	814	377
2% IS	978	213	864	327
4% IS + 22% ITC	706	485	614	577
40% PS	696	496	601	591

IS = Interest Subsidy
ITC = Investment Tax Credit
PS = Principal Subsidy

Assumptions:

o   Loan Terms = 20 years

o   Effective Property Tax = 2%

o   Effective Insurance Rate = 0.25%

o   Down Payment = 25% (15% for the grant program)

o   Commercial Lending Rate = 10%

o   Consumer Discount Rate 10% (current $)

o   Federal Discount Rate 10% (current $)

Table A.1a   TOTAL PRESENT VALUE OF CONSUMER AND FEDERAL PAYMENTS
FOR A SYSTEM VALUE OF $1000

(30-Year Term; 100% Financed)

Marginal Tax Rate of Consumer	20% ($12,000 annual income)		40% ($40,000 annual income)	
	Consumer Cost	Federal Cost	Consumer Cost	Federal Cost
No credits	1,008	204	803	409
22% ITC	808	404	503	609
30% ITC	735	477	530	682
6% IS	659	553	561	652
4% IS	765	447	632	580
2% IS	832	330	714	498
4% IS + 22% ITC	565	647	432	780
40% PS	674	538	536	676

IS = Interest Subsidy
ITC = Investment Tax Credit
PS = Principal Subsidy

Assumptions:

o   Effective Property Tax = 2%

o   Effective Insurance Rate = 0.25%

o   Down Payment = 0%

o   Commercial Lending Rate = 10%

o   Consumer Discount Rate 10% (current $)

o   Federal Discount Rate 10% (current $)

Table A.2   TOTAL PRESENT VALUE OF CONSUMER AND FEDERAL PAYMENTS
FOR A SYSTEM VALUE OF $1000

(Assuming the System is Sold After Seven Years)

20-Year Loan; 75% Financing (except as noted)

Marginal Tax Rate of Consumer	20%		40%	
	Consumer Cost	Federal Cost	Consumer Cost	Federal Cost
No credits	508	89	419	177
6% IS	352	326 (245)	306	372 (290)
22% ITC	308	289	219	377
4% IS + 22% ITC	202	453 (395)	142	513 (454)
40% PS	135	461	75	522

IS = Interest Subsidy
ITC = Investment Tax Credit
PS = Principal Subsidy

Assumptions:

o   Loan Terms = 20 years

o   Value of Property After 7 years

    as fraction of purchase value = 67%

o   Inflation Rate = 6%

o   Remaining Assumptions Same as Tabel 1

( ) = federal government recovers prepayment
     bonus from the lending institution

Table A.2a   TOTAL PRESENT VALUE OF CONSUMER AND FEDERAL PAYMENTS
FOR A SYSTEM VALUE OF $1000

(Assuming the System is Sold After Seven Years)

30-Year Loan; 100% Financed

Marginal Tax Rate of Consumer	20%		40%	
	Consumer Cost	Federal Cost	Consumer Cost	Federal Cost
No credits	481	115	366	230
6% IS	262	511 (334)	206	568 (390)
22% ITC	281	315	166	430
4% IS + 22% ITC	133	591 (463)	57	567 (539)
40% PS	120	477	43	554

IS = Interest Subsidy
ITC = Investment Tax Credit
PS = Principal Subsidy

Assumptions

o   Value of Property After 7 years
    as fraction of purchases value = 67%

o   Inflation Rate = 6%

o   Remaining Assumptions Same as Table 1a

( ) = federal government recovers prepayment
      bonus from lending institution.

Table A.3   TOTAL PRESENT VALUE OF CONSUMER AND FEDERAL PAYMENTS
FOR A SYSTEM VALUE OF $1000 FOR A PERSON NOT FILING AN
ITEMIZED TAX RETURN

(69% of All Returns)

	System Sold After Seven Years		System Maintained for 20 Years	
	Consumer Cost	Federal Cost	Consumer Cost	Federal Cost
No credits	596	0	1,192	0
6% IS	398	280	911	280
2% IS	528	100	1,092	100
40% PS	196	400	792	400

IS = Interest Subsidy
PS = Principal Subsidy
ITC = Investment Tax Credit

Assumptions are the same as those used in Tables A.1 & A.2 with the exception of the
tax status of the consumer.

Appendix to
Chapter Two
INDUSTRY

APPENDIX A

DATA ON ENERGY USE IN 1978

Data for energy use by sector (Standard Industrial Classification) and by energy service is available for 1974. A variety of scattered information, in addition to the total industrial energy consumption figure with breakdown by energy carrier (EIA), is available for the years from 1974 to 1978. One serious problem in extending data to 1978 is lack of detailed information for construction, mining and agriculture. This we do not respond to in detail because these sectors are not analyzed in detail in this study. Another serious problem is the lack of information on captive fuels and fuels used as feedstocks in the chemicals, petroleum refining and steel industries. Various sources are used

to obtain trends in these uses as mentioned in the notes to Table A.5.

The 1974 disaggregation of energy use is presented in Tables A.1, A.2, A.3 and A.4. The manufacturing data does not include losses associated with generation and transmission of electricity while the others do. It does include biomass energy (not included in the EIA total). In accordance with EIA accounting, use of gasoline is not included.

In Table A.5 the 1978 energy use by sector is determined. In Tables A.6 and A.7 this use is disaggregated by energy service.

Table A.1. CONSTRUCTION ENERGY USE, BY SERVICE CATEGORY (1974)

Service Category	Total Use[a] (10^{12} Btu)	Percent of Total Use
Space Heat	9	0.7
Mechanical Drive[b]	80	6.4
Direct Heat	13	1.0
Construction Vehicles	224	17.9
Raw Materials (asphalt)	925	74.0
TOTAL CONSTRUCTION	1,251	100.0

[a]Energy use estimates are exclusive of motor gasoline consumption. Energy losses in the generation and transmission of purchased electric power are allocated to the service categories where power is required.

[b]"Mechanical Drive" includes lighting and miscellaneous electrical overhead demand.

Source: Energy Information Administration 1977.

Table A.2. MINING ENERGY USE, BY SERVICE CATEGORY (1974)

Service Category	Total Use[a] (10^{12} Btu)	Percent of Total Use
Process Heat[b]	414	19.2
Mechanical Drive[c]	606	28.1
Electrolysis	3	0.1
Reservoir Repressuring[d]	1,137	52.6
TOTAL	2,160	100.0

[a]Energy use estimates are exclusive of motor gasoline consumption in this sector. Energy losses in the generation and transmission of purchased electric power are allocated to "Mechanical Drive" and "Electrolysis."

[b]According to information contained in the EIA Energy Consumption Data Base 74% of process heat is used at temperatures between 600°F and 1000°F. Eighteen percent is used below 600°F, and the remainder is not specified by temperature.

[c]Includes power produced on site, purchased power and mechanical drive via diesel engines. Also includes lighting and miscellaneous electrical demand.

[d]The Energy Information Administration reports a very large amount of consumption in an "other" category in EIA, 1977. Seventy-seven percent of this consumption is in the form of natural gas, nearly all of that natural gas consumption occurs in oil and gas extraction. In comparing this unspecified consumption with the amount of net production normally vented and returned to repressure reservoirs (5% of total production), it is clear that most of this unspecified consumption is for repressurization (see Energy Information Administration. 1978. Annual Report to Congress. DOE/EIA-0173/2. Washington, DC: U.S. Department of Energy pp 73-75). The remaining consumption of liquid fuels in the "other" category has been distributed in process heating.

Source: Energy Information Administration 1977.

Table A.3. AGRICULTURAL ENERGY USE, BY SERVICE CATEGORY (1974)

Service Category	Total Use, Exclusive of Motor Gasoline (10^{12} Btu)	Total Use[a] (10^{12} Btu)	Percent of Total Use
Space and Water Heat	50	50	100.0
Grain and Crop Drying	112	112	7.0
Lighting and Refrigeration	48	48	2.9
Irrigation[b]	393	404	25.0
Mechanical Drive (stationary)[c]	155	188	11.7
Vehicles and Field Machinery[d]	385	810	50.3
TOTAL	1,142	1,611	100.0

[a]Energy consumption includes the amount of energy actually lost in the conversion and transmission of energy via purchased electric power, and therefore actually exceeds the consumption reported by EIA 1977 or USDA/Economic Research Service 1976.

[b]Energy use for irrigation includes about 19 trillion Btu used in livestock production for water supply.

[c]Includes ventilation, milking and grain, egg, and feed handling machines.

[d]Includes farm trucks and autos, as well as tractors and harvesters. About 28% of the consumption was required for cultivation and tillage, about 4% for application of fertilizer and pesticides and 20% for harvesting. Most of the remaining energy consumption was for transportation-related uses. Diesel fuel dominates for the above field machinery uses, almost 3 1/2 times as much diesel fuel is used as gasoline for those purposes. On the other hand, gasoline is certainly the dominant fuel for transportation uses which comprise 48% of the demand in the "vehicles and field machinery" category.

Source: Energy Information Administration 1977, and supplemental information from Economic Research Service (USDA). 1976.

Table A.4. MANUFACTURING ENERGY USE, BY MAJOR MANUFACTURING ELEMENT AND SERVICE CATEGORY (1974)

Total Use[a] (10^{12} Btu)

		Element						All Mfg.
SIC	20	26	28	29	32	33	--	20-39
Service Category	Food	Paper	Chemicals	Petroleum	S, C, & G	Metals	Misc.	Total
Mechanical Drive[b]	105	258	450	147	153	470	673	2,256
Electrolysis			127			340		467
Process Heat	799	1,930	2,289	2,926	1,123	1,503	1,638	12,211
Space Heat	31	36	14		25	43	246	395
Chemical Feedstock			2,331			82		2413
Metallurgical Coal						2,363		2,363
TOTAL	935	2,224	5,211	3,073	1,301	4,801	2,557	20,106

[a]Energy consumption figures do not include energy losses in the generation and transmission of purchased electric power. Use of 1.4 quad of wood residues is added to the total consumption as defined by EIA. Figures shown are exclusive of motor gasoline consumption in manufacturing.

[b]"Mechanical Drive" includes lighting and other overhead electrical uses.

Source: Industrial Sector Technology Use Model (ISTUM), Energy and Environmental Analysis, Inc. 1978.

Table A.5. INDUSTRIAL ENERGY USE IN 1978 BY SECTOR (Quads)

	1978			1974 Without Losses[f]
	With Utility Losses[e]	Without Losses	Purchased Electricity	
Total Industry	30.22[a]	23.58[a]	2.73	23.65[b]
Constr., Mining, Agri.	4.45[e]	3.65[d]	0.33[d]	3.77[c]
Manufacturing	25.76	19.93	2.40	19.89[f]
Misc. Mfr.	4.46	2.60[h]	0.76[g]	2.53
Basic Mfr.	21.30	17.33[i]	1.63	17.36
Food	1.29	0.95[j]	0.14	0.93
Paper	2.56	2.17[k]	0.16	2.20
Chemicals	6.58	5.26[l]	0.54	5.15
Petroleum R.	3.71	3.44[m]	0.11	3.04
S.C. and Glass	1.52	1.25[n]	0.11	1.29
Primary Metals	5.64	4.25[o]	0.57	4.75

Source: Monthly Energy Review (MER); Industrial Sector Energy Use Model, Energy and Environmental Analysis, Inc. 1978 (ISTUM) (Table A.4); and Industrial Energy Use Data Book, Oak Ridge Associated Universities, 1980 (IEUDB). 1977 Census of Manufacturers, Part I Revised (COM).

[a]MER Dec. 1980 toal of 28.72 plus 1.50 for biomass. From the same source utility losses were 6.64.

[b]MER Dec. 1980 total less utility losses plus 1.40 for biomass.

[c]The sum from Tables A.1, A.2 and A.3 elclusive of motor gasoline, i.e., 4.55 less nominal utility loss of 0.74 (associated with estimated purchased electrical energy of 0.30 and a loss to electricity ratio of 2.46), plus 20.11 from Table A.4, or a total of 23.92, is scaled to 23.65 by a factor of 0.989. Thus for construction, mining and agriculture 0.989 (4.55 - .74) = 3.77.

[d]An overall drop of 3% 1974-78 is assumed for this category. Note: GPO for these three factors rose an aggregate 6% in the period. Purchased electrical energy is the residual of total industry less manufacturing.

[e]Losses associated with purchased electricity are taken to be given uniformly by MER Dec. 1980 ratio of 6.637/2.732 = 2.43.

[f]ISTUM results of 20.11 (Table A.4) scaled by 0.989; see note c.

[g]All manufacturing values are the 1977 (COM) figures scaled up by the 6% GPO increase in manufacturing for 1977-1978.

[h]Assumed to be the same as in 1977 (COM). Note: GPO in manufacturing grew 17% in the period.

[i]This residual is smaller than the initial sum of basic manufacturing estimates as discussed below (17.33 as contrasted with 17.69), but three SIC's are highly uncertain (because of use of capitive fuels and inconsistent reporting of feedstocks): chemicals, petroleum refining and primary metals. Therefore the initial estimates for these SIC's (of 5.41, 3.54 and 4.37, respectively) were reduced a total of 0.36, or by a factor of .973.

[j]Food industry use was 0.952 in 1977 (COM). It is assumed to be unchanged in 1978.

[k]Paper industry total energy and purchased electricity have been provided for 1978 by the American Paper Institute (IEUDB, p. 12-14) as 2.17 and 0.12, respectively. For electricity, however, the 1977 value of 0.15 (COM) is adopted as indicated by note g.

[l]Chemical energy use is made up of two parts. Feedstocks are scaled from ISTUM (1974) Table A.4, of 2.33 by the ratio of Federal Reserve Board Indices for Chemicals (1974 to 1978) of 1.24, times a correction

factor of 0.90 for physical growth compared to FRB growth. The result is feedstock use of 2.59 in 1978. Non-feedstock energy use is scaled from its 1974 value (ISTUM) of 2.88 in proportion with Chemical Manufacturer's Associated total energy use (IEUDB, p. 13-6) variation with the FRB index over the period 1972-1978, with a resulting non-feedstock energy use of 2.82 in 1978. The total of these initial estimates of 5.41 is scaled as per note i by a factor of 0.978.

[m]Petroleum refining energy is taken to be 0.707 million Btu/bbl of crude refined (V. O. Haynes, "Energy Use in Petroleum Refineries," Oak Ridge National Laboratory, 1976) times crude input of 14.74 million bpd (MER Aug. 1980) times 0.93 (reflecting as assumed 7% efficiency improvement. This initial estimate of 3.54 is scaled as per note i by a factor of 0.973.

[n]Stone, clay and glass energy use is the 1977 value of 1.25 (COM).

[o]Primary metals use is made up of two parts: metallurgical coal used in the steel industry, taken from American Iron and Steel Institute data (IEUDB, p. 16-10), is 2.16 in 1978. Other energy is purchased energy (Census of Manufacturers, IEUDB, p. 16-11) for 1977 less coke purchases, or 2.01. A correction of 0.20 is added to the latter to make 1974 figures consistent with ISTUM. These initial estimates sum to 4.37 and are scaled as per note i by a factor of 0.973.

Table A.6. INDUSTRIAL ENERGY USE IN 1978 BY SECTOR: MECHANICAL AND
            ELECTROLYTIC SERVICES (Quads, including utility losses)

	Purchased Electricity Including Losses[a]	Fuel for Internal Work[b]	Machine Drive*	Electrolysis
Mining and Construction	0.58[e]	0.12[e]	0.70(0.25)	--
Misc. Mfr.	2.62	0.01	2.63(0.77)	--
Food	0.48	0.03	0.51(0.16)	--
Paper	0.55	0.13	0.68(0.25)	--
Chemicals	1.86	0.17	1.59(0.52)	0.44[c]
Petroleum R.	0.38	0.03	0.41(0.13)	--
S.,C. and Glass	0.38	--	0.38(0.11)	--
Primary Metals	1.96	0.10	0.88(0.30)	1.08[d]
TOTALS			7.78	1.52

*The numbers in parentheses are electricity delivered to motors plus shaft work at 3413 Btu/kwh.

[a]Purchased electricity shown in Table A.5 times fuel to electricity ratio of 3.43; see note c, Table A.5.

[b]Mechanical work and electricity internally generated in 1976 converted to fuel at 7000 Btu/kWh.

[c]As chlorine production did not change substantially 1974-1978, the ISTUM value is used: .127 x 3.43 = .44. This division between electrolysis and machine drive may under estimate the former; See IEUDB, p. 6-7.

[d]Primary aluminum production fell slightly 1974-1978 (from 4.90 to 4.80 million tons) and assuming a 5% efficiency improvement the ISTUM value services: .340 x (4.8/4.9) x .95 x 3.43 = 1.08. Machine drive is hte residual after accounting, in addition for 0.1 quads for electric are furnaces.

[e]The total mechanical and miscellaneous electrical services shown in Table A.1 and A.2 for 1974 are roughly divided into purchased electricity and non-vehicle engine drive (0.55 and .14, respectively) and the electricity is increased 6% for the 1978 estimate.

Table A.7.   INDUSTRIAL ENERGY USE IN 1978 BY SECTOR: HEAT,
FEEDSTOCK AND MISCELLANEOUS SERVICES (Quads)

	"Non-Electrical" Fuel Use[a]	Process Heat[b]	Space Heat[b]	Other
Mining and Construction[c]	2.62[c]	0.41[c]	0.01	1.09[d] .21[e], .89[f]
Misc. Mfr.	1.83	1.59	0.24	
Food	0.78	0.75	0.03	
Paper	1.88	1.85	0.03	
Chemicals	4.55	2.02[g]	0.01	2.52[h]
Petroleum R.	3.30	3.30	--	--
S., C and Glass	1.14	1.12	0.02	
Primary Metals	3.68	1.47[i]	0.04	2.17[j]
TOTALS		12.51	0.38	

[a]The fuel use not associated with mechanical drive, miscellaneous electrical services or electrolysis is assigned to process heat, feedstock and other energy services in this table. The total is determined as the difference between the first column of Table A.5 and the last two columns of Table A.6.

[b]Except for chemicals and primary metals the non-electrical fuel use is assigned in proportion to the 1974 values shown in Tables A.1, A.2 and A.4.

[c]By the methods described in the notes to Table A.5, agricultural energy use including utility losses in 1978 was 1.13, so the 1978 total for construction and mining was 4.45 - 1.13 = 3.32.   The "non-electrical" fuel use of 3.32 - .70 = 2.62 is then 5% less than those uses in 1974, and all those uses are scaled accordingly.

[d]Reservoir repressuring.

[e]Non-gasoline construction vehicles.

[f]Asphalt feedstock used in construction.

[g]The residual after subtracting feedstocks.

[h]Chemical feedstocks from note l, Table A.5, the 1978 estimated use of fuel as feedstocks is 0.973 x 2.59 = 2.53.

[i]The residual after subtracting metallurgical coal and chemical feedstocks.

[j]Metallurgical coal (2.10) and chemical feedstocks (0.07).   The metallurgical coal is, from note o of Table A.5, .973 x 2.16 = 2.10.

Appendices to
Chapter Three
TRANSPORTATION

APPENDIX A
CALCULATING THE ENERGY USED IN AUTOMOBILES AND LIGHT TRUCKS

Personal Use of Automobiles and Light Trucks

Energy consumption in the year t is calculated using

$$E_t = E_{77} \frac{\gamma_p}{\gamma_m} (\gamma_f / \gamma_m)^{-.2} \left[0.41 (\gamma_g / \gamma_p)^{0.89} + 0.59 (\gamma_g / \gamma_p)^{0.23} \right]$$

where

γ_p = ratio of population in year t to that in 1977

γ_g = ratio of GNP in year t to that in 1977

γ_m = ratio of average fuel economy in year t to that in 1977

γ_f = ratio of the automotive fuel price in year t to that in 1977.

Notes

(a) The income factors in brackets are calculated from the following data for 1972-73[1]

Income Level	Household Income	Household Size	Fraction of Households	Miles Driven Per Household
Poor	2500	2.58	0.18	4200
Lower Middle	8000	2.60	0.42	10,000
Upper Middle	14,000	3.54	0.19	17,000
Well Off	24,500	3.57	0.20	20,000

by comparing per capita income with miles driven per capita. The first term in brackets gives the effect of increased per capita income on miles driven by poor and middle income households (which accounted for 41% of miles driven in 1972-73) while the second

---

[1] D.K. Newman and Dawn Day, The American Energy Consumer, a report to the Energy Policy Project of the Ford Foundation, Ballinger, 1975.

----

term gives the effect of increased income on miles driven by upper middle and well off households.

(b)   The factor $(\gamma_f/\gamma_m)^{-0.2}$ gives the impact on miles driven of the marginal cost of operating a car.   The factor γ_f/γ_m is the ratio of the fuel cost per mile in the year t to the fuel cost per mile in 1977.   The exponent -0.2 is the short-run price elasticity of demand for gasoline - a value in the mid-range of most estimates.

(c) Gasoline consumption E_{77} for personal use in autos and light trucks in 1977 estimated as follows:

i)   there were 89.5×10^6 non-fleet autos in use in 1977,[2] driven 9800 miles each[3] on the average with a fuel economy of 13.9 mpg[3].   Thus fuel use was

$$\frac{(89.5 \times 10^6 \text{ cars}) \times (9800 \text{ miles/car})}{(13.9 \text{ mpg})} \times (125{,}000 \text{ Btu/gallon}) = 7.9 \times 10^{15} \text{ Btu}$$

ii)   the percentage of light trucks in personal use historically was:[4]

1962	32.6%
1967	41.9%
1972	50.2%

---

[2]TECDB, p. 1-67

[3]Monthly Energy Review, June 1979

[4]R. Knoor and M. Millar, "Projections of Automobile, Light Truck, and bus Stocks and Sales, to the year 2000," pp. 24-25, Nov. 1979.

If this trend persisted the fraction would have been 59.2% in 1977. since there were 24.3 million light trucks in 1977[5] driven 9400 miles each at an average fuel economy of 10 mpg, fuel use was

$$\frac{0.592 \times (24.3 \times 10^6 \text{ light trucks}) \times (9400 \text{ miles/truck}) \times (125,000 \text{ Btu/gal})}{10 \text{ mgp}}$$

$$= 1.7 \times 10^{15} \text{ Btu}$$

iii)   Thus $E_{77} = 9.6 \times 10^{15}$ Btu

(d) $y_p(1990) = \dfrac{238.4}{216.9} = 1.10$

$y_p(2000) = \dfrac{254}{216.9} = 1.17$

(e) $y_g(1990) = \dfrac{1399.2}{1340.5} (1.027)^{12} = 1.44$

$y_g(2000) = \dfrac{1399.2}{1340.5} (1.025)^{22} = 1.80$

(f) $y_f(1990) = \dfrac{8.70}{5.44} = 1.60$ (w/o a tax)

$y_f(1990) = \dfrac{12.70}{5.44} = 2.33$ w/a 50¢/gallon tax)

$y_f(2000) = \dfrac{10.2}{5.44} = 1.88$ (w/o a tax)

$y_f(2000) = \dfrac{14.2}{5.44} = 2.61$ (w/a 50¢/a gallon tax)

---

[5]Barry McNutt et al, "On Road fuel Economy Trends and Impacts," USDOE Office of Conservation and Advanced Energy Systems Policy, Feb. 17, 1979, p. 52.

(g) Thus

$$E_{90} = E_{77} \left[\frac{1.10}{\gamma_m}\right]\left[\frac{1.60}{\gamma_m}\right] \times 1.15 = 1.15 \, E_{77} \, y_m^{-0.8}$$

without a tax or

$$E_{90} = 1.07 \, E_{77} \, y_m^{-0.8}$$

with a 50¢/gallon tax.

(h) Thus

$$E_{2000} = E_{77} \left[\frac{1.17}{y_m}\right] \left[\frac{1.88}{y_m}\right]^{-0.2} \times 1.25 = 1.29 \, E_{77} \, y_m^{-0.8}$$

without a tax or

$$E_{2000} = 1.21 \, E_{77} \, y_m^{-0.8}$$

with a 50¢/gallon tax.

(i) The average fuel economy m (see Table A) and y_m are as follows in 2000 for alternative new car standards m_s in 1995, assuming m_s increases linearly from 27.5 mpg in 1985:

m_s 1995	m (2000)	y_m (2000)
35	34.4	2.47
45	43.6	3.14
55	52.7	3.79
65	61.8	4.45
75	70.8	5.09

(j)   Thus E_{2000}/E_{77} is

m_s (1995)	w/o tax	w/tax
35	0.63	0.59
45	0.52	0.48
55	0.44	0.42
65	0.39	0.37
75	0.35	0.33

TABLE A

Year						Fraction of Cars in 2000	
2000	35	45	55	65	75	0.171	
1999	35	45	55	65	75	0.143	
98	35	45	55	65	75	0.126	
97	35	45	55	65	75	0.111	0.739
96	35	45	55	65	75	0.100	
95	35	45	55	65	75	0.088	
94	34.3	43.3	52.3	61.3	70.3	0.074	
93	33.5	41.5	49.5	57.5	65.5	0.058	
92	32.8	39.8	46.8	53.8	60.8	0.044	
91	32.0	38.0	44.0	50.0	56.0	0.030	
90	31.3	36.3	41.3	46.3	51.3	0.055	
89	30.5	34.5	38.5	42.5	46.5		
88	29.8	32.8	35.8	38.8	41.8		
87	29.0	31.0	33.0	35.0	37.0		
86	28.3	29.3	30.3	31.3	32.3		
85	27.5	27.5	27.5	27.5	27.5		
m (2000)	34.4	43.6	52.7	61.8	70.8		

Non-Personal Use Automobiles and Light Trucks

Energy consumption in the year t is caluclated using

$$E_t = E_{77} \frac{1}{m} (y_f / y_m)^{-0.2} y_g^{\alpha}$$

The exponent α is estimated from data for the period 1967 relating VM of travel to price and GNP (assuming constant fuel economy). Thus it is assumed that

that

$$y_v = y_f^{-0.2} y_g^{\alpha}$$

where

y_v = ratio of vehicle miles driven in year t to that in 1967.

It is estimated that $\alpha = 1$.

Notes

(a)   It is assumed that fleet cars are driven an average of 18,000 miles per year, the average for 1977.[1]

---

[1]TECDB, p. 1-70.

(b) Based on cars in fleets of 4 or more.[2]

(c) For light truck's driven 9400 miles/year. See note (c)-ii) of "Personal Use of Automobiles and Light Trucks."

(d) Thus

$$E_{90} = E_{77} \ y_m^{-0.8} \times 0.91 \times 1.445 = 1.31 \ E_{77} \ y_m^{-0.8}$$

without a tax or

$$E_{90} = 1.22 \ E_{77} \ y_m^{-0.8}$$

with a 50¢/gallon tax.

(e) Thus

$$E_{2000} = E_{77} \ y_m^{-0.8} \times 0.88 \times 1.81 = 1.60 \ E_{77} \ y_m^{-0.8}$$

without a tax or

$$E_{2000} = 1.49 \ E_{77} \ y_m^{-0.8}$$

with a 50¢/gallon tax.

(f) Assuming

---

[2]TECDB, p. 1-69

$m_s(1995)$	$m(2000)$	$y_m(2000)$
35	34.4	2.47
45	43.6	3.14
55	52.7	3.79
65	61.8	4.45
75	70.8	5.09

we obtain the following values for E_{2000}/E_{77}:

$m_s(1995)$	w/o tax	w tax
35	0.78	0.72
45	0.64	0.60
55	0.55	0.51
65	0.48	0.45
75	0.44	0.41

(g) $E_{77} = 125{,}000 \text{ Btu/gallon} \times \dfrac{0.187 \times 10^{12} \text{VM}}{13.9 \text{ mpg}} + \dfrac{0.093 \times 10^{12} \text{VM}}{10 \text{ mpg}}$

$= 2.8 \times 10^{15} \text{ Btu}$

FUEL USE BY AUTOS AND LIGHT TRUCKS
(10^{15} Btu/year)

	Personal			Non-Personal			Total		
		2000			2000			2000	
	1977	w/o Tax	w/Tax	1977	w/o Tax	w/Tax	1977	w/o Tax	W/Tax
$m_s(1995)$	9.6			2.8			12.4		
35		6.0	5.7		2.2	2.0		8.2	7.7
45		5.0	4.6		1.8	1.7		6.8	6.3
55		4.2	4.0		1.5	1.4		5.7	5.4
65		3.7	3.6		1.3	1.3		5.0	4.9
75		3.4	3.2		1.2	1.1		4.6	4.3

APPENDIX B
AIR PASSENGER TRANSPORT

Energy consumption for air passenger transport (certified and supplemental) is calculated from the expression

$$E_t = \frac{E_{77}}{y_m} \times \frac{y_r}{y_c}$$

where

y_m = ratio of fuel economy (PM/gallon) in year t to that in 1977

y_r = ratio of revenues in year t to revenues in 1977.

y_c = ratio of specific cost (¢/PM) in year t to that in 1977.

<u>Notes</u>

(a) E_{77} is calculated by noting that in 1977 the fuel economy of certified air carriers was 21.2 PM/gallon[1] and the total revenue passenger miles flown were 193.2×10^9 for certified route carriers[2] and 10.0×10^9 for supplemental carriers.[3]

Thus

$$E_{77} = \frac{203.2 \times 10^9 \text{ PM}}{21.1 \text{ PM/gallon}} = 9.58 \times 10^9 \text{ gallons} = 1.29 \times 10^{15} \text{ Btu}$$

---

[1] Statistical Abstract 1979, p. 666

[2] op. cit., p. 667.

[3] op. cit., p. 668.

399

(b) y_r is estimated by assuming the revenues (in constant dollars) increase in the future at the same average rate as in the period 1972-1978. From the following data.

		Revenue Rate (¢/PM)		Total Revenue
	10^9 RPM	(current $)	(1978$)	(10^9 1978$)
1972	162.4	6.1	9.3	15.1
1973	173.7	6.3	9.1	15.8
1974	173.8	7.3	9.6	16.7
1975	171.5	7.6	9.1	15.6
1976	187.2	7.5	8.5	15.9
1977	302.2	8.2	8.8	17.9
1978	236.8	8.2	8.2	19.4

one obtains for 1972-1978

$$\text{Revenue} = 14.46 \ (1.035)^{t-1971} \quad r = 0.83$$

Thus

$$y_r(t) = (1.035)^{t-1977}$$
$$y_r(1990) = 1.56$$
$$y_r(2000) = 2.21$$

(c)   The average fuel economy m(2000) in the year 2000 is estimated as follows:

i)   In 1976 the distribution of revenue passenger miles by trip length L_t was[4]

51.8%			$L_t \leq 1000$ miles
30.3%	1001	$\leq$	$L_t \leq 2000$
17.9%	2001	$\leq$	L_t

---

[4]A.B. Rose, "Energy Intensity and Related Parameters of Selected Transportation modes: Passenger Movements," ORNL-5506, January 979, p. 3-7.

ii)   It is assumed that advanced aircraft serve most trips less than 2000 miles. For such planes a fuel economy of 33 PM/gallon is feasible by 1990.

iii) It is assumed (following Gray) that wide body aircraft serve predominantly trips longer than 2000 miles. For such planes (e.g. 747's) a fuel economy of 48.0 PM/gallon is feasible by 1990.[5]

iv)   The average fuel economy for 2000, m(2000) is thus given by

$$y \frac{1}{m(2000)} = \frac{0.82}{33} + \frac{0.179}{48.0} \text{ or } , {}^{m}(2000) = 35 \text{ PM/gallon}$$

vi)   Thus

$$y_m (2000) = \frac{35.0}{21.2} = 1.65$$

(d)   Now

$$y_c = \frac{C_{fx}(t) + \frac{C_{fu}(t)}{m(t)}}{\frac{C_{fx}(77) + C_{fu}(77)}{m(77)}}$$

where

$C_{fx}(t)$ = fixed cost per passenger mile in the year t

$C_{fu}(t)$ = cost per gallon of fuel in the year t

$C_{fx}(77)$ = 6.3¢/mile (1977$)[6] = 6.8¢/mile (1978$)

---

[5] Verbal communication from Dr. Rose, Nasa Aircraft Fuel Conservation Technical Program, April 15, 1978, as cited in Table 9 of E.A. Allen and J.A. Edmonds, "The Future of the Personal Automobile in the U.S., IEA, September 1979.

[6] National Transportation Statistics, 1979 (NTS-1979), pp. 44 and 14-15.

Thus

$$C_{fu}(77) = (8.8-6.8)\text{¢/mile} \times 21.2 \text{ PM/gallon} = 42.4\text{¢/gallon}$$

We assume that

$$C_{fx}(t) = 6.8 (1.0028)^{t-1977}[7]$$

Also we assume that $C_{fu}(t)$ is the same as the marginal cost of diesel fuel assumed for the SERI project:

$$C_{fu}(1990) = 90\text{¢/gallon} \qquad \text{w/o a tax}$$
$$C_{fu}(1990) = 140\text{¢/gallon} \qquad \text{w/a 50¢/gallon tax.}$$
$$C_{fu}(2000) = 108\text{¢/galon} \qquad \text{w/o a tax}$$
$$C_{fu}(2000) = 158\text{¢/gallon} \qquad \text{w/a 50¢/gallon tax.}$$

Thus

$$y_c(2000) = \frac{7.3 + \frac{108}{35}}{6.8 + \frac{42.4}{21.2}} = 1.175 \text{ w/o a tax}$$

$$y_c(2000 = \frac{7.3 + \frac{158}{35}}{8.8} = 1.34 \text{ w/a 50c/gallon tax}$$

(e)   Thus

$$E_{2000}/E_{77} = \frac{2.21}{1.65 \times 1.175} = 1.593 \text{ w/o tax}$$

$$E_{2000}/E_{77} = \frac{2.21}{1.65 \times 1.34} = 1.034 \text{ w/a 50c a gallon tax}$$

Thus

$$E_{2000} = \quad \begin{array}{l} 1.59 \times 10^{15} \text{ Btu w/o a tax} \\ 1.34 \times 10^{15} \text{ Btu w/a 50¢/a gallon} \end{array}$$

---

[7]CONAES, "Alternative Energy Demand Futures to 2010, pages 145-146 and 155.

Appendices to
Chapter Four
UTILITIES

SUPPLEMENTARY TABLES

(1) Tables A-1 and A-2 provide the basis for the estimates of the potential for cogeneration made in the text.

(2) Table A-3 was used to derive the age characteristics of oil and gas burning electric generating plants used to calculate the potential for oil and gas backout.

Table A-1. PROCESS STEAM USE, POWER CONSUMPTION AND THE COGENERATION POTENTIAL IN SIX INDUSTRIES

(I) Industry	(II) Process Steam Use, in 10^{12} Btu/year[a]	(III) Process Steam Use Technically Suitable for Cogeneration, in in 10^{12} Btu/year[b]	(IV) Power Consumption in 10^9 kwh/yr.				(V) Target Congeneration Via High E/S Ratio Cogeneration Technologies in 10^9 kwh/yr. [d]
			Purch. El[c]	Onsite Mech.	Prod.[a] Elect.	Total	
Food	307	198	39.1	4.1	2.6	45.8	40
Textiles	125	75	34.8	0.6	1.5	36.9	15
Pulp and Paper	1210	859	43.5	1.8	23.4	68.7	172
Chemical	1516	1059	145.4	14.4	18.8	178.6	212
Pet. Ref.	584	364	26.3	4.1	1.5	31.9	73
Steel	366	300	54.6	10.5	10.0	75.1	60
	4108	2855	343.7	35.5	57.8	437.0	572

[a] For 1976, as given in P. Bos. "The Potential for Cogeneration Development in Six Major Industries by 1985," Report to the Department of Energy by Resource Planning Associates, Inc., 1977.

[b] See Table A-2.

[c] For 1976, as given in the Annual Survey Manufacturers 1976: Fuels and Electric Energy Consumed.

[d] Assuming the process steam use in column III is associated with cogeneration with an average E/S ratio of 200 kwh/10^6 Btu.

Table A-2. PROCESS STEAM LOAD IN 1976 TECHNICALLY SUITABLE FOR COGENERATION IN SIX MAJOR INDUSTRIES
(10^{12} Btu/year)

	Food	Textiles	Pulp and Paper	Chemical	Petroleum Refining	Steel	6 Industry Totals	
A. Process Steam Load[a]	307	125	1210	1516	584	366	4108	
Load Fluctuations	(23)	(10)	(94)	(92)	(35)	(12)	(266)	
Low Steam Load	(52)	(24)	(85)	(74)	(0)	(0)	(235)	
Low Pressure Waste Heat Steam	(8)	(0)	(0)	(143)	(132)	(35)	(318)	
Declining Demand	(26)	(16)	(172)	(148)	(53)	(19)	(434)	
B. Steam Load Technically Suitable for Cogeneration[b]	198	75	859	1059	364	300	2855	
C. Steam Load Presently Associated with Cogeneration[a]								
Electrical	27	15	236	190	15	101	584	
Mechanical	41	6	17	143	41	106	354	
Total	68	21	253	333	56	207	938	
D. Percentage of Suitable Steam Load Presently Associated with Cogeneration (100 x C/B)	34%	28%	29%	31%	15%	69%	33%	
E. Distribution of Existing Cogenerated Steam Load (Electrical)[a]								
Steam Turbine								
Coal	10.5	3.0	35.4	38.0	-	10.1	97.0	(17%)
Resid	14.6	10.8	59.0	95.0	9.8	15.2	204.4	(35%)
Waste Fuel	-	-	118.0	-	3.0	60.5	181.5	(31%)
Heat Recovery	-	-	-	9.5	0.8	5.2	15.5	(3%)
Subtotal	25.1	13.8	212.4	142.5	13.6	91.0	498.4	(85%)
Gas Turbine								
Distillate	-	0.3	-	19.0	0.7	-	20.0	(3%)
Natural Gas	1.9	0.9	23.6	28.5	0.7	10.0	65.6	(11%)
Subtotal	1.9	1.2	23.6	47.5	1.4	10.0	85.6	(15%)
Total	27	15	236	190	15	101	584	(100%)

[a]The number presented here are obtained from Peter Bos et al, "The Potential for Cogeneration Development in Six Major Industries by 1985," prepared by Resource Planning Associates, Inc., for the Department of Energy, December 1977. The various technical factors judged by RPA to reduce the size of the process steam load suitable for association with cogeneration are as follows:

* load fluctuations: arises because electric power and steam demands are out of phase. This factor was important in the RPA analysis because in that study the export to the utility grid of power produced in excess of onsite needs was not considered. Using the RPA numbers understates the cogeneration potential in the present analysis, where export is allowed, because steam and electric power demands do not have to be in phase to cogenerate when the cogeneration unit is tied to the utility grid.

* low steam load: for steam loads less than 50,000 lb. of steam per hour.

* low pressure waste heat steam.

* declining demand: based on management expectations for future reduced steam demand because of conservation, process or product changes, plant obsolescence or shutdown.

These factors collectively reduce the steam base technically suitable for cogeneration by 31% or 1253×10^{12} Btu. RPA also excluded from consideration for cogeneration another 5% or 212×10^{12} Btu of the steam base because cogeneration based on this steam base (even with steam Rankine cycles) would have resulted in more electricity than could be consumed onsite. This restriction is unappropriate for the present analysis, where export of excess electricity to the utility grid is allowed.

[b]This includes the steam load presently associated with cogeneration.

Table A.   Table A-3. OIL AND GAS FIRED GENERATING
                              CAPACITY (1977)

	Total	Completed Before 1971	Completed Before 1961
Louisiana[a]	15.6	13.5	9.8
Oklahoma[a]	7.4	4.4	3.4
Massachusetts[a]	6.6	5.3	2.3
Texas[a]	36.7	28.7	17.9
Mississippi[a]	4.7	3.3	2.0
California[a]	23.8	20.6	18.1
Florida[a]	15.9	13.2	9.2
	110.7	89.0	60.7
Total U.S.[b]	151.2		

[a]Steam-Electric Plant Construction Cost and Annual Production
Expenses 1977. Thirtieth Annual Supplement, U.S. Department
of Energy, Energy Information Administration, December 1978.

[b]National Electric Reliability Council, "8th Annual Review of
Overall Reliability and Adequacy of the North American Bulk
Power Systems," August 1978.

APPENDIX B
CALCULATION OF THE VALUE OF
SOLAR TECHNOLOGIES

This discussion will begin with a description of the techniques used to compute the cost-to-value ratio of different solar technologies. The assumptions used in the analysis will be described in some detail in the second section and tables summarizing the results of a variety of sensitivity studies are included at the end.

ANALYTICAL METHODS

The value of solar or coal equipment is computed under the assumption that no net demand growth is occurring in the utility's service area. Under these circumstances the value computed must result from fuel and operating costs saved by adding the new equipment. No credit has been assumed for the capacity which may be rendered unnecessary by the addition of new solar or coal plants.

The value of the solar investment has been computed using the diagram shown in Figure B-1. This Figure represents an approximation to the "load duration" curve of a utility. Such a curve indicates the number of hours per year when the demand on a utility equals each possible level of demand. In the simplified curve of Figure 3, it is assumed that the demand never exceeds 80% of the maximum capacity of the utility (thus allowing 20% reserve margin at all times) and never falls below 30%. The plants are dispatched to meet demand as indicated in the Figure. Hydroelectric and nuclear plants are used preferentially for the base load, coal for intermediate loads, and oil and gas for peaking.

Without storage, value of a solar device depends on when the solar energy is available. The electricity has more value if it is available when oil or gas are being used and least value if it is available only when nuclear fuel is being consumed. The availability of hydroelectric power and pumped storage or other storage equipment can be of great value in ensuring that the renewable energy is available when it is most needed and that the stability of the utility grid is maintained in spite of fluctuating output from solar equipment.

If it is assumed that the solar electricity is completely uncorrelated with demand, the energy provided by the equipment would reduce demand as

illustrated by the diagonal shaded area in Figure B-1. Since it is assumed that the net demand for electricity would not increase, the value of the solar investment is measured entirely by its ability to save fuel and operating costs. The fuel displaced depends on the mix of equipment in the utility. (For example, in Figure B-1 about 50% of the fuel displaced is oil, 25% coal and 25% nuclear.)

A coal plant could also have been added to reduce fuel demand. An advantage of a coal plant is that it could be dispatched in such a way that all of the energy it displaces is oil or gas. If coal plants operated for a relatively short period of the year, of course, the electricity they produce is expensive since the plant must be paid for whether or not it is generating electricity.

The calculations were performed using assumptions about system costs and performance characteristics summarized in Table B-2. Table B-3 indicates the mix of plant types in the system assuming that no plant additions are made after 1985. Table B-4 indicates the mix which would result if no plants are added after 1990. Both tables overestimate the amount of coal and nuclear in the nation since the data was compiled in 1978--schedules of construction and cancellations after this date have not been reflected in the tables.

REVIEW OF TECHNOLOGIES

Biomass Resources for Electricity Generation

Biomass resources that can be used for electricity generation include residues from the forest products industry, municipal solid wastes, and agricultural residues from crops and food processing. These resources can be burned directly or converted to densified solid or liquid fuels for standard turbine generators.

Several utilities are currently using wood wastes or mill residues to fire 10 to 50 MW generating units (York 1979). The forest products industry currently generates a large proportion of its total energy requirements (including steam, electricity, and direct drive) from both forest and mill wastes. In Wisconsin, the industry generates 50% of its energy requirements from these sources.

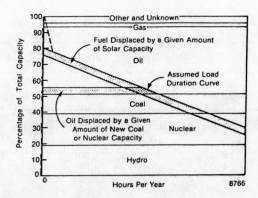

Figure B-1

**Illustrative Fuel Displacement by a
Given Penetration of Solar Capacity
or New Coal or Nuclear Capacity in
DOE Region 9.**

TABLE B-2. SUMMARY OF TECHNOLOGY AND ECONOMIC ASSUMPTIONS

General Economic Parameters (in real terms)

Capital recovery factor	0.054
Fixed charge rate	0.106
Discount rate	0.0344
Technology lifetimes	30 years

Conventional Generating Units, Characteristics for the Year 2000 (1978 $)

				Operation and Maintenance[b]	
Fuel Type	Capital Cost[a] ($/kW)	Heat Rate[b] (Btu/kWh)	30-Year Levelized Fuel Prices ($ per MBtu)	Fixed ($/kW-yr)	Variable (mills/kWh)
Nuclear	1080	10,700	1.22	3.10	1.5
Coal	820-1000	10,150	1.89	12.80	3.6
Oil	210	9,680/14,000[c]	8.40	1.40	1.5

Solar Electric Generating Units, Characteristics for the Year 2000[d] (1978 $)

Unit Type	Unit Size (MW)	Capital Cost ($/kW)	Operation and Maintenance ($/kW-yr)	Capacity Factor[e] (%)
Wind (MOD-2)	2.5	650-840	6	15-37
Solar Thermal				
Central Receiver-Stand Alone				
with storage	100	1377	27	36
without storage	100	1039	20	26
Distributed Receiver	10	685-937	13-18	15-27
Photovoltaics	100	766-1650	7	13-25
OTEC	400	2200-3200	56[f]	80

[a]Except for the higher coal cost estimate, all numbers are from Williams (1980), based on EPRI (1979). The $1000/kW coal cost estimate was used for sensitivity analysis.

[b]Source: Electric Power Research Institute (EPRI) 1979.

[c]The lower heat rate is for existing units (a combination of steam and combustion turbines). The higher heat rate is for new combustion turbines. A weighted average heat rate of 10,632 was used to calculate the value of displaced oil.

[d]Even though these costs are assumed to apply for the year 2000, they may also apply to earlier years (also in 1978 $).

[e]Capacity factors vary by region. See technology descriptions in the text.

[f]This number was calculated from a variable O&M charge of 8 mills/kWh.

Table B-3. PROJECTED GENERATING MIX FOR THE YEAR 1985 BY DOE REGION

DOE Region	Capacity (GW)	Fuel Type (% of Total Capacity)[a]					
		Nuclear	Hydro	Oil	Coal	Gas	Other & Unknown
1	25.2	31	11	53	2	0	3
2	56.3	20	10	59	11	0	0
3	87.0	19	6	25	49	0	0
4	186.1	26	8	16	47	3	0
5	149.3	21	2	13	62	1	1
6	120.9	10	2	5	32	49	1
7	45.7	9	3	11	63	12	1
8	36.8	1	19	4	75	1	0
9	73.7	15	20	45	12	2	5
10	42.0	19	70	4	5	2	1

[a]Percentages may not add to 100 due to rounding.

Note: Capacity was calculated by summing existing capacity and additions scheduled to begin operation by 1985 or before. Data ignore retirements.

Source: Shanks (1979).

Table B-4. PROJECTED GENERATING MIX FOR THE YEAR 1990 BY DOE REGION

DOE Region	Capacity (GW)	Fuel Type (% of total capacity)[a]					
		Nuclear	Hydro	Oil	Coal	Gas	Unknown and Other
1	29.4	39	9	45	4	0	3
2	72.2	31	12	46	11	0	0
3	99.0	20	8	22	49	0	0
4	210.0	28	8	14	47	3	0
5	170.0	25	2	12	60	1	1
6	132.0	11	2	5	36	45	1
7	49.7	11	5	11	61	11	1
8	43.3	1	16	3	79	1	0
9	81.9	19	19	42	14	2	4
10	47.2	28	62	3	4	2	1

[a]Percentage may not add to 100 due to rounding (Shanks 1979).

Note: Total capacity and fuel mix are those currently projected for the year 1990 (Shanks 1979). The assumption made here is that this capacity will be sufficient to the year 2000. The data do not reflect planned retirements (planned retirements to the year 1988 are 2% of 1978 generating capacity) not do they reflect the possibility of cancelling or deferring capacity currently under construction. See the text for a more thorough discussion of the assumptions made in preparing these data.

The problem with estimating the amount of electricity likely to be generated by utilities using biomass resources is that the same resource base can be used for other applications as well. Utilities will have to compete not with industrial users generating electricity and steam.

If biomass is to be a viable alternative for utilities, they will need to secure local, dedicated supplies because the resource cannot be transported economically over long distances. In addition, as the value of biomass resources becomes more widely appreciated, demand will increase and prices will rise. In the past year the cost per dry ton of wood and mill wastes has increased 20% in Eugene, Oregon where a municipal utility has been using wood or 20 years. Higher prices for wood and mill wastes will reduce the competitive advantage of biomass fuels for electricity generation relative to conventional fuels such as coal.

Hydroelectric Resources

There are currently 64 GW of hydroelectric capacity in the United States, supplying the equivalent of 2.8 quads of primary energy. Hydropower is a fully developed technology, and the maximum physical potential resources that could be developed are quite large. Hydropower resources can be separated into three categories: (1) existing capacity, (2) incremental capacity that could be developed at existing dams, and (3) underdeveloped resources, which would require constructing new dams.

The most comprehensive and recent assessment of hydropower resources in the United States is the National Hydroelectric Power Resource Study of the Army Corps of Engineers. A three-year effort undertaken at a cost of $7 million, the study (hereafter referred to as the National Hydropower Study) began by identifying 50,000 existing dams and 5000 undeveloped sites. This number is being reduced in four successive screening stages, two of which have already been completed.

The first screening--testing for sufficient storage, head and stream flow--reduced the total to 17,500 sites. The second screening, using a very relaxed economic criterion (that the benefits exceed the cost of the powerhouse equipment, which may be only 25% of the total cost) cut the inventory to 11,000 sites. The third screening which is well under way, has already reduced the inventory to 5000 sites, and the study managers expect to identify ultimately a group of 3000 sites, including 200 to 250 undeveloped sites (McDonald 1980).

Although the National Hydropower Study is well into its third stage, the results of the second stage (the 11,000 site inventory) provide the best available documentation for the study (Army Corps of Engineers 1979). This preliminary inventory estimated that the incremental and undeveloped potential of large* facilities in the U.S., excluding Alaska, is 268 GW,

which could displace roughly 6.7 quads of primary energy.** It is estimated that the incremental and undeveloped potential of small# hydropower facilities in the U.S., also excluding Alaska, is 12.4 GW, which could displace roughly 0.4 quads##.

Because of the procedures used in preparing the Preliminary Inventory, the estimated total incremental and undeveloped hydropower resources of 280 GW for the continental U.S. should be viewed as a maximum physical potential. Metz assumed that roughly half (140 GW) of this potential could be developed economically by the year 2000, ignoring land-use and environmental constraints. His assumption is consistent with the Federal Energy Regulatory Commission's (FERC) estimate of 114 GW of incremental and undeveloped large hydropower resource (Federal Power Commission 1976).

Table B-5 shows the location of existing large and small hydropower resources in the United States. Most of the existing capacity is in units larger than 15 MW, which is also true of incremental and undeveloped resources.

Table B-5.   **LOCATION OF EXISTING HYDROPOWER RESOURCES IN THE U.S. (MW)**

| | Existing Hydropower Resources | | |
DOE Region	Small (15 MW or less)	Large (more than 15 MW)	Total
1, 2, and 3	967	5,772	6,739
4	221	10,395	10,616
5	722	511	1,233
6 and 7	93	3,313	3,406
8	166	5,000	5,166
9	358	9,380	9,738
10 (excluding Alaska)	393	26,282	26,675
Total U.S. (excluding Alaska)	2,920	60,653	63,573

Source: Appendix C.

Estimating the amount of energy that could feasibly be obtained from new hydropower projects is extremely difficult, both because costs vary radically from site to site, and because environmental, social, and legal barriers prohibit the development of many sites. Table B-6 reports the range of cost estimates Metz obtained from project data. The range of costs for all types of hydropower projects is huge and illustrates the difficulty of attempting to assess the national hydropower potential. Accurate data can only be obtained from a site-by-site analysis of the type currently being conducted by the Army Corps of Engineers.

---

*   Large facilities are those whose capacity exceeds 15 MW, and that have a hydraulic head (defined as the difference in elevation between the water levels at the intake and outflow) of at least 20 meters.

** Incremental capacity is 89 GW and undeveloped is 179 GW. Alaska has an estimate 67 GW of undeveloped large hydropower resources. Alaskan resources are excluded due to their great distance from major consumption areas.

\#   Small hydropower facilities are those with a capacity of 15 MW or less.

\#\# Incremental capacity is 5.4 GW and undeveloped is 7.0 GW.

Given this caveat, Metz attempted to estimate the number of new hydropower facilities (both private and federal), which we can expect to be developed by the year 2000, given current and projected economic conditions, regulatory constraints, and existing incentives. To estimate the likely development of private facilities he used cost and resource data from national and regional hydropower studies, examined the number of applications for hydropower permits ending before FERC, and surveyed several people currently working in the field. He estimated that 21 GW of capacity is likely to be developed, of which 17 GW are additions to existing large facilities, 1 GW is new large projects, and 3 GW are additions to existing small facilities. He assumed that no small facilities at undeveloped sites would be built, largely due to their high cost. Using capacity factor data developed at FERC, he estimated that 21 GW of new hydropower facilities could displace 0.61 quads of primary energy.

Data on new federal hydropower facilities are less complete than for private facilities. Metz reports that conservative estimates of the new federal hydropower resources (all sizes) that could be developed are 5 GW (displacing 0.2 quads) and that the maximum practical development would be 25 GW (0.7 quads).

Thus, we estimate that .61 quads of new hydropower facilities are likely to be developed and that .9 quads could probably be obtained given increased incentives and a reduction in environmental constraints. Much more accurate data on hydropower potential should be available once the National Hydropower study is completed.

Wind Energy Conversion Systems

Of the three intermittent technologies, wind systems are clearly closest to commercialization for utility applications. Wind systems for irrigation and milling have been in use for many centuries (DOE 1980, p. III-51) and small turbines (1 to 10 kW) for electricity generation are commercially available today. Larger machines (greater than one MW) are most appropriate for utility applications.

Even though the wind turbine designs that appear to be best (least costly) for utility applications are not yet commercially available, a number of utilities have expressed a strong interest in wind turbines (Business Week 1979; Smith 980). Southern California Edison is currently planning to install 55 MW of wind turbines in 1986 (Western Systems Coordinating Council 1979, p. 37; Weir 1980). Wind Farms Ltd. is currently planning to build an 80 MW wind farm in Hawaii as early as 1984 (Solomon 1980).

There are several different design concepts being considered for wind turbines. The wind turbine analyzed here is DOE's Mod-2, a horizontal axis wind turbine with a rotor 300 ft. in diameter, and a rated capacity of 2.5 MW at a wind speed of 27.7 mph. Its cut-in-speed is 14 mph and cut-out speed is 45 mph. The Mod-2 was selected for analysis because it appears to be the best design for utility applications

Table B-6.   COST ESTIMATES FOR NEW HYDROELECTRIC FACILITIES IN THE CONTINENTAL U.S.

(1978 dollars)

	Unit Capacity (MW)	Cost Range Based on Actual Project Data ($/kW)
Large Facilities		
Additions to Existing Facilities	30–1000	370–1331[a]
New Facilities at Undeveloped Sites	larger than 15	542–3050[b]
Small Facilities		
Additions to Existing Facilities	15 or less	500–3000[c]
New Facilities at Undeveloped Sites	15 or less	No project data available. Experts estimate costs may be double those for additions to existing facilities.

[a]Based on data for 11 projects in the Western states.

[b]Based on data for 16 projects in the Western states.

[c]Based on data for 45 projects.

Source: Appendix C.

for which reasonable cost data are available. It is currently being developed by Boeing and the first unit should begin operation in late 1980 (DOE 1980, p. III-67).

Because the Mod-2 has not yet been produced, its cost must be projected, assuming that cost reductions through technical improvements such as mass production can be achieved. NASA estimates that the second Mod-2 produced will cost $1350/kW in 1977 dollars ($1476/kW in 1978 dollars) (Ramler and Donovan 1979, p. 5). Boeing projects a hundredth-unit cost of $686/kW installed assuming a mass production facility producing 20 units per month (Boeing 1979, p. 5-55). Boeing's detailed cost estimates are presented in Table B-7. The cost estimate used here is $840/kW (1978 dollars) which includes $686/kW installed cost, a 10% contingency fee, and $85/kW for land.* Operation and maintenance costs were assumed equal to Boeing's estimate of $15,000 per machine per year (Boeing 1979, p. 5-57).

DOE is projecting that advanced turbine designs (such as the MOD-5) should produce electricity at substantially lower cost than the MOD-2) DOE 1980, pp. III-69, III-70). For this reason, a cost estimate of $650/kW installed (including land and contingency) was also used for sensitivity analysis.

In addition to capital costs, capacity factors for the Mod-2 had to be estimated. In reality, capacity factors will vary from site to site within regions. For the purposes of this exercise, regional average capacity factors were estimated on the basis of the annual average wind resources shown in Fig. B-1. Estimates were made using an equation developed by C. G. Justus of the Georgia Institute of Technology (Justus 1978, p. 90). Table B-8 reports the average annual wind speed derived from the wind powers in

Fig. B-1, and the estimated Mod-2 capacity factors for each DOE region.

Solar Thermal Central Receiver Plants

Two types of solar thermal power plants were considered, central receivers and distributed receivers using parabolic dishes. In central receiver plants, sunlight is reflected by mirrors (heliostats) to a central receiver where the thermal energy is transferred to a working fluid. This heated fluid is used to drive a turbine-generater. Two solar thermal central receiver (STCR) concepts are repowering of existing oil- and gas-fired steam electric generating units, and new stand-alone plants coupled with three to six hours of storage capacity.

The distributed receiver systems consist of a field of parabolic dishes. Each dish system consists of a paraboloidal concentrator which focuses the solar energy onto a receiver located at the focal point. The energy is transferred from the receiver via a working fluid to a heat engine mounted on the dish.

The capital costs of both central and distributed receiver plants are estimated to be highly dependent upon the rate of production. Table B-8a shows estimated capital costs for 100 MW repowered and stand-alone STCR plants as a function of the rate of production of heliostats. These concepts are representative of the advanced solar thermal central receiver concepts using molten salt or liquid metal as the working fluid. These costs (in 1978 dollars) could be attained as early as 1985 if the production rates indicated are reached. Even though cost estimates are shown for repowered plants, it is assumed that many oil- and gas-fired units will be retired by 1995. Thus, repowering is not expected to be a major new supply option by the year 2000.

Table B-7. BOEING'S MOD-2 COST SUMMARY (HUNDREDTH UNIT)

Item	Cost (1978 Dollars)	
	Material	Manufacturing
Site Preparation	110,428	51,100
Transportation	25,320	4,000
Erection & Checkout	58,900	78,000
Rotor Assembly	150,430	178,463
Drive Train	378,527	—
Nacelle Subassembly	153,086	31,000
Tower Subassembly	270,800	—
Spares and Maintenance		
Equipment	35,000	—
Non-recurring	35,000	—
Total Initial Cost	1,217,491	342,563
10%	121,749	34,256
Total	1,339,240	376,619

Total Turnkey Cost for the MOD-2
in dollars: 1,715,859
in dollars per kW: 686

Source: Boeing (1979).

* Assumes a land cost of $3000/acre and a five-diameter spacing of the wind turbines.

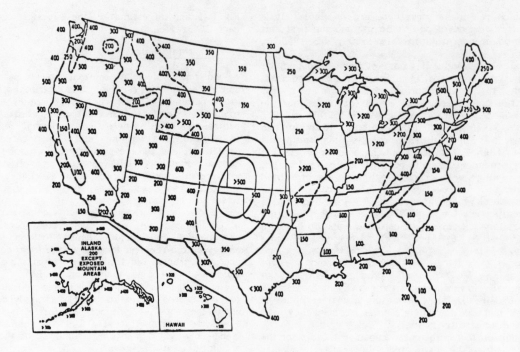

Figure B-1 **Annual Average Wind Power (W/m²) at 50m Above Exposed Areas (Elliott 1977)**

Table B-8. AVERAGE ANNUAL WIND SPEED
AND AVERAGE MOD-2 CAPACITY
FACTOR BY DOE REGION

DOE Region	Average Annual Wind Speed (meters/second)	Average MOD-2 Capacity Factor (percent)
1	8.3	33
2	7.8	30
3	7.3	26
4	6.4	15
5	7.4	26
6	8.0	31
7	8.3	33
8	8.8	37
9	8.3	34
10	8.4	35

Table B-8a. ESTIMATES CAPITAL COSTS FOR
SOLAR THERMAL CENTRAL
RECEIVER SYSTEMS
(1978 dollars)

	Year	
	1990	1995
Production Rate (heliostate per year)	15,000[a]	25,000[b]
Plant Size (MW)	100	100
Hybrid/Repowered ($/kW)		
with storage	1,545	—[c]
without storage	1,039	—
Stand-Alone ($/kW)		
with storage	1,883	1377
without storage	1,375	1039

SOURCE: L. Brandt et al. (1980). Numbers in
source are in 1980 dollars; they were
deflated to obtain 1978 dollars.

[a]At this production rate, an installed concentrator
cost of $150/m² is assumed.

[b]At this production rate, an installed concentrator
cost of $85/m² is assumed.

[c]It is assumed that many oil- and gas-fired units
will be retired by 1995. Thus, repowering is not
expected to be a major supply option by the year
2000.

Table B-8b shows the capital costs for parabolic dish systems, estimated for the years 1985, 1990 and 1995 (in 1978 dollars). The 1985 data are based on limited production of current technology, that is, an aluminum concentrator with an organic Rankine cycle heat engine. Data for 1990 and 1995 assume technological improvements as well as higher production rates. The 1990 estimates are for systems consisting of an advanced glass concentrator and a Brayton engine. The 1995 estimates are for an advanced glass concentrator and an advanced Brayton or a Stirling engine. System costs for 1990 and 1995 are given for two different production rates.

Figure B-2 shows average annual direct normal insolation for different regions. Because of the regional variability in insolation, STCRS were considered to be a viable option only in Regions 6 and 9. Midland, Texas, where the average annual direct normal insolation is 2.5 MWh/m^2, was used as the basis for calculating average capacity factors. The estimated capacity factors for STCRS with storage was estimated to be 36%; without storage was estimated to be 26% (Taylor et al. 1979).

Distributed systems were likewise considered to be most viable in Regions 6 and 9. The estimated capacity factor for Region 6 was 27% (based on Albuquerque insolation) and for Region 9 was 25% (based on Phoenix insolation). Dish systems were also assessed for Region 4 where the estimated capacity factor was 15% (based on Cape Hatteras insolation).

Photovoltaic Systems

Photovoltaic systems consist of photovoltaic cells grouped together in flat plate or concentrating configurations and encapsulated in a protective covering to form modules. Modules are mounted in structures to make photovoltiac arrays. Wiring, controls, and power conditioning equipment are the other components of a photovoltiac system.

Many materials and fabrication processes may be used to produce solar cells, the semiconductor devices that convert sunlight into direct-current electricity. Modules available commercially today primarily use single-crystal silicon cells sliced from Czochralski ingots. Polycrystalline or simicrystalline silicon cells, ribbon-grown single-crystal cells, cadmium sulfide, gallium arsenide, and several thin film cell concepts are all candidates for solar cells of the future.

Photovoltaic systems are currently too expensive to be used for central station power generation. Current R & D efforts are aimed at reducing the costs of solar cells, the major component of total systems costs. The purpose of this section is to report the range of cost estimates that might be expected for photovoltaic power plants in the year 2000 if projected technological improvements are achieved. A plant with a peak capacity rating of 100 MW (at solar insolation of 1 kW/m^2) is examined.

Photovoltaic systems in the year 2000 may use either flat plate or concentrator modules. The cost

Table B-8b. ESTIMATED CAPITAL COSTS FOR PARABOLIC DISH ELECTRIC SYSTEMS

(1978 dollars)

	1985[a]	1990[b]		1995[c]	
Module Size (rated power output (kW))[d]	20	22		29	
Production Rate (modules per year)	300	5,000	25,000	10,000	100,000
Total Installed Cost ($/kW)	3,500	1,620	1,140	937	685
Cost Components (%) of total)					
Module	72	64	58	60	52
Balance of Plant	8	16	22	20	28
Indirect	20	20	20	20	20

SOURCE: Terasawa (1980). Numbers in source are in 1980 dollars; they were deflated to obtain 1978 dollars.

[a]Costs assume 93 m^2 effective collector area per module and 22% module efficiency at 1 kW per m^2.

[b]Costs assume 93 m^2 effective collector area per module and 24% module efficiency at 1 kW per m^2.

[c]Costs assume 93 m^2 effective collector area per module and 31% module efficiency at 1 kW per m^2.

[d]Rated module size increases with assumed improvements in module efficiency. Output is rated at 1 kW per m^2.

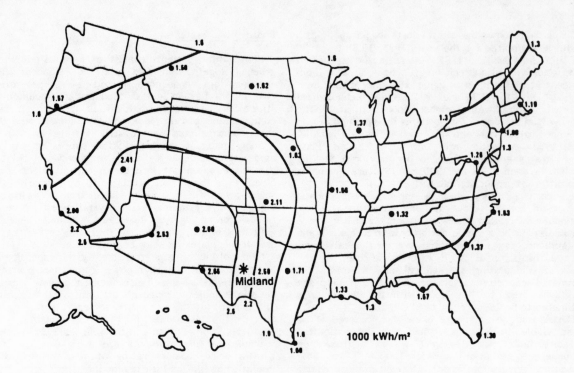

Figure B-2 **Average Annual Direct Normal Insolation (Taylor et al. 1979, p. 37; based on Sandia 1978)**

Table B-9.   CAPITAL COST ESTIMATES FOR A 100 MWp FLAT PLATE PHOTOVOLTAIC POWER PLANT IN THE YEAR 2000

(1978 dollars)

Item	Plant Costs	
	(thousand dollars)	($/kWp)
Module Cost	16,600- 66,600	166- 666
Balance of System Cost	40,000- 55,000	400- 555
Indirect Costs	20,000- 42,560	200- 425
Total Cost	76,660-164,160	766-1,642

NOTE: Assumes plant efficiency of 0.9 and peak insolation of 1 kW/m^2; module efficiency of 13.5% is used to estimate area-dependent balance-of-system costs. Although the costs shown here are for flat plate concept, the total costs could be considered representative of concentrator plant costs in the 2000 time frame. Range of cost estimates obtained from a variety of studies (See text).

projections presented in Table B-9 are based on a flat plate system concept, although they may be considered representative of the future costs of concentrator systems as well. Costs are not specific to any single solar cell technology because it is at present impossible to know which of the competing options will prove to be most cost effective. However, an efficiency range for flat plate cells of 13 to 15% is expected: 13.5% is assumed for estimates of area-dependent costs (land, structures, etc.).

A range for the capital costs of a 100 MW plant installed in 2000 is used because of the uncertainties of future photovoltaic system costs. Capital costs are divided into three categories: module costs; costs for other system components (balance of system including

array structures and installation, land costs, site preparation, and plant electrical equipment); and indirect costs, which reflect markups, engineering fees, and interest during construction. Costs are in 1978 dollars. As will be explained, the costs presented in Table B-9 represent the range of cost estimates contained in available studies but are not based on any single design or study.

Although current prices for photovoltaic modules (approximately $7 to 15 per W$_p$ in August 1979) are considerably above the costs shown in Table B-9, several studies conducted by the Jet Propulsion Laboratory indicate that the module costs are achievable (JPL 1980, pp. 4-11 - 4-16; Aster 1978). Module prices of $0.60/W$_p$ are believed to be obtainable by 1986 because of discrete technological improvements and economies of scale in either single-crystal silicon ingot or ribbon technologies. Learning and experience effects could further reduce costs to approximately $0.40/W$_p$ by 1990. If thin-film or other advanced module concepts prove successful, module prices ranging from $0.15 to .30 per W$_p$ are possible. Module prices in 2000 for 100 MW$_p$ are thus bracketed by the range $66.6 to 16.6 million. These module cost estimates shown are based on a plant efficiency of 90%. Thus 111 MW$_p$ of module capacity are needed for a plant with a peak rating of 100 MW.

Conceptual design studies have examined the balance-of-system costs for a variety of system concepts (General Electric 1977; Westinghouse 1977; Spectrolab 1977; Bechtel 1978; Bechtel 1979). The General Electric and Westinghouse studies indicate balance-of-system costs of approximately $0.40 per W$_p$. The Bechtel (1978) study estimates balance-of-system costs at $0.55/W$_p$ assuming 13.5% cell efficiency, 90% plant efficiency, and peak insolation of 1 kW/m^2. This breaks down into approximately $0.17/W$_p$ for power conditioning and controls, and $0.38/W$_p$ for land costs, array structures, and installation. A study by Motorola has examined low-cost array structures. It indicates that structure costs might be as low as $0.20/W$_p$ (Masser 1979).

Bechtel's 1979 study, which is for an experimental plant, estimated total balance-of-system costs at $1.70 to 2.90 per W$_p$. Because of the experimental nature of the plant examined, however, these costs are not believed to reflect commercial plant costs in 2000. As shown in Table B-9, balance-of- system costs are estimated at $0.40-0.55/W$_p$ here.

Indirect costs include markups, contingencies, and interest during construction. Indirect cost estimates in the conceptual design studies range rom 30% to 44% of total capital costs. A figure of 35% for indirect costs is used here.

Table B-10 shows capacity factors for flat plate photovoltaic systems in the DOE regions. The capacity factor is the average annual insolation in kWh/m^2 divided by the total number of hours per year. This is the annual proportion of peak hours of operation at peak insolation at 1 kW/m^2, or the average plant output on a 24-hour basis.

Table B-10.   CAPACITY FACTORS FOR FLAT PLATE PHOTOVOLTAIC SYSTEMS BY DOE REGION

DOE Region (City)	Annual Horizontal Insolation (kWh/m^2)	Capacity Factor
1 (Boston)	1367	.16
2 New York City)	1265	.14
3 (Washington, D.C.)	1515	.17
4 (Atlanta)	1681	.19
5 (Chicago)	1159	.13
6 (Fort Worth)	1889	.22
7 (Lincoln)	1609	.18
8 (Missoula)	1486	.17
9 (Phoenix)	2209	.25
10 (Seattle)	1155	.13

Estimates of annual operation and maintenance (O & M) costs (including insurance) are available from the General Electric, Westinghouse, and Bechtel (1978) studies cited above. Scaling the General Electric estimate to a 100 MW plant, annual O & M is equal to approximately $800,000 per year. Westinghouse O & M estimates are similar to those of General Electric. Bechtel O & M costs are given as 3.3 mills per kWh, which would be approximately equivalent to $640,000 per year using the Phoenix capacity factor of 0.22. The O & M cost estimate used in this study is $700,000 per year.

Ocean Thermal Energy Conversion (OTEC) Systems

OTEC is one of the ocean systems being developed by DOE. The others include wave energy, ocean currents, and salinity gradients as sources of power production. Of these four types of ocean systems, the OTEC concept is the most developed. OTEC systems are "based on the use of ocean temperature differences between warm surface waters and cold water depths to operate a heat cycle that generates electricity" (DOE 1980, p. III-87). OTEC plants have the advantage that they can operate as base load capacity.

The best U.S. regions for OTEC systems are tropical islands where suitable temperature gradients exist 1 to 6 miles offshore. In the Gulf of Mexico, plants would have to be located 80 to 150 miles from the coast (DOE 1980, p. III-88). The shorter distance from tropical islands reduces transmission costs not only because fewer miles of cable are required but also because alternating-current transmission can be used. Longer distances require direct-current transmission which, in turn, requires DC-to-AC conversion equipment on-shore (DOE 1980, p. III-80).

The optimum size of an OTEC plant may be around 400 MW (Shelpuk et al. 1980). On the basis of a survey of existing studies, Shelpuk et al. of SERI reports that cost estimates for the eighth OTC plant produced range between $2200 and $3200 per kW (1979 dollars) (Shelpuk et al. 1980). These cost estimates assume a 400 MW plant 80 miles off the coast of Louisiana and include the costs of the power, platform, mooring, cold water pipe, and electrical cable and DC-to-AC conversion systems. The cost of the eighth unit (compared to earlier units) excludes the cost of the shipyard needed to construct the platform system. Operation and maintenance costs have been estimated at 8 mills per kWh (1979 dollars) (Sullivan 1980).

For the purposes of this study, we assumed a capital cost of $2200 to 3200 per kW (1978 dollars), which includes the cost of the above components as well as a contingency fee of 10%. OTEC plants were also assumed to operate as base load capacity, at a capacity factor of 80%. Operation and maintenance costs were assumed to be 7 to 8 mills per kWh (1978 dollars).

APPENDIX C
HYDROELECTRIC POWER

Hydroelectric power is a renewable resource which already plays a major part in the nation's energy economy. The installed capacity of 64,000 megawatts supplies about 13 percent of the country's electrical power (the equivalent of 2.8 quads of primary energy). Hydropower capacity grew rapidly in the first half of the twentieth century. By the mid-1960s, the most favorable sites for large dams had been developed, the low cost of fossil fuels made further development less attractive, and the competition for land use and concern over environmental quality raised new barriers to implementation. The number of new hydropower projects was severely limited by these factors and the federal agencies responsible for hydropower (the Corps of Engineers and the Water and Power Resources Service, formerly the Bureau of Reclamation) were faced with concomitant restrictions. Nevertheless, additional hydroelectric power is an important energy source which is available from existing sites that are not utilized to their full potential, from select undeveloped sites for large hydroelectric systems, and from small hydroelectric sites, which have received increased attention in the past four years (1) because their minimal environmental impact circumvents many problems associated with large hydro systems.

The distribution of presently operating hydro power facilities is given in table 1. Large hydropower facilities are defined as those which have an installed capacity in excess of 15 megawatts and a hydraulic head (defined as the difference in elevation between the water levels at the intake and outflow) of at least 20 meters. Most of the large hydroelectric projects were designed to provide baseload power to the regions that they serve and four decades ago provided as much as 40 percent of the country's electrical energy. The extensive reservoirs at large hydroelectric projects provide a built-in storage capability that gives hydropower unusual flexibility. As other forms of generating capacity have taken over the bulk of the baseload supply, the role of hydropower has shifted toward supplying peaking power. Even in the northwest, where most of the hydropower generation is still baseload, this shift is occuring.

Small hydro facilities (often called low-head facilities) generally operate on a river, stream, canal, or irrigation channel in conjunction with a dam that is not designed for water storage during periods of lessened streamflow. Constrained to accept water as it arrives downstream ("run-of-the-river" operation), small hydro facilities operate with capacity factors that are closer to baseload rather than peaking values.

The U.S. Army Corps of Engineers, in its National Hydroelectric Power Resources Study (2) gives an aggregate capacity factor of 57% for existing small hydropower projects, 36% for additions of power house equipment to existing to low-head dams, and 26% for additions of new powerhouse equipment to existing large dams.

Additional Resource Availability

While it is regarded as common wisdom that most of the favorable large hydrosites have already been used up, there are significant opportunities for increasing hydropower production through upgrading existing facilities as well as developing new sites. A 90 day study (3) of existing dams done by the Corps of Engineers in 1977 identified a total of 54,600 MW of capacity from existing facilities. In particular, it cited:

o 15,800 MW of capacity gained by installing additional turbines and generators at existing dams, and

o 33,600 MW of capacity by constructing powerhouses at existing dams not presently used for hydropower.

The study also identified 5100 MW of capacity that could be gained by reworking turbines and generators at existing dams, but the improvements indicated fall more under the heading of routine maintenance and replacement of defective components than upgrading to improve the design efficiency.

National Hydropower Study Preliminary Inventory Large Hydropower
(over 15 MW based on 11,000 sites)

	Contiguous 48 States		Alaska	
Existing	61,000 MW	2.7 Q/yr	100 MW	—
Incremental	89,000 MW	2.1 Q/yr	400 MW	—
Undeveloped	179,000 MW	4.6 Q/yr	166,700 MW	4.4 Q/yr

The operating efficiency of a properly-maintained hydropower system is 86-88%. This figure could be taken to indicate that there is a margin for improvement (and increased hydropower production) through further research, but the fluid dynamics of hydraulic turbines is well understood and the operating efficiency of large generators is quite high, so further research and development to improve the efficiency of hydropower technology does not appear warranted. The reliability of hydropower equipment is also excellent.

The forced outage rate for hydro plants is only 1.5%, compared to a rate of 5 to 40% for other conventional central generating technologies. The powerhouse equipment at a hydrostation has a life expectancy of 30 years and the civil works (dam, pensitck, etc.) in excess of 50 years. Hydropower is a fully-developed technology.

The ideal efficiency of hydroelectric turbines is 88 percent and the maximum achievable efficiency for generators is 98 percent. Thus the overall efficiency of a properly maintained station is 85-86 percent. In older stations, which have worn, unbalanced or broken turbine rotors (called "runners"), the efficiency may fall to 65-70 percent. Replacement of broken runners, as well as the rewinding of generators with deteriorating dielectrics, was the task under discussion in the reference above.

The most comprehensive and up-to-date assessment of hydropower resources in the United States is the National Hydroelectric Power Resource Study of the Corps of Engineers. A three-year effort undertaken at a cost of $7 million, the study (hereafter referred to as the National Hydropower Study) began by identifying 50,000 existing dams and 5,000 undeveloped sites. This number is being reduced in four successive screening stages--two of which have been completed. The first screening--testing for sufficient storage, head and streamflow--reduced the total to 17,500 sites. The second screening, using a very relaxed economic criterion,* cut the inventory to 11,000 sites. Although the study is well into the third stage, the 11,000 site inventory published as the Preliminary Inventory (3) is the best documentation available for a work in progress. It shows the following capacity for the contiguous 48 states plus Alaska, for hydropower systems in excess of 15 MW.

These numbers represent the "maximum physical potential which could be developed, and no doubt overstate the realizable potential because the bank of 11,000 sites used to derive these numbers has not yet been closely scrutinized for economic or environmental feasibility, nor have competing water uses or legal barriers that would preclude development been considered. The third stage of screening has already reduced the inventory to 5,000 sites, and the study managers expect to ultimately identify a group of 3,000 sites, including 200 to 250 undeveloped sites (4). Economic feasibility is determined on a site-by-site basis by comparing the hydroelectric project with the most likely alternative thermal generating station (oil, gas, coal, or nuclear) under strict guidelines.*

Under those guidelines, the cost of fuel is incorporated into the comparison through "generalized power values" provided by the Federal Energy Regulatory Commission FERC), and those power values may lag far behind the market price of fuel. In a period of rising fuel prices, this lag serves to undervalue hydropower. Hydropower has also been at a competitive disadvantage in the past because fixed point-in-time costing rather than life-cycle costing has been used for purposes of comparison. Thus the reality that hydropower--with fixed capital costs, low and stable operating costs and no fuel costs--offers an inflation-proof generating system whereas the alternatives have had (and can be expected to have) costs rising faster than the general inflation rate has been overlooked. Given the historical restrictions placed upon hydropower cost comparisons, The National Hydropower Study may unduly restrict the inventory on the basis of economy alone. For the purposes of this study, it will be assumed that 50 percent of the capacity in the Preliminary Inventory could be economically developed by 2000, with an energy production of 60 percent of the figures given (because in a given sample the lowest capacity factor plants tend to be selected out first). For the purposes of this study, then, the following estimate is made:

The potential for Alaska is separated out because the undeveloped potential of this enormous resource (47 percent of the undeveloped U.S. potential) is

---

*The economic criterion was that benefits exceed the cost of the powerhouse.

---

*A new set of Principles and Standards is being issued by the National Water Resources Council, which will be applied to the final stages of the National Hydropower Study. The study will use come life-cycle cost principles, via the FERC Concept 5 model.

Assumed Amount of the Total National Large Hydropower
Potential that Could be Economically Developed by the Year 2000
Working Estimate of the National Large Hydropower Potential
(over 15 MW based on 11,000 sites)

Contiguous 48 States			Alaska	
Existing	61,000 MW	2.7 Q/yr	—	—
Incremental	44,000 MW	1.3 Q/yr	—	—
Undeveloped	89,000 MW	2.8 Q/yr	83,000 MW	4.4 Q/yr

located where it could be prohibitive to transmit the power to regions of existing demand and where the development of a large market near potential sites is problematical. The incremental potential, which could be installed at existing dams, is the most attractive target for further hydropower development, both because it would have minimal environmental and social impact and because it would generally be lower cost. A breakdown of the incremental potential by regions is included in Table 2. The undeveloped potential in the contiguous 48 states is sizeable, and perhaps not all would in fact be developed. (The Wild and Scenic Rivers Act of 1968 alone precludes 21,000 MW from development.) However, the total of incremental and undeveloped estimates is not far in excess of the figure given by FERC in 1977, namely 114,000 MW of capacity producing 4 quads per year (5). A breakdown of the undeveloped potential by regions is given in Table 4.

Resource Availability--Low Head Hydropower

Of the 50,000 dams in the country, only about 1,250 are presently used to generate power. In the past four years, thousands of low-head dams have been identified that could be used to produce hydropower. In 1977, the Corps of Engineers 90-day report ascertained that 24,000 MW was the potential from existing small dams now viewed as high. Low-head hydro was identified as one of eight mature technologies ready for commercialization by the Department of Energy, and the Small-Scale Hydroelectric Resources program was initiated. It now supports, in whole or in part, 77 small scale hydro projects, with a program budget of $28 million in 1980. Further work has reduced the initial estimate. The Preliminary Inventory of the National Hydropower Study identified 5400 MW of power producable from existing dams (excluding Alaska) with one-third of the potential in New England. An interim report by the New England River Basins Commission (6) indicated that up to 1000 MW of power could be developed at 1,750 existing non-generating dams. An in-house assessment by the Federal Energy Regulatory Commission puts the potential from small existing dams at 2,000 MW for the entire country; the DOE Small-Scale Hydro program unofficially estimates that the potential will fall between 2000 and 4000 MW,

when fully assessed. Such a capacity would yield 0.06 to 0.12 quads of energy per year.

Significant capacity could also be developed at log-head sites where no dams or impoundments exist now. The Preliminary Assessment of the National Hydropower Study identifies 7000 MW of capacity (excluding Alaska) that could produce 0.24 quads per year of energy. However, as will be laid out in the next section, such potential is likely to be the most expensive of all the categories of hydropower options. Fifty percent of the undeveloped low-head hydro potential is in the northwest, fifteen percent is in the southeast, and fifteen percent in the Great Plains. In contrast with the situation with existing dams, relatively little (7 percent) of the undeveloped potential is in New England.

Economics

Few energy technologies are more site specific than hydropower. Consequently the costs of completed projects exhibit a very large spread.

Data on large hydropower projects in the western states is available from the Bureau of Reclamation's Western Energy Expansion study, which considered (a) additions to existing Bureau of Reclamation facilities and (b) all new conventional hydroelectric facilities. Costs were determined by detailed assessment of each project. Among 31 proposed additions to existing facilities, 11 which had capacities ranging from 30 to 1000 MW had the following distribution of costs in $1978.

Additions to Existing Facilities
(1978 dollars)

$200 - $500/kW	2
$500 - $800/kW	6
$800 - $1100/kW	2
$1100 - $1400/kW	1
Total Projects	11

The range was from $370 to $1331/kW, with a median of $664/kW.

For new facilities, the WEES study developed detailed cost figures for 16 projects which would qualify as large hydro installations (over 15 MW). The distribution of these costs in 1978 dollars) was as follows:

New Facilities at
Undeveloped Sites
(1978 dollars)

$500 - $1000/kW	4
$1000 - $1500/kW	5
$1500 - $2000/kW	5
$2000 - $2500/kW	0
$2500 - $3000/kW	1
$3000 - $3500/kW	1
Total Projects	16

The range of costs was $542 to $3050/kW, with the highest capacity rating plants being the cheapest per kilowatt. (The plant in the $300-$3500 range has a proposed capacity of 28 MW.) For the purposes of this study, a figure of $1500 per kilowatt will be used.

For the costs of adding power producing facilities to existing small dams, two sources of information exist. Costs per kilowatt are available for 45 projects submitted to the DOE Small Hydro Program under PON-2. These were all proposals for adding or rehabilitating power houses at existing facilities--some only require adding generators in existing powerhouses while others require extensive work building new penstocks and tailraces. The spread of costs is shown below.

DOE Small Hydro Program-
Existing Dams and Facilities
(1978 dollars)

$500 - $1000/kW	13
$1000 - $1500/kW	18
$1500 - $2000/kW	9
$2000 - $3000/kW	5

The least expensive items on this distribution may not be conventional installations, but the addition of generating equipment to canals or water supply systems. (The $500/kW proposal is a plan to install a Pelton water wheel in the water supply system of the city of Santa Barbara.)

Another source of information is the Licensing Division of FERC, where there are 112 applications pending for permits and licenses for small hydro projects. Virtually all the projects in this group have capital costs under $200/kW, with some being as low as $1000/kW according to the director of licensing (8). For the purposes of this study, the cost of small hydro capacity installed at existing facilities will be taken as $1600/kW, in 1978 dollars.

Little interest has been shown in the construction of small hydro facilities at undeveloped sites, so there is little data available. The major difference between undeveloped and existing sites is the amount of civil engineering work required. Discussions with various experts indicates that the cost would be approximately double that for existing facilities, or $3200/kW. (Note that there are some additions that cost as much.) For the purposes of the study, a figure of $3200/kW will be used. Therefore the costs of new hydropower capacity breakdown is as follows:

Large-Scale (over 15 kW)
    Existing Facilities $600/kW
    New Facilities $1500

Small-Scale (15 kW and Under)
    Existing Facilities $1600
    New Facilities $3200

The costs of power produced vary widely, just as capital costs do. Power costs are very sensitive not only to the capital costs, but also to interest rates, tax rates and incentives, which vary markedly depending whether the project is private, municipal or federal. Short of a detailed breakdown of the pending applications and permits before FERC, the best that can be done is to say that power costs vary between 2 and 8 cents per kWh, with some license applications as high as 10¢/kWh. The only study of which the author is aware that cites resource potential as a function of economic conditions--including the return for power produced--is one by the New England River Basins Commission which will be discussed below.

Maximum Development of Hydropower Through 2000

The changing economics of energy in the past decade have already spurred a renewed interest in hydropower development. One indication of this is increased licensing activity--which has doubled yearly in the past three years. The number of applications for hydropower permits pending before FERC has gone from 18 in 1977 to 36 in 1978 to 78 in 1979 to 125 already this year with a projection of 200 at the fiscal year's end. Interest in hydropower will continue as costs of alternatives rise, and the director of the FERC licensing expects more hydropower capacity to be built in the next five year period than any other half-decade in the country's history.

The applications already on file with FERC represent 7100 MW of new conventional hydroelectric generating capacity, which will be completed by approximately 1988. Much more capacity could and probably will be built by 2000, assuming--as all signs indicate--that present trends continue. Most of the new capacity added by the end of the century can be expected to be in the category of large-capacity additions to existing sites, because of their lessened environmental effects. Given the dynamism of hydropower development, triple the 7,100 capacity now on FERC's books or 21,000 MW is likely to be built by 2000. A short survey of those experienced in the field resulted in opinions that this amount could "easily" be done, that 30,000 MW of new capacity would be "very ambitious" and only likely to be developed with strong congressional action.

The 21,000 MW figure will be adopted, with the assumption that 17,000 MW of capacity will be contributed by large incremental projects, 1,000 MW will be from large new projects, and 3,000 MW will be from small incremental projects--that is low-head hydro additions to existing dams. The 17,000 MW figure is well below the working estimates developed earlier--in fact, it is less than half the working estimate and less than one-fourth the National Hydropower Study preliminary inventory for the same category of capacity. The figure for large new projects is particularly low because the environmental social and legal barriers to them are quite complex, making the question of their development problematical; 1,000 MW is approximately twice the capacity recommended for further study in this category by the Bureau of Reclamation's WEES study.

The figure for low-head hydro is in the mid-range of the potential estimated to be available from the resource. The incentives for developing low-head hydro are particularly strong due to a number of programs. The licensing procedure has been modified for low-head projects and the time period for developing a low-head project is short--2 to 3 years. It is assumed, therefore, that most of the developable potential will be exploited by 2000.

Using the capacity factors derived from data in-house at FERC the yearly energy production of the projected capacity would be 0.61 quads per year.

Thus the total hydropower capacity of the United States in 2000 would be 85,000 MW with an annual energy production of 3.3 quads per year.

Outlook for the Future of Hydropower

The development of hydropower is already proceeding rapidly, given the inherent attractiveness of the technology and the incentives that are already in place. Particularly for small hydropower the incentives recently enacted make new developments attractive. The Windfall Profits Tax Bill has a provision that accords an 11 percent investment tax credit, over and above the general investment tax credit, to small hydro projects under 25 MW in capacity, and accords a lesser credit on a sliding scale for projects up to 125 MW. The National Energy Act authorized $300 million in low interest construction loans for small hydro facilities and also directed FERC to streamline its licensing procedures for small hydro (under 15 MW) sites with existing dams. By April 1979, FERC had streamlined its procedures not only for small hydro but for all sites with existing dams, in changes that could cut the required licensing time as much as half. The DOE small hydro program has awarded support for 54 feasibility studies and 23 demonstration projects.

Projected Additional Hydro Capacity to 2000

		Capacity Factor	Energy Production	
Incremental Large Projects (over 15 MW capacity)	17,000 MW	30%		0.45 quads/yr
New Large Projects (over 15 MW capacity)	1,000 MW	35%		0.03 quads/yr
Small Projects at Existing Dams (15 MW and under)	3,000 MW	50%		0.13 quads/yr
Totals	21,000 MW			0.61 quads/yr

Total Projected Hydro Capacity in 2000

Existing (all categories)	64,000 MW	2.8 quads/yr
Projected (all categories)	21,000 MW	0.6 quads/yr
Totals	85,000 MW	3.4 quads/yr

References

1. The President's Rural Hydroelectric Initiative of 1978 called for development of 100 small projects with a capacity of 3 MW each.
2. "Preliminary Inventory of Hydropower Resources," Vol. 1, compiled by the U.S. Army Corps of Engineers as part of the National Hydroelectric Power Resources Study, July 1979. Fort Belvoir, VA.
3. "Estimate of the National Hydroelectric Power Potential at Existing Dams," R. J. McDonald, Corps of Engineers Institute of Water Resources, July 1977.
4. R. J. McDonald, personal communication, February 14, 1980.
5. "Hydroelectric Power Resources of the United States," compiled as of January 1, 1976, Federal Power Commission, Washington, November 1976.
6. "Potential for Hydropower Development at Existing Dams in the Northeast," Volume I, Physical and Economic Findings and Methods, New England River Basins Commission, January 1980.
7. "Report of the Western Energy Expansion Study," Department of the Interior, Bureau of Reclamation, February 1977.
8. R. A. Corso, personal communication, February 26, 1980.

APPENDIX D
"NO LOSERS" RULES, EFFICIENCY,
AND UTILITY RATES
by Albert Smiley,
Princeton University, 1981

The purpose of this note is to show how, in principle, utility rates can be designed to reconcile the efficiency and equity objectives of utility conservation programs.

Suppose that the average cost of producing conventional utility energy is Ca per unit, the marginal cost is Cm, the marginal cost of saving energy is Cs per unit, and Ca is less than Cm. In addition, suppose that total consumption of conventional energy is currently Q units and that demand is expected to grow to $(1+g)Q$ units. The utility can satisfy the additional demand by producing gQ additional units of conventional energy or by installing energy saving devices at some of its customers' sites and saving gQ units of conventional energy. In the first case, the total cost of producing $(1+g)Q$ units of conventional energy is $(Ca+gCm)Q$ and the average cost per unit is Pc = $(Ca+gCm)/(1+g)$. In the second case, the total cost of Q conventional units plus gQ saved units is $(Ca+gCs)Q$. Since the cost of the saved energy is added to the cost of the conventional energy for ratemaking purposes, the average price of the Q conventional units is Ps = Ca+gCs. A customer who does not receive an onsite device is no worse off under the conservation program than he would have been in the absence of the program if the price of conventional energy is not higher under the program,

i.e. if Ps Pc

$$\text{or } (Ca + gCs) \frac{(Ca + gCm)}{(1 + g)}$$

$$\text{or } Cs \quad \frac{(Cm - Ca)}{(1 + g)}$$

Since Ca Cm, there may exist some onsite investments that will benefit all customers (i.e., those for which Cs satisifies inequality (1)). However, here may also exist some efficient onsite investments (i.e., where Cs Cm) that do not satisfy inequality (1). Thus, in order to satisfy the "no losers rule" implied by (1), some efficient onsite investments may have to be ruled out.

This conflict between efficiency and equity arises because the recipients of energy saving devices pay nothing for those devices. If this constraint is relaxed, it is possible to resolve the conflict by requiring the recipients to pay part of the cost of the devices. If an onsite investment is efficient, it must produce a cost saving.

There is clearly some price schedule for conventional utility service and some payment schedule for recipients that distributes this surplus in a way that leaves each customer no worse off than he would have been without the program. One option is to charge all customers the same price for conventional service that would have been paid in the absence of the program. This approach guarantees that nonrecipients will not be worse off, and it leaves all surpluses from the program to be distributed to the recipients. If the onsite equipment is efficient, the energy bill of the recipients will be lower than it would have been without the program.

This can be shown by assuming that each recipient of a conservation device makes a payment to the utility of

$$P = \frac{Cs\,(1 + g) + Ca - Cm}{(1 + g)}$$

for each unit of energy that is saved by the device. The total cost to the utility of producing Q conventional units and gQ saved units is now reduced by the payments from recipients:

Net Cost to Utility = $(Ca + gCs)\, Q - PgQ =$

$$\frac{(Ca + gCm)Q}{(1 + g)}$$

The average price of the Q conventional units is Pc = $(Ca + gCm)/(1 + g)$. Note that this is exactly the same as the price of conventional energy in the absence of the program calculated previously. Thus, no non-recipient is made worse off by the program.

It remains to be shown that if the onsite investments are efficient, each recipient's bill for saved energy will be lower than his bill for the same

422

amount of conventional energy. This will be the case if the price of saved energy is less than the price of conventional energy, i.e., if P Pc

or $\dfrac{Cs(1 - g) + Ca - Cm}{(1 + g)}$ $\dfrac{(Ca + gCm)}{(1 + g)}$

or Cs Cm.

Thus, if the onsite investments are efficient (i.e., if Cs Cm), the price of saved energy is less than the price of conventional energy and the recipients of onsite devices have lower total energy bills under the program.

It would also be possible for the utility to give its customers a direct grants (or equivalent loan subsidies) which will make all efficient onsite investments attractive to its customers without penalizing customers who do not participate in the program. It was shown above that if each recipient of an onsite device makes a payment to the utility of

$$P = \dfrac{Cs(1 + g) + Ca - Cm}{(1 + g)}$$

for each unit of energy that is saved, no non-recipient will be made worse off by the utility investment program and each recipient will reduce his total energy bill if the device is efficient. Clearly, a subsidy of S = Cs-P paid by the utility for each unit of energy saved by customer owned solar devices will have exactly the same effect as a payment of P by customers for each unit of energy saved by utility owned devices. This is so because the customer's net payment is Cs-S=P in both cases. Thus, the appropriate level for the subsidy is

$$S = Cs - P = \dfrac{Cm - Ca}{(1 + g)}$$

REFERENCES FOR CHAPTERS ONE THROUGH FOUR

References for Chapter One

BUILDINGS

RESIDENTIAL POTENTIAL

A.D. Little Inc. Engineering and Manufacturing Analysis in Support of Federal Minimum Efficiency Standards, February 21, 1980.

A.D. Little Inc. PG&E Assessment of Achieving Energy Conservation Potential 1980-2000, A Report to PG&E, September 1980.

AIA/RC "Life-Cycle Cost Study of Commercial Buildings" Dec. 79.

American Institute of Architects Research Corporation. "Phase One/Data Base for the Development of Energy Performance Standards for New Buildings," January 12, 1978.

Arthur D. Little, Inc. "Study of Energy-Savings Options for Refrigerators and Wate Heaters," Cambridge, MA 1977.

Batelle Pacific Northwest Laboraroty. "Economic Analysis of Proposed Building Energy Performance Standards," PNL-3044, 1979.

Berman S.M. et al. "Electrical Energy Consumption in California: Data Collection and Analysis," Lawrence Berkeley Laboratory, UCID-3847, 1976. Appendix A and Summary Report, available from the California Energy Commission Publications Office, 1111 Howe Ave., Sacramento, CA 95825.

Blue J.L. et al. "Buildings Energy Use Data Book, Edition 2," Oak Ridge National Laboratory, ORNL-5552, December 1979.

Bowman Ray. Personal Communication, Amana Inc., November 24, 1980.

Carrol W.C. "Annual Heating and Cooling Requirements and Design Day Per;formance for a Residential Model in Six Climates: NBSLD, BLAST-2, and DOE-1." Lawrence Berkeley Laboratory LBL-9270, 1979.

Design News, October, 1980.

DOE Monthly Energy Review, U.S. Department of Energy.

DOE Residential Energy Consumption Survey, DOE/EIA-0207/02 to 05, U.S. Department of Energy, 1980.

Edison Electric Institute. Annual Energy Requirements of Electrical Household Appliances, EEI-PUB #75-61 Rev. Undated.

Electric Power Research Institute. Patterns of Energy Use by Electrical Appliances, EPRI EA-682, January, 1979.

Erickson R.C. "Household Range Energy Efficiency Improvements" in "The Major Home Appliance Technology for Energy Conservation Conference," U.S. Department of Energy CONF-78-0238, Purdue University, February 1978.

Gadgil A. et al. "A Heating and Cooling Loads Comparison of Three Building Simulation Models for Residences: Two Zone, DOE-1, and NBSLD" in Changing Energy Futures (Pergammon Press, 1979).

Gadgil A. et al. Proceedings of the Fourth national Passive Solar Conference, Kansas City, MO, 1979.

General Electric Lamp Catalog.

General Electric press conference on electronic Halarc bulb, London, England, June 14, 1979.

Goldstein D.B., M.D. Levine, J. Mass. "Methodology and Assumptions for the Evaluation of Building Energy Performance Standards for Residences," Lawrence Berkeley Laboratory, LBL-9110, 1980.

Grimsrud D.T., R.C. Sonderegger, M.H. Sherman, R.C. Diamond, A. Blomsterberg, "Calculating Infiltration: Implications for a Construction Quality Standard," LBL-9416, Draft, 1980

Hirst E. "Understanding Energy Conservation" in Science, 106, 4418, November 2, 1979.

Hirst E. "Understanding Residential/Commercial Energy Conservation: The Need for Data." Minnesota Energy Agency, October 1979.

Hirst Eric et al. An Improved Engineering-Economic Model of Residential Energy Use, Oak Ridge National Laboratory, April, 1977.

Leger, E.H. and G.S. Dutt. "An Affordable Solar Home" Proceedings of the ?Fourth National Passive Solar Conference, Kansas City, MO 1979.

Levine, M.D., D.B. Goldstein, J. Mass. "Evaluation of Residential Building Energy Performance Standards." Presented at the Department of Energy/ASHRAE Conference in Thermal Performance of Exterior Envelopes of Biuldings, Orlando, Florida, 3-5 December 1979, LBL-9816.

Means Buildings Construction Cost Data, 1980, Robert Snow Means Company Inc., Kingston, MA.

NAHB, 1979 Research Results, National Association of Home Builders Research Foundation Inc., P.O. Box 1627, Rockville, MA 20850.

National Academy of Sciences. Committee on Nuclear and Alternative Energy Sources, 1979. See also Brooks and Hollander, Annual Review of Energy, 1979.

OTA 79. "Residential Energy Conservation" Vol. 1. Office of Technology Assessment, Washington, 20510.

Peart, V., DP.P De Witt, S.T. Kern. "Energy Savings Domestic Oven," in conference proceedings of Ref. 8.

Rosenfeld, A.H., D.B. Goldstein, A.J. Lichtenberg, P.D. Craig. "Saving Half of California's Energy and Peak Power in Buildings and Appliances in Long-Range Standards and Other Legislation," Lawrence Berkeley Laboratory Report, LBL-6865, 1978.

Rosenfeld, A.H., et al. "Building Energy Use Compilation and Analysis (BECA)", LBL 8912, 1979, and Energy and Buildings, 1980.

Rosenfeld, A.H. "Some Potentials for Energy and Peak Power Conservation in California," LBL 5926, 1977.

Sant, Roger W. "The Least-Cost Energy Strategy." The Energy/Productivity Center, Mellon Institute, Arlington, VA., 1979.

Schlussler L. "Final Report, The Design and Construction of an Energy-Efficient Refrigerator." The Quantum Institute, University of California at Santa Barbara, 1978.

Sears Fall/Winter 1979 Catalogue.

Sinden, F.W. "A Two-Thirds Reduction in the Space Heat Requirements of Twin Rivers Townhouse", Energy and Buildings 1, 243 (1978).

Turiel I., H. Estrada, M. Levine. "Life Cycle Cost Analysis of Major Appliances," LBL-11338, July 1980.

U.S. Department of Energy, Office of Conservation and Solar Energy, Office of Buildings and Community Systems, Notice of Proposed Rulemaking, Energy Performance Standards for New Buildings," January 12, 1978.

U.S. Department of Energy, Building Energy Performance Standards (BEPS), Notice of Proposed rules, Fed Reg. Nov. 18, 1979. The NOPR document cites 8 Technical Support Documents, and numerous technical reports.

Wright, J., A. Meier, M. Maulhardt, A.H. Rosenfeld. "Supplying Energy Through Greater Efficiency. The Potential for Conservation in California's Residential Sector," Lawrence Berkeley Laboratory, LBL-10738, December, 1980.

COMMERCIAL BUILDINGS

AIA Research Corporation and Syska & Hennessy Engineers, Inc., et al. (for the U.S. Department of Housing and Urban Development in cooperation with the U.S. Department of Energy), "Phase 1 for the Development of Energy Performance Standards for New Buildings", "Task Report, Data Analysis" January 1978.

AIA Research Corporation and Syska and Hennessy Engineers, Inc., et al. (for the U.S. Department of Housing and Urban Development in cooperation with the U.S. Department of Energy), Phase 2 for the Development of Energy Performance Standards for New Buildings", "Task Report, Commercial and Multi-family Residential Buildings," January 1979.

AIA Research Corporation, Hanscombe Associates, Inc., and Syska and Hennessy Engineers, Inc., (for the U.S. Department of Energy), "Energy Performance Standards for New Buildings: Life Cycle Cost Study of Commercial Buildings", December 1979.

AIA Research Corporation and Syska and Hennessy Engineers, Inc., (for the U.S. Department of Energy), "Energy Performance Standards for New Buildings, Analysis of ASHRAE Standard 90-75R," January 1980.

AIA/RC and Syska and Hennessy Engineers, Inc., (for the U.S. Department of Energy), "Analysis of Data Anomalies, Further Analysis of Phase 2 Buildings", March 1980.

Anderson, L. O., Bernander, K. G., Isfalt, E., and Rosenfeld, A. H., "Storage of Heat and Cool in Hollow-Core Concrete Slabs. Swedish "Building Energy Use Compilation and Analysis (BECA), An International Comparison and Critical Review, Part A: New Residential Buildings Lawrence Berkeley Laboratory (LBL-8912, EEB-BECA-79-1) November 1979.

ASHRAE, "ASHRAE Handbook and Product Directory 1977 Fundamentals", 1977.

EIA, "1977 Annual Report to Congress".

Jackson, J. R., and Johnson, W. S., "Commercial Energy Use: A Disaggregation by Fuel, Building Type and End Use, ORNL/CON-14", February 1978.

Oak Ridge National Laboratories, "ORNL Commercial Energy Use Simulation, 1970-2000", March 1980.

Swedish Experience, and Application to large, American-style Buildings. Lawrence Berkeley Laboratory LBL-8913 October 1979. Rosenfeld, A.H. W.G. Colborne, C.D. Hollowell, S.P. Meyers, L.J. Schipper, B. Adamson, B. Hidemark, H. Ross, N. Milbank, M.J. Uyttenbroeck and G. Olive.

Rudy, J. Sigworth, H., Rosenfeld, A. H., "Saving Schoolhouse Energy: Final Report", Lawrence Berkely Laboratory (LBL-9106) June 1979.

RENEWABLE ENERGY SUPPLY

AIA Research Corporation, A Survey of Passive Solar Buildings, Feb. 1979, p. 111.

AIA Research Corporation, A Survey of Passive Solar Buildings, soon to be published, Tape 114.

AIA Research Corporation, A Survey of Passive Solar Homes, soon to be published, Tape #28.

Anderson, B., testimony at BEPS hearing, March 27, 1980, Washington, D.C.

ASHRAE 1978, ASHRAE Handbook of Fundamentals, pp. 58.6-58.7, 1978.

Balscomb, J. B., Conservation and Solar: Working Together, Systems Simulation and Economic Analysis Conference, San Diego, California, Jan. 1980.

Bergquam, J., et al., A Comparative Study of SDHW Systems in California, California Energy Commission, June 1979.

Paul J. Blake and William C. Gerhert, "Load and Use Characteristics of Electric Heat Pumps in Single Family REsidence," prepared by Westinghouse Electric Corporation for EPRI. EPRI EA-793 Project Final Report, vol. I, June 1978 p.

Telephone conversation with David Block, project architect, and Laurent Hodges, project builder and owner-occupant, May 1980.

Booz-Allen and Hamilton, Inc., SHAC Evaluation Project Phase I, Final Report, Dec. 1979.

Buckles, W. E., Klein, S. A., Duffie, J. A., Analysis of Solar Water Heating Systems, ISES Annual Meeting, Atlanta, GA, 1979.

U.S. Bureau of the Census, Construction Report, Series C-25, Characteristics of New One-Family Homes, 1973, U.S. Department of Commerce, C-25-73-13, Washington, D.C., 1974.

U.S. Bureau of the Census, Current Population Reports, Series P-25, No. 873, Estimates of the Population of Counties and Metropolitan Areas: July 1, 1977 and 1978, U.S. Government Printing Office, Washington, D.C., 1980.

Cooper, J., Environmental Impact of Residential Wood Combustion Emissions and Its Implications, Wood Energy Institute Wood Heating Seminar VI, Atlanta, GA, Feb. 25-28, 1980.

Telephone conversation with Mrs. Crosley, owner-occupant, May 1980.

Davies, J. O., III, Public Attitudes and Actions Regarding Wood as a Home Heating Alternative, Wood Heating Seminar V, Sherator, St. Louis, Sept. 12, 1979.

Domestic Policy Review of Solar Energy, A Response Memorandum to the President of the United States, 1978.

Eldridge, R. R., Wind Machines, 2nd Ed., Metrek Division, the MITRE Corporation, Van Nostrand Reinhold Company, 1980.

Energy Information Administration, U.S. Doe, New and Regrofit Solar Hot Water Installations in Florida, January through June 1977, HCP/15663-01, April 1978, p. 12.

Energy Information Administration, U.S. DOE, Residential Energy Comsumption Survey: Characteristics of the Housing Stock and Households, DOE/EIA-0207/2, Feb. 1980.

Energy Information Administration, U.S. DOE, Residential Energy Consumption Survey: Conservation, DOE/EIA-0207/3, Feb. 1980.

Farrington, R., Murphy, L. M., Noreen, D. L., A Comparison of Six Generic Solar Domestic Hot Water Systems, Report SERI/RR-351-413, April 1980, Appendix B.

Telephone conversation with Carl Fike, Project design/builder, May 1980.

Telephone conversation with Charles Fowlkes, project designer, builder, owner/occupant, May 1980.

Franklin Research Center, The First Passive Solar Home Awards, 1979, p. 60.

Franklin Research Center, The First Passive Solar Home Awards, p. 170.

Franta, G., ed., Proceedings of the 4th National Passive Solar Conference, Vol. 4, American Section of the International Solar Energy Society, Inc., University of Delaware, Oct. 1979.

Telephone conversation with Michael Frerking, project designer, and Ed Pollock of Automation Industries, Inc., project monitor, May 1980.

Hapgood, W., Bemis, J. R., Report on the Performance of the Acorn Solar Heated House for the winter of 1975-1976, Acorn Structures, Inc., Concord, Mass.

Hull, J. R., Hodges, L., Block, D., The Potential for Passive Retrofit in Iowa, Proceedings of the 4th National Passive Solar Conference, Kansas City, October 3-5, 1979, p. 590.

Krawiec, F., Thorton, J., Edesess, M., An Investigation of Learning and Experience Curves, SERI, SERI/TR-353-459, April 1980.

Telephone conversation with Eugene Leger, project designer, builder, owner/occupant, May 1980.

Maxwell, T. T., Dyer, D. F., et al., Improving the Efficiency, Safety, and Utility of Woodburning Units, DOE, ORO-5552-T7, March 1979.

MITRE 1977 MITRE Corporation NETREK Division, "A System for Projecting the Utilization of Renewable Resources," prepared for US ERDA, ERHQ/2322-77/4, September 1977.

Office of Conservation and Solar Energy/Office of Buildings and Community Systems, U.S. DOE, Passive and Active Solar Heating Analysis, DOE/CS-0117, Nov. 1979.

Office of Solar Applications, Passive and Hybrid Solar Buildings Program, U.S. DOE, Passive Solar Design Handbook, Vol. 2, Passive Solar Design Analysis, DOE/CS-0127/2, Jan. 1980.

Office of Technology Assessment, Residential Energy Conservation, Vol. 1, Washington, D.C., 1979, p. 71.

The Passive Solar Institute, The 1st Passive Solar Catalog, Chapter on passive solar water heating, 1978.

Private communication, Taos Solar Energy Association and private communication, San Luis Valley Solar Energy Association.

Proceedings of the Second National Passive Solar Conference, Philadelphia, PA, 1978, p. 8.

Proceedings of the Second National Passive Solar Conference, Philadelphia, PA, 1978, p. 117.

Proceedings of the Third National Passive Solar Conference, Philadelphia, PA, 1978, p. 173.

Proceedings of the Third National Passive Solar Conference, Philadelphia, PA, 1978, p. 771.

Proceedings of the Fourth National Passive Solar Conference, Philadelphia, PA, 1978, p. 317.

Proceedings of the Fourth National Passive Solar Conference, Philadelphia, PA, 1978, p. 704.

Reid, R., Tomlinson J., Chaffin, D., Solar Energy in the TVA System: A Proposed Strategy, Vol. IV, paper 2, p. 7, Sept 1979.

Sample calculations for Program Announcement, RCS Program, 1980.

Sandia Laboratories, Passive Solar Buildings, July 1979, p. 19.

Shurcliff, W. A., Solar Heated Buildings of North America: 120 Outstanding Examples, Brick House Publishing Company, New Hampshire, 1978, p. 226.

Solar Age Magazine, Vol. 2, #11, November 1977, p. 31.

Solar Energy Industries Association, Solar Workshops Financial Incentives, under DOE contract, DE-FG-01-79cs30293, March 1980, p. 2-9.

Solar Energy Research Associates, Survey of Solar Domestic Water Heaters: Economic and Market Analysis for Arizona, 1979.

Trial audits for the RCS program conducted in Portland, Oregon by SERI.

SERI, Passive Solar: It's A Natural, soon to be published.

Stromberg, R. P., Woodall, S. O., Passive Solar Buildings: A Compilation of Data and Results, SAND 77-1204, Sandia Laboratories, Aug. 1977.

Trefil, J. S., Woodburner Efficiency--How to Measure Woodstove and Fireplace Performance, Popular Science, Jan. 1979, p. 102-104.

Trefil, J. S., Woodstoves Glow Warmly Again in Millions of Homes, Smithsonian Magazine.

Wright, B., Akers, J., eds., Proceedings of the 1st New England Site-Built Solar Collector Conference, Worchester Polytechnic Institute, Mechanical Engineering Department, 1978, p. 85.

Wornum Report, working draft, California Energy Commission, April 18, 1980.

Untitled paper sent by Susan Yanda.

Young, M. F., Bawghn, S. W., Economics of Solar Domestic Hot Water Heat in California, Systems Simulation and Economic Analysis Conference Proceedings, Jan. 1980.

Yudelson, J., Solar System Sales in California, Estimates and Predictions, Solar Cal Office, Jan. 1979.

Wood Energy Institute 1979.

BUILDINGS POLICY

Aaron, Henry J., "Shelter and Subsidies: Who Benefits from Federal Housing Policies?" The Brookings Institution, Washington, D.C. 1972.

Ashworth, John et al. "The Implementation of State Solar Incentives: A Preliminary Assessment," SERI, January, 1979.

Associated Press. "Jersey Utility: Winning Idea is Backfiring," wire story, September 13, 1974.

Balcomb, J.D. Conservation and Solar: Working Together paper submitted to the Fifth National Passive Conference, Amherst, Massachusetts, October 19026, 1980.

Baldwin, Fred D. "Meters, Bills and the Bathroom Scale," Public Utilities Fortnightly, February 3, 1977.

Barnes, Peter. "Solar Financing Through Utility Credits: A Proposal by the Solar Industry," January 30, 1980, The Solar Center, San Francisco, Calif.

Barrett, David, Epstein, Peter, and Haar, Charles M. Financing the Solar Home Lexington Books, D.C.

Heath and Company; Lexington, Massachusetts, Toronto 1977.

Battelle Pacific Northwest Laboratory. "An Analysis of Federal Incentives Used to Stimulate Energy Production," June, 1978, PNL 2410, Bruce Cone et al. and subsequent documents PNL 2609, PNL 2701.

Bernstein, Scott. "Energy Conservation in Chicago," The Neighborhood Works, February 22, 1980, Vol. 3, No. 4.

Bezdek, Roger H. "Rationale for Federal Incentives for the Development of Solar Energy," draft.

Bezdek, Roger H. and Cone, Bruce W. "Federal Incentives for Energy Development," draft, to be published in Energy - The International Journal.

Bezdek, Roger H. and Sparrow, Dr. F.T. "Are Subsidies for Solar Energy Development Justified on the Basis of Economic Efficiency," Draft, October 1979.

Bleviss, Deborah. Federation of American Scientists, "The ACEEE-FAS Residential Conservation Workshop," notes from workshop. January, 1980.

Blumstein, Carl et al. "Overcoming Social and Institutional Barriers to Energy Conservation," Energy, Pergamon Press Ltd., 1980 Great Britain.

Bohn, Robert A., et al. Energy, Environment and Resources Center, The University of Tennessee, "Solar Energy in the TVA System: A Proposed Strategy," final draft, January, 1980.

Buck, Claudia. "Save Your Energy: Balancing Consumers, Financiers," California Life, June 28, 1980.

Burger, Stephen H. "Testimony on DOE's proposed Energy Efficiency Standards for Consumer Products Other Than Automobiles" August 11, 1980.

California Energy Commission. "Nonresidential Energy Efficiency in California, Achieving a 20 Percent Improvement by 1985," staff draft 3/13/80.

California Energy Commission. "Solar Information for Local Governments," September, 1979.

California Energy Commission. Achieving Energy Efficiency in Existing Buildings, Staff Report, P110-80-003, July 1980.

California Energy Commission Staff. "Energy Service Corpora;tions: Opportunities for California Utilities" Draft November, 1980.

Carhart, Steven C., Mulherkar, Shirish S. Rennie, Sandra M. and Penz, Alton J. Creating New Choices: Innovative Approahces to Cut Home Heating and Cooling Costs. Energy Productivity Report No. 3 Energy Productivity Center, Mellon Institute, Arlington, Virginia, 1980.

Carlson, Paul T, personal communication to Mr. James R. Tanck, Office of Conservation and Solar Energy, U.S. Department of Energy August 25, 1980.

Center for Renewable Resources. "Opportunities for Informed Public Interest/Citizen Participation in the Development of Regulations by DOE and the Implementation by States of Programs Called for in Existing Legislative Initiatives Dealing with Energy Conservation and Renewable Resources," 1980.

Clear, Robert and Berman, Sam. "Cost Effective Visibility Based Design Procedure for General Office Lighting" Lawrence Berkeley Laboratory, undated.

Committee on Interstate and Foreign Commerce. Canadian Home Insulation Program, Committee Print prepared for the Use of the United States House of Representatives, November 2, 1979, Print 96-IFC41.

Conley, Craig. "How to Finance the Solar Community," Soft Energy Notes June/July 1980, Vol. 3, No. 3.

Council of State Housing Agencies. "Annual Report 1978/79," Washington, D.C.

Council on Environmental Quality. "Solar Energy: A Review of Progress and Recent Policy Recommendations," preliminary review draft, December, 1977.

Council on Wage and Price Stability. Regulatory Analysis Review Group "Department of Energy's Proposed Energy Performance Standards for New Buildings (BEPS)" April 30, 1980.

Darley, John M. et al. Scorekeeping for Retrofits: Issues Pertinent to the Management of the 1000 House Pilot Project in Lakewood, N.J. Princeton University, Center for Energy and Environmental Studies, August 1, 1980.

Department of Energy. "Solar Energy Objectives, Calendar Year 1980," April 1980.

Department of Energy. "Residential Conservation Service Program," Federal Register, Wednesday, November 7, 1979; Vol. 44, No. 217.

Department of Housing and Urban Development. "Survey of Mortgage Lending Activity," various dates. Washington, D.C.

Dickinson, Bruce, O'Regan, Brian; and Sohl, Barbara House Doctor Manual, Princeton University, August 11, 1980.

DOE. Office of Conservation and Solar Energy, "Energy Conservation Program for Consumer Products," Monday, June 30, 1980, Federal Register Vol. 45, No. 127 43976-44087.

DOE, Office of Policy and Evaluation. "Low Energy Futures" (Review Draft) April 7, 1980.

DOE. "Domestic Policy Review of Solar Energy," February 1979 and Appendices.

DOE. "Active Solar Program Strategy," draft, 1980.

DOE. "Solar Energy: Program Summary Document FY 1981, January, 1980.

DOE. "Congressional Budget Request FY 1981," January 1980.

DOE. "Energy Performance Standards for New Buildings: Proposed Rule," Federal Register Wednesday, November 28, 1979; Vol. 44, No. 230.

DOE/CS "Residential Conservation Service Program: Regulatory Analysis" U.S. Department of Energy, DOE/CS-00104/1 October 1979.

DOE/EIA Characteristics of the Housing Stock and Households: Preliminary Findings from the National Interim Energy Consumption Survey, October 1, 1979, U.S. Department of Energy, Energy Information Administration DOE/EIA-0199/P.

DOE/EIA Residential Energy Consumption Survey: Conservation U.S. Department of Energy, Energy Information Administration, February 1980 DOE/EIA-0201/3.

Doyle, Jack. Lines Across the Land, Environmental Policy Center, 1980.

Dutt, Gautum. Princeton University, personal communications.

Dutt, Gautam and Williams, Robert. "State Government Responsibilities in Residential Energy Conservation," draft, January 17, 1980.

Efford, Ian, Director Division of Energy Mines and Resources, Canada. Private communications regarding CHIP program.

Epstein, Herb and Hayes, Denis. "Strategy for Accelerated Solar Development," memorandum to Al Alm dated October 25, 1978.

Federal Home Loan Bank Board. Energy Conservation Lending Programs, pamphlet, undated.

Federal Home Loan Mortgage Corporation. The Mortgage Corporation "Energy Survey: Facts and Findings" 1980.

Federation of American Scientists "Energy Conservation and Renewable Energies: A Summary of Legislative Progress" August 1980, FAS.

GAO. April 15, 1980, "Federal Demonstration of Solar Heating and Cooling on Commercial Buildings Have Not Been Very Effective."

GAO. "The Federal Government Needs a Comprehensive Program to Curb Its Energy Use," December 12, 1979.

GAO. "Commercializing Solar Heating: A National Strategy Needed," July 20, 1979.

GAO. "Solar Demonstrations on Federal Residences--Better Planning and Management Control Needed," April 14, 1978.

GAO. "Federal Demonstration of Solar Heating and Cooling on Private Residences--Only Limited Success," October 9, 1979.

GAO. "The Solar in Federal Buildings Demonstration Program," August 10, 1979.

Gerardi, Natalie and Gers, Barbara Behrens. "What Home Shoppers Seek in Six Major Markers" Housing, October, 1978 pgs. 53-76.

Hailey, John. "Pacific Gas and Electric Company's Experiences in Energy Conservation in Residential New Construction" Prepared for the 1980 Summer Study on Building Energy Efficiency by the American Council for an Energy Efficient Economy, August 1980.

Hailey, John and Richardson, Brion C., Energy Conservation Home Program, Pacific Gas and Electric Co, January, 1980.

Harris, Jeff. "Measuring and Evaluating the Progress of Energy Conservation in Buildings: Some Design Consideration," DRAFT report March 19, 1980, California Energy Commission, Lawrence Berkeley Laboratory.

Hirst, Eric and Armstrong, John R. "Managing State Energy Conservation Programs: The Minnesota Experience," May, 1980, Minnesota Energy Office.

Hollander, Peter et al. "A National Passive Solar Communication Plan" (Review Draft) SERI, August 1979.

HUD. "Programs of HUD," Washington, D.C., U.S. GPO.

Jackson, J.R. "Energy Use in the Commercial Buildings Sector: Historical Patterns and future Scenarios" Draft Report to the Solar Energy Research Institute, November, 1980 SERI/TR-00179-1.

Johnson, Steven B. "A Survey of State Approaches to Solar Energy Incentives," July 1979, SERI.

Kahn, Edward. "Using Utilities to Finance the Solar Transition," Solar Energy Law Reporter, SERI.

Kaufman, Alvin and Daly, Barbara M. "Alternative Energy Conservation Strategies: An Appraisal," Congressional Research Service, Library of Congress, April 29, 1977.

Laughlin, Connie B. Q. "Energy Conservation: The State of the States," National Governors Association Energy Program, 1978.

Levine, Alice and Raab, Jonathan, Solar Energy, Conservation, and Rental Housing SERI/RR-744-901 October 1980.

Levine, Mark D. "Testimony on the Department of Energy Building Energy Performance Standards Before the Senate Sub-Committee on Energy Regulation of the Committee on Energy and Natural Resources" June 4, 1980.

Levine, Mark D., Goldstein, David B. Evaluation of Residential Building Energy Per;formance Standards, Lawrence Berkeley Laboratory, Energy and Environment Division, December, 1979 LBL-9816.

MacKenzie, Jim. "Summary of Remarks by Participants in Institutional Working Group" 8/14/80 summary of proceedings of the Santa Cruz ACEEE Conference.

Mattes, Martin A. "The Public Utility's Role in Residential Energy Conservation" presented December 9, 1980, Transportation Systems Center, Cambridge, Mass.

Meier, Alan. "Final Report of the Energy Conservation Inspection Service" Lawrence Berkeley Laboratory, Energy and Environment Division, March 1980, LBL-10739.

Moore, Glen J. "Solar Energy Legislation in the 95th Congress, Second Session, With a Summary of the Solar Provisions of the National Energy Act," Congressional Research Service, Library of Congress, November 14, 1978.

Moore, J. Glenn. "Solar Energy Legislation in the 95th Congress," undated, Congressional Research Service, Library of Congress.

Moore, J. Glenn. "Solar Energy Legislation in the 96th Congress, First Session," Congressional Research Service, Library of Congress, January 8, 1980.

National Association of Home Builders. "Builders Reports," Analysis of Federal and State Housing Programs, July 1979.

Office of Technology Assessment. "Analysis of the Conservation and Solar Energy Programs at the Department of Energy," May, 1980.

Office of Technology Assessment. "Residential Energy Conservation," 1979 U.S. GPO.

Office of Technology Assessment. "Application of Solar Technology to Today's Energy Needs," June, 1978, U.S. GPO.

O'Neal, Dennis L. "Energy Consumption in the Residential Sector: An Historical Analysis" Draft report to the Solar Energy Research Institute, September, 1980.

O'Neal, Dennix L. "Energy Use in the Residential Sector: An Examination of Four Possible Futures" Draft report to the Solar Energy Research Institute September, 1980.

Opinion Research Corporation. Perspective on Energy: America's Homeowners Speak Out, survey of homeowner attitudes on energy conducted for the Dow Chemical Company, January 18, 1980.

Oregon Department of Energy, Alternative Energy Development Commission. Draft Report, May 1980.

Owens, P.G.T. Energy Conservation and Office Lighting, updated paper.

Pacific Power and Light. "Zero Interest Weatherization Program, Cost Benefit Summary: Oregon" presented at the Keystone Conference on Utilities, March 1980.

Pardo, Brian. Private communication, April 8, 1980.

Partridge, Robert D. "Statement before the Subcommittee on Conservation and Credit, House Committee on Agriculture" March 19, 1980.

Penny, Susan C. "Long Term Financing Consideration for the Energy Transition," a speech at U.S. DOE/ National Consumer Research Institute Conference, April 24-25, 1980, Washington, D.C.

Perwin, Elizabeth. "Implementation of State Solar Information Outreach Programs in Selected States," SERI, January, 1980.

Peterson, Chris. "Report on Conservation Trend Indicators". Department of Energy, Office of Policy and Evaluation, Office of Conservation, Draft August 21, 1980.

Public Utilities Commission of the State of California, Decision No. 91328, February 13, 1980.

Reenie, Sandra M., Testimony Before the House Subcommittee on Energy Development and Applications, Committee on Science and Technology, September 25, 1980.

Rich, Spencer. "Taj Mahal' in New York: Sypmtom of Rent Subsidy Headache". The Washington Post, Saturday August 16, 1980, pg. A2.

Riegel, Kurt W. and Solomon, Suzanne, E. "Getting Individual Customers Involved in Energy Conservation," Public Utilities Fortnightly, November 7, 1974.

Roessner, J. David. "Implementing State Solar Financial Incentives and R, D, & D Programs" SERI, May 1980.

Roseman, Herman G. "Should Electric Utilities finance Conservation by Consumers?" Presented to American Council for an Energy Efficient Economy 1980 Summer Study on Building Energy Efficiency, August 1980.

Rosenfeld, Arthur H. and Goldstein, David B. et al. "Saving Half of California's Energy and Peak Power in Buildings and Appliances via Long-Range Standards and Other Legislation," paper submitted to California Policy Seminar, May 10, 1978.

Rosenfeld, Arthur H. Testimony on BEPS Before the Subcommittee on Energy Regulations, June 5, 1980.

Saitmen, Barry. "Commercializing Solar Energy Through Municipal Solar Utilities," American Section of International Solar Energy Society, Inc.

Scott, Eric. "Pacific Gas and Electric Company's Experiences in Energy Conservation in the Commercial Industrial Agricultural Sector" paper prepared for the 1980 Summer Study on Building Energy Efficiency sponsored by ACEEE, August, 1980.

Seaver, W. Burleigh and Patterson, Arthur H. "Decreasing Fuel-Oil Consumption Through Feedback and Social Commendation," Journal of Applied Behavior Analysis, September, 1976.

Shorey, Joan. Testimony on BEPS before the Senate Subcommittee on Energy Regulations, June 26, 1980.

Sliger, Douglas L. Tennessee Valley Authority Modular Retrofit Experiment, TVA Division of Energy Conservation and Rates, Home Insulation Program, June 17, 1980.

Southern States Energy Board. "Lighting and Thermal Efficiency Standards, An Analysis of State Initiatives," February 1980.

Spivak, Paul. "Land-Use Barriers and Incentives to the Use of Solar Energy," SERI, August, 1979.

Stern, Paul and Gardner, Gerald T. "A Revier and Critique of Energy Research in Psychology" Social Science Energy Review, Volume 3, No. 1, Spring 1980 Yale University, Program on Energy and Behavior.

Stern, Paul C., and Kirkpatrick Eileen M. "Energy Behavior: Conservation without Coercion" Environment, December 1977, ppgs. 10-15.

Stevens, James. "How Now DOE?" Appliance, July 1980 pgs 42-45.

Subcommittee on Energy and Power. Memorandum, "Department of Energy Budget--Conservation and Solar Energy," February 11, 1980.

The Bureau of Public Affairs, Inc. "State Housing Finance Agencies," November 17, 1975.

The Conservation Foundation. "BEPS GRAM," various dates.

U.S. Department of Energy. "Conservation Objectives Calendar Year 1980," Washington, D.C., January 1, 1980.

U.S. Department of Energy. Certification/Enforcement Analysis. Consumer Products Efficiency Brance, DOE/CS-0170 June 1980.

U. S. DOE Advance Notice of Proposed Rulemaking and Request for Public Comments Regarding Energy Efficiency Standards for Dishwashers, Television Sets, Clothes Washers, Humidifiers and Dehumidifiers, Office of Conservation and Solar Energy, DOE, Federal Register, Thursday, December 13, 1979; Vol. 44, No. 241 72276-85.

U.S. DOE An Analysis of the Economic Impacts on the Appliance Industry Due to DOE Testing Procedures, Preliminary Draft April 1979, under contract #EM-78-C-01-5141.

U.S. DOE Conservation and Solar Fact Book, prepared by Office of Policy, Planning and Evaluation, Assistant Secretary Conservation and Solar Energy, Department of Energy, September 23, 1980.

U.S. DOE Energy Efficiency Standards for Eight Consumer Products: Public Meeting Clarification Questions and Answers, Consumer Products division, August, 1980 DOE/CS-0185.

U.S. DOE Solar Buildings: Market Background Report, "Office of Solar Applications for Buildings, Office of the Assistanc Secretary for Conservation and Solar Energy, U.S. Department of Energy, Draft final Report, October 21, 1980.

U.S. DOE <u>Solar Energy</u>, Program Summary Document FY 1981, U.S. Department of Energy, Assistant Secretary for Conservation and Solar Energy, DOE/CS-0050 August, 1980.

U.S. DOE <u>Study of the Adverse Impacts of Improved Appliances due to Minimum Efficiency Standards</u>, Draft, November 39, 1079, under contract #DE-AC03-79-CS10757.

U.S. Dept. of Housing and Urban Development. "Compendium of Federal Programs Related to Community Energy Conservation," February 1979.

Whitcomb, John. Briefing at SERI, May 29, 1980, personal communication.

White, Sharon Stanton. "Municipal Bond Financing of Solar Energy Facilities," SERI, December 1979; SERI/TR-434-191.

White, Sharon Stanton. Private letter , Dr. Jan Laitos of SERI, July 17, 1980, regarding Windfall Profits Tax.

References for Chapter Two
INDUSTRY

A. D. Little, Inc. 1978a. Energy Efficiency and Electric Motors. HCP.M50217-01. Washington, DC: U.S. Department of Energy. April.

A.D. Little, Inc. 1978b. "Research, Development, and Demonstration for Energy Conservation, Preliminary Identification of Opportunities in Iron and Steel-making," a report to U.S. Department of Energy.

Adams, Walter, ed. 1977. The Structure of American Industry. New York, NY: Macmillan Publishing Co.

Aerospace Corp. 1978. High-Temperature Industrial Process Heat: Technology Assessment and Introduction Rationale. ATR-78(7691-03)-2. El Segundo, CA: Aerospace Corp., Energy and Transportation Division. 3 March.

Aerospace Corporation. 1979. Field Survey of High-Temperature Solar Industrial Process Heat Applications. ATR-79(4820)-1. El Segundo, CA: Aerospace Corporation, Energy and Resources Division. 15 April.

Alvis, Robert L. 1978. Solar Irrigation Program Status Report: October 1, 1977. SAND 78-0049. Albuquerque, NM: Sandia Laboratories. March.

American Iron & Steel Institute 1976, "Energy Conservation in the Steel Industry", Washington, D.C.

American Iron & Steel Institute 1980, "Steel at the Crossroads: The American Steel Industry in the 1980's," Washington, D.C., Jan.

American Metal Market. 1979. Metal Statistics 1979. New York, NY: Fairchild Publications.

American Paper Institute, 1979. As presented in Industrial Energy Use Data Book, Oak Ridge National Laboratory, 1979.

American Physical Society. Summer Study on Technical Aspects of Efficient Energy Utilization. 1974. Available as W.H. Carnahan et al. Efficient Use of Energy, A Physics Perspective, from NTIS (PB-242-773), or in Efficient Energy Use. Volume 25 of the American Institute of Physcis Conference Proceedings.

Association of American Universities 1979, "University Contribution to the National Energy Program: New Knowledge, New Talent, New Integration," a report of the Energy Advisory Committee of A.A.U., October.

Ayres, R.U.; M. Narkus-Kramer. 1976. "An Assessment of Methodologies for Estimating National Energy Efficiency." International Research and Technology. McLean, Va. June.

BDM Corporation 1980, "New Technological Options for End-Use Efficiency--Ultraviolet Radiation Curing", Mc Lean VA 22102, Jan.

Babcock, William H.; Seiegel, Steven B.; Swanson, Christina A. 1978. Energy Demands 1972 to 2000. HCP/R4024-15. Washington, DC: U.S. Department of Energy, Office of Energy Research. May.

Bain, Donald. 1978. "An Assessment of the Agricultural Wind Power Market." Wind Power Digest. Winter 1977/78; pp. 42-47.

Battelle. 1976. "Environmental Considerations of Selected Energy Conserving Manufacturing Process

Options: Vol. III, Iron and Steel Industry Report." Washington, DC: Environmental Protection Agency.

Battelle. 1979. Market Characterization of Solar Industrial Process Heat Applications. Second Quarter Progress Report. SERI/PR-353-212. Golden, CO: Solar Energy Research Institute. Dec.

Bullard, Clark W., 1974 "Energy Conservation Through Taxation", Center for Advanced Computation Document 95, University of Illinois, Urbana.

Battelle, Columbus Laboratories, 1975, "Potential for Energy Conservation in the Steel Industry", Report to the Federal Energy Administration, May.

Battelle Columbus Laboratories, 1976, "Energy Efficiency Improvement Targets for Primary Metals Industries. SIC 33, Volume 1, Target Support Document", a report to the Federal Energy Administration.

Bechtel Corp. 1978, "Resource Requirements, Impacts, and Constraints Associated with Various Energy Futures", a report to the Department of Energy, August.

BenDaniel, D.J., and E.E. David Jr., 1979. "Semiconductor Alternating-Current Motor Drives and Energy Conservation," Science Vol. 206, Nov. 16.

Berg, Charles A. 1974. "Conservation in Industry." Science p. 264, Apr. 19.

Berg, Charles A. 1978. "Process Innovation and Changes in Industrial Energy Use." Science 199. Feb. 10.

Berg, Charles A. 1979. "Energy Conservation in Industry: The Present Approach." Report prepared for the Council on Environmental Quality. May.

Berndt, Ernst R. 1978, " Aggregate Energy, Efficiency and Productivity Measurements", Annual Review of Energy, Vol. 3.

Bever, Michael B. 1976. "The Recycling of Metals-1. Ferr Metals." "The Recycling of Metals-2. Non Ferr Metals." Conservation and Recycling. Vol. 1, No. 1. Sept.

Bever, Michael B. 1978, "Recycling the 1985 Automobile," presented at Materials Substitution symposium, San Francisco, December.

Boercker, Sara W. 1979, "Characterization of Industrial Process Energy Services" Institute for Energy Analysis, Oak Ridge, Tenn.

Bosworth, B.; Duesenberry, J. S.; Carron, A. S. 1975. Capital Needs in the Seventies. Washington, DC: Brookings Institution.

Braun, Gerald W. 1979. "DOE Solar Thermal Power Systems Program." Presented at the Solar Energy Industries Association Solar Power Generation Conference." San Jose, CA. August 8-9.

Brimmer, Andrew F. and Sinai, Allen 1976. "The Effects of Tax Policy on Capital Formation, Corporate Liquidity and the Availability of Investible Funds: A Simulation Study," Journal of Finance, p. 287, May.

Brown, K. C. 1989. "The Use of Solar Energy to Produce Process Heat for Industry." SERI/TP-731-626. Golden, CO: Solar Energy Research Institute. April.

Bureau of the Census. 1978a. Annual Survey of Manufactures 1976, Fuels and Electric Energy Consumed: Industry Group and Industries. M76(AS)-4.1. Washington, DC: U.S. Department of Commerce. May.

Bureau of the Census. 1978b. Annual Survey of Manufactures 1976, Fuels and Electric Energy Consumed: States, by Industry Group. M76(AS)-4.2. Washington, DC: U.S. Department of Commerce. May.

Bureau of the Census. 1978c. Annual Survey of Manufactures 1976, Fuels and Electric Energy Consumed: Standard Metropolitan Statistical Areas, by Major Industry Group. M76(AS)-4.3. Washington, DC: U.S. Department of Commerce. May.

Bureau of the Census. 1978d. County and City Data Book 1977. 003-024-01464-5. Washington, DC: U.S. Government Printing Office. May.

Bureau of the Census. 1978e. County Business Patterns 1976. CBP-76-1. Washington, DC: U.S. Department of Commerce. August.

Bureau of the Census. 1979a. 1977 Census of Manufacturers: Fuels and Electric Energy Consumed. MC77-SR-4(P). Washington, DC: U.S. Department of Commerce. Oct.

Bureau of the Census. 1979b. 1977 Census of Manufacturers: Geographical Area Series. MC77-A-X(P). Washington, DC: U.S. Department of Commerce. Oct.

Bureau of the Census. 1979c. 1979 Statistical Abstract of the United States. Washington, DC: U.S. Department of Commerce.

Bureau of Economic Analysis. 1978. 1977 Business Statistics. Washington, DC: U.S. Department of Commerce. March.

Bureau of Economic Analysis. 1976. 1975 Business Statistics. Washington, DC: U.S. Department of Commerce. May.

Bureau of Labor Statistics. 1977a. Employment and Earnings: States and Areas 1939-1975. Bulletin 1370-12. Washington, DC: U.S. Department of Labor.

Bureau of Labor Statistics. 1977b. Productivity and the Economy. Bulletin 1926. Washington, DC: U.S. Department of Labor.

Bureau of Labor Statistics. 1978. Productivity Indexes for Selected Industries, 1978 Edition. Bulletin 2002. Washington, DC: U.S. Department of Labor. September.

Bureau of Labor Statistics. 1979. 1978 Handbook of Labor Statistics. Bulletin 2000. Washington, DC: U.S. Department of Labor.

Burchell, R. W.; Listokin, D. 1975. Future Land Use. New Brunswick, NJ: Rutgers Center for Urban Policy Research.

Business Week. 1979. "The 1980s: New Investments for the New Decade." Business Week. (No. 2618): 31 December; pp. 64-67.

Business Week. 1980. "Industrial Outlook 1980." Business Week. (No. 2619): 14 January; pp. 52-100.

Carhart, Stephen C.; et al. 1979. "The Least-Cost Energy Strategy, Technical Appendix." Energy Productivity Center. Carnegie-Mellon University Press.

Cavanaugh, H.A. 1978. "Utilities' Cleanup Cost is $19-billion." Electrical World. Vol. 190 (No. 1): 1 July; pp. 33-36.

Chapman, P.F. 1974. "Energy Conservation and Recycling of Copper and Aluminum." Metals and Materials. Vol. 8 (No. 6): June.

Chiogiogi, Melvin H. 1979. Industrial Energy Conservation. Marcel Dekker, Inc. New York.

Clark, S. H. Associates. 1978. Solar Total Energy Systems Final Technical Summary Report, Volume III: Energy Use and Price Forecasts. ATR-78(7692-01)-1. El Segundo, CA: The Aerospace Corporation. 31 March.

Committee on Nuclear and Alternative Energy Systems National Research Council (CONAES). 1979. Energy in Transition. National Academy of Sciences.

Cone, B.W., et al. 1980. "An Analysis of Federal Incentives Used to Stimulate Energy Production," Second revised report by Battelle Pacific Northwest Laboratory PNL-2410 Rev. II, Feb.

Conway, H.M.; Liston, L.L.; Saul, J. 1978. New Industries of the Seventies. Atlanta, GA: Conway Publications, Inc.

Council on Materials Science, 1979 Panel Report on Comission and Energy Systems, a report to the Department of Energy, June.

Darmstadter, Joel; Dunkerley, Joy; Alterman, Jack. 1977. How Industrial Societies Use Energy: A Comparative Analysis. Baltimore, MD: Johns Hopkins University Press.

Davidson, J. G. 1918. "The Formation of Aromatic Hydrocarbons from Natural Gas Condensate." Ind. Eng. Chem., 10, No. 11, 901-910.

Dickinson, W. C.; Brown, K. C. 1979. Economic Analysis of Solar Industrial Process Heat Systems: A Methodology to Determine Annual Required Revenue and Internal Rate of Return. UCRL-52814. Livermore, CA: Lawrence Livermore Laboratory. 17 August.

Diebold, J. P.; Smith, G. D. 1979. "Thermochemical Conversion of Biomass to Gasoline." SERI/TP-33-285. 3rd Annual Biomass Energy Systems Conference, June.

Dow Chemical Company, Environmental Research Institute of Michigan, Townsend-Greenspan and Company, Cravath, Swaine and Moore. 1975. Energy Industrial Center Study, Report to the National Science Foundation.

Drexel University. 1976. Executive Summary: Industrial Waste Energy Data Base/Technology Evaluation. CONS?2862-1. Washington, DC: Energy Research and Development Administration. Dec.

Dryden, I.G.C., ed. 1975. The Efficient Use of Energy. IPC Science and Technology Press. Guildford, Surrey, England.

Dumas, Lloyd J. 1975. The Conservation Response, Lexington Books, D.C. Heath and Co. Lexington, Mass.

Dunkerley, Joy, ed. 1978. International Comparisons of Energy Consumption. Research Paper R-10. Washington, DC: Resources for the Future.

Duscha, Rudolph A.; Masica, William J. 1979. "The Role of Thermal Energy Storage in Industrial Energy Conservation." DOE/NASA/1034-79/1. Preprint for the Conference on Industrial Energy Conservation Technology and Exhibition. Houston, TX. 22-25 April 1979. Cleveland, OH: NASA Lewis Research Center.

Economic Research Service U.S. Department of Agriculture. 1976. Energy and U.S. Agriculture: 1974 Data Base. FEA/D-76/459. Washington, DC: Federal Energy Administration. September.

Economic Research Service, U.S. Department of Agriculture. 1978. Energy and U.S. Agriculture: 1976 Data Base (DRAFT). Washington, DC: U.S. Department of Energy.

Economics, Statistics, and Cooperative Services. 1978a. Agricultural Statistics 1978. Washington, DC: U.S. Department of Agriculture.

Economics, Statistics, and Cooperative Services. 1978b. 1978 Handbook of Agricultural Charts. Agr. Handbook No. 551. Washington, DC: U.S. Department of Agriculture. Nov.

Eisner, Robert and Lawler, Patrick J. 1975. "Tax Policy and Investment: An Analysis of Survey Responses." American Economic Review. March.

Electrical World. 1979. "30th Annual Electrical Industry Forecast." Electrical World. Vol. 192 (No. 6): 15 September; pp. 70-84.

Energy and Environmental Analysis, Inc. 1978. Industrial Sector Technology Use Model (ISTUM): Industrial Energy Use in the United States. 1974-2000. DOE/FE/2344-1. U.S. Department of Energy: Washington, DC. Oct.

Energy Daily 1980, p. 3, May 12.

Energy and Environmental Analysis, Inc. 1979, "Industrial Energy Conservation and the Attainment and Maintenance of Ambient Air Quality", a report to the Environmental Protection Agency Cincinnati, Ohio.

Energy Information Administration. 1977. End-Use Energy Consumption Data Base: Serie 1 Tables (DRAFT). Washington, DC: U.S. Department of Energy. Dec.

Energy Information Administration. 1978. Annual Report to Congress 1978, Volume III: Forecasts.

DOE/EIA-0173/3. Washington, DC: U.S. Department of Energy.

Energy Information Administration. 1979a. Energy Supply and Demand in the Midterm: 1985, 1990, and 1995. DOE/EIA-0102/52. Washington, DC: U.S. Department of Energy. April.

Energy Information Administration. 1979b. State Energy Fuel Prices by Major Economic Sector From 1960 Through 1977. DOE/EIA-0190. Washington, DC: U.S. Department of Energy. July.

Energy Information Administration. 1980. Monthly Energy Review. DOE/EIA 0035/01(80). Washington, DC: U.S. Department of Energy. 21 January.

Farmer, M.H.; Magee, E.M.; Spooner, F.M. 1976. "Application of Fluidized Bed Technology to Industrial Boilers." Report to the Federal Energy Administration. Exxon Research and Engineering.

Fish, J. D. "Solar Industrial Process Heat Markets for Central Receiver Technology." SAND/80-8214. Livermore, CA: Sandia Laboratories. April.

Foster, George H.; Peart, Robert M. 1976. Solar Grain Drying: Progress and Potential. Agr. Information Bulletin No. 401. Washington, DC: U.S. Department of Agriculture. Nov.

Forrester, Jay W. 1979. "A self-regulating Energy Polciy", Astronautics and Aeronautics. p. 40, July/August.

Friedman, Stephen J. 1980. Testimony in Hearings on Federal Securities Laws and Small Business Legislation before the Subcommittee on Securities of the Senate Committee on Banking, Housing and Urban Affairs. May 16.

Fromm, G. Ed., 1971. Tax Incentives and Capital Spending, Brookings Inst., Washington, DC.

General Accounting Office 1979. "Nuclear Power Costs and Subsidies," June 13.

General Energy Associates 1979 (H. Brown and B. Hamel), Chem Hill, N.J. private communication.

Gordian Assoc., 1975, "The Potential for Energy Conservation in Nine Selected Industries, Vol. 6, Steel," U.S. Government Printing Office.

Gray, Jerry; Sutton, George W.; Zlotnick, Martin. 1978. "Fuel Conservation and Applied Research." Science 200. 135. Apr. 14.

Grumman Data Systems Corp. 1979. Energy Price Distribution Study. SERI Contract No. AM-9-8355-1. Bethpage, NY: Grumman Data Systems Corp. Dec.

Gyftopoulos, Elias P., Lazaridis, Lazaros J., Widmer, Thomas F. 1974. Potential Fuel Effectiveness in Industry, a report to the Energy Policy Project of the Ford Foundation, Ballinger, Cambridge, Mass.

Hamel, B.; Brown H.; Ross M.; Hedman B.; Sweeney J.; Smith S.; Koluck M. 1979. Study of the Second Las of Thermodynamics as Related to Energy Conservation, a report by General Energy Associates, Inc. to the Department of Energy. November.

Hamel, B.B.; Brown, H.L. 1980. Total Energy, Steam Use and Boiler Profile for the Industrial Manufacturing Sector (SIC 20-39). Philadelphia, PA: Drexel University. Jan.

Hannon, Bruce; Herendeen, Robert; Penner, Reser 1979. "An Energy Conservation Tax: Impact and Policy Implications", report to the Council on Environmental Quality of the Excutive Office of the President. July.

Harvey, Douglas 1980. Private communication.

Hatsopoulos, G. 1979. testimony at "Hearings before the Coinmittee on Ways and Means, House of Representatives on the Tax Restructuring Act of 1979". 14 Nov.

Hausz, Walter. 1979. "Thermal Energy Storage and Transport." P-835. Preprint for the 1979 ASME Winter Annual Meeting. New York, NY. 3-7 December 1979. Santa Barbara, CA: General Electric Co., Center for Advanced Studies. October.

Heckler, Ken 1980. Toward the Endless Frontier (a history of the House Committee on Science and Technology). U.S. Government Printing Office, Stock No. 052-070-05133-7.

Heizer, E. F., Jr. 1980. Testimony before the Subcommittee on Securities of the Senate Committee on Banking, Housing and Urban Affairs, Hearings on Federal Securities Laws and Small Business Legislation. Apr. 29.

Hill, C.T.; Overby, C.M. 1979. "Improving Energy Productivity Through Recovery and Reuse of Wastes." Energy Conservation and Public Policy. John C. Sawhill, ed. Prentice-Hall.

Hitch, Charles; Charles J., Ed. 1977. Modeling Energy-Economy Interactions = Five Approached.

Research Paper R-5, Resources for the Future. Washington, DC:

Holland, Daniel M. 1977. "The Role of Tax Policy," in Capital Productivity and Jobs, Eli Shapiro and William L. White, Eds., The American Assembly, Prentice-Hall, Inc., Englewood Cliffs, NJ.

Hollomon, J.H. 1979. "Government and the Innovation Process, Technology Review, May.

Industrial Advisory Subcommittee 1979. (on Direct Federal Support of Research and Development) in "Industrial Advisory Committee to the Domestic Policy Review on Industrial Innovation," Department of Commerce.

Industry and Trade Administration. 1979 U.S. Industrial Outlook. Washington, DC: U.S. Department of Commerce. January 1979.

Industry and Trade Administration. 1980. 1980 U.S. Industrial Outlook. Washington, DC: U.S. Department of Commerce. Jan.

International Pulp Bleaching Conference. 1979. from several latitudes, Technical Section, Canadian Pulp and Paper Association. Montreal, Quebec.

InterTechnology. 1977. Analysis of the Economic Potential of Solar Thermal Energy to Provide Industrial Process Heat. 00028-1. Warrenton, VA: InterTechnology Corp. Feb.

Jack Faucett Associates. 1973. Development of Capital Stock   by Industry Sector.

Jayadev, T.S.; Roessner, D. 1980. Basic Research Needs and Priorities in Solar Energy: Vol. II. SERI/TR-351-358. Golden, CO: Solar Energy Research Institute. Jan.

Ketels, P.A.; Reeve, H.R. 1979. Market Characterization of Solar Industrial Process Heat Application. Progress Report. SERI/PR-52-212. Golden, CO: Solar Energy Research Institute. April.

Kuester, J. L. 1979. "Liquid Hydrocarbon Fuels from Biomass." ACS Sumposium on Biomass as a Non-Fossil Fuel Source. Honolulu, HA: April.

Kutscher, C.; Davenport, R. 1980. Performance Results and Experience of the Operational Industrial Process Heat Field Tests. SERI/TR-333-385. Golden, CO: Solar Energy Research Institute.

Lockeretz, William, ed. 1977. Agriculture and Energy. New York, NY: Academic Press.

Lund, Robert T. 1977. "Making Products Last Longer." Technology Review. Jan.

Mansfield, E.; et al. 1977. The Production and Application of New Industrial Technology. New York, NY: Norton and Company.

Mallon, G. M.; Finney, C. S. 1972. "New Techniques in the Pyrolysis of Solid Wastes." AICHE 73rd National Meeting in Minneapolis, MN: 27-30 August.

May, E. K. 1980. "Solar Energy and the Oil Refining Industry." SERI/TR-733-592. Golden, CO: Solar Energy Research Institute. March.

Miernyk, William H.; Giarratani, Frank; Socher, Charles, F. 1978. Regional Impacts of Rising Energy Prices. Cambridge, MA: Ballinger Publishing Co.

Murgatroyd, Walter; Wilkins, E.C. 1976 "Efficiency of Electric Motive Power in Industry." Energy. Vol. 1 p. 337.

National Academy of Science. 1975. National Materials Policy. Washington, DC: National Academy of Sciences.

National Association of Securities Dealers. 1980. "Small Business Financing", Testimony by J. Stephen Putnam in Hearings on Federal Securities Laws and Small Business Legislation before the Subcommittee on Securities of the Senate Committee on Banking, Housing and Urban Affairs. Apr. 29.

National Science Foundation. 1977. "National Patterns of R&D Resources, Funds and Manpower in the United States, 1953-1977." NSF 77-310.

Nelklin, Dorothy 1971. The Politics of Housing Innovation, The Fate of the Civilian Industrial Technology Program. Cornell Univ. Press.

Nelson, W. G. 1930. "Waste-Wood Utilization by the Badger-Stafford Process." Ind. Eng. Chem., 22, No. 4, 312-315.

Office of Conservation and Solar Energy. 1980. Solar Energy: Program Summary Document FY1981. Washington, DC: U.S. Department of Energy. Jan.

Office of Energy Technology. 1979. Solar, Geothermal, Electric and Storage Systems: Program Summary Document. DOE/ET-00102. Washington, DC: U.S. Department of Energy. July.

Office of Industrial Programs, Department of Energy, 1979, "Industrial Energy Efficiency Program, Annual Report, July 1977 through December 1978". December.

Office of Technology Assessment. 1980. "Technology and Steel Industry Competitiveness, U.S. Government Printing Office.

Page, Talbot. 1977. Conservation and Economic Efficiency: An Approach to Materials Policy. Baltimore, MD: Resources for the Future, Johns Hopkins University Press.

Phillips, Almarin, Ed., 1975. Promoting Competition in Regulated Markets, The Brookings Institution, Washington, D.C.

Pilati, David A. 1980 (Brookhaven National Laboratory), private communication, April 15.

Rauch, James, ed. 1976. The Kline Guide to the Paper and Pulp Industry. Fairfield, NJ: C. H. Kline & Co., Inc.

Reay, D.A. 1979. Industrial Energy Conservation, A Handbook for Engineers and Managers. 2nd ed. Pergamon Press.

Resource Planning Associates, Inc. 1977. (P. Bos, et al). The Potential for Cogeneration Development in Six Major Industries by 1985. Report to the Federal Energy Administration.

Russ, Mare H. and Williams, Robert H. 1981. Our Energy: Regaining Control. McGraw Hill Book Co. N.Y., N.Y.

Sant, Roger 1979. "The Least Cost Energy Strategy". Carnegie Mellon University Press.

Sant, Roger 1980. "Coming Markets for Energy Services. Harvard Business Review, May-June.

Senate Budget Committee. 1975. Tax Expenditures: Briefing Paper on Tax Incentives for Business Investment.

Shaker, W. M., Ed., 1976. Electric Power Reform: The Alternatives for Michigan, Institute of Science and Technology, The University of Michigan, Ann Arbor.

Shafizadeh, F. 1968. "Pyrolysis and Combustion of Cellulosic Materials." Adv. in Carbohydrate Chem., 23, 419-75.

Stamm, A. J. 1956. "Thermal Degradation of Wood and Cellulose." Ind. Eng. Chem., 48, No. 3, 413-417.

Sternlieb, George; Hughes, James W., ed. 1975. Post-Industrial America: Metropolitan Decline and Inter-Regional Job Shifts. New Brunswick, NJ: Rutgers Center for Urban Policy Research.

Stigler, George J.; Kindahl, James K. 1970. The Behavior of Industrial Prices. New York: National Bureau of Economic Research.

Szekely, Julian, ed. 1975. The Steel Industry and the Energy Crisis. New York, NY: Marcel Dekker, Inc.

TAPPI 1979. Future Technical Needs and Trends in the Paper Industry ***. Technical Association of the Pulp and Paper Industry, Atlanta GA: Robert W. Hagemeyer, Ed.

Thermo Electron Corp. 1976. A Study of Inplant Electric Power Generation in the Chemical, Petroleum Refining, and Paper and Pulp Industries. PB-255-659. Washington, DC: U.S. Department of Commerce (NTIS). June.

Thermo Electron Cor. 1979. "Impact on Energy Conservation of Automatic Control Systems Utilization in the U.S. Pulp and Paper Industry." Report to the Department of Energy. June.

Treadwell, G. W.; Grandijean, N. R.; Biggs, F. 1980. "An Analysis of the Influence of Geography and Weather on Parabolic Trough Solar Collector Design." SAND/79-2032. Albuquerque, NM: Sandia National Laboratories. March.

Vigerstad, T.J.; Sharp, J.M. 1979. Assessment of the Availability of Wood as a Fuel for Industry in South Carolina. Columbia, SC: South Carolina Energy Research Institute. Dec.

Ward, G. M.; Knox, P. L.; Hobson, B. W. 1977. "Beef Production Options and Requirements for Fossil Fuel." Science. Vol. 198 : pp. 265-271. 21 October.

Wall Street Journal. 1980. "Alcoa is Planning to Lift Capacity for Aluminum." Wall Street Journal. Vol. CII (No. 46): p. 5. 6 March.

Wall Street Journal. 1980. "Manufacturers in Japan Seen Raising Spending to Increase Capacity." Wall Street Journal. Vol. CII (No. 46): p. 21. 6 March.

Washington Scientific Marketing, Inc. 1977. Department of Energy Program and Objectives: Energy Conservation in Agricultural Production. DOE/CS-0004. Washington, DC: U.S. Department of Energy. December.

Williams, J.H.; Whitney, S.C.; Ball, H. 1980. Solar-Augmented Applications in Industry. Final 1979 Report, Phase I. Los Angeles, CA: Insights West, Inc. Feb.

Williams, R.H. 1978. "Industrial Cogeneration." Annual Reviews of Energy. Vol. 3.

Williams, Robert H. 1980. private communication.

Wishart, Ronald 1978. Science

Woolf, P.L., 1975. "Improved Blast Furnace Operation,: in Symposium Papers on Efficient Use of Fuels in the Metallurgical Industries, Institute of Gas Technology, Chicago.

BIOMASS

Adelman, H. R, R. Pefley, et al., End Use of Fluids From Biomass as Energy Resources in Both Transportation and non-Transportation Sectors, contractor report to OTA, January 1979.

American Paper Institute, Raw Materials and Energy Division. U.S. Pulp, Paper and Paperboard Industry Estimated Fuel and Energy Use (brochure). New York NY 1978.

Boffinger, Testimony before the U.S. Senate Committee on Energy and Natural Resources, Hearings on Wood Energy Development, Hanover, N.H., February 1980.

DOE, "Briefing on the Commercialization of Wood Energy," Washington D.C. 1979.

DOE 1979a, The Report of the Alcohol Policy Review, Washington, DC 1979.

Fege, A. S., et al. "Energy Farms in the Future" Journal of Forestry. Vol. 77, No. 6 June 1979.

Glidden, Report to the Solar Energy Research Institute, Wood Energy Supply and Availability, Dartmouth College, 1980.

Bethel, J., et al Energy from Wood: A Report to the Office of Technology Assessment. University of Washington, Seattle, Washington 1979.

OTA, U.S. Congress. Energy from Biological Processes Washington D.C. 1980.

OTA 1980a, Energy from Biological Processes, U.S. Congress. Draft, Jan. 1980.

Reed, T. B., Efficiencies of Methanol Production from Gas, Coal, Waste or Wood. Presented at the 171st Meeting of the Division of Fuel Chemistry of the American Chemical Society Symposium on Net Energies of Integrated Synfuel Systems, April 1976.

Personal communication from T.B. Reed to M. Gibson, Solar Energy Research Institute, February 1981.

SAF 1979, "Forest Biomass as an Energy Source" Study Report of SAF Task Force. Journal of Forestry, Vol. 77, No. 8, August 1979.

Ernest, R. K., et al. Mission Analysis for the Federal Fuels from Biomass Program Vol. III: Feedstock Availability. SRI Internationl, Menlo Park, California 1979.

Tillman, David. Wood as an Energy Resource, Academic Press, New York, NY 1978.

Tyner, Report to the Solar Energy Research Institute, Bioenergy from Agriculture task force report, Purdue University, 1980.

USDA, 1978. The Federal Role in the Conservation and Management of Private Nonindustrial Forest Lands, USDA 1978.

Holmes, An Analysis of the New England Pilot Fuelwood Project, 1980 USDA/FS, S&PF, October 1980.

USFS 1979a. An Assessment of the Forest and Range-land Situation in the United States. USDA, Washington, DC 1979.

USFS 1979b. A Report to Congress on the Nation's Renewable Resources: RPA Assessment and Alternative Program Directions. USDA, Washington, DC 1979.

USFS 1973. The Outlook for Timber in the United States. USDA, Washington, DC 1973.

USFS 1978. Forest Statistics of the U.S. 1977 Review Draft. USDA, Washington, DC 1978.

Wan E., et al. Technical Economic Assessment of the Production of Methanol from Biomass Vol. II: Assessment of Biomass Resource and Methanol Market Science Applications, Inc., McClean, VA 1979.

Gallup Poll conducted during June-July, 1979 for the Wood Energy Institute and Gardenway Publishing. Washington, DC 1979.

Williams, Robert. Special report to the Solar Energy Research Institute, Princeton University, December 1980.

IPH

Casamajor, A. B., and Wood, R. L. 1978, "Limiting Factors for the Near Term Potential of Solar Industrial Process Heat", Lawrence Livermore Laboratory Report UCRL-52587, October 1978.

Casamajor, A. B. 1980, personal communication.

Dickenson, W. L., and Brown, K. C. 1979, "Economic Analysis of Solar Industrial Process Heat Systems" Lawrence Livermore Laboratory Report UCRL-52814.

DOE 1978. Annual Report.

DRI. Energy Review, Update, 4/79.

General Electric Co., TEMPO, "Combined Thermal Storage and Transport for Utility Applications, EPRI EM-1175.

Insights West, Inc. 1980, "Solar Augmented Applications in Industry--Phase 1", prepared by Insights West, Inc., 900 Wilshire Blvd., Los Angeles California 90017, for Gas Research Insitute, contract number 5011-343-0105

May, K., SERI, 11/6/80 IOM.

Kutscher, C. F., and Davenport, R. L. "Preliminary Operational Results of the Low-Temperature Solar Industrial Process Heat Field Tests", SERI/TR-632-385, June 1980.

Rabl, A. "Solar Industrial Process Heat," SERI December, 1980.

Rabl, A. ,1980, "Yearly Average Performance of the Principal Solar Collector Types," to be published in Solar Energy.

Shell Oil Company, The National Energy Outlook, 1980-1990 Feb. 1979.

The Pace Company Consultants & Engineers Inc., The Pace Energy & Petrochemical Outlook to 2000, Winter 1978.

References for
Chapter Three
TRANSPORTATION

Airline Transport Report - Airline Fuel Efficiency January 1980.

Personal Correspondence from E.L. Thomas, ATA, to Charles Gray, EPA, April 1980, May 1980.

An Inexhaustable, Low-Cost Fuel? Fleet Owner. Feb. 1980: p. 147.

Another Look at Methanol. Automotive Engineering. Vol. 83; No. 4. april 1975: p. 38.

AAR Intermodal Policy Studies Group Staff Memorandum 79-8 F.L. Smith, Jr. An Efficiency Assessment of the Highway User Charge System, Oct. 1979.

A Study of Consumer Behavior Towards Fuel Efficient Vehicles, Task D: Focus Group Analysis. Market Facts - Washington. Report Submitted to United States Department of Transportation. (DOT-HS-7-01781). May 1978: p. 16.

Atlanta Journal, the Atlanta Constitution, Bike Firm Experimenting with 60-mph Commuter Cycle, Nov. 27, 1980.

Automotive News, New Fuel Gains Linked to Turbo-compounding, June 30, 1980, p. 8.

Barber, Ton. Personal Communication, JPL, January 1981.

Barry, E. G., et al. If Autos Go to Diesel Fuel. in Hydrocarbon Processing. May 1977: p.111.

Barth, Edward A. and Kranig James M., "Evaluation of Two Turbocharged Diesel Volkswagen Rabbits," (EPA Office of Mobile Source Air Pollution Control, Emission Control Technology Division, Technology Assessment and Evaluation Branch, Oct. 1979).

Benninton, et al. Stable Colloid Additives for Engine Oils-Potential Improvement in Fuel Economy. SAE Paper 750677. June 1975.

Big Fleets Move to Radials. Diesel Equipment Superintendent. June 1979: pp. 80-82.

Bolt, J. A., Engine Modification For Use of Methanol, DOE #E(29-2) 3682, 1979.

Bond, Langhorne (Administrator, Federal Avaiation Administration) Aviation doesn't guzzle much gasoline Washington Star 3/19/80.

Booz, Allen, Personal Communication with John Wiegman, December 1980.

Booz, Allen and Hamilton, Inc. Energy Use in the Marine Transportation Industry. (prepared for DOE Transportation Energy Conservation Division). 11 Jan. 1977.

Glenning, Bower, Tire Testint for Rolling Rinstance and Fuel Economy, Tire Sciency and Technology, Vol. 2 and 4.

Broman, et al. Testing of Friction Modified Crankcase Oils for Improved Fuel Economy. SAE Paper 780597. June 1978.

Bunch, Howard M. "The Small Car May Be Dangerous to Your Health! - The Consequences of Downsizing." A Paper Presented for the Symposium on Technology, Government and the Future of the Automobile Industry. Harvard Business School. 19-20 Oct. 1978.

U.S. Autos Losing a Big Sement of the Market - Forever? An Article in Business Week, 24 March 1980.

CBO Energy Use in Urban Transity Systems, 1977.

CBO, Guidelines for a Study of Highway Cost Allocation, Feb. 1979.

CBO, Fuel Economy Standards for New Passenger Cars After 1985. Dec. 1980.

CQ, Congress Clears Major Bill Cutting Trucking Regulation, Judy Sarasohn, 6/28/80.

Cheslow, Melvyn (ERC), The Effect of Changing Household Composition on the Size Mix of New Automobile Sales: 1979-2000, May 1980 Evaluation Research Corporation, Vienna Virginia.

Clark, S. K. Geometric Effects on the Rolling Resistance of Pneumatic Tires.

Congressional Record, 6/19/80, Conference Report, Energy Security Act of 1980.

Corporate Tech Planning, Inc., "Review of Improved Rolling Resistance of Tires," in Automotive Fuel Economy Contractors' Coordination Meeting Summary Report, April 24-26, 1978 (US DOT, HS-803 362, 1978), p. II-A-21, quoted in Von Hipple.

Costle, D.M., letter to Jack Watson, Vehicle Fuel Efficiency Program, Jan 30, 1980.

Coussens, T. G. and Tullis, R. H., Propfan Propulsion for Commercial Air Transports, SAE paper #800733.

Crowley, A. et al. Methanol-Gasoline Blends Performance in Laboratory Tests and In Vehicles. SAE Paper No. 750419.

Increased Truck Size and Weight, the Impact for Highways, Safety and Energy, Central States Resource Center, March, 1980.

D'Eliseu, Prof. P. N. Biological Effects of Methanol Spills into Marine, Estuarine, and Freshwater Habitats. Presented at the International Symposium on Alcohol Fuel Technology, Methanol and Ethanol. Wolfsburg, FRG: 23-23 Nov. 1977. CONF-771175, UC 61, 90, 96.

Dependency Dilemma: U.S. Gasoline Consumption and America's Security. Harvard University. 232 March 1980.

Department of Commerce (DOC). Statistical Abstract of the United States 1978 Bureau of Cencus.

Big Fleets Move to Radials, Diesel Equipment Superintendent, June 1979.

Detroit Free Press Vanpools Can Find Wide Open Highways, 18 Feb. 1980.

Dickson, Edward M.; Bowers, Raymond. The Video Telephone, Impact of a New Era in Telecommunications. Cornell University. 1973.

Big Fleets Move to Radials. Diesel Equipment Superintendent. June 1979: pp. 80-82.

Difiglio, Carmen, (DOE), Robert Dulla, and K.G. Duleep (EEA) Cost Effectiveness of 1985 Automobile fuel Economy Standards, presented at the Society of Automotive Engineers, October 104, 1979, Warrendale, PA.

Difiglio, Carmen. The Urban Transportation Sector: A Preliminary Energy Conservation Strategy Assessment, March 1979.

DOE, State of Competition in Gasoline Marketing, Office of Competition, as required by Title III of the Petroleum Marketing Practices Act of 1977, January 1981.

Allocation of Highway Cost Responsibility and Tax Payments, Bureau of Public Roads, FHWA, DOT May 1970.

DOE, Study of Potential for Motor Vehicle Fuel Economy Improvement, Truck and Bus Panel Report, DOE, EPA, Jan 1975.

DOE/CS. Near Term Electric Test Vehicle ETV-1, Phase II Final Report, S1294-01, Oct. 1980.

DOE, Reducing U.S. Oil Vulnerability, DOE/PE-0021, November 10, 1980.

DOT, Commercial Vehicle Post-1980 Goak Study, DOT, EPA, FEA, ERDA, May 1976.

DOE, Reducing U.S. Oil Vulnarability: Energy Policy for the 1980's. P&E of D&E 11/10/80.

DOE, Light Duty Diesel Engine Development States and Engine Needs, OTP/CS/52184-01, Aug. 1980.

DOT 1977, Information on Integrated Research VW, SAE Research Contract Report, DOT-TSC-NHTSA/ 77-31.

Department of Transportation (DOT). "Highway Statistics, 1977".

A Study of Consumer Behavior Towards Fuel Efficient Vehicles, Task D: Focus Group Analysis. A Report of Market Facts - Washington. Submitted to the U. S. Department of Transportation (DOT-ITS-7-01781). May 1978.

Department of Transportation "The US Automobile Industry, 1980".

Department of Transportation, Report to the President from the Secretary of Transportation, The U.S. Automobile Industry, 1980, January 1981.

The US Department of Transportation's New Car Crash Protection Assessment Program. The Fourth Car, The Dodge Magnum, Is a Mid-size Car. US/ DOT Fact Sheet. 28 Feb. 1980.

Douglas, Dave, Personal Communication, EPRI, January 1981.

EEA (Energy and Environmental Analysis Inc.), Medium and Heavy-Duty Truck Fuel Demand Module, 1980. Note that our definition of "medium trucks includes the two classes "medium" and "light-heavy" defined in the EEA report.

DPA, Transportation Task Force Report to SERI, Charles Gray, May 1980.

Analysis of Oregon's Inspection and Maintenance Program, Becker & Rutherford, EPA, APCA #79-7.3

Emmenthal, K. D., Hagemann G., and Walzer, P., "Fuel Economy Improvements by Turbocharging," Proceedings of the First International Automotive Fuel Economy Research Conference, (U.S. DOE, NHTSA, 1979) p. 195, quoted in Von Hipple.

EPRI, Journal, Electric Vehicles, Sept. 1979.

Effects of Goodyear Unisteel Radial Ply Tires on Fuel Economy, Goodyear, Inc. 1975. p. 16.

Ford, Personal correspondence from T.J. Galbreath, Environment and Safety Engineering Staff, to Charles Grey, EPA, April 1980.

GAO, The Federal Government Should More Actively Promote Energy Conservation By Heavy Trucks (EMD-80-40, 1980).

General Accounting Office, Report EMD-80-40, March 31, 1980.

Giles, W. L. Expanded Applications Wide Base Radial Truck Tire, Society of Automotive Engineers. Paper No. 791044.

Gleming and Bowers. Tire Testing for Rolling Resistance and Fuel Economy, SAE Paper No. 750457. 1975: pp. 13-14.

Gleming, D. A.; Bower, P. A. Tire Testing for Rolling Resistance and Fuel Economy, Tire Science and Technology. TSTCA Vol 2.: No. 4. Nov. 1974.

GM Corporation Presentations to Environmental Protection Agency at the GM Proving Grounds. Milford, MI: 22 Feb.-26 July 1978.

GM, letter to Charles Gorey, EPA, from Gerald Stofflet, May 1980.

Goodsen, Eugene R. Electric Vehicles - How Do They Compare With Synthetic Liquid Fuels EPRI EM-1326. Feb. 1980.

Goodwin and Haviland, Fuel Economy Improvements in EPA and Road Tests With Engine Oil and Rear Axle Lubricant Viscosity Reduction, SAE Paper 78059. June 1978.

Gorman Heitman, A Comparison of Costs for Automobile Energy Conservation is Synthetic Fuel Production, presented at 5th International Automotive Propulsion Systems Symponuim, April 1980.

Grey Advertising, Inc. Marketing Plan to Accelerate the Use of Vanpools, Publication No. HCP/M60437-01. Dec. 1977. (prepared for DOE).

Conservation and Alternative Fuels in the Transportation Sector, Transportation Task Force of the SERI Solar Conservation Study, Charles Gray, EPA, Chairman.

Gray, Charles and Frank VonHipple "The Fuel Economy of Light Vehicles" (to be published).

Hagey, G. et al. Methanol and Ethanol Fuels - Environmental, Health and Safety Issues, Presented at the International Symposium on Alcohol fuel Technology, Methanol, and Ethanol. Wolfsburg, FRG: 21-23 Nov. 1977 CONF-771175, UC 61, 90, 96.

Haviland Merril L. and Goodwin, Malcolm C., "Fuel Economy Improvements With Friction-Modified Engine Oil in Environmental Protection Agency Road Tests," (SAE Paper #790945, 1979).

Hayden A.C.A., "The Effects of Technoloagy on Automobile Fuel Economy Under Canadian Conditions," (SAE Paper #790229, 1979.

Heaton, George U.S. Automobile Regulation: An Examination of the Foreign Experience and Its Implications for U.S. Policy, Third Automotive Fuel Economy Research Contractor' Coordination Meeting, Washington, D.C. December 1-2, 1980.

The Highway Loss Reduction Status Report, A Publication of the Insurance Institute for Highway Safety. Vol. 12: No. 7. 9 May 1977.

Hill, R. "Alcohol Fuels." Popular Science. Mar. 1980.

Hulsing, K. L., Diesel Engine Design Concepts for the 1980s, SAE paper #790807.

Hurter and Lee. A Study of Technological Improvements to Optimize Truck Configurations for Fuel Economy, DOT-TSC-OST. 1975: pp. 5-43.

Jennings, Lloyd R., The Police Chief. "A Study of Road Damage as a Result of Truck Weight" (hereafter "A Study of Road Damage"). Jan. 1979: p.42.

JPL, Three State-of-the-Art Individual Electric and Hybrid Test Reports Volume II, EC-77-A-31-1011 NASA/JPL/DOE, November 1978.

John, Richard R. (DOT). Transition to the Post-1985 Motor Vehicle. First International Automotive Fuel Economy Research Conference. Washington, DC: 31 Oct.-1 Nov. 1979.

Kelderman, Jake. "Tough Standards for Small Diesels Delayed Until '85." Automotive News. 25 Feb. 1980: p. 1.

Klamp, W. K., Power Consumption of Tires Related to How They Are Used, Proceedings of the 1977 SAE-DOT Conference. Tire Rolling Losses and Fuel Economy - An R&D Planning Workshop.

Klineberg, John, The NASA Aircraft Energy Efficiency Program, paper presented at the Royal Aeronautical Society, Joint RAeS/AIAA Conference on Energy and Aerospace, December 5-7, 1978.

Knight, R. E., Tire Parameter Effects on Truck Fuel Economy, SAE paper #791043, 1979.

Knights, R. E. Tire Parameter Effects on Trucks Fuel Economy. Society of Automotive Engineers. Paper No. 791043. 1980.

Knorr, R.; Millar M. Projections of Automobile, Light Truck, Bus Stocks and Sales to the Year 2000, Nov. 1979.

Kollen, J.; Garwood, J. Travel/Communications Tradeoffs: The Potential for Substitution Among

Business Travelers, Business Planning. Bell Canada, Montreal, Canada: 1975.

Lane, L. Lee. The Rail Industry and the Growing Crisis in Highway Finance, Association of American Railroads. 16 Feb. 1980 (hereafter "The Rail Industry. . .").

Lee; Martin, E. H.; Glover, Matthew F.; Eavy, Paul W. Differences in the Trip Attributes of Drivers With High and Low Accident Rates, Society of Automotive Engineers Paper 800384 (SP-461). 25-29 Feb. 1980.

Linn, Richard, Future Aircraft Requirements, Presentation at University of Maryland seminar on Air Transportation for th 1980s.

McNutt, Barry et al. On Road Fuel Economy Trends and Impacts, US/DOE Office of Conservation and Advanced Energy Systems Policy, 17 Feb. 1979.

Malen, Personal Communication with Gerald Maden, EPRI, January 1981.

Meuller, H. G.; Wouk, Victor. Efficiency of Coal Use, Electricity for EUS Synfuels for ICE, Society of Automotive Engineers, Inc. Congress and Exposition. Detroit,MI: 25-29 Feb. 1980: p. 800109.

Moriarity, Dr. A. J. Toxicological Aspects of Alcohol Fuel Utilization, Presented at the International Symposium on Alcohol Fuel Technology, Methanol, and Ethanol. Wolfsburg, FRG: 21-23 Nov. 1977. CONF-771175, UC 61, 90, 96.

Your Market Place, Modern Tire Dealer, Feb 7, 1980.

The State Perspective of Highway Tax Desing, Loyd Henion, Oregon Dept. of Transportation, October 1979.

Mott, W. J.; Longwell, J. P. "Single Cylinder Engine Evaluation of Methanol - Improved Energy Economy and Reduced NOx." SAE Paper No. 750119. Feb. 1975.

Motor Vehicle Manufacturers Association Motor Vehicle Facts and Figures. 1979.

Nagalingam, et al. "Surface Ignition Initiated Combustion of Alcohol in Diesel Engines - A New Approach." SAE Paper No. 800262. Feb. 1980.

NASA/Lewis, Tweleve State-of-the-art Omdovodia; Electric and Hybrid Test Reports Volume II, EC-77-A-31-1011 NASA/JPL/DOE, Nov. 1978.

National Journal, Hard Times for the Highway Trust fund May Mean Trouble for Highway Repair, Rochelle Stanfield, 8/16/80.

National Transportation Statistics, 1979; p. 120.

Nationwide Personal Transportation Survey, 1969-1970. Bureau of the Cencus Table P-8 (unpublished).

Nored, Donald, Propulsion, Astronautics and Aeronautics, July/August 1978.

N.Y. State Assembly Scientific Staff, Guidelines for Using Operating Characteristics in the Evaluation of Public Transit Service, William G. Allen, et al Prepared for NSF June, 77.

Ostrouchor, N., "Effect of Cold Weather on Motor Vehicle Emissions and Fuel Consumption-II," (SAE Paper #790229, 1979).

Pefley, End Use of Fluids from Biomass as Energy Sources in Both Transportation and non-Transportation Sectors, contractors report to Offices of Technology Assessment, U.S. Congress, Jan. 1979.

Pefley, R. K. An Alcohol Fuel Alternative, Mechanical Engineering. Nov. 1979.

Pope, G. G., Prospects for Reducing the Fuel Consumption of Civil Aircraft, paper presented at the Royal Aeronautical Society, Joint RAeS/AIAA Conference on Energy and Aerospace, December 5-7, 1978.

Pratsch, L., Starling, R. Vanpooling: An Update, DOE Sept. 1979.

Reilly, J.; Sandrock, G. Hydrogen Storage in Metal Hydrides, Scientific Americaan. Feb. 1980.

Personal Communications with Diane Ryding. Feb. 1980. Michigan Bell Telephone Company, Detroit, MI:

Reister and Chamberlin. A Test Track Caomparison of Fuel Economy Oils, SAE Paper 790213. Feb. 1979.

Rettie, Jan Aerodynamic Design, presentation for Univ of Maryland Seminar, Air Transportation for the 1980s, June 1975.

Reuyl, J. Private Communication. 1980.

Ripple, Wallace, Personal Communication, JPL, January 1981.

Rolling Resistance Measurement Procedure for Passenger Car Tires, Society of Automotive Engineers Recommended Practices. SAE J1269. Nov. 1979.

Rose, A. B. "Energy Intensity and Related Parameters of Selected Transportation Modes: Passenger Movements." ORNL-5506 Jan. 1979: pp. 3-7.

The SBS Digital Communications Satellite System. Satellite Business Systems. McLean, VA: 1979.

Schneider, William. Public Opinion and the Energy Crisis, A Presentation for "The Dependency Dilemma: U.S. Gasoline Consumption and America's Security." Harvard University. 22 March 1980.

Sears, P.M. Vanpooling The Three Major Approaches, DOE Aug. 1979.

Seiffert, U., and P. Walzer, (Research Division, Volkswagenwerk AG. Wolfsburg Germany), "Development Trends for Future Passenger Cars" presented at the 5th Automotive News World Congress. Prepring received August 28, 1980.

Seiffert, Ullida, Walger, et al. Improvements in Automotive Fuel Economy, VWAG Paper presented at the First International Automotive Fuel Economy Conference, Washington, DC, Nov. 1979.

Shackson, Richard H., and H. James Leach, Using Fuel Economy and Synthetic Fuels to Compete with OPEC Oil", The Energy Productivity Center, Mellon Institute, Arlington, VA. August 18, 1980.

Shavell, Richard S., Technological Development of Transport Aircraft--Past and Future, Journal of Aircraft 17, 1980, p. 67.

Smith, F. L. Jr., An Efficiency Assessment of the Highway User Charge System, Staff memo 79-8, AAR Inter model Policy Studies Group, Oct. 16, 1979.

Smith, Fred. Alternatives to Motor Fuel Taxation Weight Mileage Taxes, Oct. 23, 1980, presented to North American Gasoline Tax Conference.

Sorge, Marjorie, TECO Systems Saves Fuel, Automotive News, June 30, 1980, p. 8.

South Coast Technology Inc., "Review of Improved Lubricants," in Automotive Fuel Economy Contractors' Coordination Meeting Summary Report, April 24-26, 1978, (U.S. DOT Report # DOE HS-803 362) p. II-A-19.

South Coast Technology Inc., Unregulated Diesel Emissions and Their Potential Health Effects, in US DOT, Automotive Fuel Economy Contractors Coordination Meeting, 11-13 Dec. 1978. Summary Report (DOT-HS-803-706): p. 1-4-1.

Spenny, C.H. (U.S. DOT) An Update of the Costs and Benefits of Railroad Electrification, April 8, 1980.

SRI International, Evaluation of the FEA Vanpool Marketing and Implementation Demonstration Program, Publication HCP/J60438-01, April 1978. (prepared for DOE).

SRI International, Railroad Electrification in America's Future: An Assessment of Prospects and Impacts, January 1980.

Steers, Louis L. and Saltzman, Edwin J., Reduced Truck Fuel Consumption Through Aerodynamic Design, Journal of Energy 1 , 1977, p. 312.

Stofflet, Gerald F. letter to Henry Kelly of SERI, dated July 11, 1980

Stofflet, Gerald F. letter to Henry Kelly of SERI, dated February 25, 1981.

Swihart, J. M. and Minnick, J. L., Why the New Air Fleets Challenge the Designer, Astronautics and Aeronautics, January 1979, p. 26.

National Journal, A New Plan to Target Transit Aid-But will It Get you From Here to There? By Rochelle L. Stanfield 8/2/80.

Transportation Energy Conservation Data Book: Edition 3, ORNL, Nov. 1978.

Transportation Energy Conservation Data Book: Edition 4, Oak Ridge National Laboratory, Sept. 1980.

Texas U. at Hustin, Council for Advanced Transportation Studies, An evaluation of Promotional Tactics and Utility Measurement Methods for Public Transportation, Mark Alpert, for DOT, 1980.

Thomas, E.L. (Vice President Air Transport Association) Letter to Charles Greg April 23, 1980.

Thompson, G. D.; Burgeson, R. N. Determination of Tire Energy Dissipation Analysis and Recommended Practice, US Environmental Protection Agency. April 1978.

Thompson, G. D.; Torres, M. Variations in Tire Rolling Resistance - A Real World Information Need, Proceedings of the 1977 SAE-DOT Conference. Tire

Rolling Losses and Fuel Economy - An R&D Planning Workshop, 1977.

Timourian, H.; Milanovich, F. Methanol as a Transportation Fuel: Assessment of Environmental and Health Research, Lawrence Livermore Laboratory. 18 June 1979. UCRL-52697.

Traffic Quarterly, The Use of Performance-based Methodologies for the Allocation of Transit Operating Funds. Oct. 1980.

Traffic Quarterly. Creating an Upward Cycle in Urban Transit Ridership: A Case Study, Tschangho Joh Kim, and William L. Volk, 1980.

Transportation Association of America, Transportation Facts and Trends 1979, p. 18.

Transportation and Energy: Energy Crises Requires Expansion of U.S. Railroads, United Transportation Union, 1975.

Tyler, M. et al. The Contribution of Telecommunications to the Conservation of Energy Resources. Office of Telecommunications, Department of Commerce. Washington, DC: July 1977.

The Rail Industry and the Growing Crisis in Highway Finance, L. L. Lane, Association of American Railroads, February 1980.

Uniroyal, correspondence from W.J. Woehrle to Charles Grey, EPA, May 1980.

U.S. Federal Highway Administration, 1977 National Personal Transportation Study, as presented in MVMA Motor Vehicle Facts and Figures 1980.

CBO Urban Transportation and Energy: The Potential Savings of Different Modes. A Report of the Congressional Budget Office to the Congress of the United States. Dec. 1977.

Model Shifts in Short-Hand Passenger Travel and the Consequent Energy Impacts. Report of the United Technologies Research Center for the Dept. of Energy, March 1978.

United Transportation Union, Transportation and Energy: Energy Crisis Requires Expansion of U.S. Railroads, 1975.

U.S. Senate, Heatings Before the Committee of Energy and national Resources, Potential for Improved Automobile Fuel Economy Between 1985 and 1990, April 30, 1980.

U. S. Autos Losing a Big Segment of the Market - Forever? An Article in Business Week, 24 March 1980.

Vantine, H. C., et al., The Methanol Engine: A Transportation Strategy for the Post-Petroleum Era, Lawrence Livermore Laboratory. 25 Mar. 1976. UCRL-52041.

Von Hippel, Frank, "Forty Miles a Gallon by 1995 at the Very Least: Why the U.S. Needs a New Fuel Economy Goal," in Yergin, D., Ed., The Dependence Dilemma: Gasoline Consumption in America's Security, Part III, pp. 89-108. Cambridge: Harvard University, Center for International Affairs, #43, 1980.

Von Hippel, Frank "Is it Cost Effective to Push Automotive Fuel Economy Beyond 4 Ompg" Feb. 16, 1980 (Draft).

VW 1979, VW Comments to EPA Light Duty Diesel Particulate Reulmaking, Appendix 3, April 1979.

Waddey, et al. Improved Fuel Economy Via Engine Oils. SAE Paper 780599. June 1978.

WAO, Personal Communication with U.P. Wao, Argonne, January 1981.

Webster and Wag, Progress and Forecast in Electric Vehicle Batteries, proceed. 30th Annual Vehicular Technology Society, Dearborn, Mich., Sept. 1980.

Weidemann B. and Hofbauer, P. 1978, "Data Base for Light-Weight Automotive Deisel Plants, " (SAE Paper #780634) quoted in Von Hipple.

Wigg, E.; Lunt, R. Methanol as a Gasoline Extender - Fuel Economy, Emissions, and High Temperature Driveability. SAE Paper No. 741008. Oct. 1974.

Williams, Robert. A $2.00 A Gallon Political Opportunity, Report PU/CEES #102, Center for Energy and Environmental Studies, Princeton, 1980.

Yurko, J., Tire Related Effects on Vehicle Fuel Economy. US Environmental Protection Agency Technical Report. SDSB 79-27.

References for
Chapter Four
UTILITIES

Alley, Weible M., Manager, Energy Applications, Arkansas Power and Light Co. Testimony before the Subcommittee on Energy Development and Applications. Committee on Science and Technology, U.S. House of Representatives. July 22, 1980.

Army Corps of Engineers. 1979. Preliminary Inventory of Hydropower Resources. Vol. 1. Preliminary Report of the National Hydroelectric Power Resources Study. Fort Belvoir, VA: U.S. Army Corps of Engineers. July.

Aster, Robert W. 1978. Economic Analysis of A Candidate 50¢/W$_p$ Flat Plate Photovoltaic Manufacturing Technology. Pasadena, CA: Jet Propulsion Laboratory. December.

Averch, Harvey and Leland Johnson "Behavior of the Firm Under Regulatory Constraint" American Economic Review. Vol. 52 No. 3, December 1962 p. 1052.

Bauer, Douglas; Hirshberg, Alan S. 1979. "Improving the Efficiency of Electricity Generation and Usage." John C. Sawhill, ed. Energy Conservation and Public Policy. Englewood Cliffs, NJ: Prentice-Hall, Inc.

Bechtel National, Inc. 1978. Terrestrial Central Station Life Cycle Analysis Support Study. Pasadena, CA: Jet Propulsion Laboratory. August.

Bechtel National, Inc. 1979. Requirements Definition and Preliminary Design of a Photovolatic Central Station Test Facility. Albuquerque, NM: Sandia Laboratories. April.

Beck, R.W. and Associates, "Analysis of Expansion of the Pacific Northwest Southwest Energy Intertie System, California Energy Commission, July 1979.

Boeing Engineering and Construction Co. 1979. MOD-2 Wind Turbine System Concept and Preliminary Design Report. DOE/NASA CR-159607. Draft Report. Seattle, WA: Boeing Engineering and Construction Co. January.

Bossong, Ken. "The Case Against Public Utility Involvement in Solar/Insulation Programs, Washington D.C.: Citizen's Energy Project 1978 Report, Series no. 19.

Brandt, L.; Edelstein, R.; et al. 1980 (Aug 21). "Solar Thermal Cost Goals: Methodology and Preliminary Results." Presentation to the Department of Energy at the first Report of the Solar Thermal Cost Goal Committee. Golden, CO: Solar Energy Research Institute.

Business Week. 1979. "New England Electric System: Embracing the Conservation Ethic." December 10; pp. 119-120.

Business Week. May 28, 1979, p. 108. "A Dark Future for Utilities."

California Public Utilities Commission. "Financing the Solar Transition." San Francisco CA. CPUC January 1980.

Cantor v. Detroit Edison Co. 428 U.S. 579,596 (1976) quoted in Satlow 1981.

Cicchetti, Charles J.; Gillen, William J.; Smolensky, Paul. 1977. The marginal Cost and Pricing of Electricity, An Applied Approach. Cambridge, MA: Ballinger Publishing Co.

Coleman, L. R. et. al. Initial Brief of the United States Department of Energy Before the City of Los

Angeles Department of Water and Power PURPA proceedings.

Considine, Timothy J. 1979. "Planning Models Used for Assessing Utility Applications of Solar Electric Technologies." Theresa Flaim, et al. Economic Assessments of Intermittent, Grid-Connected Solar Electric Technologies, A Review of Methods. SERI/TR-353-474. Draft. Golden, CO: Solar Energy Research Institute. November.

California Public Utilities Commission Decision No. 91272, Docket No. 00I 42 January 1980, p. 6.

Department of Energy. Answers to questions posed by the House Committee on Science and Technology dated October 22, 1979.

Department of Energy 1980. "Conservation and Renewable Energy Activities for PMA's July 10, 1980.

Department of Energy. 1980. Solar Energy Program Summary Document FY-1981. Washington, D.C.: U.S. Department of Energy, Assistant Secretary for Conservation and Solar Energy. January.

Department of Energy/OGC Docket no. U-14495 before the Louisiana Public Service Commission. Guld States Utility Company, October 6, 1980.

Doane, J. W.; O'Toole, R. P.; Chamberlain, R. G.; Bos, P. B.; Maycock, P. D. 1976. The Cost of Energy from Utility-Owned Solar Electric Systems, A Required Revenue Methodology for ERDA/EPRI Evaluations. ERDA/EPRI-1012-76/3. Pasadena, CA: Jet Propulsion Laboratory. June.

EUS, Inc. 1978. Survey of Utility Load Management and Energy Conservation Projects. ORNL/Sub-77-13509/4. Oak Ridge, TN: Oak Ridge National Laboratory. December.

Electric Power Research Institute. "Under-Over cost Assessment" 1979.

Electric Power Research Institute. 1979. Technical Assessment Guide. EPRI PS-1201-SR. Palo Alto, CA: Electric Power Research Institute. July.

Elliot, D. L. 1977. Synthesis of National Wind Energy Assessments. BNWL-2220 WIND-5. Richland, WA: Pacific Northwest Laboratory. July.

Energy Information Administration. 1978. Annual Report to Congress. Vol. 2. DOE/EIA-0173/2. Washington, D.C.: Department of Energy, Energy Information Administration.

Energy Security Act, 1980, Joint Explanatory Statement of the Committee of Conference.

Federal Power Commission. 1976. Hydroelectric Power Resources of the United States as of January 1, 1976. Washington, DC: Federal Power Commission. November.

Federal Power Commission "Proposed Generating Capacity Additions 1977-1986" as reported by the Regional Electric Reliability Councils Under Docket R-362, Order 383-4, September 6, 1977.

Fellner, William. "The Influence of Market Structure on Technological Progress." Readings in Industrial Organization and Public Policy. American Economic Association, p 287.

Feuerstein, Randall J., "Utility Rates and Solar Commercialization. Solar Law Reporter July/August, 1979, p. 326.

Flaim, Theresa; Timothy J. Considine; Robert Witholder; Michael Edessess. 1979. Economic Assessments of Intermittent, Grid-Connected Solar Electric Technologies: A Review of Methods. Draft Report. SERI/TR-353-474. Golden CO: Solar Energy Research Institute. November.

Ford Boyd references

General Accounting Office "Construction Work in Progress" EMD-8075.

Garfield, Paul and W.F. Lovejoy. Public Utility Economics Prentice Hall Inc., Englewood Cliffs, N.J. 1964.

General Electric Co. 1977. Conceptual Design and Systems Analysis of Photovoltaic Systems. Washington, D.C.: Department of Energy.

General Motors. 1979. Heliostat Production Evaluation and cost Analysis. SERI/TR-8052-1. Golden, CO: Solar Energy Research Institute. December.

Graban, William. 1980. Personal communication to Theresa Flaim (SERI). Washington, D.C.: Economic Regulatory Administration, Power Supply Planning Branch. February 26.

Hamilton, Michael S.; Norman, I. Wengert 1980. Environmental Legal and Political Constraints on Power Plant Siting in the Southwestern United States: A Report to the Los Alamos Scientific Laboratory. Ft. Collins, CO: Department of Political Science, Colorado State University.

House Conference Report No. 96-1104 96th Congress, 2nd Session on S 932

Iwler, Louis, "Conflict Continues over CWIP in the Rate Base," Electrical World, August 15, 1980, p. 21.

JBF Scientific Corp. 1978. Wind Energy Systems Application to Regional Utilities. Draft. Wilmington, MA: JBF Scientific Corp. September.

Jet Propulsion Laboratory. 1980. Federal Policies to Promote the Widespread Utilization of Photovoltaic Systems. Washington, D.C.: Department of Energy. February.

Joint Explanatory Statement of the Committee of Conference, Report No. 95-1750 p 98 quoted in Lock 1980

Joskow, Paul L. "Applying Economic Principals to Public Utility Rate Structures: The Case of Electricity," in Studies in Electric Utility Regulation edited by Charles J. Circhetti and John L. Jurewitz, Ballinger 1975.

Justus, C. G. 1978. Winds and Wind System Performance. Philadelphia, PA: The Franklin Institute Press.

Justus, C. G. and W.R. Hargraves. "Wind Energy Statistics for Large Arrays of Wind Turbines (Great Lakes and Pacific Coast Regions), Georgia Institute of Technology RLO/2439-77/2, May 1979.

Kahn, Alfred. 1970. The Economics of Regulation: Principles and Institutions. Vol. 1&2, Economic Principles. New York, NY: John Wiley and Sons, Inc.

Kahn, Edward. "Bankruptcy Risk in the Electric Utility Industry: Policy Issues" LBL Draft May 1980.

Kahn, Edward. "Using Utilities to Finance the Solar Transition", Solar Law Reporter, Vol 2 No. 2.

Kahn, Edward, and Stephen Schultz. "Utility investments in on-site solar: Risk and Return Analysis for Capitalization and Financing@ September 1978 (LBL-7876 p. 21).

Kahn, Edward, Leonard, Ross, Peter Berenson, and Chevvy, James, Utility Solar Finance: Economic and Institutional Analysis, Lawrence Berkeley Laboratory, 1979.

Kahn, Edward, Berenson, Peter, Brown, Burnett, Commercialization of Solar Energy by Regulated Utilities Economic and Financial Risk Analysis, Lawrence Berkeley Laboratory 1980.

Laitos, Jan; Feuerstein, Randall J. 1979. Regulated Utilities and Solar Energy. SERI/TR-62-255. Golden, CO: Solar Energy Research Institute. June.

Lindley, C.A.; Melton, W.C. 1979. Electric Utility Application of Wind Energy Conversion Systems on the Island of Oahu. Draft. ATR-78 (7598)-2. El Segundo CA: The Aerospace Corp. February 23.

Lock, Reinier H.J.H. "Encouraging Decentralized Generation of Electricity: Implementation of the New Statutory Scheme" Solar Law Reporter 2(4) 1980.

Loose, Verne, W. and Theresa Flaim. "Economies of Scale and Reliability: The Economics of large versus small generating units." Energy systems and policy volume 4 no. 1-2 pp 37 (1980).

Lowe, John (Director of Wind Energy programs), Boeing Engineering and Construction Company, Statement to the House Science and Technology Committee Subcommittee on Energy Development and Applications. Washington, D.C. 9/26/79, Lerner, California Energy Commission.

Marsh, W.D. 1979. Requirements Assessment of Wind Power Plants in Electric Utility Systems. EPRI-ER-978. Palo Alto, CA: Electric Power Research Institute. January.

Masser, P. 1979. Low-Cost Structures for Photovoltaic Arrays. SAND 79-7006. Albuquerque, NM: Sandia Laboratories. August.

Maulden, Jerry, (President Arkansas Power and Light Co.) "Energy Conservation and Resource Development: The Need for Public Policy Changes" June 20, 1980.

McDonald, R.J. 1980. Personal communication to William Metz. Fort Belvoir, VA: U.S. Army Corps of Engineers, Institute of Water Resources. February 14.

Mitchel, Bridger M. and Jan Paul, Action, Peak Load Pricing in Selected European Electric Utilities, Rand Corp 1977.

New Hampshire Public Utilities Commission Fifth Supplemental Order No. 14, 280. June 18, 1978.

Office of Power Marketing. "Activities of the Power Marketing Administration in FY80 to conserve Energy and to Develop the Use of Renewable Energy Sources."

Oklahoma Stat. Ann. tit. 18 section 1.27(West), quoted in Satlow 1981.

Pacific Gas and Electric Company, "Annual Report to the Federal Energy Regulatory Commission for the Year Ended December 31, 1979" (FERC Form), p 228-229, Quoted in Praul and Marcus.

Praul, Cynthia and William B, Barcus "Energy Service Corporations: Opportunities for California's Utilities" California Energy Commission, Draft report November 1980.

Primeaux, Walter J. JR., "The Monopoly Market in Electric Utilities" in Promoting Competition in Regulated Markets, Almarin Phillips, Editor The Brookings Institution, Washington D.C. 1975.

Ramler, J. R.; Donovan, R. M. 1979. "Wind Turbines for Electric Utilities: Development Status and Economics." DOE/NASA/1028-79/3. Cleveland, OH: National Aeronautics and Space Administration. Paper prepared for the Terrestrial Energy Systems Conference, Orlando, FL: June 4-6.

Reddoch, Thomas W. 1980. Personal Communication to Theresa Flaim (SERI) regarding the TVA/DOE study in progress, "Analysis of the Operation of An Electric Power System With and Without Wind Generation." Knoxville, TN: University of Tennessee. April.

Reference on deregulating generation

Sandia Laboratories. 1978. Availability of Direct Total and Diffuse Solar Radiation for Fixed and Tracking Collectors in the U.S.A. SAND-77-0885. Albuquerque, NM: Sandia Laboratories. January 30.

Satlow, Barry " The Energy Security Act and Public Utilities: A Yellow Light for Utilitys Solar Financing and Marketing," Solar Law Reporter, vol. 2 no. 5, January/February 1981.

Schlueter, Robert; Park, Gerald; Modir, Hassan; Dorsey, John; Lotfallian, Mohson. 1979. Impact of Storm Fronts on Utilities with WEC'S Arrays. COO/4450-79/2. East Lansing, MI: Michigan State University, Division of Engineering Research. October.

(Seplow 80)

Seplow, Kenneth quoted in Electrical Week 3/24/80, p. 5.

Shanks, Kenneth J. 1979. Inventory of Power Plants in the United States - April 1979. DOE/EIA-0095.

Washington, D.C.: U.S. Department of Energy, Energy Information Administration.

Shelpuk, Ben; Don Petty; Peter Davidoff. 1980. "Ocean Energy Systems." Working paper prepared in support of the Utilities Task Force of the SERI Solar/Conservation Study. Golden, CO: Solar Energy Research Institute. February.

Sillman, Sanford. 1980. "Electric Storage Technologies." Working paper prepared in support of the Utilities Task Force of the SERI Solar/Conservation Study. Golden, CO: Solar Energy Research Institute. February.

Smiley, Albert "Solar Energy and Public Utilities-- Institutional Issues" Report (To be published in SERI's "Solar Energy in Review"), August, 1980

Smith, Jeffrey R. 1980. "Wind Power Excites Utility Interest." Science. Vol. 207. February 15; pp. 739-742.

Solomon, Burt. 1980. "Windmills Move into the Market". The Energy Daily. March 27.

Solar Energy Research Institute. 1978. Annual Review of Solar Energy. SERI/TR-54-066. Golden, CO: Solar Energy Research Institute. November.

Spann, Robert M. in Electric Power Form; Alternatives for Michigan, edited by William H. Shaker and Wilbert Steffy, Ann Arbor 1976.

Spectrolab, Inc. 1977. Photovoltaic Systems Concept Study. Washington, D.C.: Department of Energy.

Sullivan, Robert. 1980. Personal communication to Theresa Flaim (SERI) regarding the OTEC cost estimates used in the Florida Power Corp. study sponsored by DOE. Gainesville, FL: University of Florida, Department of Electrical Engineering. February 25.

Taylor, Roger; John, Day; Brian, Reed; Mike, Malone. 1979. Solar Thermal Repowering Utility Value Analysis. SERI/TR-8016-1. Golden, CO: Solar Energy Research Institute. September.

Terasawa, K. 1980 (Aug. 21). "Cost Goals of Solar Thermal Parabolic Dish Systems." Presentation to the Department of Energy at the First Report of the Solar Thermal Cost Goal Committee. Pasadena, CA: Jet Propulsion Laboratory.

Trade Reg. Rep. (CCH) section 71,134 (D.N.J. 1956) (Consent judgment)

Turvey, Ralph; Dennis, Ander. 1977. Electricity Economics, Essays and Case Studies. Baltimore, MD: The Johns Hopkins University Press.

Turvey, Ralph, Dennis, Anderson, Electricity Economics, Johns Hopkins Univ. Press, Baltimore 1977.

TVA Moves Further Along Solar Path." Solar Law Reporter. 1979.

Uhler, Robert G. 1977. Rate Design and Load Control, Issues and Directions. A Report to the National Association of Regulatory Commissioners. Palo Alto, CA: Electric Utility Rate Design Study. November.

United States v. Western Electric and AT&T, 13 Rad. Reg. (P-H) section 2143, 1956

Vankuiken, J.C.; et al. 1978. Reliability Energy and Cost Effects of Wind Integration with Conventional Electric Generating Systems. Draft. Argonne, IL: Argonne National Laboratory. November.

Weir, Roger. 1980. Personal Communication to Theresa Flaim (SERI). Salt Lake City, UT: Western Systems Coordinating Council. February.

Weiss, Leonard W. "Antiturest in the Electric Power Industry" In Promoting Competition in Regulated Markets, Almarin Philips, Editor The Brookings Institution, Washington D.C. 1975.

Western Systems Coordinating Council. 1979. Ten-Year Coordinating Plan Summary, 1979-1988. Salt Lake City, UT: Western Systems Coordinating Council.

Westinghouse Electric Corp. 1977. Conceptual Design and Systems Analysis of Photovoltaic Power Systems. Washington, D.C.: Department of Energy.

Wood, Archie L. "The Social Cost of Imported Oil", TRW Energy Systems Planning Division prepared for the department of energy November 28, 1979.

York, Wendy L. 1979. Electric Utility Solar Energy Activities, 1978 Survey. EPRI-ER- 66-SR. Palo Alto, CA: Electric Power Research Institute. May.